Advances in Urban Flood Management

BALKEMA – Proceedings and Monographs
in Engineering, Water and Earth Sciences

Advances in Urban Flood Management

Editors

R. Ashley
Department of Civil and Structural Engineering, Pennine Water Group, Sheffield, UK

S. Garvin
Building Research Establishment, East Kilbride, UK

E. Pasche
River and Coastel Engineering, Hamburg-Harburg University of Technology, Hamburg, Germany

A. Vassilopoulos
Geocultural Park of Eastern Aegean, Athens, Greece

C. Zevenbergen
UNESCO-IHE, Sustainable Urban Infrastructure, Delft, The Netherlands &
Dura Vermeer Group NV, Hoofddorp, The Netherlands

Taylor & Francis
Taylor & Francis Group

LONDON / LEIDEN / NEW YORK / PHILADELPHIA / SINGAPORE

Cover photo: Courtesy of Rob Lengkeek

COST is supported by the EU RTD Framework programme

ESF provides the COST Office through an EC contract

Taylor & Francis is an imprint of the Taylor & Francis Group, an informa business

Typeset by Charon Tec Ltd (A Macmillan Company), Chennai, India
Printed and bound in Great Britain by Bath Press Ltd (CPI-group), Bath

Published by: Taylor & Francis/Balkema
P.O. Box 447, 2300 AK Leiden, The Netherlands
e-mail: Pub.NL@tandf.co.uk
www.balkema.nl, www.taylorandfrancis.co.uk, www.crcpress.com

ISBN: 978-0-415-43662-5 (Hbk)
ISBN: 978-0-203-94598-8 (Ebook)

Table of Contents

Editors

Professor Richard Ashley
Departement of Civil and
Structural Engineering
Pennine Water Group
Mappin Street
S1 3 JD Sheffield
United Kingdom
Tel: +44 114 222 5766
Fax: +44 114 222 5700
Email: r.ashley@sheffield.ac.uk

Dr. Stephen Garvin
Building Research Establishment
Scotland
Kelvin Road
East Kilbride
G75 ORZ Glasgow
United Kingdom
Tel: +44 1355 576242
Fax: +44 1355 576210
Email: garvins@bre.co.uk

Professor Erik Pasche
River and Coastel Engineering
Hamburg University of Technology
Denickestraße 22
21073 Hamburg
Germany
Tel: +49 40 42878 3463
Fax: +49 40 42878 2802
Email: pasche@tu-harburg.de

Dr. Andreas Vassilopoulos
Geocultural Park of Eastern Aegean
Fliass 13 & V. Ipeirou,
151 25, Athens
Greece
Tel: +30 210 8056448
Fax: +30 210 8056091
Email: vassilopoulos@gcparks.com

Dr. Chris Zevenbergen
UNESCO-IHE
Sustainable Urban Infrastructure
PO Box 3015
2601 DA Delft
The Netherlands
Tel: +31 15 2151854
Fax: +31 15 2122921

and

Dura Vermeer Group NV,
PO Box 3098
2130 KB Hoofddorp
The Netherlands
Tel: +31 23 569 2380
Fax: +31 23 569 2332
Email: c.zevenbergen@
    duravermeerdiensten.nl

Foreword

Global climate change is increasing the variability of extreme flood events and cyclones. Current measures to mitigate flood impacts specifically for urban environments no longer provide optimal solutions for previously planned flood risk intervals. Being prepared for uncertain changes and extreme flood events requires a paradigm shift in current strategies to avoid and manage flood disasters. A major rethink of current planning and flood management policies and practices on different spatial and temporal scales is required to reverse the trend of increasing impacts of urban floods. In November 2004 an International Expert Meeting on Urban Flood Management was held to discuss these challenges. The outcome of this meeting resulted in a book on Urban Flood Management which was edited by A. Szöllösi-Nagy and C. Zevenbergen and published by A.A. Balkema Publishers in 2005. This International meeting also accompanied the launch of COST[1] action C22 on Urban Flood Management in March 2005. This European Concerted Action C22 brings together a large number of European research groups. Its main objective is to increase knowledge required for preventing and mitigating potential flood impacts in urban areas by exchanging experiences, developing integrated approaches, and by promoting the diffusion of best practices in Urban Flood Management. In the action, four working groups are active in the following areas:

Flood modelling and probability assessment: Computational methods to determine flood hydrographs of urban watercourses (hydrological models); flood propagation in storm water pipe and open channel networks (hydraulic models); and the assessment of the probability of extreme floods (statistical methods). Special reference is given to data management by making use of web-services, the modelling of flash floods; SUDS (sustainable drainage systems); the "daylighting" of urban streams by removing culverts and restoring open space for flood storage; and the assessment of climate change effects.

Flood risk assessment and mapping: Methods used to determine the impact of floods in urban environments. Special reference is given to the assessment and mapping of risk, and to methods to assess the impact of flood mitigation through structural and non-structural measures.

---

[1] Founded in 1971, COST is an intergovernmental framework for European Co-operation in the field of Scientific and Technical Research, allowing the co-ordination of nationally funded research on a European level. COST Actions cover basic and pre-competitive research as well as activities of public utility.

Flood resilience: Methods used to reduce flood damage for the built environment. Special reference is given to the techniques of flood resilient repair and dry and wet-proofing; the capacity building of residents and the prerequisites for efficient flood resilience.

Integrative concepts of urban flood management: Strategies to integrate technological and non-technological aspects into an holistic approach to flood risk management with special reference to non-structural solutions and the relevance of policy, insurance, risk awareness, communication, consistency management and the socio-cultural environments.

This book comprises initial results from C22 and aims to address the aforementioned issues by highlighting and analyzing the state of the art and best practices in this emerging field of urban flood management. Our thanks go to the contributing authors, whose efforts have made this book a useful reference in setting the direction for the future in urban flood management.

Chris Zevenbergen
Chair COST Action C22

Erik Pasche
Vice-chair COST Action C22

January 2007

1

Challenges in Urban Flood Management

Chris Zevenbergen[1,2] & Berry Gersonius[1]

[1] *UNESCO-IHE, SUIS, Delft, The Netherlands*
[2] *DuraVermeer Business Development BV, Hoofddorp, The Netherlands*

1.1 INTRODUCTION

Floods are the most common type of natural disaster in Europe (EEA, 2005). During the past five years floods have affected an estimated land area of one million square kilometres. According to CRED[1] in the period between 1900 and 2006 about 415 major damaging floods occurred in Europe. Over this period the number of flood events per year increased, while the number of deaths exhibited a decrease, although the economic losses increased substantially. The decreasing loss of life is likely to be due to improved warning and rescue systems. The dramatic and consecutive flood disasters in 2002 which occurred in countries such as Austria, the Czech Republic, Germany, Hungary and the Russian Federation have triggered several initiatives to improve preparedness and response to extreme weather events and enforced the need for a concerted European action (Barredo et al., 2005). The countries most affected by floods between 1998 and 2002 are in central and eastern Europe. Severe floods in 2005 further reinforced the need for concerted action.

In Europe, urban land expanded by 20% during the period 1980 to 2000, while the population increased by 6% (EEA, 2005). The building stock is mainly aging and there is much heritage, although overall, new building means the total stock is slowly increasing. Within 30 years, some one third of the European building stock will be renewed (ECTP, 2005). European cities are faced with a continuing demand for high quality, low density housing areas in the proximity of metropolitan areas. For example (re)development plans have been made for 12,000 new waterfront habitations (dwellings) in Hamburg Hafencity, Germany by 2010; there are allocations for 150,000 homes in the Thames Gateway, England by 2016, and 550,000 new houses are planned in the Randstad (The Netherlands) by 2010. Almost 80% of the original floodplains along the Elbe and Rhine rivers have been lost due to urban development and agricultural use. Both urban growth and the restructuring and transformation of existing urban areas result in an intensification of land use

---

[1] Centre for Research on the Epidemiology of Disasters, UCL, Belgium

and subsequently an increase of the level of investments in flood prone areas. Moreover, Europe has thousands of levees, dams and canal systems, some of which stem from Victorian times, in which the structural integrity is questionable and needs to be better understood.

For many centuries in most European countries, flood management policies have favoured engineering structural solutions over other complementary non-structural measures. The aim of these flood defence measures was to modify the flood and make it easier to cope with by eliminating the highest probability floods. Such measures gave a false sense of safety: the reduction in perceived vulnerability to flood loss was largely offset by the rapid growth of urban areas in flood-prone areas combined with an even larger increase in investment levels. In addition, the people in the cities were (and often still are) unaware of the risks associated with floods and accordingly the responses by individuals for reducing losses are limited. As a consequence of these policies, flood exposure and potential social and economic damage has increased substantially. In addition, reducing losses from high frequency floods has also caused a greater risk of disastrous consequences when more extreme events have occurred. The threat posed by climate change together with continuing urbanisation of flood plains has triggered many European countries to reconsider their flood management policies as part of their overall responses (Europa, 2006).

In response to the 2002 floods, the European Commission developed in 2005 a proposal for a Directive on flood risk management. On 18 January 2006, after extensive consultations with stakeholders and the public the Commission adopted its proposal for a Directive of the European Parliament and of the Council on the assessment and management of floods. Its aim is to reduce and manage the risks that floods pose to human health, the environment, infrastructure and property. Under the Directive member states first need to carry out a preliminary assessment to identify the river basins and associated coastal areas at risk of flooding. For such zones they then require drawing up flood risk maps and then flood risk management plans focused on prevention, protection and preparedness. Since most of Europe's river basins are shared by more than one country, concerted action at a European level is expected to enhance effective management of flood risks. A binding legal instrument will ensure flood risks are properly assessed, coordinated protection measures taken and the public properly informed. Some European countries such as The Netherlands and Germany are already taking initiatives in further developing their flood risk maps and the UK has them already implemented.

1.2   TOWARDS A BETTER UNDERSTANDING OF URBAN FLOOD GENERATION AND IMPACTS

With this book, COST action C22 intends to support this new legal framework of flood risk management by presenting a review of best practice of flood risk management for urban environments. Due to the special nature of the flood regime, watercourses in urban catchments need special consideration and their own particular strategies for flood risk management. Currently, each city or district embarks on its own mitigation strategy – with some being more successful than others. For this reason there is a strong need to share current approaches to flood risk management for these urban catchments at a European level and to identify which are the good solutions in practice, and also highlight improved ways of managing and coping with urban water courses.

1.3   FLOOD MODELLING AND PROBABILITY ASSESSMENT

This book has selected this issue as a key topic. It begins with a characterisation of urban streams and urban flooding (paper by Douglas *et al.*). The impact of the contribution of runoff to flood risk in urban areas is described and how imprudent management practices and climate change are exacerbating urban flooding.

As flood risk is the product of probability of occurrence multiplied by its impact, the procedure for flood risk assessments is composed of three tasks: the determination of the inundation areas, the exploration of the vulnerability, and the assessment of probable damage. The determination of inundation areas (flood analysis) must include a hydrological and a hydraulic analysis. The former gives the design hydrographs and flood probability of occurrence. The latter determines the flooded areas for these design hydrographs. The paper by Pasche gives a state-of-the-art in hydrological and hydraulic modelling for urban environments. It concludes that for hydrological modelling, deterministic rainfall-runoff models, based on semi-distributive lumped approaches are appropriate when combined with hydraulic modelling that is 1- and 2-dimensional hydrodynamic water surface models. Together these may be used for the determination of the flood plains and the flood-prone areas requested by the EC Water Directive. It shows that within the hydraulic models the traditional roughness approach of Manning and a constant eddy assumption in the presence of vegetation, could lead to considerable errors in predictive mode, despite a good calibration using historical data. This deficiency can be avoided by using the Darcy-Weisbach-Equation and by applying special routines for modelling the effect of woodlands.

The paper by Oberle and Merkel addresses the modelling techniques of decision support systems. It concludes that for operational applications, in which a fast response and reliable numerical results are most relevant, 1-dimensional hydraulic models are still state-of-the-art. But they also show, that by integrating the models in a GIS and simplifying the theoretical approaches, 2-dimensional modelling techniques combined with 1-dimensional models can achieve a level of robustness and performance that allow a real-time simulation to be provided in time scales which are appropriate for disaster management, even for large rivers liker the Elbe.

The relevance of good knowledge of the spatial and temporal distribution of precipitation in storm events is shown by Kobold. She demonstrates the sensitivity of hydrological models for these input data and that the type of rainfall model has only minor relevance to the quality of the results as long as sufficient calibration events are available. Statistical models are needed to evaluate the probability of flooding. The paper by Frances gives an overview of the state-of-the-art in flood frequency analysis. It concludes that non-systematic information characterising the observed flood events is preferable to the present practice of annual maximum floods as input data. Hence, historical and paleo-flood information can be used for flood frequency analysis, which is of great use in the assessment of flood probability. Only few statistical concepts consider the non-stationarity of the long term flooding process, which is necessary for a better estimation of future flood hazards in a changing climate. Frequency analysis of precipitation data is highlighted by Koutsoyiannis. He demonstrates that the Extreme Value Type 2 distribution should replace the Gumbel distribution for probabilistic modelling of extreme rainfall.

The two papers by Koutsoyiannis and Jovanovic deal with sustainable measures to reduce the probability of flooding. The paper by Tourbier and White shows that sustainable urban drainage systems (SUDS) reduce the flood volume and peak flow only for minor storms. However, by expanding SUDS to include exceedence conveyance systems, which route the excess flood flow, the efficiency of SUDS can be extended to help cope with extreme floods. In his paper Jovanovic gives examples of technical solutions, which ensure the preservation of the riparian wildlife and the ambient quality of the stream corridor in addition to efficient flood attenuation.

Despite the high standard of the available mathematical methods to model and assess flood probability, some major deficiencies and gaps of knowledge have still been detected which should be addressed in the near future. Further experimental studies are required on hill slope hydrology in order to gain a better understanding of the contribution of subsurface flow to flood flows. This knowledge will support on-going research activities dealing with the coupling of hydrological (rainfall-runoff models) and hydraulic models (riverine flow models and groundwater models) by providing the necessary empirical input to model the mass and momentum transfer between the different flow regimes. The runoff models should be directly coupled with meteorological models to study, at a global scale, the effect of changes in evapotranspiration, which is a result of climate change caused by warmer periods with less rain. Floods in urban areas result from interaction between the storm water pipe network, the overland flow and flow in small watercourses (the minor and the major systems respectively). These can result in flash floods with extreme spatial and temporal variability. The modelling capabilities need to be extended by coupling pipe flow models with 2-dimensional (at least) overland flow models and developing robust numerical schemes for discontinuous and transient flow. In these flow domains erosion has a strong impact on flood damage.

Thus modelling techniques should be able to simulate any sheet erosion on green areas, scouring at buildings and the transportation and deposition of sediment in the drainage system and in the urban environment. They can then be used to help to develop efficient erosion-sediment-control plans.

Inputs to and the results from all modelling tools used in flood risk assessment and management are still of great uncertainty. Probabilistic methods should be included in deterministic models to estimate the ranges of uncertainty. Paleo-hydrologic and paleo-climatic data should be included in the frequency analysis of extreme events. However, the methods used to derive the flood characteristics of these historic events need further improvement.

Due to the deficiencies of today's meteorological models in forecasting small-scale storm events, radar data should be further analyzed to explore the possible predictability of precipitation. However, with climate changing the traditional view of what may be predictable held in the past will have to be revised.

Sustainable drainage systems need to be extended in their capacity to attenuate extreme floods. The possibilities to convey excess water through open space (e.g. parks, green areas, roads) should be further explored and methods to better evaluate the attenuation flood waves in such areas should be developed.

Decision support tools need a standardisation of the pre- and post-processing information stages. This would create the possibility of switching easily between different mathematical models allowing a direct comparison of the model qualities and an assessment of the uncertainty of the numerical results. The complexity of the flow in urban environments needs high spatial and temporal resolutions of models. These requirements cannot be fulfilled with current operational models. Performance improvements making use of parallel computing and grid technology can help to overcome these limitations.

A prerequisite for the reliable application of hydrological and hydraulic models is the availability of topographic, geomorphologic and hydrometeorological data of good quality and appropriate spatial and temporal resolution. The technological advancements in geo-data sampling by remote sensing (LIDAR-technology, satellite imaging) should be used to develop a geo-processing service which provides digital terrain models, numerical grids and the spatial distribution of vegetation and surface texture structured in hydraulic roughness classes for flood plains.

1.4 FLOOD RISK ASSESSMENT AND MAPPING

In order to establish a well defined theoretical and practical framework on flood resilience, it is vital to address the concept of vulnerability in flood prone areas. Although flood impact assessment methods have been around for many years (e.g. Penning-Rowsell, 1976), rapid changes in the contemporary urban environment require the continuous adaptation and development of theory, methods and applications in vulnerability assessment. Although, the concept of vulnerability is strongly related to susceptibility (e.g. the protection level provided by primary flood defence systems), quantitative assessment starts with impact estimation. Due to the increasing availability of data this holds true not only for retrospective assessment but to a large extent also for prospective cases in which potential flood impact assessment can address disruption of society's socio-economic backbones and their effects on regional, national and even trans-national networks. This applies especially to methods used to evaluate the indirect effects of flood impact, which are a relatively new research area in which econometric models are used to measure the effect local impacts have on a larger scale. While many of the models used have originated as methods of assessing impacts of policy changes (e.g. taxation), their adaptation for use within the hazard impact domain offers exciting new prospects that will put a new perspective on impact assessment, economic vulnerability and ultimately the implementation of design methods to improve resilience.

Increasing precision and the estimation of both direct and indirect flood impacts is vital for the evaluation (e.g. cost-benefit analysis) of impact reduction measures and flood risk management. The potentially large impact of extreme events caused by climate change further increases the urgency for further development of assessment models and the need for vulnerability indicators and localization of 'hotspots' of vulnerability. Furthermore, flood risk maps indicating direct and indirect damage can be used to raise awareness of both policy makers and citizens to influence preparation for flooding and adaptive responses. An essential factor in successful application is therefore, the accessibility of information for various groups of stakeholders. This puts emphasis on the development of a better presentation of the often complex results now only accessible for experts.

In addressing the theoretical framework of direct- and indirect flood loss estimation, Veerbeek describes the state-of-the-art in indirect flood loss estimation models. The paper covers Unit-Loss models, Input-Output Analysis models and the increasingly popular Computable General Equilibrium models within the flood impact domain. Computable General Equilibrium models in particular, provide a rich framework in which dependencies, individual agent behaviour as well as various multiplier effects can be modelled to closely resemble the behavioural response of economic systems as a result of local disruptions. Due to the relative novelty and complexity, the use of these models within actual case studies is so far limited. Veerbeek adds that further modifications by implementing GIS within the statistical framework of these models might further increase their application value and predictive qualities.

The paper by Kron deals with requirements for flood and flood risk mapping in the context of the shift from classical flood protection to flood risk management. Flood risk maps can support stakeholders in prioritising their investments and increase the awareness level of both authorities and citizens, possibly resulting in a more effective response. The increasing level of detail in flood risk maps results in better flood impact and risk assessments, and paves the way for a flood resilient planning approach. Models include assessment for direct and indirect damage for the whole range of flood events from the occurrence of first damage to catastrophic events with a low recurrence probability (10^{-3} or lower). Kron also acknowledges that with the coupling of hydraulic (hazard) simulation models and Flood Damage Estimation tools, risk assessment can be derived with a high level of detail. However, the paper concludes by questioning the effectiveness of detailed risk assessment, given the inherent uncertainties in integrated flood risk management.

1.5   INCREASED COMPLEXITY AND UNCERTAINTY

The challenges caused by climate change and the potential increase in the likelihood of extreme events are now widely recognized as a major challenge and risk for flood management approaches. Although research on climate change impacts has improved knowledge about certain variables, with increasing certainty, such as for the likely rise in sea level – in many situations the projections are still rather uncertain. These impacts include peak flows, extreme rainfall, extreme waves and winds. Here, there is a need for further evidence and research to reduce local and regional variations and variations between impact models. The uncertainties resulting from these projected changes imply that conventional risk-centered approaches can no longer be maintained as a robust strategy to ensure future safety and development. This means that seeking optimal solutions, in a cost-benefit approach based on statistical analysis of flood frequency, is no longer sufficient or feasible. In New Orleans, for example, a decision has been made as to the level of flood protection (to withstand storm surges from category 4 or 5 hurricanes) based on a cost-benefit analysis taking into account climate change and other human related disruptions of the environment and second order impacts: Here, two evaluations reached opposite conclusions (Hallegatte, 2005). This illustrates the deficiencies associated with decision making based on a cost-benefit analysis, which in turn is based on assumptions underlying a probability distribution.

Furthermore, the problems facing the future of urban flood management are set within the whole sphere of human activity. They are as much the result of planning and development, stakeholder and actor behaviour, and economic systems as they are of physical factors. Many challenges have to do with the complexity of the system to be managed and the uncertainties in socio-economic developments influencing the performance of future responses. The paper by Ashley et al. describes these challenges and recognizes that 'compared with entire river basins, urban areas are more complex in terms of the physical, institutional and scale dimensions'.

1.6   BUILDING FLOOD RESILIENT URBAN AREAS

To deal with the challenges of increased uncertainty and complexity, efforts have to be made to increase the resilience of the urban environment both before and after disasters occur. Resilience reflects the amount of change a system can undergo and still retain the same control on function and structure, the degree to which a system is capable of self-organisation and the ability to build and increase the capacity for learning and adaptation (Holling, 1973 cited in Pahl-Wostl et al., 2005). Hence, the concept of resilience represents a profound shift in conventional perspectives, which attempt to control system changes that are assumed to be stable, to a more realistic viewpoint aimed at sustaining and enhancing the capacity of complex systems to adapt to uncertainty and surprise (Adger et al., 2005).

It should be noted that the delivery of resilient urban areas is not only a technical exercise. Flood resilient planning and building has two dimensions; on one hand it is a technical approach, and on the other it has to do with many other aspects, including policy, regulation, decision making and engagement. It has therefore become increasingly necessary to engage with stakeholders and communities in general so that these are better prepared for flooding, but so that they can also take an active role in decision making for where, when and how investments and measures are taken. The papers by Ashley et al. and Tippett & Griffiths recognize that resilience in a 'soft' sense can only be developed by creating the conditions within which resilience will emerge – and that resilience cannot as yet be provided (in a non-structural sense).

In addition, the challenge of building resilience will involve multiple scales, and hence the overall performance is dependent on the ability to understand and take advantage of different initiatives at different levels (Allenby, 2005). For example, designing a flood-proofed building is useful, and a number of such buildings in the urban environment will enhance the area's overall resilience against floods. But flood-proofing will not substitute for an early warning system, and neither for the flood relief effort that the city as a whole will need to mount.

This multi-dimensional nature of the concept of resilience makes it difficult to put into operation. What constitutes resilience in a complex system, such as flood risk management, is as yet poorly defined and the concept seems to remain elusive other than as an aspirational concept or principle, although ecological theories and principles of resilience have been around for a long time. How urban areas should be planned, designed or managed to contribute to resilience, however, has not yet been defined. The results of the work done so far by COST C22 support the adoption of more resilient approaches to urban flood management by presenting the current state-of-the-art in technological and non-technological measures to increase the resilience of the urban environment.

1.7   TECHNOLOGICAL MEASURES TO INCREASE FLOOD RESILIENCE

The technological dimension of resilience refers to the ability of physical systems (including components, their interconnections and interactions, and entire systems) to perform to acceptable levels when subject to floods, as well as developing and implementing, to the maximum extent possible, dual-use measures that offer societal benefits even if anticipated disasters never occur (Bruneau et al., 2003; Allenby and Fink, 2005).

The book also discusses means of improving the performance of buildings in a flood. Current guidance on flood resilient repairs involves not just the use of materials or a change of design, but the whole process of investigating the extent of flood damage to a building, the drying and decontamination process and the use of risk assessment to set the standard of repair required. It is clear from recent repair standards that many of the issues regarding building technology and flood protection products are already well known, but the scope for further research to enhance flood resilience of the urban fabric is also identified. Areas in which research and innovation are required include, for instance, performance of material related to flood conditions, the role of planning and building regulations, and knowledge of the benefits of improved flood resilience.

The lack of information on the economics of using flood proofing measures for domestic buildings has also been identified in this book. The paper by Zevenbergen et al. illustrates a benefit-cost analysis carried out to enhance knowledge about application of these technologies. The three flood proofing techniques that have been investigated are dry flood proofing, wet flood proofing and building an elevated structure. For this study, functional relations between hypothetical flood damage and inundation depth were established for different types of flood proofing of newly built dwellings and buildings under construction.

In addition to taking measures at the building level, (residual) flood risk can only be sensibly managed through adopting a holistic approach, aimed at creating robustness and redundancy. In this sense, technical measures within the urban area and the wider catchment involve building permanent and semi-permanent flood protection structures. Some of these measures are highlighted in the papers by Salagnac and Bjerkholt and Lindholm, respectively.

The paper by Brilly provides an overview of local flood defence measures in the urban environment. The local flood defence systems are closely connected to the urban drainage system and the related pollution this leads to. These systems have specific characteristics, contrary to the regional and national systems. The local authority usually carries out maintenance of local systems and often these systems are not integrated into the national system. Due to these arrangements information on the performance of these local systems is generally lacking. This paper provides a useful insight describing the policies in local flood protection in a number of countries.

Notwithstanding advances in flood protection measures, urban planning as part of a holistic approach needs to avoid residential developments in areas of particularly high residual risk. Particular caution should be exercised in allocating essential infrastructure, vulnerable uses and developments that attract large numbers of people. In this respect, the move from a strategy of flood defence to one of flood risk management is very significant. The paper by Kelly and Garvin deals with new approaches which focus on managing floods in a more sustainable manner. These include 'making space for water' and creating areas which can be used to accommodate flood water during and after an event.

1.8   OUTLOOK

There is growing recognition in Europe that flood management strategies need to change. This particularly holds true for the urban environment. Living with Flood, Space for Water and 'resilience' are becoming key issues in national and European policies. Support for this is based on a general trend towards sustainable development in urban society. Climate change has become an additional incentive. We must recognize that increasing climate variability is now unavoidable so we need to start adapting to the potential impacts. If we succeed in coping with current extreme flood events, it will help us to deal better with gradual changes associated with climate change.

For the future, a twin approach should be adopted in which (i) precautionary investments need to be made in responding to the changes we think we can predict reasonably well and (ii) an adaptive and incremental approach in which we are less certain about future changes. Such an approach has recently been specified for the different aspects of flood risk management in the UK (Defra, 2006). To achieve this, governments must have access to relevant information and they must involve stakeholders more effectively and be as open as possible about the flood risks involved. The scientific community should develop proper analytical instruments and more cooperation is needed between various disciplines such as (urban) planners and water managers. The private sector and more specifically the building industry has little understanding of the challenges and threats at stake and consequently their participation in research and development in this domain is still much too limited. Ultimately, integrated water management approaches will emerge, but this seems a long way off at the moment, with barriers at each stage: governance, policy, private sector, institutions, and communities (Ashley, 2006). It is expected that this book will not only encourage further relevant research and development in urban flood management, but also help to inform and engage stakeholders in promoting integrated and cooperative approaches in water management in general.

REFERENCES

Adger W.N., Hughes T.P., Folke C., 3 Carpenter S.R., Rockström J. (2005) Viewpoint: Social-Ecological Resilience to Coastal Disasters. Science Vol. 309. no. 5737, pp 1036–1039

Allenby B., Fink J. (2005) Viewpoint: Toward Inherently Secure and Resilient Societies. Science Vol. 309. no. 5737, pp 1034–1036

Ashley R. (2006) The future for water and flood risk management in highly susceptible urban areas; 9th Inter-Regional Conference on Environment-Water, EnviroWater 2006, Delft, the Netherlands

Barredo J.I., Lavalle C., De Roo A. (2005) European flood risk mapping. European Commission, DG-Joint Research Centre, Ispra, Italy

Bruneau M., Chang S.E., Eguchi R.T., Lee G.C., O'Rourke T.D., Reinhorn A.M., Shinozuka M., Tierney K., Wallace W.A., von Winterfeldt D. (2003) A Framework to Quantitatively Assess and Enhance the Seismic Resilience of Communities, Earthquake Spectra, Vol. 19, no. 4, pp 733–752, Earthquake Engineering Research Institute

Cumming G.S., Barnes G., Perz S., Schmink M., Sieving K.E., Southworth J., Binford M., Holt R.D., Stickler C., Van Holt T. (2005) An Exploratory Framework for the Empirical Measurement of Resilience. Ecosystems (2005) 8: 975–987

Defra (2006) Flood and Coastal Defence Appraisal Guidance FCDPAG3 Economic Appraisal. Supplementary Note to Operating Authorities – Climate Change Impacts. October

Europa (2006) Climate Change and the EU's response. Europa press release. 6th November 2006 (MEMO/06/406)

European Construction Technology Platform (ECTP), *Strategic Research Agenda for the European Construction Sector* (2005)

EEA. Climate change and river flooding in Europe. EEA Briefing 2005/01, European Environment Agency, Copenhagen

Hallegatte, S. "The Cost-Benefit Analysis of the New Orleans Flood Protection System" (Center for Environmental Sciences and Policy, Stanford University, and Centre International de Recherche sur l'Environnement et le Développement, Ecole Nationale des Ponts-et-Chaussées, 2006)

Pahl-Wostl C., Downing T., Kabat P., Magnuszewski P., Meigh J., Schlueter M., Sendzimir J., Werners S. (2005) Transition to Adaptive Water Management; The NeWater project. Water Policy. NeWater Working Paper X., Institute of Environmental Systems Research, University of Osnabrückklimate change and river flooding in Europe. EEA Briefing 2005/01, European Environment Agency, Copenhagen

Pahl-Wostl C., Möltgen J., Sendzimir J., Kabat P. (2005) New methods for adaptive water management under uncertainty – The NeWater project. Paper accepted for the EWRA 2005 conference

Penning-Rowsell E.C. (1976) The effects of flood damage on land use planning, Geographica Polonica, Vol. 34, pp 135–153

2

Sustainable Measures for Flood Attenuation: Sustainable Drainage and Conveyance Systems SUDACS

Joachim T. Tourbier
Dresden University of Technology, Dresden, Germany

Iain White
Centre for Urban and Regional Ecology (CURE), School of Environment and Development, University of Manchester, UK

ABSTRACT: On-site stormwater management, or sustainable drainage systems (SUDS) have been conceived to satisfy the ecologic, social and economic aspects of sustainability. In certain European countries SUDS are supported by institutional arrangements, covering aspects of runoff attenuation, quality improvement, infiltration, and detention. It is being shown that SUDS often only control the volumes and peaks of minor storms and only reduce flood risk to a limited extent. This situation can be improved if SUDS include conveyance systems for the routing of drainage that exceeds the capacity of measures traditionally considered to be part of the "treatment train". It is recommended to expand the concept of SUDS to make them "Sustainable Drainage and Conveyance Systems" (SUDACS) to fully include routing of exceedance flows. With such an expansion SUDS could more adequately address the needs of flood management in Europe, which needs to cover the prevention, protection, preparedness, emergency response and recovery from floods.

2.1 INTRODUCTION

In the United States a first manual on Best Management Practices for on-site stormwater management was issued under the Clean Water Act (WPCA, 1972) in the mid-nineteen-seventies (Tourbier and Westmacott, 1974) and ordinances have been passed for water quality control, infiltration and on-site detention up to the 100 year frequency storm (Tourbier and Walmsley, 1994). "Low Impact Development" (LID) in stormwater management is being widely advocated on the east coast of the US (www.lid-stormwater, 2006). In Germany the IBA Emscherpark International

13

Construction Exhibit (1989–99) sparked off an approach of "Near Natural" stormwater management, including infiltration that is being advocated by the "Emscher-Lippe Verband" in the Ruhr Valley (Sieker, Kaiser, Sieker 2006). In the United Kingdom there has been a considerable initiative through (CIRIA), the "Construction and Industry Research and Information Association" (www.ciria.org, 2006) advocating Sustainable Drainage Systems (SUDS). In 2001 a first planning advisory note on SUDS was published in Scotland, to be followed by the consultation document *Framework for Sustainable Urban Drainage Systems (SUDS) in England and Wales* in 2003 (NSWG, 2003). In 2004 an *Interim Code of Practice for Sustainable Drainage Systems* (NSWG, 2004), was published by the National SUDS Working Group. It lead up to the *SUDS Best Practice* manuals as well as documents to aid implementation. CIRIA has now published "Designing for Exceedance in Urban Drainage – Good Practice" (CIRIA C635) considering floods that exceed system capacity. The ongoing UK project *AUDACIOUS* will be modelling such drainage exceedance www.eng.brad.ac.uk/audacious (2006). Despite the potential benefits, on-site stormwater conveyance of major storms is not yet a standard approach in the UK, nor is the practice of SUDS yet generally accepted on the European Continent.

2.2   THE CONCEPT OF SUSTAINABLE DRAINAGE

Throughout Europe there has been an increase in the frequency and depth of flooding with the related property damage and loss of life. The causes are the spread of urbanisation and the affiliated impervious surfaces exacerbated by climate change, which affects the intensity and frequency of precipitation. In Germany for example, built up urban areas have more than doubled in the last 70 years, from 5.1% to 12.3% (Kaiser, 2005). Land use changes through urbanisation cause substantial changes in the volume and peak flows of stormwater runoff. The volume of runoff generated by impervious surfaces can be as much as 20 times that of an undeveloped site (Lampe, 2005). Fig. 2.1 shows typical hydrograph changes for both an urban area and one with natural ground cover. In an undeveloped environment 10% of rainfall forms runoff, 40% evapotranspirates and 50% infiltrates into the ground, whereas in a site with 75–100% paved surfaces 55% of the rainfall runs off, 30% evaporates and only 15% infiltrates.

Since the nineteen-seventies on-site stormwater management follows the approach to internalise the responsibility for stormwater control, by charging those who cause problems with the responsibility of solving them, rather than to have all taxpayers pay for corrective actions (Heaney *et al.*, 1975). However, despite the construction of ever larger and more expensive federally funded flood control dams in the US flood losses were mounting. Progressive municipalities there started to implement ordinances that require post-development stormwater

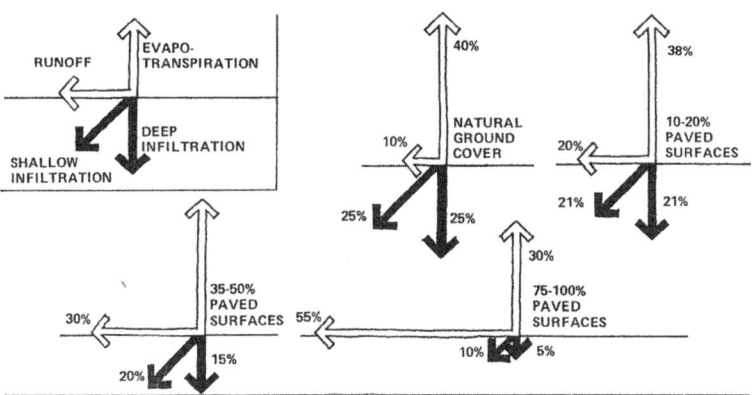

Figure 2.1. Typical hydrograph changes due to increases in impervious surface areas (*Source*: Tourbier and Westmacott, 1981).

discharge peaks not to exceed pre-development peaks. The 1992 Earth Summit in Rio, which produced Agenda 21, also internalised responsibilities by encouraging participating countries to develop sustainable development strategies at local and national government levels.

In the UK the Office of the Deputy Prime Minister put forth the *Planning Policy Guidance Note 25: Development and Flood Risk* (DTLR, 2001), for application in England. It requires that runoff from development should not increase flood risks elsewhere in a catchment area and that sustainable drainage systems be incorporated wherever practicable. It is also being argued in Britain that Integrated Urban Water Resources Management (IUWRM) should be practiced as part of Integrated Catchment Management, as set forth in the EU Water Framework Directive. In the UK, there is also widespread encouragement for SUDS and to modify the current right of connection of piped surface water drainage into the existing sewer (Tyson, 2004).

In Germany there are several legal provisions with implications on SUDS. A new law for improved preventive flood protection (GVVH, 2005) as well as the federal working commission on water (LAWA, 1995) hold, among others, that it is advantageous to detain and to infiltrate floodwater where it is generated. Many German cities charge separate stormwater fees in an effort to make the determination of water and sewer fees more transparent. The city of Berlin charges land users app. 1.30 Euro/year per square meter of impervious surface, unless there is an on-site disposal for runoff, such as infiltration practices. This provides a substantial incentive for the implementation of SUDS.

SUDS are a range of flexible drainage techniques that alter the focus of drainage design, practice, construction and maintenance to facilitate a higher consideration for society in general and the receiving environment (CIRIA, 2000a). There are

Preventative measures	The first stage of the SUDS approach is to prevent or reduce pollution and runoff. This may include good housekeeping, to prevent spills and leaks, storage in water butts, rainwater harvesting systems, and alternative roofs (i.e. green and brown roofs).
Pervious surfaces	Surfaces that allow inflow of rainwater into the underlying construction or soil.
Green roofs	Vegetated roofs that reduce the volume and rate of runoff and remove pollution.
Filter drains	Linear drains consisting of trenches filled with a permeable material, often with a perforated pipe in the base of the trench to assist drainage, to store and conduct water; they may also permit infiltration.
Filter strips	Vegetated areas of gently sloping ground designed to drain water evenly off impermeable areas and to filter out silt and other particulates.
Swales	Shallow vegetated channels that conduct and retain water, and may also permit infiltration; the vegetation filters particulate matter.
Basins, ponds and wetland	Areas that may be utilised for surface runoff storage.
Infiltration devices	Sub-surface structures to promote the infiltration of surface water to ground. They can be trenches, basins or soakaways.
Bioretention areas	Vegetated areas designed to collect and treat water before discharge via a piped system or infiltration to the ground.
Filters	Engineered sand filters designed to remove pollutants from runoff.
Pipes and accessories	A series of conduits and their accessories normally laid underground that convey surface water to a suitable location for treatment and/or disposal. (Although sustainable, these techniques should be considered where other SUDS techniques are not practicable.)

Figure 2.2. Summary of SUDS components (National SUDS Working Group, 2004).

a number of different types of SUDS, which can be used either as an individual system or an integrated network of techniques. Whilst each SUDS measure might only bring a small amount of benefit, the cumulative effects over an entire catchment can be significant. Fig. 2.2 provides an overview of the most popular SUDS components[1].

---

[1] Note that although SUDS use predominately "soft" engineering techniques, they can also use hard engineering solutions.

SUDS are more sustainable than conventional drainage techniques, offering the following benefits:

Flood related benefits:
Attenuation of runoff prior to concentration
- Reduction of runoff peaks
- Reduction of runoff volumes
- Reduction of flood related erosion and deposition in channels and reservoirs

Water quality benefits:
- Through a passive level of treatment the quality of runoff is improved (CIRIA, 2000a; SEPA and EA, 1999)

Groundwater benefits:
- Pre-development groundwater recharge rates can be maintained through infiltration

River related benefits:
- Reduction in Floodwater flows that cause channel degradation through erosion of stream bed and banks
- Minimization of adverse flood impacts on the environment

Social and economic benefits:
- Reduction of flood damage to property
- Reduction of flood related public health and safety problems
- Creation of visual enhancement and amenity
- Passive recreation
- Employment opportunities in construction and maintenance

The storage of stormwater has been identified as one of the key mechanisms to protect against damaging floods (Defra, 2004; Evans *et al.*, 2004). Yet, in practice there are a number of significant barriers inhibiting the widespread use of SUDS (White and Howe, 2004b). White (2005) demonstrated that the impact of seemingly robust SUDS planning policies is severely undermined by a lack of wider information and guidance, as many developers continue to manage runoff using conventional engineering approaches.

Detention or retention basins can store precipitation and so increase the level of rainfall needed before flash flooding occurs. Utilisation of SUDS may also reduce the volume of surface water that the drainage infrastructure has to manage and so lessen the possibility of sewer flooding due to storm events. A higher utilisation of basins in a catchment also reduces the demand on flood defenses downstream, changing the emphasis from a protectionist to preventionist outlook and help to adapt to the challenges of climate change (Evans *et al.*, 2004; White and Howe, 2002).

Finally, on the EU level the "Directive of the European Parliament and of the Council on the Assessment and Management of Floods" {SEC 92006 66} was published in 2006. It covers the steps of "prevention, protection, preparedness, emergency response, recovery and review". SUDS, when expanded to include conveyance systems will help to meet those demands.

2.3   DESIGN AND PLANNING APPROACHES

A range of studies has been conducted to address a widespread concern about the performance of SUDS. A comparison of pre- and post development discharge peaks of newly constructed residential sites in Germany (Kaiser, 2005) show that SUDS consistently bring a reduction of peak flows. The comparison to peaks generated by conventional drainage is substantial (see Fig. 2.3).

In a recent study US and UK researchers conducted *Post-Project Monitoring of BMP's/SUDS to determine Performance and Whole-Life Costs* (Lampe, 2005). The relative hydrologic performance of individual facilities was determined through computer modelling for observed events, design storms, and five years of stochastically generated rainfall data. The study also provided valuable data on maintenance costs.

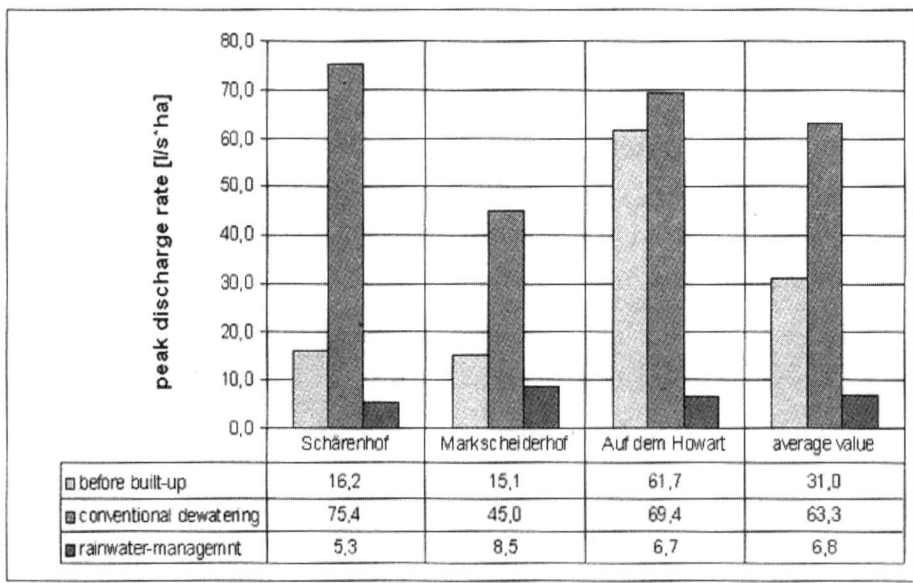

	Schärenhof	Markscheiderhof	Auf dem Howart	average value
before built-up	16,2	15,1	61,7	31,0
conventional dewatering	75,4	45,0	69,4	63,3
rainwater-managemnt	5,3	8,5	6,7	6,8

Figure 2.3. Pre- and post development runoff peaks of four residential developments in comparison to conventional drainage (*Source*: Kaiser, 2005).

Design criteria for the sizing of on-site stormwater management facilities can vary. In the US many municipalities that have adopted a view that stormwater management ordinance requires that post-development discharge peaks for the 2, 5, 10, 20, 50 and 100-year frequency storms should be equal to the pre-development peak. However, in areas adjacent to large bodies of water such detention may not be meaningful. There also is a trend in the US to primarily control the 25- to 50-year storm, and to create flood routing depressions and passages to handle larger storms (Tourbier, 2003). A further concern in the US, where on-site detention is routinely used, is that delayed releases may contribute to flooding when the longer duration discharges from detention basins coincide with discharges of upstream sub areas. This point reinforces the need for integrated catchment management.

2.4   A PLANNING PROCESS FOR SUSTAINABLE DRAINAGE AND CONVEYANCE SYSTEMS (SUDACS)

The concerns associated with the current concept of SUDS provides a strong argument that their remit should be expanded to include both Sustainable Drainage and Conveyance Systems (SUDACS) to increase their potential to mitigate flood risk. In practice SUDS are rarely designed to detain runoff volumes and peaks that exceed the 30-year frequency storm. This means that facilities overflow during exceedance of these storms, not protecting against flood potential. Fig. 2.4 below shows a "treatment train" of SUDS, addressing the runoff volume generated by a design storm. Volume reduction can occur through attenuation, quality improvement measures, infiltration and detention. The 30, 50, 100, and 200-year frequency storms will generate exceedance flows. Their management should become an integrated component of SUDS.

Figure 2.5 shows stormwater conveyance measures for channel stabilisation through transverse and flow parallel structures, point protection, stabilisation of upper bank areas, and measures for emergency floodways and flood damage control, providing benefits in addition to conventional SUDS.

The planning procedure for such systems should start with (1) an analysis of pre-development condition and should consider (2) goals and objectives to be defined

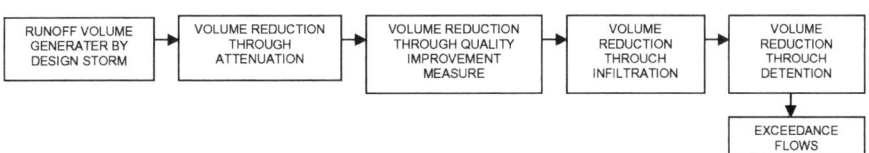

Figure 2.4.   The sustainable urban drainage and conveyance system.

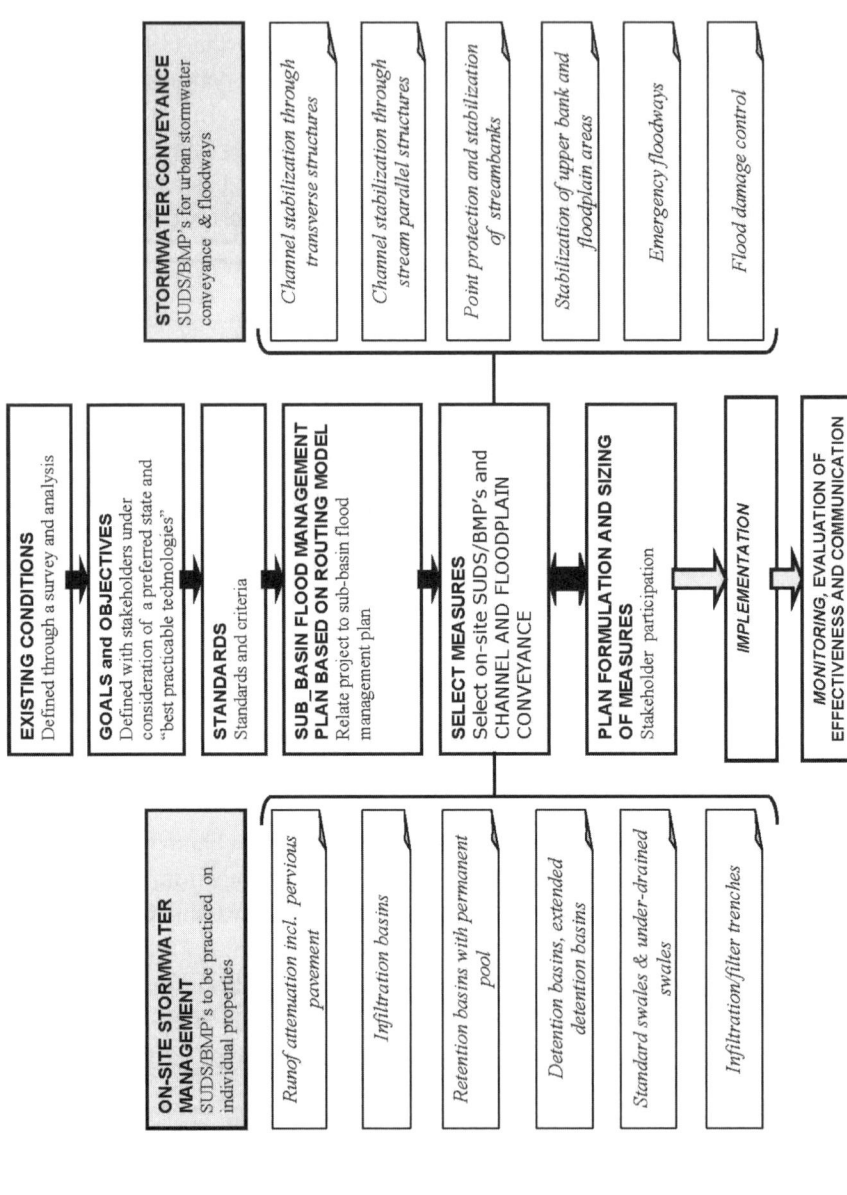

Figure 2.5.   Flow diagram SUDACS planning procedure (*Source:* Tourbier, URBEM, 2005).

under stakeholder participation, leading to the definition of (3) standards, criteria and performance controls. A site concept should then be related to a (4) sub-basin flood management plan that includes a flood routing model. Steps 1–4 should be completed before (5) on-site SUDS and stormwater conveyance measures are defined. The plan (6) to be formulated should also provide for exceedance flows, leading to (7) implementation and (8) monitoring and evaluation of effectiveness and the communication of results.

2.5   DESCRIPTION OF MEASURES

Section 6 describes and discusses the various measures that could be employed for on-site stormwater management and conveyance. It is anticipated that this description will enable the SUDACS process to be understood in more depth.

2.5.1   Runoff attenuation including pervious pavement

The differing aspects of surface cover have a substantial effect on the rate and volume of runoff. Porous pavement, gravel road surfaces and modular porous pavers can all provide for infiltration where precipitation falls, whilst extensive roof covers and evaporation through landscaped areas can both intercept precipitation and allow evapotranspiration back to the atmosphere (Fig. 2.6). Furthermore, exceedance flows can be directed to additional measures. Evapotranspiration is being increasingly considered as important mechanism in reducing runoff rates. The Bagrov-Glugla model (Glugla *et al.*, 2003), as recommended by Harlass and Herz (2006) can be used to calculate the water balance in urban areas with consideration to evapotranspiration losses. It considers the following:

$$\frac{d\overline{ET_a}}{d\overline{P_{korr}}} = 1 - \left(\frac{\overline{ET_a}}{ET_{\max}} \right)^n$$

P_{korr} Precipitation (mm/a) after correcting systematic errors of measurement
ET_a Actual Evapotranspiration
n Bagrov Parameter of Effectiveness
$ET_{\max}$ Maximum Evapotranspiration

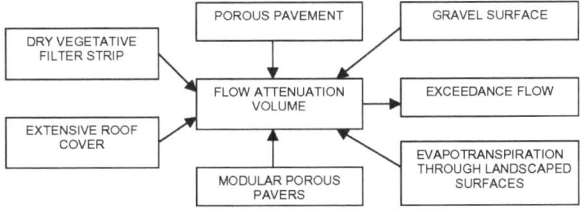

Figure 2.6. Runoff attenuation measures.

$$\overline{ET}_{\max} = \overline{ET_0} * f_I$$

with ET_0 Evapotranspiration of Standard Gras (mm/a)
       f_I Factor Depending on Type of Land Use, Slope and Exposure

Another example of runoff attenuation is pervious pavement, which can be installed to include a reservoir course with a void ratio designed for runoff detention. A sample cross section is shown in Fig. 2.7 below.

2.5.2   Water quality improvement

Runoff pollution can be defined by the type of surface that the runoff is generated from. Treatment can be achieved through biological treatment ponds, extended detention basins, constructed wetlands, or dry vegetative filter strips. The 2 year or 5 year frequency storm is usually used as the standard design, after which exceedance flows will be passed on.

2.5.3   Infiltration

Distance to groundwater table, infiltration rates of soil and the consideration of hazardous wastes in soil can all influence the selection of below ground or above

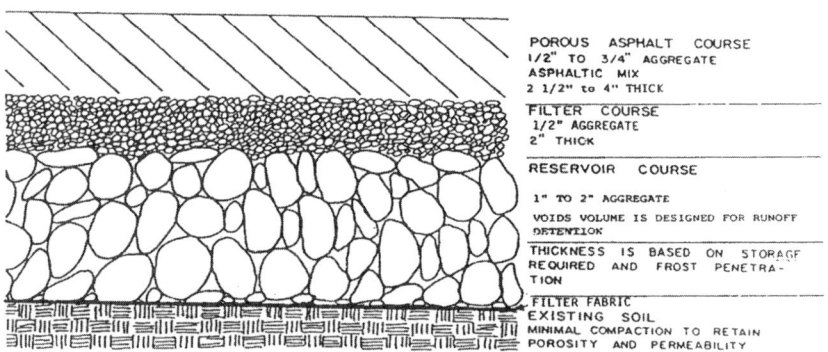

Figure 2.7. Pervious asphalt (*Source*: Tourbier, 2003).

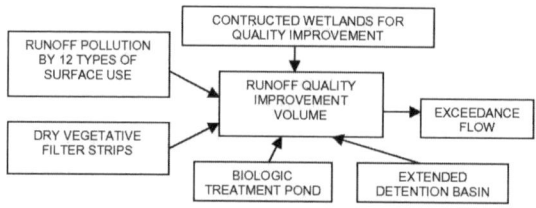

Figure 2.8. Water quality improvement measures.

ground infiltration measures and their respective infiltration volumes. It is considered that infiltration from a two-year frequency storm may be sufficient to maintain pre-development base flow of streams. However, although infiltration devices can divert runoff there may still be exceedance flows during major storm events. An example of measures with limited recharge potential are infiltration swales, shown in Fig. 2.10 below, while infiltration basins, shown in Fig. 2.11, have the benefit of considerable storage, but tend to have problems with clogging over time and need regular inspection (ASCE, 1992).

2.5.4   Detention

Detention basins and detention areas hold back, and slowly release runoff. They may be wet – or dry detention basins. Surface area, spillways and dam width have

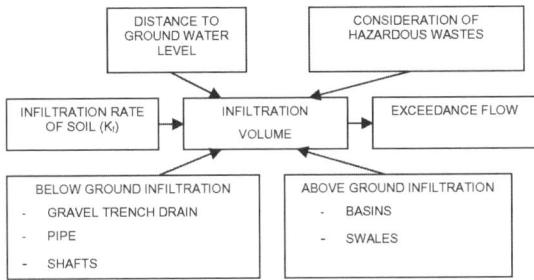

Figure 2.9.   Infiltration measures.

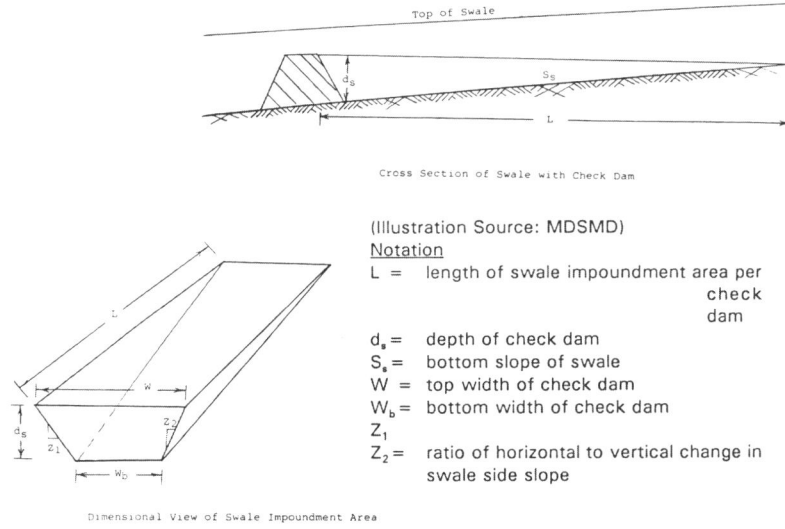

(Illustration Source: MDSMD)
Notation
L   =   length of swale impoundment area per
                                                check
                                                dam
d_s =   depth of check dam
S_s =   bottom slope of swale
W =   top width of check dam
W_b =   bottom width of check dam
Z_1
Z_2 =   ratio of horizontal to vertical change in
            swale side slope

Figure 2.10.   Infiltration swales (*Source*: Tourbier, 2003).

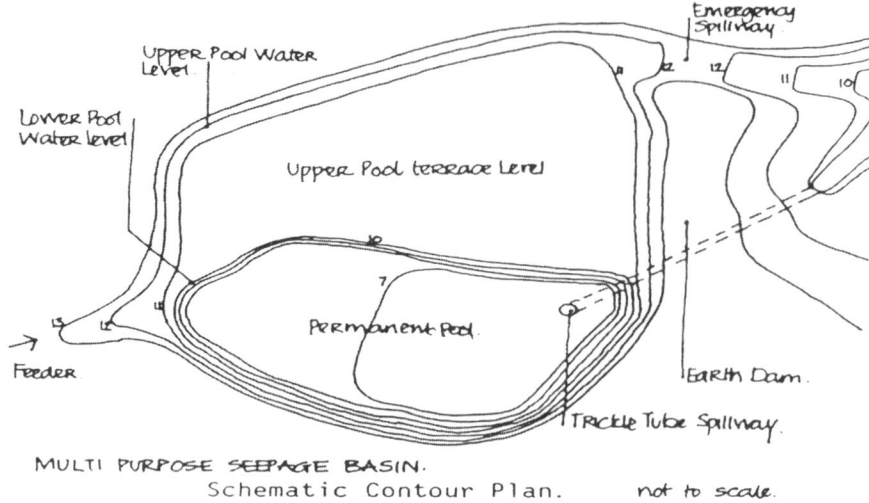

Figure 2.11. Infiltration basin (*Source*: Tourbier & Westmacott, 1981).

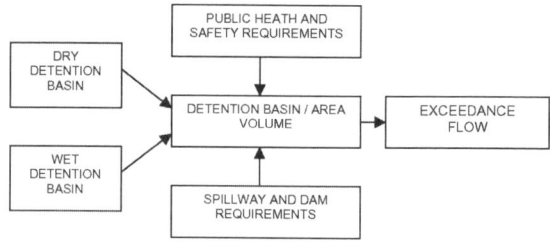

Figure 2.12. Detention facilities.

considerable space requirements. There also are public health and safety concerns, limiting the steepness of side slopes and affecting the size of basins in urban areas where space is at a premium. Although detention basins can hold back precipitation, dependant upon design of the mechanism there may also be exceedance flows during a major storm event.

2.5.5   Surface conveyance

Fig. 2.13 above shows how a digital terrain model and runoff model that subtracts storage in SUDS will yield above ground excedence volumes and flows that need to be conveyed through a site. Conveyance may be accommodated in vegetation areas, on pedestrian ways, roads and parking lots. The objective of this conveyance

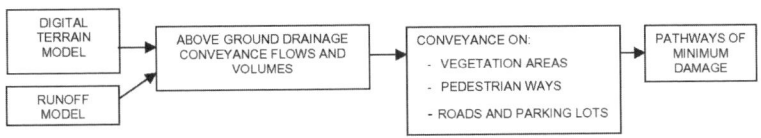

Figure 2.13. Surface conveyance systems.

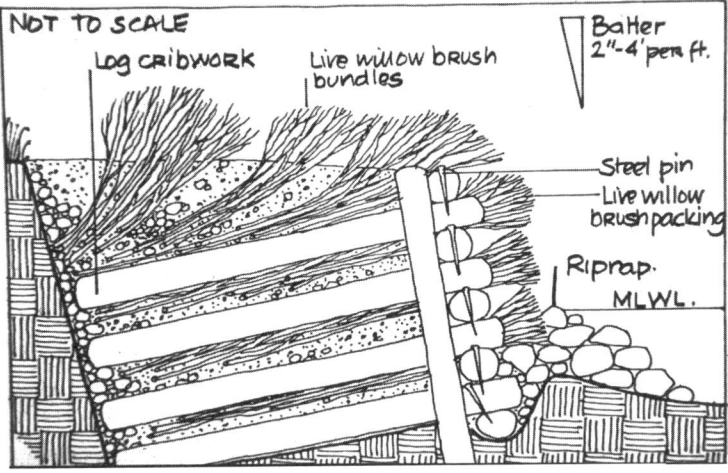

Figure 2.14. Life cribwall for point protection of emergency floodways in parks (*Source*: Tourbier and Westmacott, 1981).

method is to define a pathway of minimum potential damage. Channel conveyance can be calculated by using the Manning Equation:

$$Q = \frac{1}{n} A_C R^{\frac{2}{3}} s^{\frac{1}{2}}$$

where
Q = discharge, m^3/s
N = Manning roughness value
R = hydraulic radius = NP
S = slope (decimal)
A_C = cross-sectional area of flow, m^2
P = wetted parameter

Surface channels for exceedance storm flows should have non-erosive surfaces. Soil-bioengineering techniques that combine mechanical and vegetative techniques to protect critical areas on banks against erosion can included vegetated brush packing, wing deflectors, brush matrices and life crib walls (Fig. 2.14) that can be integrated into the landscaping of emergency flood channels.

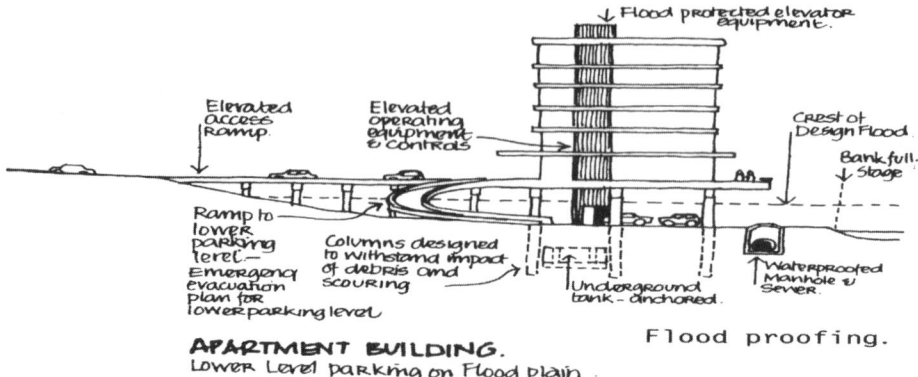

Figure 2.15. Flood proofing and emergency access on floodplains (*Source*: Tourbier and Westmacott, 1981).

Prevention, protection and preparedness, can be achieved through the careful routing of emergency foodways through urban areas, while response, recovery and being prepared for emergencies are aided through flood proofing and emergency access provisions (Fig. 2.15). All of these components should be part of stormwater management planning efforts.

In summary, we argue that the potential of future problems caused by conventional drainage could be increased due to the impact of climate change and the rising demand for housing, presenting environmental planning and society in general with considerable challenges. Therefore, alternative approaches to managing surface water can help to mitigate these impacts and achieve a more sustainable outcome, a view supported by recent research (DEFRA, 2004; Evans *et al.*, 2004).

Sustainable Drainage Systems as a drainage approach are inspired by natural processes and gaining in sophistication (CIRIA, 1992, 1996, 2000a, 2000b), having the potential to prevent stormwater from having a negative influence on society (CIRIA, 2000a; Howe and White, 2001). Yet, their benefits to flood mitigation could be improved. We recommend that the conveyance of exceedance flows needs to become an integrated component of sustainable drainage systems (SUDACS) and that this new emerging method should become an integrated component of both, city planning and urban design.

REFERENCES

ASCE, American Society of Civil Engineer, *Design and Construction of Stormwater Management Systems*, New York NY, 1992

CIRIA, Construction Industry Research and Information Association, *Scope for Control of Urban Runoff*, Report 124, Vol. 1,2,3, Construction Industry Research and Information Association, London, 1992

CIRIA, Construction Industry Research and Information Association (2000a) *Sustainable Urban Drainage Systems – Design Manual for England and Wales*, London, 2000

CIRIA, Construction Industry Research and Information Association (2000b) *Sustainable Urban Drainage Systems – Best Practice Manual*, London, 2000

CIRIA, Construction Industry Research and Information Association (1996) *Infiltration Drainage – manual of good practice*, London, 1996

CIRIA Construction Industry Research and Information Association, C635 Balmforth D., Digman C., Kellagher R.B.B., Butler D., Designing for Exceedance in Urban Drainage – Good Practice. ISBN 0-86017-635-5, 2006

COST, Memorandum of Understanding, Technical Annex, March 2000

DTLR (2001) *Planning Policy Guidance Note 25: Development and Flood Risk*, DTLR, London

Geiger, W., Dreiseitl, H. *Neue Wege für das Regenwasser*. IBA Emscherpark GmbH. 2001

GVVH (2005) Gesetz für einen verbesserten vorbeugenden Hochwasserschutz der Bundesrepublik Deutschland, 2005

Glugla, G., Jankiewicz, P., Rachimow, Lojek, K., Richter, K., Fürtig, G., Krahe, P. *Wasserhaushaltsverfahren zur Berechnung vieljähriger Mittelwerte der tatsächlichen Verdunstung und des Gesamtabflusses* – BFG Bericht 1342. Koblenz, Bundesanstalt für Gewässerkunde, 2003

Harlass, R., Herz, R. *Water Balance in Urban Development – Accounting for Evapotranspiration*. Paper presented at the COST C13 End of Action Conference on "Sustainable Urban Development" in Bergen, Norway, Aug. 24/25, 2006

Heaney, J.P. *et al. Urban Stormwater Management Modelling And Decision Making*, Gainesville, FL: University of Florida, 1975

Kaiser, M. *Near Natural Stormwater Management as a Model in the Sustainable Development of Settlements*, in Tourbier, J.T., Schanze, J. (Eds.) *Urban River Rehabilitation Proceedings – International Conference on Urban River Rehabilitation*. Technische Universität Dresden & Leibniz Institute of Ecological and Regional Development, Dresden, ISBN 3-933053-29-3, 2005

Lampe, L. *et al. Post-Project Monitoring of BMP's/SUDS to Determine Performance and Whole-Life Costs*. Alexandria, VA: Water Environment Federation, 2005

LAWA (1995) Länderarbeitsgemeinschaft Wasser und Abfall, 1995

National SUDS Working Group (2003) *Framework for Sustainable Drainage Systems (SUDS) in England and Wales*. Copy available from www.environment-agency.gov.uk/yourenv/consultations/

National SUDS Working Group, *Interim Code of Practice for Sustainable Drainage Systems*. ISBN 0-86017-904. Copy available from www.ciria.org/suds/, 2004

Office of the Deputy Prime Minister (2005) *Consultation on Planning Policy Statement 25:Development and Flood Risk* ODPM, London

Department of the Environment, Food and Rural Affairs (2004) *Making space for water: developing a new Government strategy for flood and coastal erosion risk management in England*, Department of the Environment, Food and Rural Affairs, London

Sieker, F., Kaiser, M., Sieker, H. Dezentrale Regenwasserbewirtschaftung im privaten, gewerblichen und kommunalen Bereich. Stuttgart, Fraunhofer IRB Verlag, 2006

Tourbier, J., Westmacott, R. *Water Resources Protection Measures in Land Development – A Handbook.* For the Office of Water Resources Research of the U.S. Department of the Interior. University of Delaware Water Resources Center. Recommended reading in EPA's "Section 208 Guidelines for Area Wide Waste Treatment Planning" 1974

Tourbier, J.T. Westmacott, R. *Water Resources ProtectionTechnology*, Washington, DC: Urban Land Institute, 1981

Tourbier, J.T. in *Urban River Basin Enhancement Methods (URBEM), Work Package 8, New Techniques for Urban River Rehabilitation*, 2006

Tourbier, J.T. *Blue Green Technologies, Integrated Practices to Manage Stormwater as an Asset*, Geraldine R. Dodge Foundation, USA, Madison, NJ: Great Swamp Watershed Association, 2003

Tourbier & Walmsley, London, *Grove Township Stormwater Management Ordinance*, Avongrove, Pa., 1994

Tyson, J.M. *A Review of Current Knowledge – Urban Drainage & the Water Environment: a Sustainable Future?* Marlow: Foundation for Water Research, 2004

WPCA 1972, Water Pollution Control Act Amendment of 1972, PL 92–500

www.lid-stormwater.net/intro/background.htm, 30.10.2006
www.cira.org/suds/background.html, 30.10.2006
www.eng.brad.ac.uk/audacious/, 30.10.2006

3

Characterisation of Urban Streams and Urban Flooding

Ian Douglas[1], Mira Kobold[2], Nigel Lawson[1], Erik Pasche[3] &
Iain White[1]
[1] *University of Manchester, UK*
[2] *University of Ljubljana, Slovenia*
[3] *Technical University Hamburg-Harburg, Germany*

ABSTRACT: This chapter explores a number of issues concerned with urban flooding. It firstly examines the broad problems of rural and urban flooding before focusing on the classification of the different types of urban watercourse, from major rivers to sewers to minor streams, found in urban areas. Secondly, the paper explores in depth the impact of runoff in contributing to flood risk in urban areas, providing both a broad discussion and a specific investigation of flow processes in water bodies. Section four expands on the issue of runoff by providing on overview of the influence of land use on urban flood risk. Sections five and six attempt to link the information discussed thus far by examining the different types of flood event that these runoff regimes, flow processes and land uses may create and introduces two brief case studies describing how cities in the UK have suffered from flooding. It concludes that unwise management practices and climate change are exacerbating urban flooding and that there is a need for a holistic approach towards improving our understanding of the complexities of urban flood risk.

3.1   INTRODUCTION

Floods are natural recurring hydrological phenomena that have been affecting human lives from time immemorial. They become disasters because of their impacts on people and ecosystems, particularly in terms of human life and property damage (Smith and Ward, 1998; WMO, 2004). Munich Reinsurance data (Bruen, 1999) show that floods cause nearly one third of natural catastrophes. With rivers providing water for human use and industrial production and valley-floor alluvium often producing best agricultural land, urban settlements grew on river banks. They gradually organised protection against flooding by building dikes and embankments. However, streams carrying eroded material from the upper parts of the watershed deposited some of their load, gradually raising the natural riverbed. Eventually levees develop along the river raising the height of the banks, but leaving lower land further away from the river liable to flooding.

Floods are associated with extreme natural events that happen on a geographical area, such as a river basin, a catchment area or a watershed. These areas can be rural and urban, the former commonly being much larger than the latter. Hence, flooding can be rural and urban. An extreme natural event only becomes a disaster when it has an impact on human settlements and activities. At the European scale, 5 types of urban rivers and watercourses can be envisaged, each causing differing problems associated with flooding:

1. Major rivers, such as the Danube, Elbe and Rhine, adjacent to which urban areas have developed;
2. Rivers rising in adjacent mountains or uplands, such as the Drava and Mura in Slovenia and the Ribble and Mersey in England, which descend relatively steep courses from the highlands and then enter urban areas;
3. Streams whose entire catchment area, or watershed, lies totally within the built-up area, such as the Irk in Manchester and the Brent or Wandle in London;
4. Small urban streams, that an adult can jump across, that have been totally or partially culverted, such as the Cornbrook and Baguley Brook in Greater Manchester.
5. Sewers and the total artificial urban drainage network.

These five broad types of watercourses can cause flooding in both rural and urban areas. The basic cause of rural or river basin flooding is heavy rainfall or rainfall combined with snowmelt, followed by slow development of flood flows, which exceed the capacity of natural waterways. Other causes of rural floods are:

- surcharge in water levels due to natural or man-made obstructions in the flood path (bridges, gated spillways, weirs)
- sudden dam failure
- landslide
- mud flow
- inappropriate urban development (excessive encroachment in the floodway)
- ice jam
- rapid snowmelt
- deforestation of the catchment basin

Rural floods are river-basin events, whereas urban floods can have both area-wide and local origin, and can be accompanied by water pollution problems. As the proportion of the stream catchment area that is urbanized increases, and as the degree of management, modification and channelisation increases, so the flashiness of floods tends to increase.

From an urban perspective, flooding has become increasingly severe in most cities and usually occur in built-up areas as a result of heavy rain and the large amount of rain water that subsequent runoff. Rivers overflowing do not always cause floods; they can also be caused by high rain intensities over the city combined with inappropriate sewer systems and diverse urban land cover. However, generally, flooding in cities originates from extreme high flows and stages in major neighbouring rivers as a result of severe regional meteorological disturbances, or from local high intensity thunderstorms occurring over parts of the urban area. Thus, management of urban floods requires both the knowledge of physical characteristics of specific flood events and understanding of urban hydrometeorological issues. Urban flooding therefore requires special attention due to its sheer complexity incorporating a host of social, economic, institutional and technical factors within both rural and urban environments.

Furthermore, it should be noted that flood occurrences are not bound by local administrative areas because storm water drainage and protection facilities are part of an environmental system that is usually larger than an incorporated city territory (Andjelkovic, 2001). Thus, effective management of risk can present difficulties due to the disparity between natural and administrative boundaries and also between the differing strategies in place within neighbouring local, regional or national areas (White and Howe, 2002, 2004a). This is especially the case within urban areas as the policy framework becomes increasingly complex and strategies to research and mitigate flooding need to understand the broader drivers of risk, such as runoff, discussed in section 2.

3.2   THE IMPACT OF RUNOFF

3.2.1   Processes acting on the formation of surface runoff

In an urban area the management of surface runoff has been highlighted as a key driver on flood risk (White and Howe, 2004b). Horton's well-known concept (Horton, 1933) has been used widely to explain the surface runoff in terms of the precipitation water that exceeded the infiltration capacity of the topsoil layer (infiltration-excess water). This concept has been subsequently been developed and we now understand the surface runoff generation process in much more depth. Surface runoff (Figure 3.1) is not only generated by infiltration excess flow but also on saturated topsoil layers, and on water bodies (lakes, rivers and streams) as saturation flow (Dunne, 1978). Subsurface runoff can be generated by rapid throughflow of newly infiltrated storm water within macropores and soil pipes feeding directly into the stream flow. In case of saturation of the soil matrix (micropores) and the macropores pre-storm soil water returns to the surface through the additional water pressure created by the blocked runoff within the macropores at the depression zones. Thus this return runoff mainly occurs close to rivers and is significant on concave hillslopes and in wide, flat river valleys (Dunne *et al.*, 1975; Tanaka *et al.*, 1988). Furthermore, surface crusts can develop on loamy soils by the direct impact of heavy rain drops on the surface. The raindrop impact creates a sealing effect

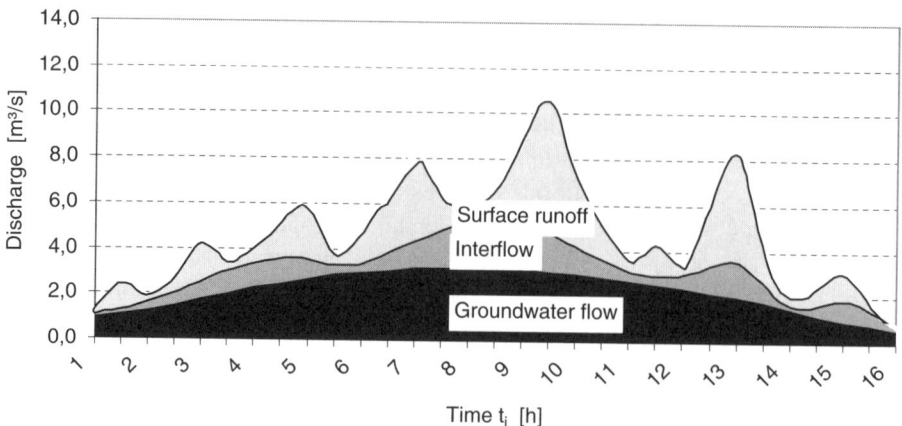

Figure 3.1. Runoff components of a flood hydrograph.

on the soil surface. Water can only infiltrate through the remaining cracks (Roth, 1992). Such situations commonly arise on farm land that lacks a vegetation cover.

3.2.2   Processes acting on the formation of subsurface flow

Interflow, or hypodermic flow, (Figure 3.1) forms in the unsaturated soil layer and flows more or less directly through this soil layer to the river. It can be divided into slow and fast interflow. While slow interflow results from long flow paths in the soil layer, fast interflow occurs in areas with a distinct topographic gradient close to rivers and in unsaturated soil zones with decreasing permeability with depth, or a layered soil profile in which the permeability of the lower layer is much less than the upper one (by power of 2 to 3 for sandy soil and power of 1 for loamy soil (Peschke *et al.*, 1999)). Zuidema (1985) showed that the interflow can occur in the network of macropores, soil pipes and cracks or through porous medium along an impervious layer (piston-flow). In steep regions (e.g. at hillsides) gravity can drive free soil water through the soil matrix, called matrix-throughflow (Kirkby and Chorley, 1967). Groundwater ridges can block interflow draining into wide flat valleys and direct the groundwater into the river by lifting the groundwater table (Blowes and Gillham, 1988).

In urban areas interflow is reduced by various people-made environmental changes. First, because the sealing of the surface reduces the infiltration of the precipitation, less free water is available in the unsaturated soil layer. Furthermore, much of the original natural soil layer may have been removed or have been consolidated by construction activities. These processes reduce the number and size of macro- and micropores, so decreasing the formation of interflow.

Because urban soils are often developed on composite materials derived from previous uses and exogenous sources, spatial heterogeneity is a typical feature. Their evolution is controlled almost exclusively by humans, who impose very rapid transformation cycles compared with those occurring in less disturbed areas. However, there is a continuum from the natural soils to the extensively disturbed soils, and their basic functions are essentially the same. As a result of their origin and uses, urban soils may contain pollutants that can find their way into interflow and groundwater and thus into rivers.

3.2.3   Processes acting on the formation of groundwater flow

Groundwater is subsurface water which fills totally the pores and cavities of the lithosphere (saturated zone). It comprises about 97% of the fresh water on earth. Although the groundwater aquifers can be regarded as huge water reservoirs, the groundwater table shows a distinct pressure gradient through which the water in these reservoirs is subjected to continuous flow and exchange processes with rivers, lakes and the unsaturated zone. Between 60 and 80% of the long-term flow volume in rivers is a result of groundwater discharge. Compared to interflow and surface runoff this flow process is much slower and can take years before the infiltrated precipitation water can reach the river. Due to these hysteresis effects, the outflow of groundwater into rivers persists even during long periods of no rain. Thus the groundwater is responsible for the dry-period flow in rivers (also named baseflow). However, during floods, rapid changes in groundwater outflow can be observed, which are not due to lateral flow processes through the aquifer. In confined aquifers a distant rise of the groundwater table through infiltrated precipitation water can increase the groundwater pressure close to the river, similar to the principle of communicating tubes in hydraulics. This process is often called piston flow, because the new water pushes out the old water. Old groundwater outflows along the river, while new groundwater enters distant aquifer regions.

Like interflow, groundwater in urban areas is affected by land surface changes. Surface sealing reduces groundwater recharge. The overall amount of groundwater in urban watersheds declines and the groundwater table is lowered, reducing the groundwater discharge to rivers. In dry periods, rivers draining urban catchments have lower runoff than rural streams of comparable size and at times may become totally dry.

The characteristics of groundwater baseflow may have a significant influence on the quality of urban surface watercourses. Transport of dissolved phase contaminants from the aquifer to the river will take place across the groundwater/surface water interface where processes are governed by the rapid change in physical and chemical conditions. Mixing of groundwater and surface water occurs at variable depths below the riverbed within the hyporheic zone where sorption and degradation may be significant with high levels of organic matter and microbial activity. The spatial distribution of flux through the riverbed is complex and sudden changes in river water levels may lead to reversals in flux direction.

Research to assess the impact of groundwater on the quality of the River Tame draining an urban catchment ($408\,km^2$) containing the industrial city of Birmingham above an underlying Triassic Sandstone Aquifer found the river to be typically 8–12 m wide, 0.2–2 m deep with average dry weather flow velocities of $0.1–0.8\,m\,s^{-1}$. The river more than doubles its mean discharge across the 24 km study reach between the gauging stations at Bescot, 182 mega litres per day ($Ml\,d^{-1}$) and Water Orton $397\,Ml\,d^{-1}$. Work on a 7.4 km section receiving $\sim6\%$ of its total baseflow (60% of which is groundwater) from the underlying Triassic Sandstone aquifer and flood-plain sediments provided surface water and groundwater flow, head and physical/chemical data. Field data and supporting computer modelling indicated positive piezometric heads in the riverbed and the convergence of groundwater flows from the sandstone/drift deposits and variable discharge to the river (0.06 to $10.7\,m^3\,d^{-1}\,m^{-1}$, mean $3.6\,m^3\,d^{-1}\,m^{-1}$). The data reveal the discharge of organic (chlorinated solvents) and inorganic groundwater contaminant plumes through the bed of the river (Ellis *et al.*, 2000).

3.3   FLOW PROCESSES ACTING ON FLOOD PROPAGATION IN WATER BODIES

The concentration and formation of runoff ends in rivers, storm water pipes, channels or any other water body. Within these transport elements, the runoff is subjected to translation and retention. The translation corresponds to the travel time of the water, which is dependent on the flow velocity and the length of the water course (Figure 3.2). The flow velocity is influenced by the gradient of the bed of the water course, the water depth and the flow resistance.

Retention, the second process contributing to flood propagation in water bodies, is caused by temporary storage of water within the water course. This occurs either on the rising stage, when the river itself is filled with water, or when overbank flow occurs inundating flood plains. Especially in natural rivers the retention has a strong effect on the propagation of the flood wave. Here wide flood plains, meandering river beds and wooden vegetation cause high flow resistance and vast areas of inundation. They have a positive effect on the retention at flood by dampening the flood wave and attenuating its peak. However they are very complex and not fully understood. Retention is strongly influenced, not only by the size of the flooded water body, but also the variation of the flow velocity along the wetted perimeter within this water body. If the velocity is nearly constant, all water particles reach the downstream end of the river section at the same time. In this case no retention occurs. The more the velocity profile varies over the cross-section the more the water will be retained in the slower sections of the river. For no flow in inundation areas the retention can be easily assessed. But in general on flood plains flow still occurs and returns back to the river through momentum exchange or transverse flow

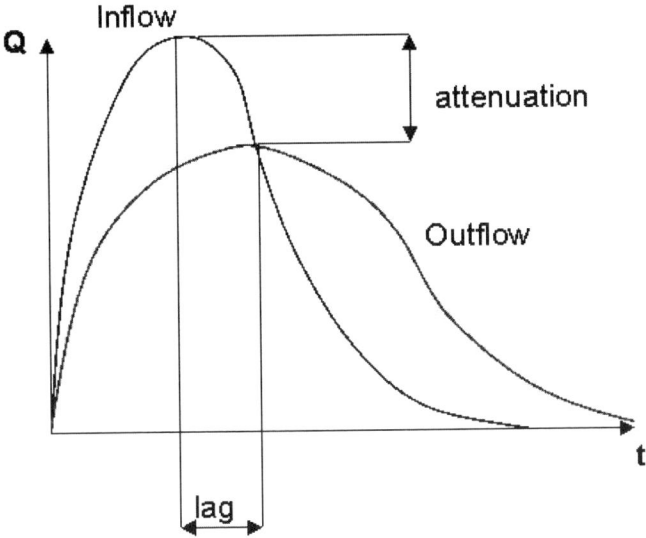

Figure 3.2. Inflow/outflow hydrograph of a flood wave.

at converging flood plains. As the flow velocity varies considerably in space and time, the retention effect can not be easily evaluated or determined for natural rivers. Often the retention effect in natural rivers is overestimated. Pasche and Plöger (2004) demonstrated that merely widening flood plains and the meandering of the main channel does not give a substantial attenuation of the flood peak, although these measures have led to higher water depth and larger inundation volume.

However, for flooding it is not merely a case of evaluating land cover, it is important to characterise hydraulic constraints and obstacles to stormwater flows. Yu and Lane (2005) have used high-resolution data obtained from airborne remote sensing to represent small-scale structural elements (e.g. walls, buildings) in complex floodplain systems using two-dimensional (2D) models of flood inundation. and to determine patterns of fluvial flood inundation in urban areas. Their model shows that even relatively small changes in model resolution have considerable effects on the predicted inundation extent and the timing of flood inundation. Timing sensitivity would be expected, given the relatively poor representation of inertial processes in a diffusion-wave model. Sensitivity to inundation extent is more surprising, but is associated with: (1) the smoothing effect of mesh coarsening upon input topographical data; (2) poorer representation of both cell blockage and surface routing processes as the mesh is coarsened, where the flow routing is especially complex; and (3) the effects of (1) and (2) upon water levels and velocities, which in turn determine which parts of the floodplain the flow can actually travel to. The combined effects of wetting and roughness parameters can compensate in part for a coarser mesh resolution. Nevertheless, the coarser the resolution, the poorer the ability to control the inundation process, as these parameters not only affect the speed, but also the direction of wetting. Thus, high-resolution data will have to be coupled to a more sophisticated representation of the inundation process in order to obtain effective predictions of flood inundation extent. Using roughness parameters to represent sub-grid-scale topography inadequately reflects the effects of structural elements on the floodplain (e.g. buildings, walls), as such elements not only act as momentum sinks, but also have mass blockage effects. These can be extremely important in floodplains within built-up areas. By using high-resolution topographic data to precisely represent sub-grid-scale topographic variability in terms of the volume of a grid cell that can be occupied by the flow and the effect of that variability upon the timing and direction of the lateral fluxes, significantly better prediction of fluvial flood inundation in urban areas than traditional calibration of sub-grid-scale effects using Manning's *n* can be obtained.

In addition to built topographic elements, vegetation affects flows in many urban channels. Davenport *et al.* (2004) have developed an urban river survey (URS), from the River Habitat Survey (RHS) which is applied routinely to UK rivers. The URS recognizes that most urban channels have been engineered and that management decision-making has to take account of previous "channel improvements". Urban river stretches are identified for survey according to their engineering type (a combination of planform, cross-sectional form and level of reinforcement). The URS is then applied to stretches of a single engineering type and incorporates recording of (i) additional variables to the RHS that are particularly relevant to urban channels (e.g. indicators of pollution); (ii) improved resolution in the recording of some variables in comparison with the RHS (e.g. habitat features); and (iii) separation of layers of information that relate to the engineered (e.g. artificially introduced materials) and more natural (e.g. bank materials and morphological features) channel properties so that the interaction between these properties can be identified. The URS has the potential to improve channel characterization and to enable dynamic vegetation communities to be incorporated into flood flow monitoring.

Retention is only of minor relevance in stormwater pipes and drainage channels, where compact cross sections produce a nearly constant flow velocity. In addition, the travel time of the water is reduced by a straight channel and a smooth surface along the wetted perimeter. This acceleration of flow reduces the translation time with the effect that flood waves, especially in storm water pipe networks and channels, reach the downstream end of their system without attenuation and much faster than flood waves in natural rivers. The more catchments are developed and drained by sewers and channels the more the flood waves will overlap with the negative effect of increasing the flood peaks in the central draining rivers. But this wave interference can be very complex especially in rivers where the flood waves from sewers and natural drains come together. Figures 3.3a and 3.3b show a flood wave separated into a slow moving part coming from the natural watershed and a fast one from an urban area released through the sewer pipe network. They show all

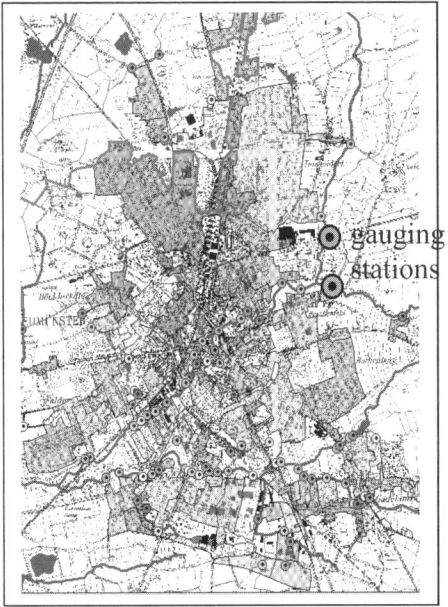

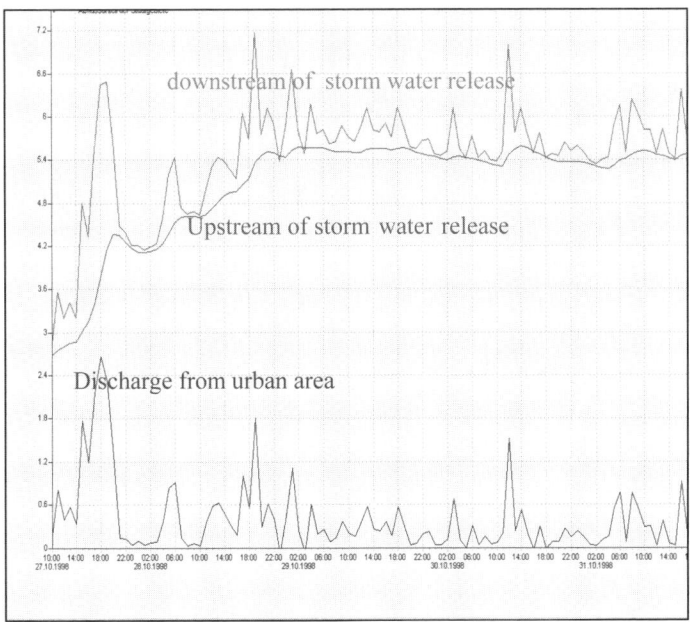

Figure 3.3. (a) Urban area draining into a stream. (b) Calculated discharge hydrograph upstream and downstream of an stormwater outlet.

the characteristics and differences between flood waves from natural watersheds and urban areas. They overload the pipe networks, flooding streets, cellars and the low-lying parts of urban areas. The flooded area however is restricted to the surrounding of the pipes and seldom comprises extensive urban areas. In natural watersheds, long lasting rain events with high total precipitation volumes but relatively low intensity cause the most critical floods. The larger the river basin, the longer the flood wave lasts. For large rivers the floods can last several weeks as observed at the Elbe flood and Danube flood in 2006.

Due to these distinctive differences in the flood waves, their accumulation in the river is hard to determine. Therefore it is difficult to assess effect on flood attenuation in the receiving urban rivers of retention measures in urban drainage networks, such as detention in green roofs, ponds and reservoirs, and infiltration to the subsurface through drains, porous pavements and ponds. Pasche *et al.* (2004) showed that for increasing flood events the attenuation effect of sustainable retention measures in urban areas is decreasing. In extreme floods, only reservoirs that retain large volumes of flood water clearly exhibit an impact on flood magnitude.

3.4   THE INFLUENCE OF LAND USE

3.4.1   The influence of land cover

The great variety of ground cover in urban areas leads to far greater complexity than the simple characterization of urban surfaces as permeable or impermeable suggests. Attention has now been paid to detailed multiple land-cover mapping. However, despite the value of this approach, many are attracted by a single "index" variable that characterizes the magnitude of urban development in a watershed. Patterns can be readily displayed, correlations are simplified, and communication between scientists and planners is enhanced. Yet urban development comes in many styles, occurs on many different types of landscapes, and is accompanied by a variety of mitigation measures designed to reduce its negative consequences on downstream watercourses. So any simple correlation between any single measure of urbanization and aquatic-system condition are unlikely to be precise. Past efforts to quantify the degree of urban development have not been consistent. Thus, the green infrastructure can be viewed as consisting of corridors, patches, and the overall matrix (Figure 3.4) (Gill *et al.*, 2006).

Vegetated areas are not actually examined closely in urban flood hydrology. However, the components of the green infrastructure play varying roles in influencing runoff. For example, flood storage is especially important in corridors, but also has some importance as SUDS in the patches. In Greater Manchester, for example, green spaces (golf courses, nature reserves, etc.) alongside the River Mersey are used as flood storage basins at times of high river flow. On the other hand, the matrix is especially important when it comes to rainwater infiltration, as

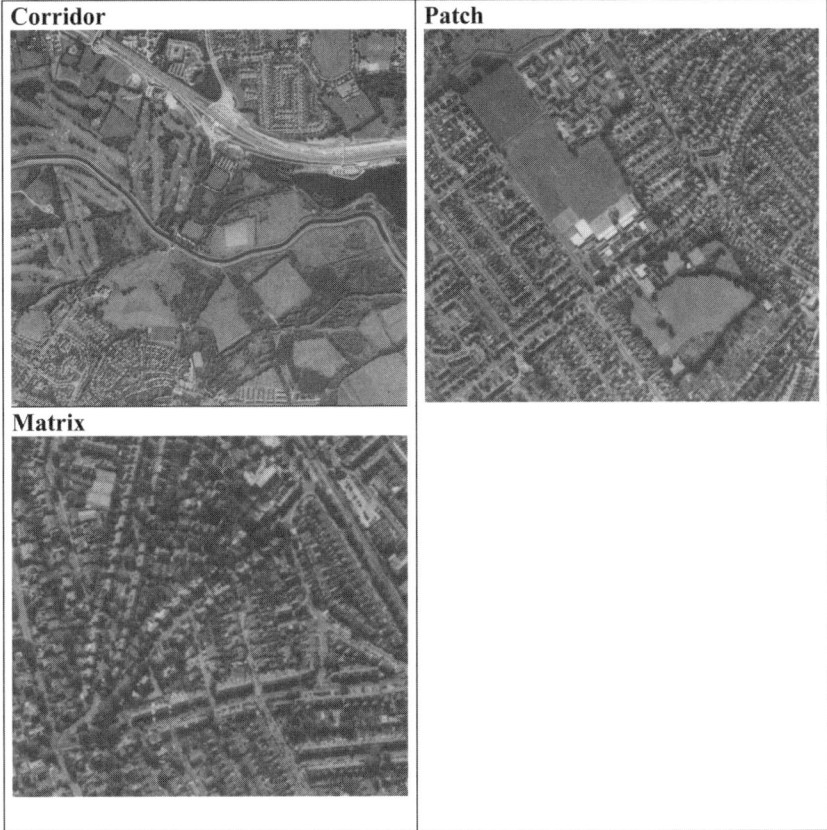

Figure 3.4. Green corridor, patch and matrix.

are patches (linking back to the work on surface runoff modelling). There may be a case for restricting infill development in lower density residential areas with high infiltration capacity.

The key element in urban runoff is the sealing of the ground surface, or the extent of the impermeable area. The extent of sealing of the surface varies considerably within urban areas. Although nearly 100% of precipitation, less that lost by evaporation, will runoff from impermeable areas, many sealed areas are not connected to a storm water pipe drainage network. The more they are surrounded by gardens and green spaces, the more the sealed areas shed rainwater laterally into depressions, hollows or drains where it can infiltrate into the ground (Sieker, 1999). Therefore Pasche *et al.* (2004) distinguish the sealing rate of urban areas in term of both the total sealed area and the proportion connected to drainage pipes. This has been expressed by Schueler (1995) as the distinction between the *total impervious area* (TIA) and the *effective impervious area* (EIA).

TIA is the "intuitive" definition of imperviousness: that fraction of the watershed covered by constructed, non-infiltrating surfaces such as concrete, asphalt, and buildings. Hydrologically, this definition is incomplete for two reasons. First, it ignores nominally "pervious" surfaces that are sufficiently compacted or otherwise so low in permeability that the rates of runoff from them are similar or indistinguishable from pavement. For example, Wigmosta and others (1994) found that the impervious unit-area runoff was only 20 percent greater than that from pervious areas, primarily thinly turfed lawns over glacial till, in a western Washington residential subdivision. Clearly, this hydrologic contribution cannot be ignored entirely.

The second limitation of TIA is that it includes some paved surfaces that may contribute nothing to the storm-runoff response of the downstream channel. For example, the only hydrologic change a shelter in the middle of parkland would cause would be highly localized increases of soil moisture at the drip line of its roof. Less obvious, but still relevant, will be the contrasting impacts of roofs that drain a) into piped storm-drain systems directly discharging into natural streams, or b) onto splashblocks that disperse the runoff onto the garden at each corner of the building.

The first of these TIA limitations, the production of significant runoff from nominally pervious surfaces, is typically ignored in the characterization of urban development because it is difficult to identify such areas and estimate their contribution, even though such tasks can be done with simple field methods and the resulting hydrologic insights are often valuable (Burges *et al.*, 1989). Furthermore, the degree to which pervious areas shed water as overland flow should be related to the amount of impervious area. Where construction and development is more intense and covers progressively greater fractions of the watershed, the more likely that the intervening green spaces have been stripped and compacted during construction and only imperfectly rehabilitated for their hydrologic functions during subsequent "landscaping."

The second of these TIA limitations, inclusion of non-contributing impervious areas, is formally addressed through the concept of effective impervious areas, defined as the impervious surfaces with direct hydraulic connection to the downstream drainage (or stream) system. Thus, any part of the TIA draining onto pervious (i.e. "green") ground is excluded from the measurement of EIA. This parameter, at least conceptually, captures the hydrologic significance of imperviousness. EIA is the parameter normally used to characterize urban development in hydrologic models.

The direct measurement of EIA is complicated, requiring direct, independent measurements of both TIA and EIA (Alley and Veenhuis, 1983; Laenen, 1983; Prysch and Ebbert, 1986). The results can then be generalized either as either a correlation between the two parameters or as a "typical" value for a given land use. Dinicola (1989) used a typical land-use value approach to compile the findings

Table 3.1. Presumed relationship between imperviousness and land use
(Dinicola, 1989).

Land use	TIA (%)	EIA (%)
Low density residential (1 unit per 1–1.5 ha)	10	4
Medium density residential (2.5 unit per ha)	20	10
"Suburban" density (10 units per ha)	35	24
High density (multi-family or 20+ units per ha)	60	48
Commercial and industrial	90	86

of earlier studies to recommend a single set of impervious-area values based on
five landuse categories for use in studies of catchments in western Washington,
USA (Table 3.1).

Use of satellite imagery facilitates detailed analysis of land cover, especially the
high resolution of IKONOS, but often cost and time constraints force people to
use coarser resolution data such as 30-m pixels which can only detect land-cover
differences permitting evaluation of only *total* imperviousness.

3.4.2   The influence of wetlands

Urban wetlands occur along drainage ditches, in storm water detention ponds,
irrigation canals and at the edges of small ponds and lakes. Many wetlands contain
plants such as grasses, sedges and bulrushes. These areas can also be critical
habitat for wildlife and are important both for recreation and minimizing urban
flood damage. Unfortunately such urban wetlands are rapidly disappearing from
cities and towns as land use intensifies. Nevertheless, efforts are being made to
create urban wetlands along river channels by removing or setting back existing
defences, or during channel widening works. However, unless they involve a large
area, such wetlands will have a neutral impact on storage processes within the
floodplain but can increase the capacity of the channel to pass flood flows. Such
schemes often provide opportunities for environmental enhancements, especially
in urban parks e.g. on the River Quaggy, south London.

As an alternative to wetland construction on enhancement, an offline flood storage basin on existing sports fields is being constructed as part of the ongoing River
Irwell Flood Alleviation Scheme at Salford, Greater Manchester, which also relies
on raised flood defences. Significant under drainage is required to enable the basin
to continue to be used for recreational purposes. Remotely operated sluices control
inundation and evacuation, which requires high standards of forecasting during
an event. The reservoir is sited close to the main areas of flood risk so, with good
knowledge of the shape of the flood hydrograph, best use can be made of the storage
capacity of the reservoir and the flow capacity of the channel through the risk area.
Catchment characterization needs to take account of these management practices.

3.4.3   The influence of floodplains

Commenting on the 1998 Easter floods in Britain, Horner and Walsh (2000) noted the importance of adequate characterization of floodplains and stressed the importance of examining flood issues in a holistic manner. They suggested that identification of floodplains is an essential prerequisite to effective flood-risk management. However, despite increased attention there is still much to do, including:

- Establishing national consistency in the estimation of flood probability associated with the extents of observed floodplain inundation;
- Using computational hydraulic modeling for floodplain mapping only where theoretically or empirically based analyses are valid and there are adequate data to calibrate and verify the models;
- Investigating alternative approaches to flood plain mapping that take account of hydrological and hydraulic uncertainties;
- Enhancing hydrometric standards to obtain more reliable estimates of rainfall and flood magnitude and probability, including improving the ability of flow-measuring stations to gauge larger rare events.

The current understanding of national flood risk is represented by the Indicative Floodplain Map (IFM), which are held by the Environment Agency and published on the Internet (www.environment-agency.gov.uk). This dataset is a compilation of: (i) historic flood outlines; (ii) the results of detailed local studies; and (iii) a comparatively low resolution nationwide flood outline produced in 1996 by the then Institute of Hydrology (now Centre for Ecology and Hydrology). This nationwide outline is known as IH130, after the report describing its generation (Morris and Flavin, 1996). The relatively coarse spatial resolution has resulted in areas shown at flood risk that are in fact well above the river channel. Similar problems result from use of some of the historical outlines which were also based on records held on relatively low resolution mapping (e.g. 1:50,000). In addition, the return period of the historical outlines was not always known (Bradbook et al., 2005) and the affect of existing flood defences were not taken into account.

Improved estimates can be made using the following types of data (Hall et al., 2005), but resolution is too coarse to accurately obtain flood depths:

- Indicative Floodplain Maps (IFMs) are the only nationally available information on the potential extent of flood inundation. The IFMs are outlines of the area that could potentially be flooded in the absence of defences in a 1:100-year return period flood for fluvial floodplains and a 1:200-year return period flood for coastal floodplains.
- 1:50,000 maps with 5 m contours. The methodology has been developed in the absence of a national topographic dataset of reasonable accuracy. Topographic information at 5 m contour accuracy has only been used to classify floodplain types as it is not sufficiently accurate to estimate flood depths.
- National map of the centreline of all watercourses.

Table 3.2. Levels of protection and flood return period.

Type of development	Minimum level of protection to 2050
All residential development	200 year return period (0.5%)
Sheltered housing	1,000 year return period (0.01%)
Bungalows and ground floor flats	500 year return period (0.2%)
Property where the scope for warning is limited (e.g. flash flooding)	500 year return period (0.2%)

- National Flood and Coastal Defence Database provides a national dataset of defence location, type and condition.
- National database of locations of residential, business and public buildings.
- Land use maps and agricultural land classification.

As an outcome of the inquiries into the 1998 UK floods, specific degrees of protection for residential areas were recommended (Table 3.2).

3.5   TYPES OF FLOOD EVENT

The differing types of watercourse, runoff regimes, flow processes and land uses all combine with physical geography to create various types of flood event. Broadly speaking, flood events affecting urban areas in Europe stem from 6 main causes, each of which are discussed in sub sections below. These are:

1) In mountain areas, the effects of heavy, warm spring rains on winter snow causing sudden melting.
2) Regional weather systems that are blocked by high pressure systems and produce widespread heavy rain over large sections of major river catchments.
3) Flash floods in hilly and mountainous regions.
4) Short duration, high intensity thunderstorm driven local flooding on small streams entirely within the urban areas.
5) Sewer flooding associated with blocked sewer overflows to larger rivers or surcharging through manhole covers.
6) Groundwater flooding can be a result of prolonged heavy rainfall in certain geological conditions.

3.5.1   Rain and snow melt in mountain areas

Exceptional conditions in the French Alps have caused extreme flood events such as the millennial inundation of the Guil River in 1957 (Tricart, 1975), which induced a runoff coefficient greater than 1 due to rain falling on snow cover. More recently, the flooding of the Ouvèze river at Vaison-la-Romaine was the subject of a number of papers on the damage caused to vineyards (Wainwright, 1996) as well as stream beds and banks (Piégay and Bravard, 1997). The Guil valley floods affected whole

Figure 3.5. Flood in Celje, November 5, 1998 (Photo: J. Uhan).

villages that had been built on alluvial fans where tributaries discharged into the Guil. Normally the tributary flowed along a stream through the middle of the village. In the extreme flood it became active across virtually the whole of the alluvial fan, sweeping away nearly all the buildings in the settlement.

3.5.2   Regional, widespread floods

Regional widespread floods resulting from prolonged heavy rain, such as those on the Rhine and Elbe mentioned above, have major impacts on the towns and cities along the affected rivers. Catastrophic widespread flooding occurred in Slovenia in 1990 and 1998 (Kolbezen, 1991; Polajnar, 2000). Both events were assessed as 1 in 100-floods. They had similar causes, occurring in November when the soil moisture content was high and the permeability of the ground surface was low. After a long period of modest rainfall, there was a storm causing an immediate increase in runoff. Flood damage was great in both cases, the cities of Celje and Laško suffering enormous economic damage to infrastructure and buildings (Figure 3.5). The floods caused an exceptional amount of stream bank erosion and a large number of landslides.

3.5.3   Flash floods in hilly and mountainous regions

Stream draining upland areas and descending into urban areas often cause serious damage to urban infrastructure. This is particularly the case in the European Alps

Figure 3.6. The trace of the peak flood level on a building in Sevnica in August 21, 2005 (Photo: J. Uhan).

where, for example, analysis of the records reveals that a few individual areas of Slovenia are affected by flash floods every 5 to 6 years. However, flash floods are not evenly distributed in time and space. The last catastrophic flash floods in Slovenia occurred in August 2005 (Figure 3.6), a wet month with higher discharges than usual. Although the whole country had a higher monthly rainfall than usual, the situation was worst during form 20th to 23rd August in southeastern Slovenia and the Posavje region where more than 100 mm of rain fell on August 21st and 22nd (Figures 3.7 and 3.8). The August rainfall was two to three times above the mean 1971–2000 August precipitation. Heavy rains with showers and long duration rainfall caused numerous landslides and flash floods. The runoff from small catchments exceeded periodic maximum discharges (Figure 3.9), while the discharges of large rivers were usual. The return period of peak discharges was 50 years and more in the most threatened area.

3.5.4 Short duration, high intensity thunderstorm driven local flooding

All cities experience the impacts of occasional extreme high intensity thunderstorm rains. These events are predicted to grow in frequency due to increased development and the intensification of storm events associated with the effects of climate change (Evans *et al.*, 2004; IPCC, 2001). They are also especially difficult to defend against effectively as they can occur within a very short time period and there is little time

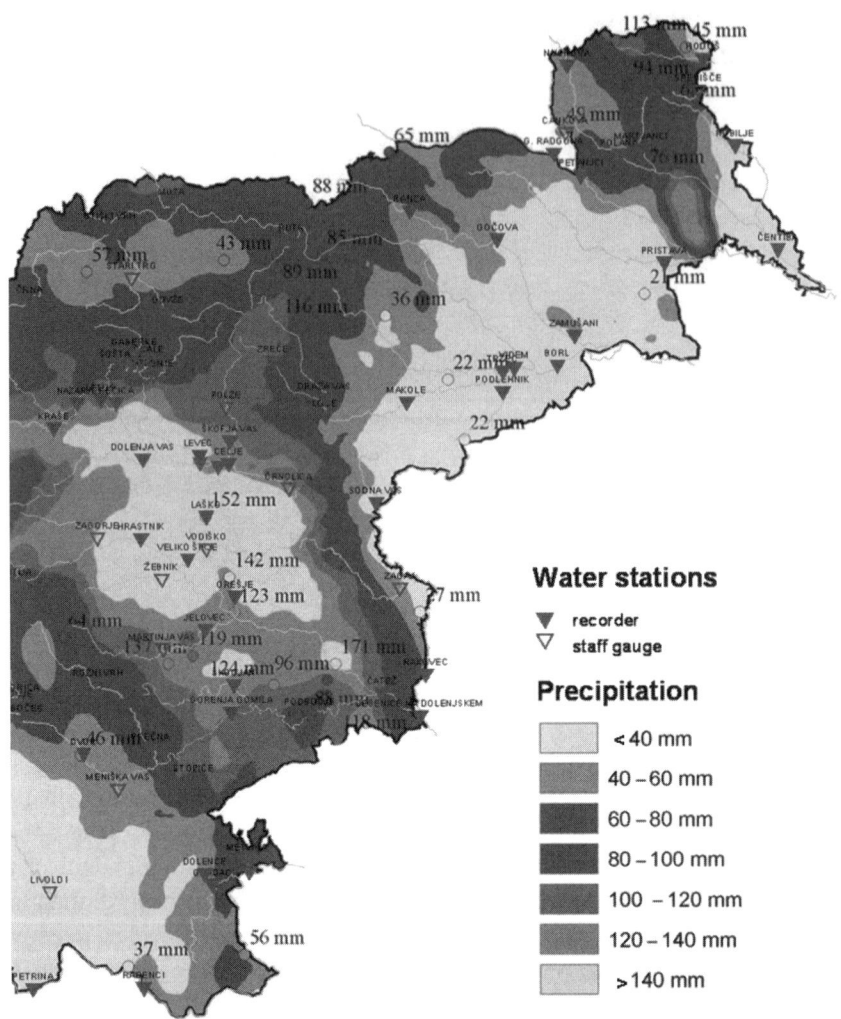

Figure 3.7. Cumulative 2-days precipitation between 20th and 22nd August 2005.

to issue warnings or enact predefined management strategies. The Manchester case study discussed in section 6.2 explores a practical example of this phenomenon in more depth.

3.5.5   Sewer flooding

Flooding from sewers occurs when the rivers into which stormwaters overflow are so high that the stormwater cannot escape or when the sewers are not large enough to cope with all the water flowing into them from upstream. Quite commonly, low

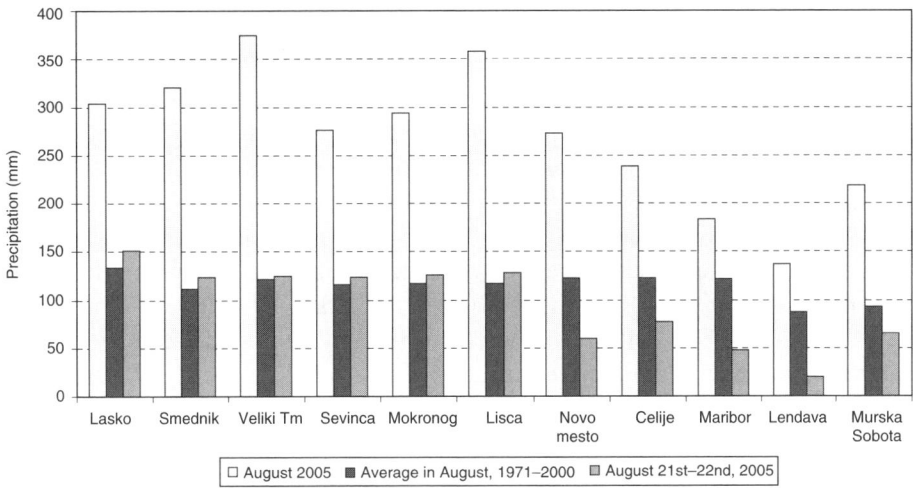

Figure 3.8. August amount of precipitation in eastern and northeastern Slovenia.

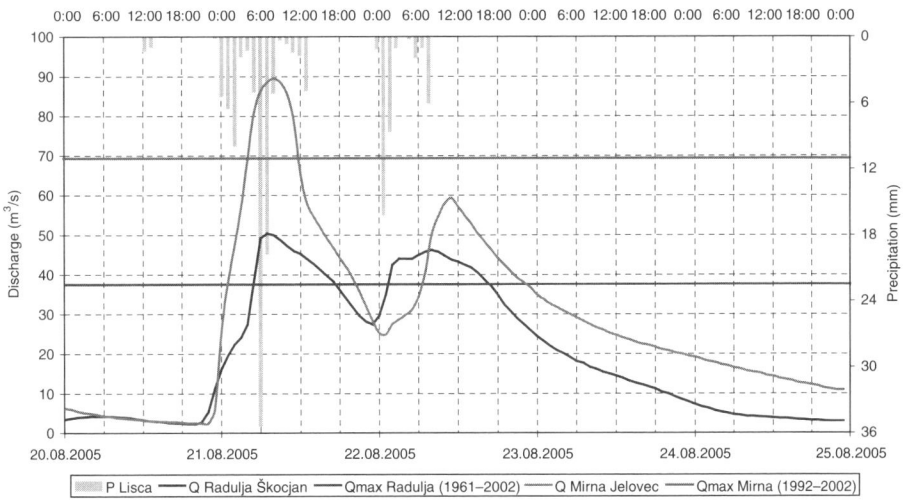

Figure 3.9. Hydrographs of two streams between 20th and 25th August 2005, periodic maximal discharge Qmax and hourly rainfall intensity from the nearest raingauge station on Mount Lisca.

lying urban areas close to rivers suffer from the overflowing of sewers, rather than the overtopping of river banks. This happened in Lower Kersal, Greater Manchester in 1980, when prolonged regional rainfall caused flooding throughout the Mersey Basin and the high levels in the Irwell tributary caused sewers to back up and overflow through housing areas.

3.5.6   Groundwater flooding

Under certain geological conditions, particularly on the Chalk of northwestern Europe, groundwater flooding may be a real problem. In the UK, 83,000 properties located on the exposed chalk aquifers of southern England where ground-water levels fluctuate widely are vulnerable to groundwater flooding. Unlike surface water flooding events, which are usually of relatively short duration, groundwater flooding may last several weeks or even months. Such flooding may follow an unusually wet period when water tables rise so much that flows from normally dry springs and winterbournes inundate roads and overwhelm all drainage systems. The British Geological Survey (BGS) provides data on ground-water flooding that is the product of integrating several datasets: a digital model of the land surface, digital geological map data and a water level surface based on measurements of groundwater level made during a particularly wet winter. This dataset provides an indication of areas where groundwater flooding may occur (see: http://www.bgs.ac.uk/britainbeneath/haz_flooding.html).

3.6   CASE STUDIES

These differing types of flooding are now discussed within a practical city context by two brief case studies. Although these two cities are situated within the UK, their experiences may be relevant towards all major European urban areas. These examples demonstrate how flood risk within specific urban areas can be varied and even smaller events can have significant localised effects.

3.6.1   London case study

London is a major European city threatened by three main source of flood risk:
- tidal flooding such as the 1953 storm surge;
- fluvial floods on the river Thames and its many tributaries;
- high intensity, convective storms coupled with inadequate storm and sewer drainage systems.

The land cover throughout London continues to be greatly modified by urbanization, changing the response to rainfall. Flood peaks are higher, and thus the extent of floods of a given probability is larger, and the tributary rivers are flashier than formerly providing shorter flood warning lead times. For example, new housing and industrial estates developed in the Silk Stream catchment in north London have generated a threefold increase in flood peaks (Hollis, 1986). Recent investigations of rainfall patterns over London concluded that storm and sewer drainage systems now have insufficient capacity for today's runoff volumes (Moore *et al.*, 1993). By the 1990s there appeared to have been an increase in flash flooding owing to intense rainfall (Parker and Tapsell, 1995).

The main response to fluvial flood risk in the London region is an on-going and largely reactive programme of river channel improvements (including large flood relief channels) and structural defences which are becoming more difficult to implement because of fierce opposition from those living close to rivers and environmentalists. Reactive flood defence is encouraged by the current central government grant-aid system for flood defence projects. Typically flood plains become developed before it is possible to demonstrate that the benefits flowing from such developments warrant the costs of protection.

Because London is an old city where replacement of infrastructure has not kept with the rate of its decay, the urban area has become increasingly vulnerable to severe disruption and this is now being frequently highlighted by flooding. The capacity of London's old combined storm drain and sewer systems is now often insufficient to take the volumes of water generated by thunderstorms, especially when they become stationary. Severe flash flooding results either along river valleys or almost anywhere (including on hills) the drains are inadequate. Commuters depend on the underground and surface rail networks which are vulnerable to both tidal and flash flooding from streams and inadequate storm water drainage.

A storm on the 11th August 1994 led directly or indirectly to the closure of 28 stations (Figure 3.10) illustrating the vulnerability of the London underground rail network to flash flooding. Train services were shut down for at least five hours and about 60,000 passengers were stranded at stations. At the same time, road congestion caused by localized flooding spread rapidly over a large geographical

Figure 3.10. Map of underground stations affected by flooding.

area limiting the effectiveness of alternative routing (Parker and Tapsell, 1995). Furthermore, the August 2002 flash flood in London caused major disruptions to mainline surface rail services out of Euston and Kings Cross underground station was closed at the height of the rush hour.

3.6.2   Manchester case study

Manchester, the archetypal industrial city, is surrounded to the north and east by peat covered hills which catch the wet westerly airstreams from the Atlantic and so feed rain to the two rivers that meet downstream at the city, the Irwell and the Mersey. Average annual rainfall in the hills at the top of the catchments at 5/600 m above sea level is 1,450 mm, dropping to 825 mm in the city centre at some 20 m above sea level. The Environment Agency is responsible for improving flood protection along the rivers but the effectiveness varies greatly. Along the Mersey south of the city centre, protection a 1:200 year event exists as a result of the creation since 1970 of flood storage basins on parkland and golf courses. On a tributary of the Irwell to the north-east of the city centre, flood protection for a 1:100 year event has been obtained by constructing by-pass channels, making channel adjustments, installing reinforcements and storm-water tanks, and by increasing drainage and sewage capacity. However, on the Irwell itself, immediately northwest of the city centre, flood protection for only a 1:75 year event will be obtained even after the completion of a 650,000 m^3 flood storage basin. Climate change is predicted to increase peak flood flows and thus the Environment Agency is giving further work to protect this part of the Irwell floodplain the highest priority (Environment Agency, 2005).

Flooding from 19th century sewers and urban drains originally designed to cope with 1:5 year floods (Read, 1986), has gradually been reduced since the 1989 privatisation of the water supply companies with large scale sewer reconstruction and the building of storm-water tanks. However, the combined sewers can still back-up and overflow during intense thunderstorms. Whilst water companies accept responsibility for the adequacy of the sewer system, culverts are either the responsibility of local authority highway departments or the owners of land under which they pass; and relatively little information on the flood capacity of these culverts exists. An updated version of Ashworth's (1987) map shows how extensive the buried stream channel network is in Manchester (Figure 3.11).

3.7   CASE STUDIES

Streams in culverts have caused much localised flooding in many parts of the city and two recent examples are presented here.

Broadheath and Timperly on the low lying flat lands of south Manchester is intersected by a series of small urban streams and has a long history of localised

Figure 3.11. Small urban catchments and culverts in Manchester.

flooding. Following a 1981 storm when the insufficient capacity of culverts resulted in 80 houses being inundated by up to 1.75 m of water in the Brunswick Road area, the then Rivers Division of the North West Water Authority produced a comprehensive plan designed to end flooding in the area. The 1983 Timperley, Baguley and Sinderland Brook Improvement Scheme involved the use of playing fields as a 5,000 m³ flood basin, increasing the capacity of the culverts to cope with a 1:35 year event by inserting a new culvert next to the original 1772 culvert where

Figure 3.12. Timperley Brook bank reinforcement, culvert modification with debris, flood regulator.

Figure 3.13. Pilsworth Rd. Heywood 2nd July 2006 (Courtesy Heywood Advertiser).

Timperley Brook passes under the Bridgewater Canal; channel straightening: the reinforcement and raising of banks; and undertakings by the authorities to control debris and limit illegal tipping (Figure 3.12).

However, residents in the Timperley Brook area continued to witness regular sewer flooding and a July 2006 storm resulted in a sewer overflow and further flooding in Brunswick Road for the fourth time in two years (Griffen, 2006a), with the utility company undertaking to carry out detailed studies (Griffen, 2006b).

Heywood is on the northern edge of the city under the surrounding hills at 130–140 m above sea level. The Pilsworth Road area of Heywood experienced severe flooding following summer storms in 2004 and 2006 (Figure 3.13). On July 2nd 2006 the rain gauge at Cowm, Whitworth, recorded a total of 50.4 mm in 3.5 hours. The first hour saw 33.2 mm of rainfall and the peak rainfall intensity was 17.2 mm in 15 minutes (=68.8 mm/hr). Excessive surface water and back-up of water from old 1.08 m and 1.25 m wide combined sewerage systems and the now culverted Wrigley Brook (Figures 3.14 and 3.15) under the area resulted in 200 properties being inundated with up to 700 mm of sewage infected water and 40 homes had to be evacuated. The same area was also hit by a storm on August 3rd 2004 when 37.5 mm of rain fell in 1.5 hours and whist the drainage systems were not designed to cope with such extreme weather, grids blocked with silt greatly

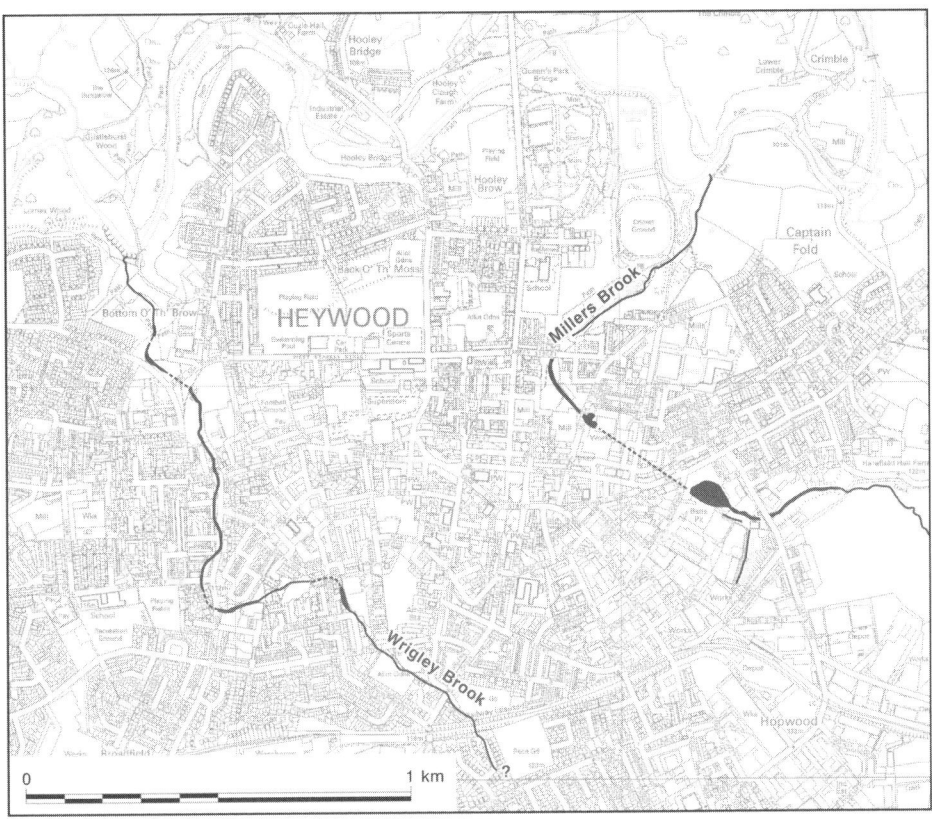

Figure 3.14. 1898 course of streams in Heywood (now in culverts).

increased the amount of flood damage incurred. Several areas of Heywood are now flood prone, not least because it is primarily still drained by the original 19th century system designed when there were 4,000 homes where there are now 13,000 (Lambert, 2006).

3.8   FUTURE CHALLENGES

Because several different agencies are responsible for the provision and maintenance of urban drainage systems in the UK many small events go unrecorded and a comprehensive analysis of the scale and consequences of sewer and culvert flooding in Manchester is not possible, but it is noticeable that this type of urban flood has become more prevalent since the 1980s.

Under existing socio-economic conditions the National Appraisal of Assets at Risk of Flooding suggests that by 2050, with median climate change predictions, average flood damage could exceed £80 million per year in the northwest of England

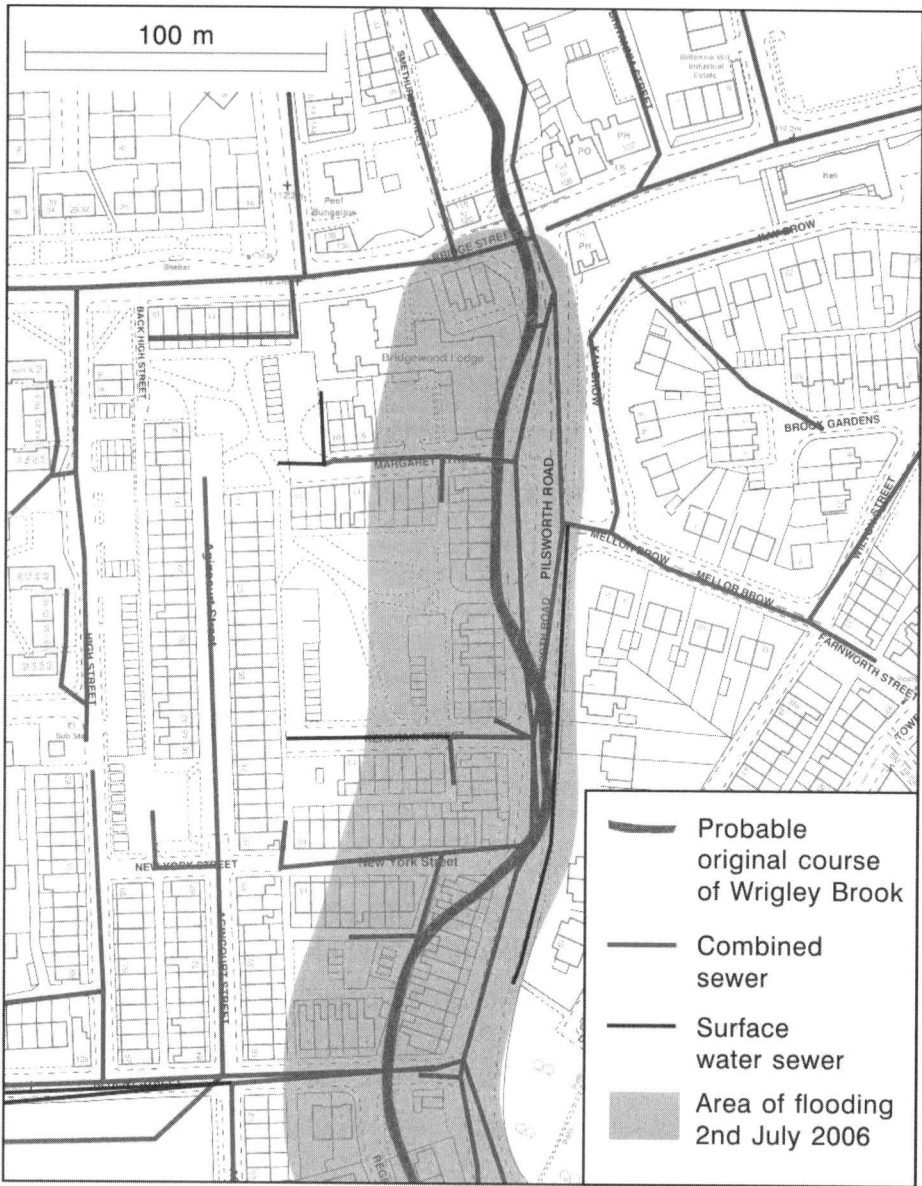

Figure 3.15. Pilsworth Rd. Heywood: area flooded 2nd July 2006, sewer system and probable original course of Wrigley Brook.

(Holman *et al.*, 2002). The climate in the North West is set to get warmer and wetter, weather patterns could become more extreme and this will increase the risk of storms and flooding (Environment Agency, 2006) with both average and extreme rainfall likely to increase by a factor of 1 to 1.25 in the region by the end of the 21st century (STARDEX, 2005). In addition to these climate related predictions increased urban infilling and intrusions into natural floodplains mean that the hydraulic capacity of the now largely antiquated drainage systems in Greater Manchester requires urgent amelioration.

Without co-ordinated widespread participation in flood mitigation by all stake-holders ranging from individual home owners to utility providers and governmental organisations, localised flooding in Manchester will continue to be a problem.

3.9   CONCLUSION

The use of land is fundamental to the alleviation of flood risk (Parker, 1995; White and Howe, 2004a) and climate change, combined with escalating development demands, will intensify the need for effective long term mitigation and adapta-tion strategies in this field (Evans *et al.*, 2004; IPCC, 2001; White and Howe, 2004b). The impact of flooding on urban areas is perhaps the most damaging due to the high population densities and consequent threat to life. Throughout history, unwise management practices and growing human vulnerability, together with the global climate change observable for the last 20 to 30 years, have led to increasing flooding and corresponding growing economic losses. According to the European Environmental Agency (EEA, 2001) floods are the most common natural disasters in Europe and, in terms of economic damage, the most costly ones. Flooding and its impacts are often influenced by a combination of natural factors and human inter-ference, including changes in the catchment and in flood-plain land use, population growth, urbanisation and increasing settlement, roads and railways, and hydraulic engineering measures. The London and Manchester case studies demonstrate the necessity to give greater consideration to the inadequacy of drainage systems in old established urban areas to cope with extreme events and climate change. There is a need to holistically address the most sustainable and least disruptive technical solutions to reduce the impact of urban flash floods, to better understand cost ver-sus benefits, and to identify the roles and responsibilities of both public agencies and private stakeholders. Therefore a contemporary challenge for the scientific community is to engage effectively with the complexities of urban flood risk to inform and enhance flood risk mitigation and adaptation strategies with evidence based, cutting edge scientific research.

REFERENCES

Alley, W. A. and Veenhuis, J. E. 1983 Effective impervious area in urban runoff modeling. *Journal of Hydrological Engineering, ASCE*, 109(2), 313–319.

Andjelkovic, I. 2001 Guidelines on non-structural measures in urban flood management. *IHP-V, Technical Documents in Hydrology*, No. 50, UNESCO, Paris.

Ashworth, G. 1987 *The lost rivers of Manchester.* Willow Publishing, Altrincham, Cheshire.

Blowes, D.W. and Gillham, R.W. 1988 The generation and quality of streamflow on inactive uranium tailings near Elliot lake, Ontario. *Journal of Hydrology*, 97, 1–22.

Bradbook, K., Waller, S. and Morris, D. 2005 National Floodplain Mapping: Datasets and Methods – 160,000 km in 12 months. *Natural Hazards*, 36, 103–123.

Bruen, M. 1999. Some general comments on flood forecasting. *Proceedings of the Euro-Conference on Global Change and Catastrophe Risk Management: Flood Risk in Europe*, June 1999, IIASA, Laxenburg, Austria.

Davenport, A., Gurnell, A.M. and Armitage, P.D. 2004 Habitat survey and classification of urban rivers. *River Research and Applications*, 20, 687–704.

Decroix, L. and Mathys, N. 2003 Processes, spatio-temporal factors and measurements of current erosion in the French Southern Alps: a review. *Earth Surface Processes and Landforms*, 28, 993–1011.

Dinicola, R.S. 1989 Characterization and simulation of rainfall-runoff relations for headwater basins in western King and Snohomish Counties, Washington state. *U.S. Geological Survey Water-Resources Investigation Report* 89-4052, 52 pp.

Dunne, T. 1978 Field studies of hillslope flow processes, *Hillslope Hydrology*, Ed.: M.J. Kirkby, John Wiley, Chichester.

Dunne, T., Moore, T.R. and Taylor, C.H. 1975 Recognition and prediction of runoff producing zones in humid regions. *Hydrological Sciences Bulletin*, 20, S. 305–327.

EEA 2001. *Sustainable water use in Europe, Part 3: Extreme hydrological events: floods and droughts. Environmental issue report No 21*, European Environmental Agency, Copenhagen.

Ellis, P.A., Rivett, M.O. and Mackay, R. 2000 Assessing the impacts of a groundwater pollutant plume on the River Tame, West Midlands.

Environment Agency 2005 *River Irwell Catchment Flood Management Plan Pilot Study-Consultation Draft April 2005.* The Environment Agency, Warrington.

Environment Agency 2006 *Climate change in the North West.* www.environment-agency.gov.uk accessed 24/9/2006.

Evans, E., Ashley, R., Hall, J., Penning Rowsell, E., Saul, A., Sayers, P., Thorne, C. and Watkinson, A. 2004 *FORESIGHT. Future Flooding. Scientific Summary Volume 1 and 2*, OST, London.

Gill, S., Handley, J. and Ennos, R. 2006 *Greenspace to adapt cities to climate change.* Unpublished paper for the Engineering and Physical Science Research Council's Adaptation Strategies to Climate Change in the Urban Environment (ASCCUE) project.

Griffen, C. 2006a *Flash floods bring sewage.* Sale and Altincham Messenger, July 13th 2006.

Griffen, C. 2006b *Call for action on sewage threat.* Sale and Altincham Messenger, September 13th 2006.

Hall, J.W., Sayers, P.B. and Dawson, R.J. 2005 National-scale Assessment of Current and Future Flood Risk in England and Wales. *Natural Hazards*, 36, 147–164.

Hollis, G.E. 1986 Water management. In: Clout, H., Wood, P. (eds.), *London: Problems of Change.* 101–110. Longman, Harlow.

Holman, L.P., Loveland, P.J., Nicholls, R.J., Shackley, S., Berry, P.M., Rounsevell, M.D.A., Audsley, E., Harrison, P.A. and Wood, R. 2002 *REGIS-Regional Climate Change Impact Response Studies in East Anglia and North West England.* Defra, London.

Horner, M.W. and Walsh, P.D. 2000 Easter 1998 floods. *Water and Environmental Management*, 14, 415–418.

Horton, R.E. 1933 The role of infiltration in the hydrological cycle. *EOS Transactions American Geophysical Union*, 14, 460–466.

IPPC (2001) *Climate Change 2001: The Scientific Basis*, Cambridge University Press, Cambridge.

Kirkby, M.J. and Chorley, R.J. 1967 Throughflow, overland flow and erosion. *Bulletin International Association of Hydrological Sciences*, 12(3), 5–21.

Kolbezen, M. 1991. Flooding in Slovenia on November 1, 1990. *Ujma* 5, Ljubljana, 16–18.

Laenen, A. 1983. Storm runoff as related to urbanization based on data collected in Salem and Portland, and generalized for the Willamette Valley, Oregon. *U.S. Geological Survey Water-Resources Investigations Report*, 83-4238, 9 pp.

Lambert, C. (Leader Heywood District Council) 2006. Personal communication.

Moore, R., Hitchkiss, D. and Black, K. 1993 *Rainfall patterns over London.* Institute of Hydrology, Wallingford, Oxon.

Morris, D.G. and Flavin, R.W., 1996, *IH Report No. 130: Flood Risk Map for England and Wales*, Institute of Hydrology, Wallingford, UK.

Parker, D.J. 1995 Floodplain development policy in England and Wales. *Applied Geography*, 15, 341–363.

Parker, D. and Tapsell, S. 1995 Hazard transformation and hazard management issues in the London megacity. *GeoJournal*, 37, 313–328.

Pasche, E., Brüning, C., Plöger, W. and Teschke, U. 2004 Möglichkeiten der Wirkungs-analyse anthropogener Veränderungen in naturnahen Fließgewässern, *Erschienen in Proceedings zum Jubiläumskolloquium 5 Jahre Wasserbau an der TUHH Amphibische Räume an Ästuaren und Flachlandgewässern*, Hamburger Wasserbau-Schriften, Heft 4, Hrsg. Erik Pasche, Hamburg.

Peschke, G., Etzberg, C., Müller, G., Töpfer, J. and Zimmermann, S. 1999 Das wis-sensbasierte System FLAB – ein Instrument zur rechnergestützten Bestimmung von Landschaftseinheiten mit gleicher Abflussbildung. *IHI – Schriften, Internationales Hochschulinstitut Zittau*, Heft 10, 122 Seiten.

Piégay, H. and Bravard J.P. 1997 Response of a mediterranean riparian forest to a 1 in 400 year flood, Ouvèze river, Drôme-Vaucluse, France. *Earth Surface Processes and Landforms*, 22(1), 31–43.

Polajnar, J. 2000 High waters in Slovenia in 1998 *Ujma* 13, Ljubljana, 143–150.

Prysch, E.A. and Ebbert, J.C. 1986 Quantity and quality of storm runoff from three urban catchments in Bellevue, Washington. *U.S. Geological Survey Water-Resources Investigations Report* 86-4000, 85 pp.

Read, Geoffrey F. 1986 *The Development, Renovation and Reconstruction of Manchester's Sewerage System.* Proceedings of the Manchester Literary and Philosophical Society 1895-86,14–30.

Roth, C. 1992 Die Bedeutung der Oberflächenverschlämmung für die Auslösung von Abfluss und Abtrag. *Bodenökologie und Bodengenese*, Heft 6, Inst. f. Ökologie, TU Berlin.

Schueler, T. 1995 The importance of imperviousness. *Watershed Protection Techniques*, 1(3): 100–111.

Sieker, F., Bandermann, S., Holz, E., Lilienthal, A., Sieker, H., Stauss, M. and Zimmermann, U. 1999 *Innovative Hochwasserreduzierung durch dezentrale Maßnahmen am Beispiel der Saar* – Zwischenbericht.- Deutsche Bundesstiftung Umwelt, DBU, Projekt AZ 07147, Osnabrück.

Smith, K. and Ward, R. 1998. *Floods, Physical Processes and Human Impact*, John Wiley & Sons Ltd, Chichester, England.

STARDEX (Statistical and Regional dynamic Downscaling of Extremes for European regions) 2005. *STARDEX scenarios information sheet: How will the occurrence of extreme rainfall events in the UK change by the end of the 21st century?*. www.cru.uea.ac.uk/projects/stardex/deliverables/ accessed 24/9/2006.

Tanaka, A., Yasuhara, M., Sakai, H. and Marui, A. 1988 The Hachioji experimental basin study – Storm runoff processes and the mechanism of its generation. *Journal of Hydrology*, 102, 139–164.

Tricart, J. 1975 Phénomènes démesurés et régime permanent dans des bassins montagnards (Queyras et Ubaye, Alpes Françaises). *Revue de Géographie Alpine*, 23, 99–114.

Wainwright, J. 1996 Infiltration, runoff and erosion characteristics of agricultural land in extreme storm events, SE France. *Catena* 26(1–2), 27–47.

White, I. and Howe, J. 2002 Flooding and the role of planning in England and Wales: A critical review. *Journal of Environmental Management and Planning*, 45(5), 735–745.

White, I. and Howe, J. 2004a Like a fish out of water: The relationship between planning and flood risk management. *Planning Practice and Research*, 19(4), 415–425.

White, I. and Howe, J. 2004b The Mismanagement of Surface Water. *Applied Geography* 24(4), 261–280.

Wigmosta, M.S., Burges, S.J. and Meena, J.M. 1994. Modeling and monitoring to predict spatial and temporal hydrologic characteristics in small catchments. *Report to U.S. Geological Survey, University of Washington Water Resources Series Technical Report No.* 137, 223 pp.

WMO, 2004 *Water and disasters. WMO-No. 971*, Geneva, Switzerland.

Yu, D. and Lane, S.N. 2005 Urban fluvial flood modelling using a two-dimensional diffusion-wave treatment, part 1: mesh resolution effects. *Hydrological Processes*, 20, 1541–1565.

Zuidema, P.K. 1985 Hydraulik der Abflussbildung während Starkniederschlägen.- *Mitteilungen der Versuchsanstalt für Wasserbau, Hydrologie und Glaziologie*, Nr. 79, ETH Zürich.

4

Flood Modelling in Urban Rivers – the State-of-the-Art and Where to Go

Erik Pasche
Technical University Hamburg-Harburg, Hamburg, Germany

ABSTRACT: This paper gives an overview about today's mathematical modelling instruments of the hydrological processes participating at the formation of flood in urban environments. They are divided into hydrological and hydraulic models. While the first group of models addresses the flow processes in the catchment, the second group simulates the transport processes of runoff in channels and rivers. The overall trend in rainfall-runoff modelling is directed towards more refined physically based modelling. The lumped and distributed models are more and more approaching each other. For the flood routing in rivers hydraulic methods are nowadays dominating compared to hydrological approaches. 1-dimensional flow models are still the preferred hydraulic modeling tools but due to better availability of topographic data by remote-sensing technique and more robust numerical methods 2-dimensional models are advancing.

The paper compiles the most promising mathematical approaches to an integrative, the whole water cycle covering modelling approach for stormwater-induced floods in urban environments. It demonstrates that for the hydrological modelling deterministic rainfall-runoff models based on semi-distributive lumped approaches and for the hydraulic modelling 1- and 2-dimensional hydrodynamic water surface models are appropriate instruments for the determination of the design floods and the corresponding inundation on the flood plains. It shows that within the hydraulic models the traditional roughness approach of Manning's and a constant eddy assumption could lead in the presence of vegetation to considerable errors despite a good calibration. This deficiency can be avoided by using the Darcy-Weisbach-Equation and by applying special routines for modelling the effect of wooden vegetation.

The study presents some important deficiencies in the knowledge and modelling capabilities of urban floods. Hillslope flow processes still need further research. Especially the subsurface flow is not understood in all details. Sustainable drainage systems need to be further studied concerning their capability to retain runoff for extreme flood. The combination of different model components like hydraulic groundwater models and riverine flow models or semi-distributed hillslope models with hydraulic models has to be further studied and techniques of integration by better linkage have to be refined. Also meteorological models should be connected with hydrological models to study the interaction of hydrological and meteorological processes. The models need to be extended with better user interfaces, which support the parameter assessment and the definition of scenarios by an expert system and further integration of GIS-functionality. New technologies of remote sensing are not sufficiently tested and evaluated in their use for flood models. Especially the high resolution and diversity of satellite images open a large potential for application in hydrological and hydraulic models.

4.1   INTRODUCTION

Stimulated by the increasing occurrence of extreme floods in the last years worldwide the flood research has been intensified leading to a better understanding about the phenomenon of flood. The physical processes contributing at its generation have been analysed and new refined mathematical methods have been developed. While at the beginning floods were regarded only as surface runoff in the meantime we understand that various flow processes on the surface and in the underground contribute to its formation. The impact of anthropogenic changes on the water balance and flood flow became aware and has been studied in many national and international research programmes. In the meantime, climate change has been detected as the main uncertainty in determining the probability of flood. Meteorological models have been used to develop different scenarios of future climate condition. They show an increase of extreme weather conditions like droughts and heavy storm events. These scenarios have been used as input to hydrological models to determine the effect of climate change on the water balance in catchments and on the flood situation in rivers. Still research is at the beginning on that field. More refinement of modelling capabilities will be needed like the coupling of hydrological and meteorological models.

It is the objective of this paper to present today's knowledge about the mathematical modelling of the hydrological processes participating at the formation of flood. They are divided into hydrological and hydraulic models. While the first group of models addresses the flow processes in the catchment, the second group simulates the transport processes of runoff in channels and rivers. At the end deficiencies of today's knowledge on floods and modelling instruments will be discussed and needs for research are formulated.

4.2   MODELLING OF FLOODS IN URBAN RIVERS

Since the 1930s the science of hydrology has concentrated on the development of mathematical models to predict a flood hydrograph in catchments (deterministic hydrology). It started with the unit hydrograph model in 1932 of Sherman and Horton's simplistic infiltration approach in 1933. Although further refined these two concepts were the only modelling approach for more than 30 years. They are still today the theoretical basis of the conceptual models in which the catchment is regarded as a system in which the input (precipitation) is transferred by a system function (unit hydrograph) to a runoff hydrograph at the outlet of the catchment. These rainfall-runoff models are also called "lumped models". In the 1970s the science of hydraulics joined the research field of runoff modelling. At the beginning they concentrate on the prediction of the changing magnitude, speed and shape of a flood wave as it propagates through rivers and channels (flood routing). They resolve the flow path by dividing the stream channel in profiles and applying the equations of motion (Mahood/Yevjevich, 1975). Also at the end of the 1970s the Hillslope Hydrology (Kirkby, 1978) has provided new knowledge on the relationship between precipitation input and hillslope discharge output in terms of the spatial soil distribution and terrain contour. With the availability of powerful computers these physically based theoretical concepts (Physical Hydrology) have stimulated the development of a new generation of rainfall-runoff models, the physically based distributed models (Beven, 1985; Abbot et al., 1986). They subdivide the catchment in a regular grid of cells. For each cell they solve the equations of motion to get the runoff pattern of overland and subsurface storm flow. At the beginning these models have been restricted to one-dimensional vertical simulations. Later they were extended to 2- and even 3-dimensional approaches to give the spatial pattern of overland flow, interflow and groundwater flow. The most sophisticated and applied distributed models are the SHE-model (Abbot et al., 1986) and the Hillflow-3d model (Bronstert/Plate, 1996)). All of them use a one-dimensional approach to simulate the flow in the unsaturated soil layer. While the SHE-model and Hillflow-3d models solve the one-dimensional Richards equation the Topmodel (Beven, 2001a) applies a simple exponential function of water content in the saturated zone. The latest version of the SHE-model, MIKE SHE (DHI, 1998) provides a three-dimensional groundwater model and calculates the overland flow on the basis of a 2-dimensional implicit finite difference model based on the kinematic wave theory. The channel flow is described by using MIKE 11 (DHI, 1998) or MIKE 21 (DHI, 1998), river-modelling systems, which solve the full 1-dimensional Saint-Venant equation or the 2-dimensional shallow water wave equations (dynamic wave). This system of models has reached the highest grade of hydraulic components to simulate the rainfall-runoff process. It comes closest to the target to describe fully the flow path of precipitation from the moment it reaches the earth surface to the outlet of the catchment at the end of the river.

Parallel to the development of the distributed models the lumped models have been further refined and improved to more physically based models which fully cover the whole hydrological cycle of land-bound water movement and refine its spatial resolution by subcatchments and hydrotopes. The last ones are areas of equal hydrological characteristic with respect to the vertical processes of interception, infiltration, evapotranspiration and groundwater recharge. Some of the lumped models use the same mathematical approaches as the distributed models to simulate these vertical processes, like the Richard's equation or exponential functions (ARCEgmo, Becker et al., 2002). The subcatchments are composed of hydrotopes, in which the runoff of each hydrotop is aggregated and according to the terrain contour and stormwater network discharged into the stream channel. The horizontal flow components of overland flow and interflow are modelled by own system functions, making use of the concept of parallel reservoir cascades (Pasche/Schröder, 1994). The translation of surface-runoff and interflow is covered by a time-area-histogram, which presents the histogram of the travel time of runoff within the catchment and thus represent the drainage characteristics of the subcatchments. Another very popular representative of these refined lumped models is the Topmodel (Beven/Kirkby, 1979; Beven, 2001). It also uses a transfer-function, which is derived from raster-based values of the topography (hillslope of each raster cell) and morphology. Topmodel assumes a close correlation between mean soil moisture, topography, the pattern of saturated areas and the hillslope runoff formulation. In dependence of the resolution of subcatchments, raster cells and hydrotopes today's conceptual models can reach a degree of spatial resolution, which comes close to the distributed models. Therefore they are often described as semi-distributed models.

The overall trend in rainfall-runoff modelling is directed towards more refined physically based modelling. The lumped and distributed models are more and more approaching each other. Still the enormous amount of computing time is one of the major drawbacks of a distributed model. Realistic simulations need a high resolution of the numerical grid. A very fine resolution is needed in the unsaturated soil zone to get the variation of the soil profile over the depth and to resolve the distribution of macropores. Despite the difficulty to get these data the three-dimensional application of the Richards equations will not be feasible for large watersheds. Thus aggregation methods like subdividing the soil profile in homogeneous layers (Ostrowski, 1982) and regarding them as hydrological units (lump approach) seem to be more realistic than introducing simplified and guessed vertical soil profiles in the Richards equations. The same difficulty arises in the modelling of the groundwater flow and the surface runoff. Structures of the mezo scale (with typical sizes of 0.10 m to 5.0 m) can have an important effect on the flow pattern and the hillslope runoff hydrograph. Neither the base data nor the computing capacity is available to resolve these structures in groundwater and surface runoff models. Additionally in urban areas the flow path is often diverting from the natural hillslope through urban drainage measures. Therefore the flow path cannot be determined purely on the basis of terrain data. Distributed models introduce dispersions terms to include these deficiencies in their models. They can be hardly determined on the basis of physical parameters but need to be found through model calibration. Another difficulty arises at the interface between the river and the groundwater aquifer (collimation layer). Micro and mezzo structural elements influence the permeability of this layer, which are hardly to resolve in distributed models. These effects are considered by pedo-transfer functions, which give the leakage through this layer in dependence of the hydraulic gradient (Giebel, 2002). The leakage factor stands for the structure and composition of the collimation layer. Our knowledge is not sufficient to quantify these parameters on the basis of physical conditions but have to be determined by model calibration. The numerical grids of the different runoff zones (groundwater zone, the non-saturated zone, the surface runoff area and the channel) are varying in spatial and temporal resolution. Thus the coupling of the different models of hillslope, groundwater and channel flow needs the introduction of special boundary elements with matching sides at which the primary variables are varying continuously and the mass conservation is preserved over these boundaries. Wohlmuth (2000) has developed special coupling elements, the Mortar elements which satisfy these inner boundary conditions between the models.

The review of today's techniques of flood modelling has demonstrated that no longer a clear difference exists between distributed and lumped models. In general hydraulic methods are replacing hydrological methods. Especially for the flood routing in natural rivers with wide flood plains and wooden vegetation these methods improve the quality of modelling results. They are able to reproduce the different transport capacities between rising and falling floods due to the hysterese effect (Sacher et al., 2006) and the non-linear retention process during the flooding of riparian land and encroached river sections (Teschke/Pasche, 2004). Also by coupling 3-dimensional groundwater models with 2-dimensional riverine flow models the interaction of the water body in the groundwater aquifer and in the river could be modelled more realistic than by lumped models (Ruf et al., 2006). But the demand for computing resources is extremely high restricting these models to small-scale applications. Mezo-scale and macro-scale analysis of medium and large catchments are not feasible yet. Also the Richards equations for interflow modelling and the kinematic wave equations for surface runoff modelling can be applied for small-scale studies like hillsides studies. In medium and large catchments the required high-resolution of the numerical grid cannot be accomplished that these methods are not improving the modelling results compared to the lumped models.

Taking into account the strengths and weaknesses of the hydraulic and hydrological methods of flood modelling the following chapters try to compile the most promising approaches of the distributed and lumped models to an integrative, the whole water cycle covering modelling approach for stormwater induced floods in urban environments. Due to the limitations of distributed approaches a semi-distributed approach is applied for modelling the horizontal flow processes on the surface and within the unsaturated soil zone (Hillslope Hydrology). For the groundwater flow a lumped approach as well as a hydraulic approach based on the 1-dimensional Darcy equation is suggested. The flood routing in channels and rivers will be modelled purely on the basis of hydraulic approaches.

4.2.1   Modelling instruments in hillslope hydrology

In semi-distributed hillslope hydrology the whole catchment is subdivided into smaller units, the subcatchments. They cover only parts of a whole water shed area but represent still drainage units in which the hillslope runoff (surface and subsurface) confluents to one point, the drainage point of the subcatchment. All sub-catchments are linked through channel sections. Their ends match with the drainage point of the subcatchment and drain into the adjacent channel section (fig. 4.1). Still in each subcatchment the vegetation cover, land use and the hydrogeology might vary. Therefore the subcatchments are further subdivided in hydrotopes. They are areas in which homogeneous conditions with respect to the vertical runoff processes occur. In general this is given for areas with homogeneous land use, pedology and geology. They can be easily determined within a GIS by overlaying and intersecting these three thematic maps.

In each hydrotop the vertical flow process is subdivided into a chain of reservoirs. The central element of this vertical chain is the unsaturated soil zone, which com-prises the root zone and further underlying zones of unsaturated soil (fig. 4.2). It can be modelled by any number of reservoir layers to get the vertical variation of the soil structure. For each layer the mass conservation is balanced by the 1-dimensional continuity equation of soil water

$$\frac{d(m \cdot sw_i(t))}{dt} = inf_{micro,i}(t) - (perc_i(t) + intf_i(t)) - evt_i(t) + ca(t) + q_B(t) \quad (4.1)$$

The change of water content $sw(t)$ in each soil layer is the sum of direct infiltration of the precipitation water into the soil matrix, the micropores ($inf_{micro}(t)$ [mm/h]), the capillary uprise $ca(t)$, the interflow $intf(t)$ [mm/h], the evapotranspiration $evt(t)$ [mm/h] and the exfiltration of soil water out the layer called percolation $perc(t)$. In the last layer of the unsaturated soil zone the percolation rate is replaced by the rate of groundwater recharge $q_B(t)$ [mm/s]. Due to rising and falling of the groundwater table the size m [mm] of the last layer of the unsaturated zone has to be adjusted. In case of high groundwater table soil layers need to be totally eliminated.

The challenge lies in the determination of the infiltration and exfiltration rates. As they are mainly a function of the soil water content $sw(t)$. Based on an approach of Ostrowski (1982) and the infiltration theory of Horton and Green-Ampt the potential infiltration rate into the soil matrix can be described by the following equation:

$$inf_{micro,pot}(t) = k_Z(\psi(t))_i \cdot \frac{sw_{max} - sw(t)}{sw_{max} - wp} \quad (4.2)$$

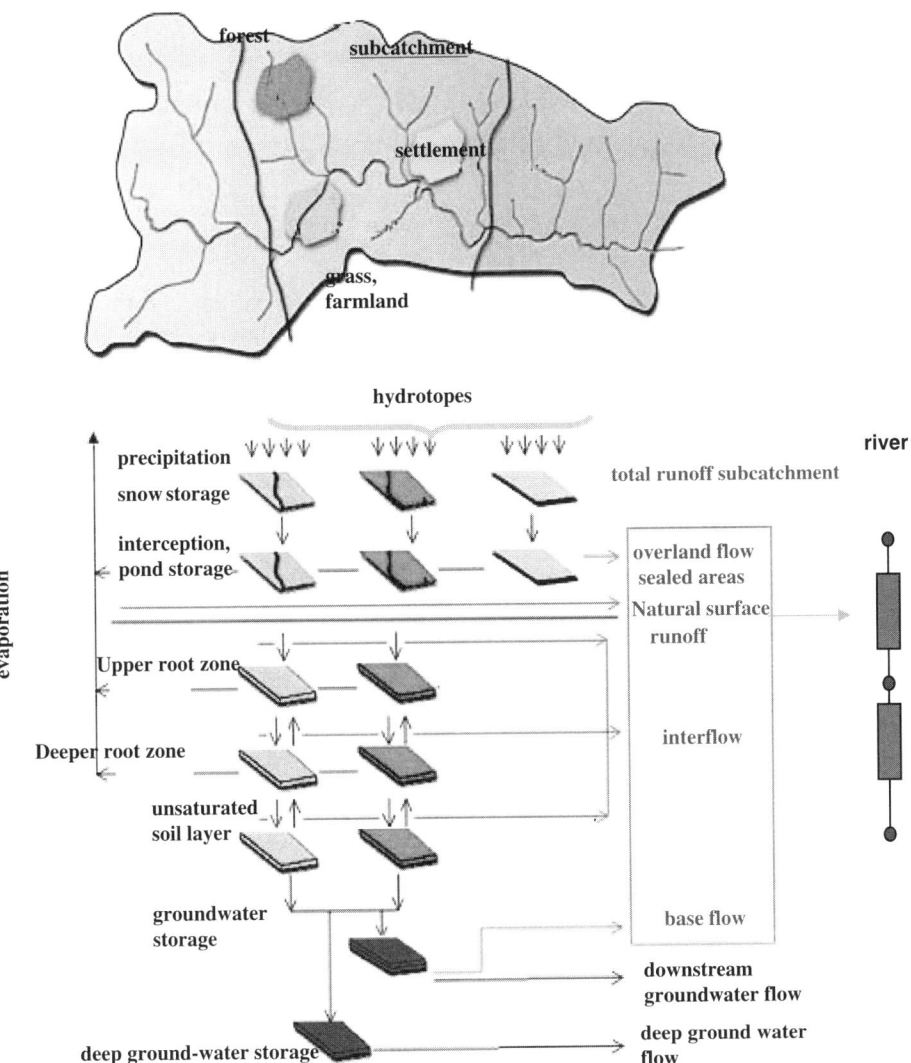

Figure 4.1. Structure of the semi-distributed hillslope model.

with

$k_Z(\psi(t))_i$ = hydraulic conductivity in dependence on the soil water tension [mm/h]

$\psi(\overline{sw(t)})$ = mean soil water tension within the micropores of the root layer [mm]

wp = wilting point in [mm]

fc = field capacity [mm]

sw_{max} = maximum soil water content at saturation in [mm].

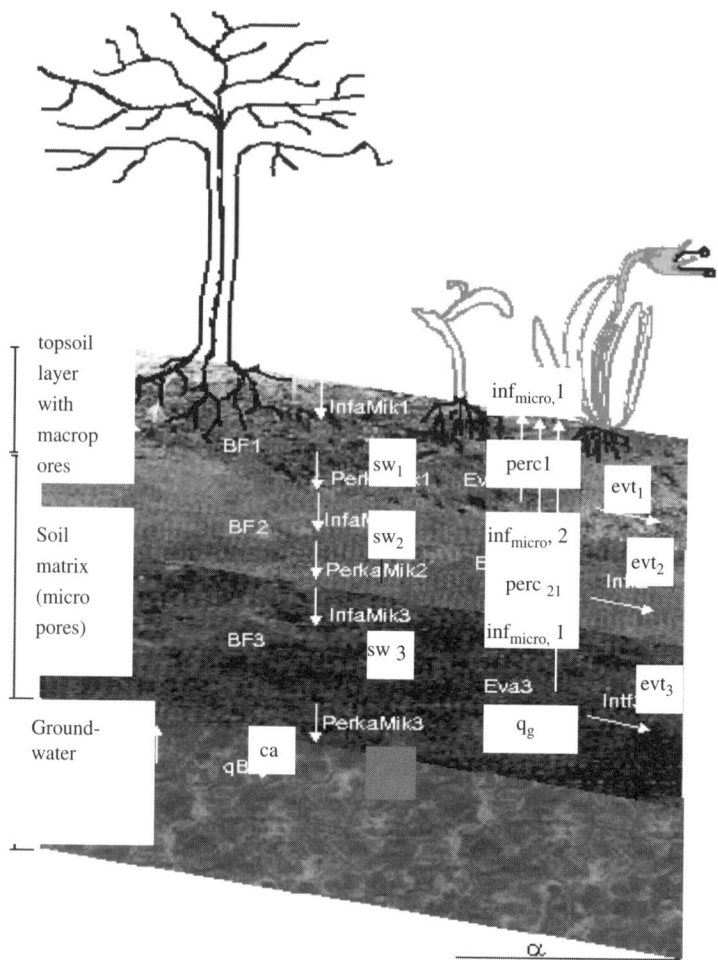

Figure 4.2. Soil water dynamics.

The real infiltration $inf_{micro,i}(t)$ into the micropores is the minimum of the potential infiltration rate and the maximum available water on the surface which corresponds to the intensity of the precipitation above surface $inf_{max}(t)$ in [mm/h].

The hydraulic conductivity $k_Z(\psi)$ for unsaturated soil can be estimated by a method from Van Genuchten (1980).

$$k_Z(\psi(t)) = k_F \cdot \frac{[1 - (\alpha|\psi(t)|)^{n-1}(1 + (\alpha|\psi(t)|)^n)^{-m}]^2}{[1 + (\alpha|\psi(t)|)^n]^{m/2}} \qquad (4.3)$$

with $\alpha, n, m =$ Van-Genuchten parameter of the pF-curve (AG-Boden, 1999) and $k_F =$ hydraulic conductivity at saturation of the soil water.

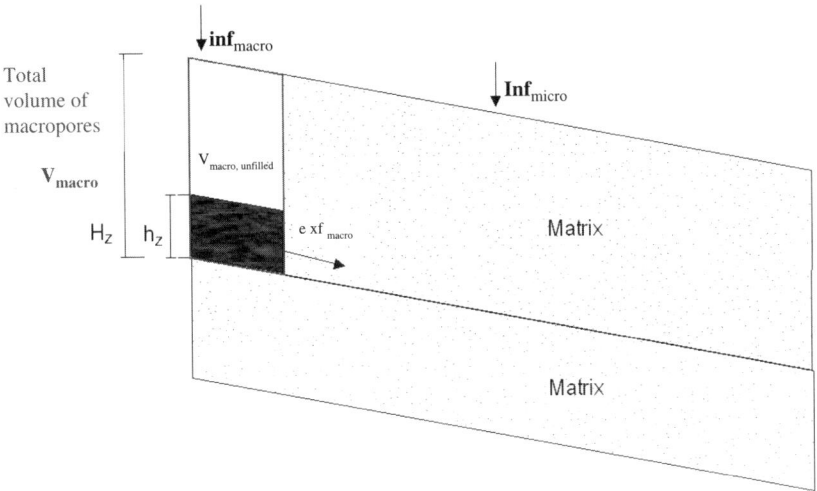

Figure 4.3. Distribution of dual pore system within topsoil layer.

Beven/Germann (1982) refined this infiltration approach by adding an additional component which includes the exfiltration $exf_{macro}(t)$ [mm/h] out of the macropores into the micropores.

$$\frac{d(m \cdot sw_i(t))}{dt} = inf_{micro,i}(t) + exf_{macro}(t) - (perc_i(t) + intf_i(t)) - evt_i(t)$$
$$+ ca(t) + q_B(t) \tag{4.4}$$

On the basis of this macropore concept Bronstert (1994) developed an analytical method for the determination of the exfiltration $exf_{macro}(t)$. He assumes that the dual poresystem of micro- and macropores occurs only in the root zone with the thickness H_Z [m] (fig. 4.3). The filling of the macropores starts not before the infiltration capacity of the micropores is reached and the macropores are filled without any time lap.

With these assumption the potential infiltration $inf_{macro,pot}(t)$ into the macropores is only a function of the volume $V_{macro,unfilled}(t)$ of the unfilled macropores:

$$inf_{macro,pot}(t) = \frac{V_{macro,unfilled}(t) \cdot H_Z}{\Delta t} \tag{4.5}$$

$V_{macro,unfilled}(t)$ (related to the total soil volume of the root zone [-]) needs to be updated for each time step Δt by:

$$V_{macro,unfilled}(t) = V_{macro} \cdot \frac{(H_Z - h_z(t))}{H_Z} \cdot \Delta t \tag{4.6}$$

with $V_{macro}(t)$ = total share of the macropores within the root zone and $h_Z(t)$ = height of the macropores, already filled with water in [mm]

$$h_z(t) = h_z(t - \Delta t) + [inf_{macro}(t - \Delta t) - exf_{macro}(t - \Delta t)] \cdot \Delta t \cdot V_{macro} \quad (4.7)$$

For the determination of the real infiltration $inf_{macro}(t)$ the potential infiltration has to be compared with the maximum available water for infiltration ($inf_{max}(t)$):

$$inf_{macro}(t) = \begin{cases} inf_{macro,pot}(t) & \text{for } inf_{max}(t) - inf_{micro}(t) \geq inf_{macro,pot}(t) \\ inf_{max}(t) - inf_{micro}(t) & \text{for } 0 < inf_{max}(t) - inf_{micro}(t) < inf_{macro,pot}(t) \\ 0 & \text{for } inf_{max}(t) - inf_{micro}(t) = 0 \end{cases}$$

$$(4.8)$$

The exfiltration $exf_{macro}(t)$ out of these macropores into the micro pores is for filled macropores a function of the soil water content $sw(t)$ and the conductivity $k_{macro}(t)$ between the macropores and micropores:

$$exf_{macro}(t) = k_{macro}\psi(sw(t)) \cdot \frac{sw_{max} - sw(t)}{sw_{max} - wp} \quad (4.9)$$

But at each time step the exfiltration rate has to be compared with potential exfiltration rate which is a function of the maximum available water in the macropores.

For the percolation $perc(t)$ of the soil water into the deeper soil layer it is assumed that only the micropores contribute to this exfiltration of water into deeper layers.

The potential percolation $perc_{pot}(t)$ is calculated in dependence on the soil water tension and the amount of free soil water ($sw(t) - fc$):

$$perc_{pot}(t) = k_Z(\psi(t)) \cdot \frac{sw(t) - fc}{sw_{max} - fc} \quad \text{for } sw(t) \geq fc \quad (4.10)$$

If the potential percolation $perc(t)$ of a soil layer i exceeds the infiltration rate of the following soil layer $i + 1$, interflow is generated within the layer i:

$$intf_i(t) = perc_i(t) - inf_{micro,i+1}(t) \quad (4.11)$$

According to Germann (1990) the interflow is also generated at a hillslope due to gravity. It is mainly fed by water from the macropores. Its intensity $intf_{gravity}(t)$ is limited by the maximum infiltration rate $inf_{macro}(t)$ and the exfiltration rate $exf_{macro}(t)$ out of the macropores into the soil matrix:

$$intf_{gravity}(t) = inf_{macro}(t) - exf_{macro}(t) \quad (4.12)$$

The transport capacity and the retention of the macropores within the vadose soil zone is influenced by many parameters, like the soil properties, the diameter of the macropores, the length of continuous pathways through the macropores, the slope of the hillside and the antecedent moisture within the tubes of the macropores. This dependency cannot be expressed on the basis of the Darcy law, as no piezometric pressure gradient exists in the vadose zone and the horizontal conductivity of the macro pores is not uniform within the soil layers. As stated by Germann (1990) storage, flow and the driving energy of the interflow need to be decoupled. He could show that kinematic wave models can be successfully applied to macropore flow in hillslopes. Despite this successful application the mechanisms of runoff generation by macropore flow is not fully understood. It is unclear to what degree water is indeed flowing through macropores and to what extend energy waves simply push water out at the downstream ends of the macropores. In the meantime many researchers have taken up the ideas of Germann and developed the concept of saturation-excess overland flow. They assume that in catchments with hillslopes, decreasing hydraulic conductivity with increasing infiltration depth and wide valley floors fast subsurface flow transports new stormwater along the lateral macropores to the valley floor where the macropores end in capillary fringes which are filled with old soil water. In case of saturation this prestorm water will be replaced by new stormwater coming form the macropore channels. In narrow valleys without flat valley floors the flow of the macropore channels is directly released into the river channel. Bronstert (1994) compares equations of Eaglson (1970), Beven (1982) and German (1992) for the estimation of the velocity of the flow within the macropore channels. They are based on the kinematic wave approach in which the friction slope within the macropores is equal to the slope of the terrain. According to Germann (1992) the velocity $v_{macro}(t)$ within the macropore channels is a function of the hillside slope I_o, the height $h_z(t)$ and the friction parameter k_{macro} of filled macropores:

$$v_{macro}(t) = k_{macro} \cdot I_o \cdot [h_z(t)]^{2.5} \tag{4.13}$$

For the friction parameter k_{macro} Germann determined values between 0.2 and $10.0 \, \text{m}^{-3/2}/\text{s}$. This equation has been evaluated for hillslopes of $10°$. The validity of this function at other slopes is unclear. Unsatisfactory are the physical meaningless dimensions of the friction parameter. Thus the equation should be improved by making use of hydrodynamic approaches.

As mentioned above the stormwater hydrograph at the outlet of a subcatchment is mainly a function of the infiltration-excess overland flow, the interflow (saturation-excess overlandflow) and the groundwater baseflow.

In semi-distributed models these runoff components are determined by a lump model approach. In the simplest way this is a linear reservoir, which has been extended to linear reservoir cascades or parallel reservoir cascades. However the empirical parameters (retention coefficient k and the number n of the reservoirs) are empirical parameters with little physical background and have to be determined by calibration. In ungauged catchments these methods cannot be applied reliably as the validity of the empirical parameters determined at other catchments is in general not given. A better physical basis is given for the method of isochrones. In this approach the runoff process is divided in a translation and retention process. The first one is described on the basis of the isochrones, which represents lines of the same transport time of the runoff within the catchment. The area between two lines is arranged in a time-area-diagram corresponding to the transport time of the downstream and upstream isochrone. Assuming over the whole subcatchment a unit infiltration-excess runoff of 1 mm the time-area-diagram represents the unit translation hydrograph u_t, which can be calculated for discrete time steps $n\Delta t$:

$$u_t(n\Delta t) = \frac{\sum_{i=1}^{n} A(i\Delta t) - \sum_{i=1}^{n-1} A[(i-1)\Delta t]}{\Delta t \cdot A_e} \qquad (4.14)$$

with $A(i\Delta t)$ = area of the subcatchments that has already drained at the moment $i\Delta t$ and A_e = total area of the catchment.

For the determination of the isochrones Geographical information systems have been applied very successfully. Based on a Digital Terrain Model (DTM) the flow paths on the overland can be determined. The total transport time t_c (concentration time) along this flow path can be derived by segmenting and determining the flow velocity v_i within each segment i of length s_i.

$$t_c = \sum_{i=1}^{n} \frac{s_i}{v_i} \qquad (4.15)$$

In most publications the kinematic wave equation and the Manning's formula are used for the determination of the flow velocity. Pasche/Schröder (1994) showed that the sheet flow around vegetation is better described on the basis of the Darcy-Weisbach law and a resistance law derived from flow around cylindrical bodies:

$$v = \sqrt{2 \cdot g \cdot I_o \cdot \frac{a^2}{d_P \cdot c_{WR}}} \qquad (4.16)$$

with I_o = hillside slope, d_P = diameter of the vegetation element [m], a = distance of the vegetation elements [m].

The drag coefficient c_{WR} can be assumed according to DVWK (1993) to $c_{WR} = 1.5$.

When the sheet flow reaches the micro channels (rilles and ephemeral gullies) a boundary layer flow is generated, in which the mean velocity is dependent on the water depth. This shows that the transport time in catchments is not a constant but varies in dependence on the intensity of the available infiltration-excess runoff. Consequently the time-area diagram cannot be Unit Hydrograph for all precepitation events but has to be adjusted. Due to the uncertainties in hillslope hydraulics and the lack of detailed information about the interrill areas, structure of rilles and ephemeral gullies the Unit Hydrograph of the time-area diagram is sent through a linear reservoir u_R. It also considered other retention effects, which are not included in the isochrone calculation, e.g. trapping in ponds and hollows.

$$u_{TR}(n\Delta t) = \sum_{i=1}^{n} u_t(i\Delta t) \cdot u_R[n\Delta t - (i - 1)\Delta t] \tag{4.17}$$

with

$$u_R[n\Delta t - (i - 1)\Delta t] = \frac{1}{k} \cdot e^{-[n\Delta t - (i-1)\Delta t]/k} \tag{4.18}$$

and k = retention coefficient of the surface runoff. This parameter is compensating all the deficiencies in your time-area-diagram and thus is less a physically based parameter but more a calibration parameter. It was found to vary between $k = 0.5$ h to 5.0 h for surface runoff in natural environments.

On the basis of this translation-retention Unit Hydrograph $u_{TR}(t)$ the hydrograph of the surface runoff at the exit of the subcatchment can be calculated by the lump model approach:

$$Q_a(n\Delta t) = \frac{1}{3.6} A_e \sum_{i=1}^{n} i_{eff}(i\Delta t) \cdot u_{TR}[n\Delta t - (i - 1) \cdot \Delta t]\Delta t \tag{4.19}$$

with $i_{eff}(i\Delta t)$ = generation rate of infiltration-excess runoff in [mm/h] at moment $i\Delta t$.

This isochrone model can be also applied for modelling the subsurface run-off (interflow). As the flow velocity within the macropores was assumed to be dependent on the slope of the hillside the flow paths within the macropores can be assumed to be parallel to the ones of the surface runoff. Then the transport time of the macropore flow along each flow path can be calculated on the basis of eqn. (13). Also the retention coefficient of the linear reservoir has to be adjusted to the subsurface flow conditions and needs to be found by calibration. For the determination of the groundwater induced baseflow into the river channels the lumped model approach has been successfully applied in semi-distributed rainfall-runoff models (Pasche et al., 2004). In this approach the groundwater aquifer is approximated by a reservoir with horizontal groundwater table. The mass balance of this reservoir is given by the 1-dimensional continuity equation of the groundwater body:

$$\frac{dh_{GW}(t)}{dt} = \frac{1}{3.6} qg(t) + [Q_{GWIN}(t) - Q_{Base}(t) - Q_{GWOUT}(t)$$
$$- Q_{Deep}(t)]/(A_E \cdot 10^6) \tag{4.20}$$

The changing rate of the ground water stage $h_{GW}(t)$ is balanced with the ground-water recharge $qg(t)$, the groundwater inflow $Q_{GWIN}(t)$ [m³/s], the baseflow

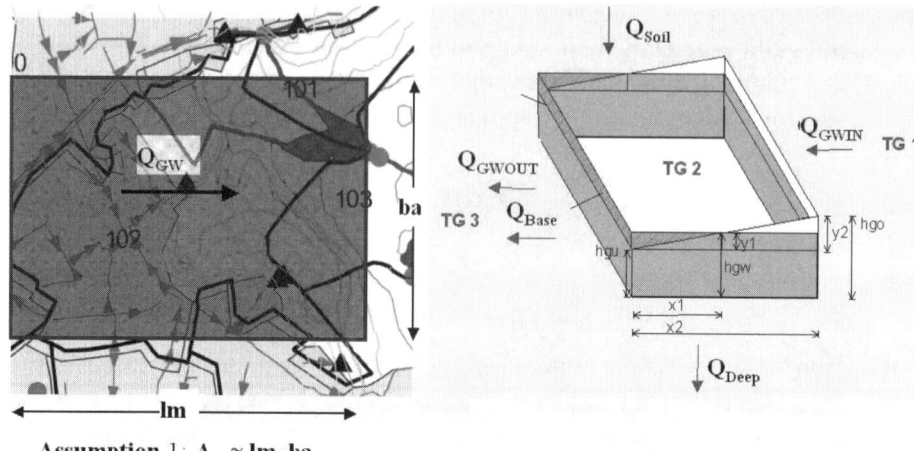

Assumption 1: $A_E \approx \text{lm} \cdot \text{ba}$

Figure 4.4. Concept and definition sketch of the groundwater reservoir.

$Q_{Base}(t)$, the outflow $Q_{GWOUT}(t)$ of the aquifer to the downstream side and the outflow $Q_{Deep}(t)$ into a second deeper groundwater aquifer. Based on the linear reservoir theory the groundwater outflow $Q_{GWOUT}(t)$ and the baseflow $Q_{Base}(t)$ are calculated in dependence on the stored volume $S(t)$ in the groundwater aquifer. As an example the resulting equations are given for groundwater stages between the bottom height h_{GU} [in m above bottom of aquifer] and the highest elevation h_{GO} [in m above bottom of aquifer] of the channel:

$$Q_{Base}(t) = \frac{0.5}{k_{Base}} \cdot \frac{h_{GW}(t) - h_{GU}}{h_{GO} - h_{GU}} \cdot (h_{GW}(t) - h_{GU}) \cdot n_e \cdot A_E \qquad (4.21)$$

$$Q_{GWIN}(t) = \frac{1}{k_{GW}}(h_{GW}(t) - h_{GU})\left[1 - 0.5\frac{h_{GW}(t) - h_{GU}}{h_{GO} - h_{GU}}\right] \cdot n_e \cdot A_E \qquad (4.22)$$

with A_e = area of the subcatchment in [m^2], n_e = effective porosity of the groundwater aquifer [-].

The retention coefficients k_{Base} [h for baseflow] and k_{GW} for the groundwater outflow are empirical coefficients, which need to be determined by calibration. The elevations h_{GO} and h_{GU} stand for the thickness of the aquifer layer. The difference between these two parameters stands for the part of the aquifer, which participates at the generation of the baseflow. Although Pasche et al. (2004) could show that these parameters can correspond with the physical parameters (fig. 4.5), they are nonidentical and have to be verified by calibration. Especially for low lands with mild topographic gradients and strong backwater effect within the groundwater aquifer they need to be reduced. Thus Pasche et al. suggested for these areas to

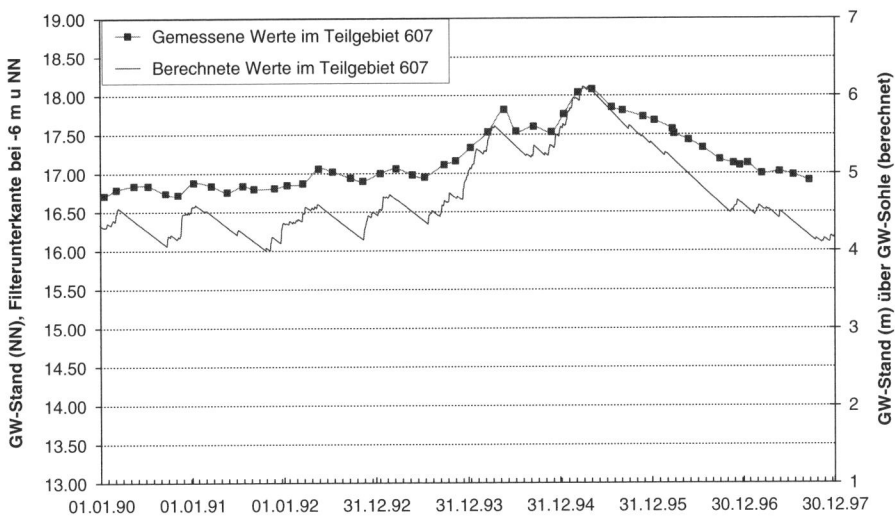

Figure 4.5. Mean groundwater stages within a subcatchment.

determine the groundwater flow on the basis of the 1-dimensional Darcy law:

$$\frac{\partial}{\partial x}\left(h_{GW}(t) - b \cdot k_F \frac{\partial h_{GW}(t)}{\partial x}\right) + q(t) = n_e \frac{\partial h_{GW}(t)}{\partial t} \qquad (4.23)$$

with k_F = hydraulic conductivity at saturation [m/s], $q(t)$ = sink- and source term [m/s] and b = lateral width of the aquifer in [m].

4.2.2 Hydraulic modelling instruments for flood routing in rivers

Due to still increasing computing performance and increased demand for reliable computer simulations hydraulic approaches will substitute the lumped models in riverine flood routing. Thus this chapter will restrict its state-of-the-art debate on presenting the most promising hydrodynamic approaches in riverine flood routing

Good consensus exists between engineers in the formulation of the basic flow equations for rivers. For one-dimensional flood-routing the theoretical basis is the well-known Saint Venant equation (Pasche et al. 2004)

$$\frac{1}{gA}\frac{\partial Q}{\partial t} + \frac{1}{gA}\frac{\partial(\alpha' Q^2/A)}{\partial x} + \cos\Theta\frac{\partial z_{WSP}}{dx} + I_R - \cos\varphi\frac{v_e q_e}{gA} = 0 \qquad (4.24)$$

with Q = total discharge [m³/s], A = cross section [m²], q_e = inflow/outflow [m³/sm], z_{WSP} = water surface elevation [m a.O.], I_R = friction slope [-], α' = energy coefficient [-], Θ = longitudinal slope of river bottom [°], φ = angle between inflow/outflow and main flow [°], $i, i+1 =$ downstream, upstream profile [-], $x =$ length of river along thalweg [m].

For two-dimensional flood-routing the shallow water equations are theoretical basis which are based on the assumption of hydrostatic water pressure and are derived by integration of the Reynolds stress equations over the depth. In rivers the Coriolis force and the shear stress at the water surface are of minor relevance that the equations can be further simplified to the following form.

$$\frac{\partial u_i}{\partial t} + u_j \frac{\partial u_i}{\partial x_j} = -g \frac{\partial}{\partial x_i}(z_{WSP}) + \frac{1}{h}\frac{\partial}{\partial x_j}\left[h\left(\frac{1}{\rho}\tau_{L,ij} + \frac{1}{\rho}\tau_{t,ij} - \widetilde{u_i}\widetilde{u_j}\right)\right] - \frac{1}{h}\frac{\tau_{so,i}}{\rho} \quad (4.25)$$

with $u =$ local velocity component [m/s], $h =$ local water depth [m], $\tau_L =$ viscous shear stress [N/m²], $\tau_{t,ij} =$ turbulent shear stress component [N/m²], $\widetilde{u_i}\widetilde{u_j} =$ dispersion terms [m²/s²], $\tau_{So} =$ bed shear stress [N/m²], $\rho =$ density of water [kg/m³], $i, j = 1, 2$ (x-, y-component and index of the Einstein summation convention).

Both models, the one- and two-dimensional flood routing model, need empirical parameters. The most relevant parameter for the one dimensional equations is the friction slope I_R and the energy coefficient α'. In the shallow water equation the friction losses are covered by the bed shear stress τ_{So}. Further parameters are needed to quantify the turbulent shear stresses $\tau_{t,ij}$ and the dispersion term $\widetilde{u_i}\widetilde{u_j}$ which are considering the effect of secondary current.

4.2.2.1 Parameter estimation in one-dimensional flow models

The simplest approaches use only one parameter to quantify the friction slope IR in the one-dimensional equation (4.24) and the bed shear stress in the two-dimensional equation (4.25). It includes all flow losses and represents the variable roughness along the wetted parameter by a mean value. In dependence of the flow formula this parameter is named Manning's n (GMS-formula) or Darcy Weisbach coefficient λ (DW-formula).

$$I_R = \frac{1}{8g}\frac{\lambda}{r_{hy}}\frac{Q^2}{A^2} \quad (4.26) \qquad I_R = \frac{n^2}{r_{hy}^{4/3}}\frac{Q^2}{A^2} \quad (4.28)$$

$$\tau_{S0,i} = \frac{\lambda}{8}\rho u_i \sqrt{u_i^2 + u_j^2} \quad (4.27) \qquad \tau_{S0,i} = \frac{\rho g n^2}{h^{1/3}} u_i \sqrt{u_i^2 + u_j^2} \quad (4.29)$$

Darcy-Weisbach-formula $\qquad\qquad$ Gauckler-Manning-Strickler-formula

Only the DW-formula is physically based which can be seen by the units of the empirical parameters. While the Darcy-Weisbach-coefficient λ is dimensionless the Manning's n parameter has the physically senseless units of $[s^2/m^{1/3}]$. Making use of the boundary layer theory and the theory of flow around bodies physically based equations can be derived in which the Darcy-Weisbach-coefficient λ is expressed in terms of directly determinable geometric parameters or equivalent parameters, like the equivalent sand roughness k_s in the Colebrook-White formula.

$$\frac{1}{\sqrt{\lambda_{So}}} = -2.03 \cdot \log \left(\frac{2.51}{fRe\sqrt{\lambda_{So}}} + \frac{k_s}{f\,14.84R} \right) \tag{4.30}$$

with $Re = \frac{UR}{v}$ = Reynold's number, R = hydraulic radius, v = kinematic viscosity and f = form factor.

BWK (1/1999) has shown that on this basis nearly all relevant friction losses can be well quantified. Pasche (1984) has shown that especially for natural rivers the DW-formula has advantages as the flow losses caused by non-submerged wooden vegetation and by momentum transfer at the interface between river and flood plain can be well described by the following formulas

$$\lambda_p = \frac{4 \cdot h_P \cdot d_P}{a_x \cdot a_y} \cdot c_{WR} \cdot \cos(\alpha_{lat}) \tag{4.31}$$

$$\frac{1}{\sqrt{\lambda_T}} = -2 \cdot \log \left[0.07 \cdot \left(\frac{c \cdot b_m}{b_{III}} \right)^{1.07} \cdot \Omega \right] \tag{4.32}$$

$$\Omega = \left[0.07 \cdot \frac{a_N^L}{a_x} \right]^{3.3} + \left[\frac{a_{NB}}{a_y} \right]^{0.95} \tag{4.33}$$

with a_x, a_y = distance of vegetation elements in both horizontal directions [m]; h_P = water depth in front of the vegetation element [m]; d_P = diameter of vegetation element [m]; α_{lat} = lateral inclination of bottom [°]; c_{WR} = drag coefficient of wooden vegetation [-] according to Pasche (1984); a_{NL}, a_{NB} = wake length and wake width [m] according to BWK 1/1999, c = form factor [-], b_m, b_{III} = contributing width of vegetation zone and river [m] according to BWK (1/1999).

The vegetation parameters and the contributing vegetation width are representing real parameters, which can be directly determined in nature or derived on this basis. In contrary the GMS-formula uses a parameter, which can be only quantified on the basis of tables, own experience or by calibration. Chow (1959) and some newer publications have tried to correlate Manning's n to photos in order to make the estimation of this parameter more reliable. It can be easily seen that the effort to evaluate the parameters in the GMS-formula is much less than in the DW-formula what explains the persistence of this formula in the engineering world.

Pasche et al. (2005) showed that on the basis of the GMS-formula the quality of the flood modeling could be substantially improved. All parameters of the DW-formula have been estimated on the basis of a field survey in which the corn fraction of the bed material and the geometry of the wooden vegetation has been

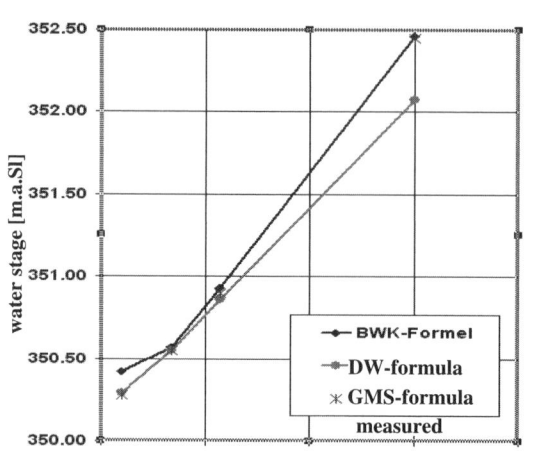

Figure 4.6. Water stages of the river anube at gauge Neustadt.

determined and transferred into equivalent sand roughness and average vegetation parameters a_x, a_y, and d_P according to the recommendations of BWK (1999). Only little adjustments were necessary in the calibration procedure. With one parameter set the whole range of monitored flow events could be well reproduced (fig. 4.6). In contrary one parameter set of the GMS-formula did not give a good fit over the whole range of observed flow events. Especially in rivers with wooden vegetation on the bank the GMS-formula underestimated the water stage with increasing deviation for rising water stages. This deviation has an easy explanation. A constant Manning's n parameter will always result in increasing flow velocities with rising water stages. For non-submerged wooden vegetation the flow velocity however stays more or less constant. This characteristic of flow in vegetated rivers is included in the formula (4.31) of the DW roughness concept that the good approximation of the DW-formula was not by accident but due to a good physical soundness of the method. This effect has been observed in the application of both momentum equations, the Saint Venant equation and the shallow-water-equations.

Considerable improvement of the quality of the model has been accomplished by using high-resolution remote-sensing data of the type Quickbird to determine the spatial distribution of the roughness condition on the flood plain (fig. 4.7) (Rath, 2006). The roughness distribution was much more refined than the field survey and detected errors like the classification of large parts of the flood plains as short cut grasses instead of natural wetlands with weed and long grass. This refined roughness classification had a strong effect on the calculation. The water stages increased by 20 cm and reduced the error to the measurements considerably.

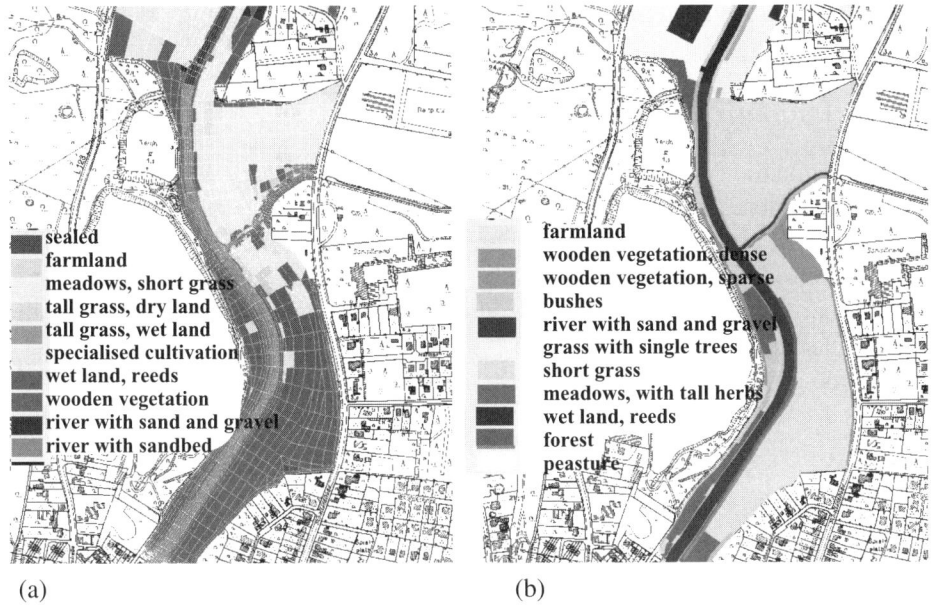

sealed
farmland
meadows, short grass
tall grass, dry land
tall grass, wet land
specialised cultivation
wet land, reeds
wooden vegetation
river with sand and gravel
river with sandbed

farmland
wooden vegetation, dense
wooden vegetation, sparse
bushes
river with sand and gravel
grass with single trees
short grass
meadows, with tall herbs
wet land, reeds
forest
peasture

(a)   (b)

Figure 4.7. (a) Roughness distribution gained from analysis of QuickBird satellite images; (b) roughness distribution determined by field survey.

The momentum coefficient α' results from the integration of the square of the velocity $v(y,z)$ over the cross-section A divided by the square of the mean velocity v times the total area A:

$$\alpha' = \frac{1}{v^2 A} \iint\limits_{A} v(y,z)^2 \, dA \qquad (4.34)$$

In compact cross-sections its value is nearly 1.0. But in natural rivers with wooden vegetation on the bank and adjacent flood plains the momentum coefficient can not be neglected. Separating the whole channel into the three independent stream tubes of left flood-plain (fl,l), main channel (ch) and right flood-plain (fl,r) the momentum coefficient can be calculated by:

$$\alpha' = \frac{A[A_{fl,l}[R_{fl,l}/\lambda_{fl,l}] + A_{ch}[R_{ch}/\lambda_{ch}] + A_{fl,r}[R_{fl,r}/\lambda_{fl,r}]]}{[A_{fl,l}[R_{fl,l}/\lambda_{fl,l}]^{1/2} + A_{ch}(r_{ch}/\lambda_{ch})^{1/2} + A_{fl,r}[R_{fl,r}/\lambda_{fl,r}]^{1/2}]^2} \qquad (4.35)$$

This approach underlies the assumption, that the flow velocity within each stream tube (left and right flood-plain, stream channel) can be approximated by a constant velocity. On flood plains with extreme changes of land use cover or water depth this assumption is no longer valid and equation 35 needs to be extended by dividing the stream tubes in smaller units. Additionally in compound channels the friction slope IR has to adjusted in a similar way as the energy coefficient:

$$I_R = \frac{1}{[A_{fl,l}(R_{fl,l}/\lambda_{fl,l})^{0.5} + A_{fl,r}(R_{fl,r}/\lambda_{fl,r})^{0.5} + A_{ch}(R_{ch}/\lambda_{ch})^{0.5}]^2} \cdot \frac{Q^2}{8g} \quad (4.36)$$

The determination of the friction slope based on the Darcy Weisbach law can require considerable computer resources especially for unsteady flow simulation in rivers with compound channel geometry and composite roughness. Teschke (2004) showed that the computing time could be substantially reduced by using polynomial functions for the friction slope, which have been determined on the basis of a steady state calculation.

Assuming that for unsteady flow the flow resistance is the same as for steady flow the friction slope I_R, steady of steady flow can be transferred into the friction slope IR, unsteady of unsteady flow by the following equation:

$$I_{R,unsteady}(h) = \frac{I_{R,steady}(h) \cdot Q(h)^2_{unsteady}}{(Q_{steady}(h))^2} \quad (4.37)$$

with Q_{steady} is the discharge at steady flow for the specific water depth h and determined from the polynomial function.

He tested several rivers with this method like the river Danube in Bavaria/Germany and the tidal section of the river Stör in North Germany. The agreement with the observed discharge hydrograph was very satisfactory (fig. 4.8). In the meantime extended this method to structures, like bridges and weirs, where he approximates the upstream-downstream-water stage-discharge relationship by a 2-dimensional polynomial function (Teschke, 2006).

4.2.2.2 *Parameter estimation in two-dimensional flow models*
For evaluating the turbulent shear stresses τ_t in 2-dimensional hydrodynamic models only the Boussinesq approach is applied in the engineering world:

$$\tau_{t,ij} = \rho \left(\nu_T \left(\frac{\partial u_i}{\partial x_j} + \frac{\partial u_j}{\partial x_i} \right) - \frac{2}{3}k\delta_{ij} \right) \quad (4.38)$$

with k = turbulent kinetic energy [m^2/s^2], ν_T = turbulent viscosity [m^2/s], δ_{ij} = Kronecker delta [-].

The simplest models assume that the turbulent viscosity ν_T is a constant for a given flow domain (constant-eddy-viscosity-approach). In deed it could be shown that this assumption leads to good results but only if the parameter could be calibrated. Theoretical and experimental analysis of turbulence however demonstrates that the turbulent viscosity is not a constant but varies in dependence on the flow situation. These characteristics of the turbulent viscosity are well considered by the mixing length approach (term 1 in eq. 4.39) in combination with the bed shear

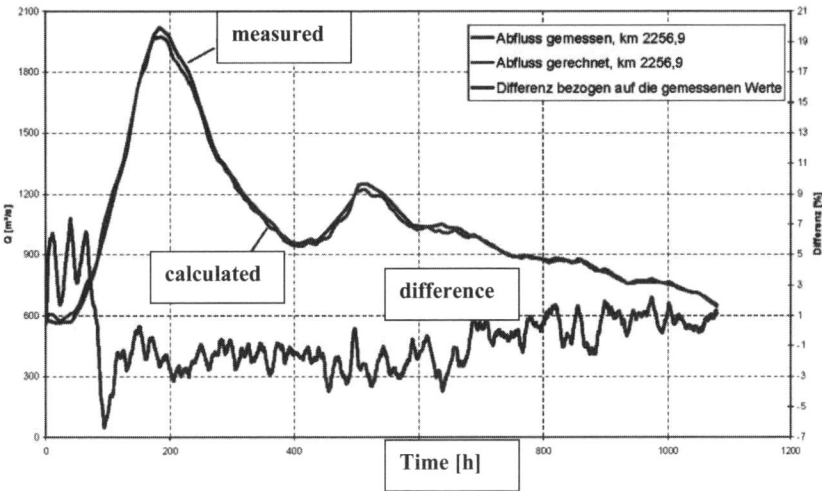

Figure 4.8. 1-dimensional flood simulation on the basis of polynomials for friction slope assessment.

stress approach (term 2 in eq. 4.39) (Lippert, 2005):

$$v_{t,ij} = l_m^2 \left[2 \left(\frac{\partial u_i}{\partial x} \right)^2 + 2 \left(\frac{\partial u_j}{\partial y} \right)^2 + \left(\frac{\partial u_i}{\partial y} + \frac{\partial u_j}{\partial x} \right)^2 \right]^{1/2} + e^* \sigma_t h \sqrt{\frac{\lambda}{8}} \sqrt{u_i^2 + u_j^2}$$

(4.39)

with $e^* \sigma_t = 0.15$.

The mixing length l_m can be quantified on the basis of a free shear layer approach from Rodi (1984).

$$l_m = 0.09 b_S$$ (4.40)

with b_S = mean width of the free shear layer (Free-Shear-Layer-Approach). In an approach of Smagorinsky (1963) the mixing length is correlated to the geometry of the numerical grid to compensate the numerical diffusion which occurs in most numerical grids (Malcherek, 2001).

$$l_m = 0.197 \Delta$$ (4.41)

with Δ = equivalent length of one element side of the numerical grid.

Pasche et al. (2006) demonstrated at the same section of the river Danube as the test runs for the DW-formula that indeed a constant eddy viscosity parameter is not able to accomplish an acceptable fit to all observed flow events. The deviations were in some cases more than 40 cm. Considerable improvement has been accomplished by the mixing length approach in combination with the bed shear stress approach (eq. 4.39). Both the Smagorinsky equation (eq. 4.41) and the free shear layer equation (eq. 4.40) do not show the deficiencies of the constant eddy viscosity

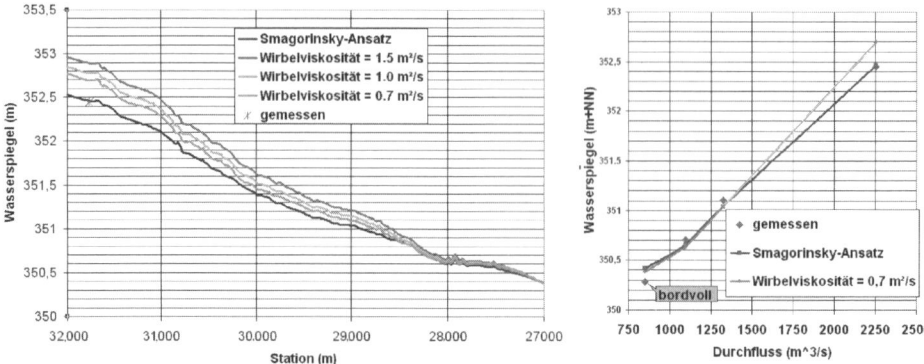

Figure 4.9. (a) Water surface profile of the Danube at the flood in 1999; (b) water stages at the gauging station Neustadt.

approach (fig. 4.9). While the Smagorinsky approach does not need any parameter estimation the free shear layer approach contains one parameter, the width of the shear layer, which is dependent on the flow domain and thus needs to be taken from the numerical result. Free shear layers occur at the interface between flood plain and main channel. Thus they are only relevant at flood. At flow below bankful the turbulence is dominated by bed shear stress, as the velocity gradient in the vertical exceeds the one in the horizontal direction by far. This characteristic of turbulence in rivers is well covered by equation 4.39.

Inadequate results produced the Smagorinsky approach in grids in which the grid resolution had a discontinuity. While in the section with the coarse grid the best fit has been accomplished with the Smagorinsky parameter (0.197) an adjustment to 0.100 was necessary in the section with the refined grid. The capability of the turbulence approaches to reproduce the depth-averaged velocity distribution have been further studied on the basis of experimental data from Pasche (1984). They give the depth-averaged velocities in compound channels with and without vegetation on the flood-plain (fig. 4.10). The physical model of the laboratory flume was algorithm of Rath (2006) leading to unstructured Finite Element nets.

Due to numerical diffusion the calculated depth-averaged velocity distribution showed the free shear layer a strong dependency from the degree of refinement which was more or less the same for all turbulence approaches given above (fig. 4.10b). This sensitivity diminished for the ratio $ds/b_s < 0.15$ with $ds =$ length of the lateral element side and $b_s =$ width of extending this relationship by the velocity difference dv of the free shear layer and the maximum velocity v_{max} in the main channel a parameter could be found which can be regarded as an indicator of the intensity of numerical diffusion in free shear layers:

$$c = ds/b_s(dv/v_{max}) \qquad (4.42)$$

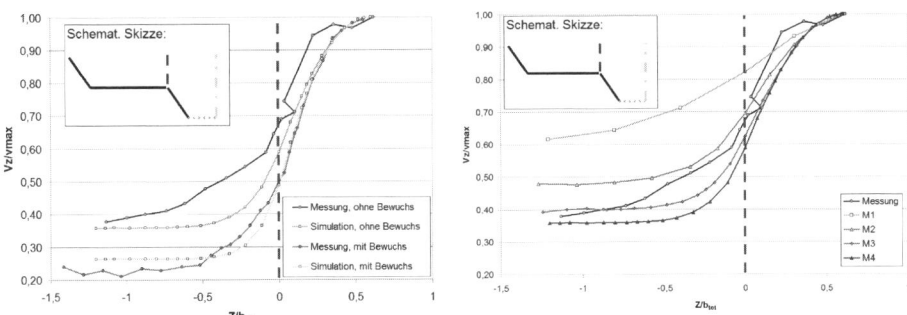

Figure 4.10. (a) Depth-averaged velocities with Smagorinsky-approach; (b) depth-averaged velocities for differentgrid resolutions.

For the analysed laboratory flume this parameter was found to be $c = 0.07$. More studies at free shear layers in different rivers however are necessary to validate this parameter.

Unstructured grids behaved more stable in the numerical calculation then structured grids in which the elements are oriented parallel to the main flow direction, which can be explained by a better dampening of the numerical wiggles in the Finite Element solution procedure. For the unstructured and diffusion free grid the different turbulent approaches were used to show the importance of the eddy viscosity on the calculated velocity distribution in the free shear layers. Overestimated eddy viscosities produce unrealistic mild gradients of the depth-averaged velocity in the free shear layer. Consequently the constant eddy viscosity approach needs a good guess to reproduce this velocity gradient whereas the turbulent approaches based on the mixing length theory were directly giving an acceptable result without modification of the Smagorinsky-parameter or by taking the free shear layer width from the numerical result (fig. 4.10a). Thus these two turbulent approaches can be applied in a predictive way. A first verification of the free shear layer approach has been accomplished for the river Lippe and Stör. Without modification of the empirical parameters and just by adopting the shear layer width to the numerical result a very satisfactory velocity distribution was determined (fig. 4.11). At the end of the iterative process the free shear was determined to 7–10 m in the river Stör and to 23 m in the river Lippe, which corresponds very well with the observation.

More refined turbulence models as the k-ε-model have no relevance in the engineering world, as the numerical effort is considerably more than in analytical approaches without improving the quality of result substantially (Pasche, 1984).

As the shallow-water-equations cannot simulate secondary currents directly, they need to be included by analytical approaches. Most often this effect is included in the turbulent viscosity, which is without physical meaning. A more realistic and physically based approach is the quantification of the dispersion terms $\tilde{u}_i \tilde{u}_j$ by

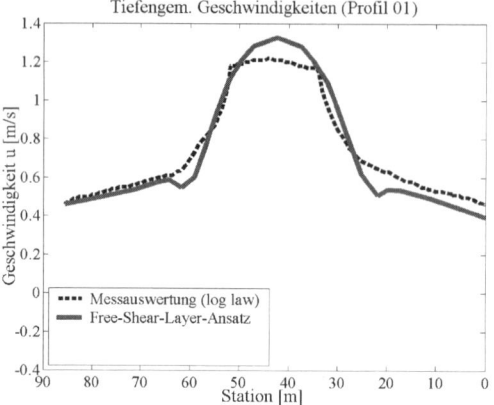

Figure 4.11. Depth averaged velocity of the river Lippe at flood.

analytical equations which Lien et al. (1999) developed for meandering rivers and has been further modified by Lippert (2005):

$$\overline{\tilde{u}_1 \tilde{u}_1} = -\rho u_1^2 h \left(\frac{\sqrt{\lambda}}{\kappa \sqrt{8}} \right)^2$$

$$\overline{\tilde{u}_1 \tilde{u}_2} = \overline{\tilde{u}_2 \tilde{u}_1} = -\rho \left[u_1 u_2 h \left(\frac{\sqrt{\lambda}}{\kappa \sqrt{8}} \right)^2 + \frac{u_1^2 h^2}{\kappa^2 R_K} \frac{\sqrt{\lambda}}{\kappa \sqrt{8}} \cdot F_1(\zeta) \right]$$

$$\overline{\tilde{u}_2 \tilde{u}_2} = -\rho \left[u_2^2 h \left(\frac{\sqrt{\lambda}}{\kappa \sqrt{8}} \right)^2 + \frac{2 u_1 u_2 h^2}{\kappa^2 R_K} \frac{\sqrt{\lambda}}{\kappa \sqrt{8}} \cdot FF1 + \frac{u_1^2 h^3}{\kappa^4 R_K^2} \cdot F_2(\zeta) \right] \quad (4.43)$$

with κ = von Karman constant, R_K = radius of the meander bend and

$$f_m(\zeta) = 1 + \frac{\sqrt{\lambda}}{\kappa \sqrt{8}} + \frac{\sqrt{\lambda}}{\kappa \sqrt{8}} \ln \zeta \quad (4.44)$$

$$F_1(\zeta) = \int_{z*}^{1} \frac{\ln \zeta}{\zeta - 1} d\zeta \quad (4.45)$$

$$F_2(\zeta) = \int_{z*}^{1} \frac{\ln^2 \zeta}{\zeta - 1} d\zeta \quad (4.46)$$

ζ represents the normalized flow depth

$$\zeta = (z - z_o)/h \quad (4.47)$$

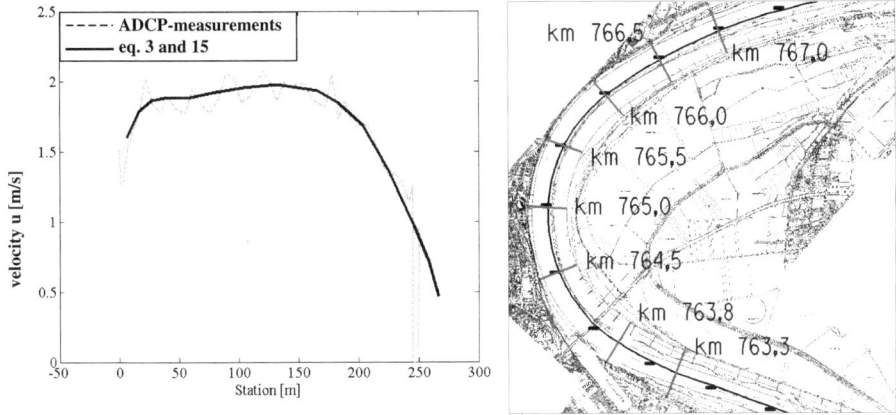

Figure 4.12. (a) Depth-averaged velocities at Rhine-Km 766.5; (b) investigated river bend at the Lower Rhine.

This approach has been tested by Lippert at a bend of the river Rhine close to Wesel (fig. 4.12). The curvature radius was derived from maps and determined to 1100 m. The velocities have been measured with an ADCP in eight characteristic cross-sections. The application of the shallow water equations (eq. 4.25) without special consideration of the dispersion terms did not accomplish an acceptable reproduction of the depth-averaged velocity distribution at all. However, with the dispersion model of Lien et al. (1999) the result could be improved substantially (fig. 4.12). Surprising was the observation that the longitudinal gradient of the water surface was not effected by the dispersion indicating that for large width-to-depth-ratios the secondary current does not increase the backwater-effect in meanders. Thus in these rivers no special consideration of the dispersion term is necessary if only the water stages are requested.

4.3   CONCLUSIONS

The new EU water policy calls for a paradigm change in the mitigation of urban flood problems: From "Fighting against flood" to "Living with flood". Flood Management has to start with flood risk assessment and rising the risk awareness at all stakeholders. The focus lies on the capacity development of stack holders and more sustainable urban development by retaining the stormwater. For the support of this water policy sophisticated mathematical models are needed to study the effect of urban development and sustainable stormwater management in a predictive way. Thus refined mathematical models play today an important role in Urban Flood Management.

The state-of-the-art analysis has shown that within the last two decades our knowledge about the mechanisms of flood formation and the available mathematical models have already reached a very high level. But this study also presented some important deficiencies in our knowledge and modelling capabilities. Together with the availability of new technologies in data mining and computer based graphical data management they mark the future issues of research on the hydrology of floods, which can be summarized as follows.

4.3.1   Process studies

Hillslope flow processes still need further research. Especially the subsurface flow is not understood in all details. Macropores have been detected as an important soil structure, which contributes substantially at the generation of runoff. But still it is difficult to quantify the different components, the rapid throughflow and the displacement of old prestorm water in flat valleys by new storm water coming mainly from the macropores.

A second aspect is the relevance of sealed surfaces and its connectivity with the stormwater pipe systems on the generation of extreme floods. Especially with the implementation of sustainable drainage systems, which foster the retention and infiltration of stormwater on site their contribution at the formation of runoff is not fully understood. For each element of sustainable drainage the infiltration and retention capacity can be evaluated but combinations of measures and whole system with their complex interaction need further analysis.

4.3.2   Refinement of modelling tools

First the models need to be refined with respect to the sustainable drainage components. The new methods of stormwater retention and infiltration into the ground need to be modelled more realistic than by aggregating them on a subcatchment level. The processes of green roof water storage, pond storage, and infiltration into the ground and its drainage into the groundwater and stormwater pipe system for the case of overload need to be included in the models.

The combination of different model components like hydraulic groundwater models and riverine flow models or semi-distributed hillslope models with hydraulic models has to be further studied and techniques of integration by better linkage have to be refined. Also meteorological models should be connected with hydrological models to study the interaction of hydrological and meteorological processes.

The models need to be extended with better user interfaces, which support them in the parameter assessment and the definition of scenarios. Expert systems and further extension of GIS-functions can be appropriate techniques. They should be extended to a decision support system, which provide optimization tools, economic evaluation methods and multi-criteria analysis to evaluate the conflict potential.

Due to the still increasing complexity of the models there is a need for increasing the performance of the models. To make use of remote computing and clustering of computers the models should support parallel computing which has an effect on the solution algorithms applied in the models.

4.3.3   Data mining and management

New technologies of remote sensing are not sufficiently tested and evaluated in their use for flood models. Especially the high resolution and diversity of satellite images open a large potential for application in hydrological and hydraulic models. The classification of land use and surface cover should be totally feasible on the basis of these data. Additionally the images can be used for the detection of surface and subsurface flow paths, the spatial distribution of soil moisture and for monitoring the spatial distribution of a snow cover and vegetation index.

The quality of LIDAR data has made obsolete classical topographic surveying methods. But the huge amount of data requires sophisticated methods for data reduction and data filtering. Here some progress is to be observed but needs more activity.

REFERENCES

Abbott, M.B., Bathurst, J.C., Cunge, A., O'Connell, P.E., Rassmussen, J. (1986a): An introduction to the SHE, 1: History and philosophy of a physically-based, distributed modelling system. Journal of Hydrology, 87, 45–59.

Abbott, M.B., Bathurst, J.C., Cunge, A., O'Connell, P.E., Rassmussen, J. (1986b): An introduction to the SHE, 2: History and philosophy of a physically-based, distributed modelling system. Journal of Hydrology, 87, 61–77.

AD-HOC-Arbeitsgruppe Boden (1996): Bodenkundliche Kartieranleitung, Bundesanstalt für Geowissenschaften und Rohstoffe, Hannover.

ASCE Task Committee on Turbulence Models in Hydraulic Computations (1988): Turbulence Modelling of Surface Water Flow and Transport. Journal of Hydr. Eng., 114, 9.

Becker, A., Messner, F., Wenzel, V. (2002): Integrierte Analyse der Auswirkungen des Globalen Wandels auf die Umwelt und die Gesellschaft im Elbgebiet (GLOWA-Elbe), Statuskonferenz München.

Berlekamp, L.-R., Pranzas, N. (1992): Erfassung und Bewertung von Bodenversiegelungen unter hydrologisch-stadtplanerischen Aspekten am Beispiel eines Teilraumes von Hamburg, Dissertation Universität-Hamburg.

Beven, K.J., Kirkby, M.J. (1979): A physically-based variable contribution area model of basin hydrology. Hydrological Sciences Bulletin, 24, S. 43–69.

Beven, K.J. (1985): Distributed models. In Anderson, M.G. and Burt, T.P. (eds), Hydrological Forecasting, John Wiley, Chichester, 405–435.

Beven, K.J., Freer, F. (2001): A dynamic TOPMODEL. Hydrol. Proc., 15, 1993–2011.

Blowes, D.W., Gillham, R.W. (1988): The generation and quality of streamflow on inactive uranium tailings near Elliot lake, Ontario. Journal of Hydrology, 97, S. 1–22.

Bronstert, A. (1994): Modellierung der Abflussbildung und der Bodenwasserdynamik von Hängen, Mitteilungen des Instituts für Hydrologie und Wasserwirtschaft an der Universität Karlsruhe; Dissertation.

Bronstert, A., Plate, E.J. (1996): Modelling of runoff generation and soil moisture dynamics for hillslopes and micro-catchments. Journal of Hydrology, 198 (1997), 177–195.

Bronstert, A. (1999): Capabilities and limitations of detailed hillslope hydrological modelling. Hydrological Processes, 13, S. 21–48.

Bronstert, A. (2005): Abflussbildung – Prozessbeschreibung und Fallbeispiele, Forum für Hydrologie und Wasserbewirtschaftung; Heft 13.05; Hennef.

BWK (1999): Hydraulische Berechnung von naturnahen Fließgewässern. Merkblatt 1, Bund der Ingenieure für Wasserwirtschaft, Abfallwirtschaft und Kulturbau e.V., Düsseldorf.

Chow, V.T. (1988): Open Channel Hydraulics. Classical Textbook Reissue, MC Graw Hill.

Deutscher Verband für Wasserwirtschaft und Kulturbau (DVWK) E.V. (1996): Ermittlung der Verdunstung von Land- und Wasserflächen. Nummer 238 in Merkblätter zur Wasserwirtschaft. Wirtschafts- und Verl.-Ges. Gas und Wasser, Bonn.

Deutscher Verband für Wasserwirtschaft und Kulturbau (DVWK) E.V. (1991): Hydraulische Berechnung von Fließgewässern. Merkblatt 220, Verlag Paul Parey, Hamburg.

Dunne, T., Moore, T.R., Taylor, C.H. (1975): Recognition and prediction of runoff producing zones in humid regions. Hydrological Sciences Bulletin, 20, S. 305–327.

Dunne, T. (1978): Field studies of hillslope flow processes, Hillslope Hydrology, Ed.: M.J. Kirkby, John Wiley, Chichester.

Evertz, T., Pasche, E., von dem Bussche, M. (2002): Simulation der Abflusskonzentration auf der Oberfläche und in der oberflächennahen Bodenzone in einem ebenen, von landwirtschaftlichen Drängräben durchzogenen Untersuchungsgebiet. In: Wittenberg, H., Schöniger, M. (Hrsg.), Wechselwirkungen zwischen Grundwasserleitern und Oberflächengewässern, Beiträge zum Tag der Hydrologie 2002.

Germann, P. F. (1990): Macropores and hydrologic hillslope processes. In: Process Studies in Hillslope Hydrology. M.G. Anderson and T.P. Burt. (Eds). Chapter 10, 327–363. John Wiley & Sons Ltd., New York.

Giebel, H. (2002): Austauschvorgänge zwischen Fluss- und Grundwasser: Grundlegende Untersuchungen im Neuwieder Becken und ergänzende Untersuchungen im Rheinabschnitt Bonn/Köln, Wechselwirkungen zwischen Grundwasserleitern und Oberflächengewässern, Beiträge zum Tag der Hydrologie 2002, Forum für Hydrologie und Wasserbewirtschaftung, Heft 01.02.

Horten, R.E. (1933): The role of infiltration in the hydrological cycle. EOS Trans. AGU, 14, S. 460–466.

Kirkby, M., Chorley, R. (1967): Throughflow, overland flow and erosion. IAHS Bull., 12 (3), S. 5–21.

Kirkby, M.J. (Ed.) (1978): Hillslope Hydrology, John Wiley, Chichester.

Lien, H.C., Hsieh, T.Y., Yang, J.C. (1999): Bend-Flow Simulation Using 2D Depth-Averaged Model. Journal of Hydr. Eng., 125, 10, 1097–1108.

Lippert, K. (2005): Analyse der Turbulenzmechanismen in naturnahen Fließgewässern und ihre mathematische Formulierung für hydrodynamische Modelle, Hamburger Wasserbau-Schriften, Heft 4, Hrsg. Erik Pasche, Hamburg.

Mahmood, K., Yevjevich, V. (1974): Unsteady Flow in Open Channels: Littleton, Colo., Water Resources Publications, 29–62.

Malcherek, A. (2001): Hydrodynamik der Fließgewässer, Habilitation, Institut für Strömungsmechanik und Elektronisches Rechnen im Bauwesen der Universität Hannover, Bericht Nr. 61/2001.

Moore, I.D., Foster, G.R. (1990): Hydraulics and overland flow. In: Process Studies in Hillslope Hydrology. Hrsg. M.G. Anderson & T.P. Burt. John Wiley, S. 215–254.

Ostrowski, M.W. (1982): Ein Beitrag zur kontinuierlichen Simulation der Wasserbilanz. Mitteilungen 42. Institut für Wasserbau und Wasserwirtschaft, Rheinisch-Westfälische TH Aachen, Aachen.

Pasche, E. (1984): Turbulenzmechanismen in naturnahen Fliessgewässern und die Möglichkeiten ihrer mathematischen Erfassung, Mitteilungen Institut für Wasserbau und Wasserwirtschaft, RWTH Aachen.

Pasche, E., Schröder, G. (1994): Erhebung und Analyse hydrologischer Daten auf der Basis geographischer Informationssysteme. Seminarunterlagen zum 18. DVWK-Fortbildungslehrgang Niederschlag-Abfluss-Modell für kleine Einzugsgebiete und ihre Anwendung, Karlsruhe, DVWK.

Pasche, E., Björnsen, G. (1994): Umweltverträglicher Hochwasserschutz am Beispiel Albungen/Werra, Proceedings zum Wasserbaulichen Symposium 1994, Mitteilungen der Universität Gesamthochschule Kassel, 1995.

Pasche, E., Wyrwa, J., Lippert, K.-U. (1995): Die Untersuchung der Hochwasserhydraulik am Rhein auf der Basis zweidimensionaler Strömungsmodelle, erschienen in Proceedings zur Informationsveranstaltung "Der Rhein im Wandel", Ministerium für Umwelt und Gesundheit, Rheinland-Pfalz.

Pasche, E. (1997): Model-Based Monitoring and Assessment of Flow and Hydrochemics in Rivers, Proceedings International Workshop on Informative Strategies in Water Management, Nun-speet, Netherlands.

Pasche, E., Brüning, C., Plöger, W., Teschke, U. (2004): Möglichkeiten der Wirkungsanalyse anthropogener Veränderungen in naturnahen Fließgewässern, Erschienen in Proceedings zum Jubiläumskolloquium 5 Jahre Wasserbau an der TUHH "Amphibische Räume an Ästuaren und Flachlandgewässern", Hamburger Wasserbau-Schriften, Heft 4, Hrsg. Erik Pasche, Hamburg.

Pasche, E., Plöger, W. (2004): "Retention effect by natural rivers with riprarian forest", Proceedings of 6th International Conference on Hydroinformatics 2004, Singapore.

Pasche, E., Teschke, U. (2004): "A new Approach for One-Dimensional Unsteady Flow Simulation in natural Rivers with Flood Plains and Vegetation", Proceedings of 6th International Conference on Hydroinformatics 2004, Singapore (Vol. 1, pp. 399–406).

Peschke, G., Etzberg, C., Müller, G., Töpfer, J., Zimmermann, S. (1999): Das wissensbasierte System FLAB – ein Instrument zur rechnergestützten Bestimmung von Landschaftseinheiten mit gleicher Abflussbildung.- IHI- Schriften, Internationales Hochschulinstitut Zittau, Heft 10, 122 Seiten.

Rath, S. (2002): Entwicklung eines hybriden CVFEM-Ansatzes zur Lösung der zwei-dimensionalen tiefengemittelten Flachwassergleichungen. Arbeitsbereich Wasserbau, TU Hamburg-Harburg, Diploma-Thesist.

Rath, S. (2003): Automatic Description of Fluvial Topography and Relief for Hydrodynamic Flood Wave Simulation In: Festschrift 5 Jahre Wasserbau an der TU Hamburg-Harburg, Hrsg.: E. Pasche.

Rath, S. (2006): Model Discretisation in 2D Hydroinformatics based on High Resolution Remote Sensing Data and the Feasibility of Automated Model Parameterisation. Dissertation, Institute für Wasserbau, Technische Universität Hamburg-Harburg.

Rodi, W. (1993): Turbulence Models and Their Application in Hydraulics. IAHR-AIRH Monograph, A.A. Balkema, Rotterdam, 3. ed.

Roth, C. (1992): Die Bedeutung der Oberflächenverschlämmung für die Auslösung von Abfluss und Abtrag.- Bodenökologie und Bodengenese, Heft 6, Inst. f. Ökologie, TU Berlin.

Rouvé, G. (1987): Hydraulische Probleme beim naturnahen Gewässerausbau. VCH-Verlag, Bonn.

Ruf, W., Perona, P. Molnar, P. Burlando, P. (2006): Kopplung eines hydrodynamischen Strömungsmodells und eines Grundwassermodells, Dresdner Wasserbauliche Mitteilungen Heft 32, Dresden.

Sacher, H., Fritz, M., Johann, G., Pfeiffer, E. (2006): Betrachtung von Auswirkungen der Hystereses auf die Verwendung von Abflusstafeln mittels Messwertanalyse und instationärer 1D/2D-Modellierung, Dresdner Wasserbauliche Mitteilungen Heft 32, Dresden.

Sherman, L.K. (1932): Streamflow from rainfall by unit-graph method. Eng. News Record, 108, S. 501–505.

Sieker, F., Bandermann, S., Holz, E., Lilienthal, A., Sieker, H., Stauss, M., Zimmermann, U. (1999): Innovative Hochwasserreduzierung durch dezentrale Maßnahmen am Beispiel der Saar – Zwischenbericht. Deutsche Bundesstiftung Umwelt, DBU, Projekt AZ 07147, Osnabrück.

Smagorinsky, J. (1963): General circulation experiments with the primitive equations. Monthly Weather Review, 91, 3, 99–165.

Stein, C.J. (1990): Mäandrierende Fließgewässer mit überströmten Vorländern – Experimentelle Untersuchungen und numerische Simulation. Mitteilungen des Lehrstuhls und Instituts für Wasserbau und Wasserwirtschaft, RWTH Aachen, Heft 76.

Tanaka, A., Yasuhara, M., Sakai, H., Marui, A. (1988): The Hachioji experimental basin study – Storm runoff processes and the mechanism of its generation. Journal of Hydrology, 102, S. 139–164.

Teschke, U. (2003): Zur Berechnung eindimensionaler instationärer Strömungen natürlicher Fließgewässer mit der Methode der Finiten Elemente. Dissertation, Technische Universität Hamburg-Harburg, Arbeitsbereich Wasserbau.

Teschke, U., Pasche, E. (2003): Grenzen der Modellierung – Wellenablauf, Workshop der BafG, Koblenz, November 2003.

Van Genuchten, M. TH. (1980): A closed-form equation for predicting the hydraulic conductivity of unsaturated soils. Soil Sci. Soc. Am. J., 44, 892–898.

Von dem Busche, M. (2002): Entwicklung eines Programms zur Simulation der ober-
flächennahen Abflußkonzentration in durch Drainage entwässerten Gebieten und
Prüfung des Programmes in einer vollständigen Niederschlag-Abflußsimulation,
Diplomarbeit an der TU-Harburg.

Wackermann, R. (1981): Eine Einheitsganglinie aus charakteristischen Systemwerten ohne
Niederschlag-Abfluß-Messungen. Wasser und Boden, Parey-Verlag, Heft Nr. 1.

Wohlmuth, B. (2000): A Mortar Finite Element Method Using Dual Spaces for Lagrange
Multiplier. SINUM, 38, 989–1014.

Zuidema, P.K. (1985): Hydraulik der Abflussbildung während Starkniederschlägen.- Mitteilungen der Versuchsanstalt für Wasserbau, Hydrologie und Glaziologie, Nr. 79, ETH
Zürich.

Internet links:

AG-Boden at: www.bgr.de/saf_boden/AdhocAG/Ergaenzungsregel_1_18.pdf www.bgr.
de/saf_boden/AdhocAG/Ergaenzungsregel_1_20.pdf.

MIKEShe and MIKE11 at: DHI,2006: http://www.dhigroup.com/Software/Water
Resources.aspx.

5

Urban Flood Management – Simulation Tools for Decision Makers

Peter Oberle & Uwe Merkel
Universität Karlsruhe (TH), Germany

ABSTRACT: Hydrodynamic-numerical simulation of free-surface flow is a key element for the implementation of flood management concepts in urban areas.

In this context, Geographic Information Systems (GIS) turned out to be a fertile platform for flood related data analysis and essential for model setup as well as further processing of hydraulic results (e.g. flood-hazard mapping). In recent years, there have been various ambitions to integrate flood relevant geodata and simulation components into online flood information systems. Effects of technical measures, precautionary sanctions and operational short-term strategies can be more comprehensibly illustrated.

Since many river systems are characterized by meandering waters courses, spacious floodplains and hydraulic complex confluence of river streams, where the hydraulic behaviour is changing and different for every single hydrological event, flood-risk assessment in urban areas remains a challenging task, especially regarding the uncertainties related to extreme events and the impacts of failures of flood control measures. In addition, a high accuracy is bought with long setup and calculation times. Sometimes too long for urgent threats, sometimes too complex for fast universally valid solutions. Simplifications towards operational usability and automated workflows are the subject of current research efforts.

This report gives a review on established and promising simulation methods and discusses the current state of their fields of application as decision support systems. Methods and programs are illustrated with two actual projects. The supra-regional IKoNE-Project at the Neckar River in Southern Germany uses GIS technology with directly connected one-dimensional hydraulic simulations. Upgrades and refinements will be included for operational purposes in a present medium-scale case-study at the Middle Elbe in Eastern Germany. For this project, calculation times of high-resolution two-dimensional simulation techniques could be reduced to an extent that suits disaster management time scales.

5.1 INTRODUCTION

Extreme flood events with catastrophic effects on the affected population keep reminding us of the urgent need for the implementation of effective protection concepts. These comprise, apart from infrastructural protection facilities, especially the consistent implementation of precautionary measures in flood-plain areas as well as strategies for operational management during a flood event.

The interdisciplinary assessment, optimisation and implementation of lasting protection concepts require tools that enable decision-makers to simulate the course of a flood event, including its effects, as a function of the specific conditions of the river valley. The quantification procedure can be subdivided into three major steps (Büchele et al., 2006):

- Regional estimation of flood discharges (basin-, site-specific hydrological loads)
- Estimation of flow characteristics in potential inundation areas (local hydraulic impacts)
- Estimation of the resulting damages (area- or object-specific risk assessment)

An overview of suitable approaches for these steps of hazard and vulnerability assessment is given in Table 5.1 (appendix). The left column states minimum requirements on data and methods for a standard quality of hazard and risk assessment on a local scale. The right column lists more sophisticated approaches, which require more spatial information and more complex calculations up to fully dynamic simulations of unobserved extreme flood situations.

To determine the hazard of floods and quantify their risks on a local or rather regional scale for urban water courses, the relevant hydraulic processes of stream flow need to be analysed. Therefore, predicted flood discharges and discharges with a certain recurrent interval have to be transformed into detailed hydraulic parameters like water levels, inundation depths or flow velocities by means of hydrodynamic-numerical (HN) models. For the further processing of the calculation results through interdisciplinary stakeholders, GIS-supported functionalities and interfaces are useful for visualisation (e.g. floodplain mapping) and risk analysis. Increasingly, public flood-hazard maps are available on Internet platforms.

In recent years, there have been various ambitions to integrate the numerical simulation component into online flood information systems. The goal is to enable decision makers in water management administrations, local authorities and disaster control management to autonomously access the up-to-date geo-datasets and models and thus to provide a basis for decisions regarding both middle- and long-term precautionary measures and for short-term analyses during or just before an incident.

Depending on the extent and the complexity of the area of interest, one- or multi-dimensional models have to be used. With higher complexity, the computing time is rising rapidly. For precautionary planning, the computing time is subordinate. But

Table 5.1. Overview of methods and data for high-resolution flood-risk mapping in Germany (Büchele et al., 2006).

	Basic approach (minimum requirements)	More detailed approach (higher spatial differentiation, inclusion of dynamic aspects)
Regional estimation of flood discharges	– Extreme value statistics (mainly for gauged basins) – Regionalisation of flood parameters (for ungauged basins)	– Rainfall-runoff simulation for extreme flood events – Long-term simulation of flood variability (probabilistic evaluation, e.g. climate trends)
Estimation of local hydraulic impacts	– Documentation of historical water levels/inundation zones (varying availability/quality) – Calculation of water levels and inundation zones/depths based on hydraulic models and digital terrain models (local scale, e.g. 1:5,000) – Simplified approaches (only large-scale, e.g. 1:50,000)	– assessment of flow situation based on 2D HN models (e.g. spatial distribution of flow velocities/directions) – unsteady hydrodynamic simulation of extreme flood scenarios (e.g. impact of dyke failures)
Damage estimation	– stage-damage curves for aggregated spatial data (e.g. ATKIS-data)	– stage-damage curves for individual buildings or land-units (e.g. ALK-data) – consideration of further damage-determining factors (e.g. flow velocity, precautionary measures, warning time)

for operational use during a flood event, it is a basic necessity to make the computing results available very quickly. Therefore new approaches are based on a simplified two-dimensional processing and on a combination of one- and two-dimensional modeling techniques.

5.2   HYDRAULIC FLOOD SIMULATION IN URBAN AREAS

5.2.1   Objectives

The following basic objectives for the use of hydrodynamic-numerical models in connection with GIS-technologies in the context of urban flood management can be outlined:

- Examination of objects' degree of protection and evaluation of protection deficits
- Examination and revision of legally binding inundation areas (flood-hazard mapping)
- Assessment of the impacts of construction measures within the river valley on the flood flow (back water effects, acceleration of flood propagation)
- Prognosis of flood damages in endangered areas (risk analysis) as a basis of precautionary measures
- Formulation and monetary evaluation of flood protection concepts (cost-benefit analysis)
- Refinement of local and regional warning and emergency plans
- System application in operational mode as basis for decision-making during a flood event
- Visualisation and animation of flood relevant data and hereby raise of public danger awareness in potentially endangered areas

5.2.2   Methods and tools

Over the last few decades, hydrodynamic-numerical (HN) methods based on physical equations (laws of conservation of mass and momentum) are available for the simulation of flood events. Target parameters of the numerical flow simulation are the calculation of water levels and flow velocities as well as the quantification of flood propagation including retention effects and wave superposition with tributaries. Furthermore, HN models can be used for analyzing erosion and sedimentation processes as well as the dispersion of pollutants. By quantifying the flow parameters in endangered areas (intensity and duration of inundation, flow velocity), in combination with land use data the damage potential can be determined. Besides the reasonable choice of modeling method (spatial discretisation, time dependency, etc.) and model resolution, it is especially the calibration of the model that is of particular importance for flood simulation due to the great uncertainties of the topographical and hydrological input data as well as the need for empiric hydraulic parameters.

The probability of flooding can be determined by combination of hydrological modeling and the determination of statistical information to recurrent intervals of peak discharges. This makes it possible to conduct risk assessments, compile priority lists for the implementation of protection concepts, quantify cost-benefit ratios and determine design parameters for technical protection measures. Furthermore, the results can be used as a basis for warning and emergency plans. By interacting with operational services of flood forecast centres, they can decisively enhance the coordination of disaster operations.

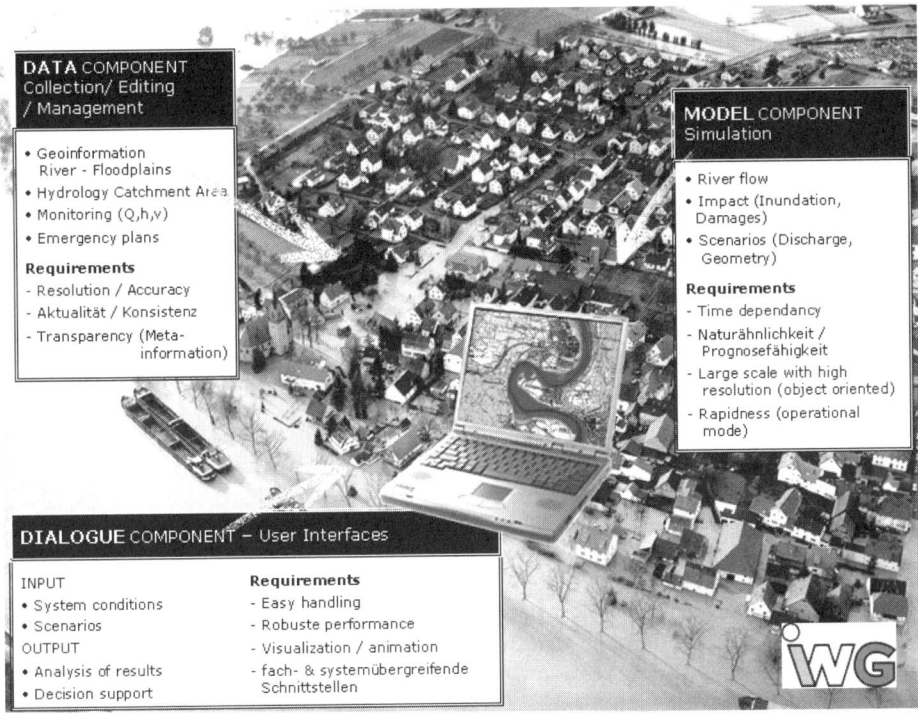

Figure 5.1. System components of information and planning instruments for flood management.

For the pre- and post-processing of the model components, among others, Geographic Information Systems (GIS) are highly effective. They serve the compilation, management, visualisation and analysis of spatial input and output data. However, modern GIS-software systems can be used considerably more extensively, since they integrate a large – cross-component – potential of functionalities. In addition, some Geographic Information Systems can be upgraded to additional functions and algorithms by internal programming environments. Thus, their field of application is not restricted to the data component, but comprises also areas of the model component.

In addition to the various fields of application as data and model component, some GIS also allow the integration of graphical interfaces that are adapted to the special requirements of the user and can be used as effective dialogue components within the framework of a flood information system. The modular system component of a GIS-based information and planning instrument is illustrated in Figure 5.1.

5.2.3   Model requirements and challenges

The necessary accuracy for calculated flood parameters normally results from the requirement of a reliable analysis of potentially endangered objects (<2 decimetres

Figure 5.2. Aerial photo of the urban region of Heilbronn (Neckar river) and GIS-based analysis of flooding scenario due to failure of protection measures (Oberle, 2004).

for calculated water levels; land parcel size for calculated inundation borders). Taking into account the oftentimes high uncertainty of the topographical and hydrological input data as well as the bases of calibration (observed water levels and inundation areas), this demand can often not be met. This holds true especially for urban areas with complex hydraulic boundary conditions. For the simulation of failure scenarios in dykeleveed areas, for example (Fig. 5.2), assumptions have to be made regarding the time and characteristic of an overflow or breach formation. In addition, the impact of dyke breaches on flood hydrographs along a river course must be quantified.

The prognosis of the current situation within the urban developed area can be especially vague due to the complexity of the topological and topographical situation, the dynamic alterations in case of flooding (flow through road networks, temporary flood barriers like cars or containers, blocking of undercrossings) as well as the lack of calibration data. In addition, the sewage network can affect the local flood situation. On the one hand, the volume of the sewage network can store a certain amount of water, on the other hand, areas without direct interconnection with the river can be inundated through the sewage network. Assessment of modeling the exchanges between sewage network and surface water during flooding is a topic of recent studies (Renouf et al., 2005). Thus hydraulic models can be used to help planning mitigation measures also in a dense urban area (Oberle, 2004; Mignot, et al., 2006).

For the application of simulation tools in operational flood management, the demands for model accuracy are made against simulation methods. Hence they cannot be fulfilled with common numerical calculation methods. In this regard the crucial point is the time of computation. In 1D-models (see below), calculation results are available within a short time (few minutes running time of CPU), whereas complex or changing hydraulic situations cannot be analysed with required

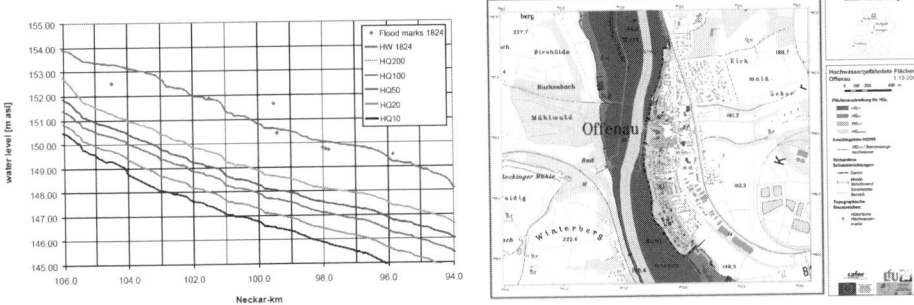

Figure 5.3. Left: Maximum water levels along the Neckar river for statistical flood events (HQT with T = 10, 20 . . . 200 years) and the historical event 1824 (HW1824), the latter as reconstructed from water-level marks. Right: Flood-hazard map of the dyked community Offenau (Neckar-km 98.0; design discharge HQ_{100}).

accuracy. In contrast to 1D-models, the results of 2D-models (solving the complete shallow water equations) show a high temporal and spatial resolution (water depths, flow velocity vectors), but take a relatively long calculating time, from several hours to days. It makes them inefficient for operational mode or analyses of spacious river systems (several hundred river kilometres). Simplified 2D approaches and a rough approximation of the physical processes are a necessary compromise.

5.2.4   Modeling of extreme floods

In most cases, HN-models are applied to documented flood events from the last decades (in order to assess the present hydraulic conditions) or to statistical flood events with recurrence intervals up to 100 or 200 years (e.g. for delineation of inundation zones or as design events for flood protection measures). Because of an increasing process complexity, for instance in case of an overflow or even destruction of a flood protection structure, and the lack of measurements for calibration and validation of model parameters, the application of HN-models to larger floods is rarely practiced (AE). However, despite of the uncertainties, it is necessary to apply HN-models to floods that exceed the 100- or 200-year level, as these are the most relevant situations in terms of residual risks, causing severe damages and fatalities. In particular with respect to residual risks, it is obvious that model parameters should also be valid for extreme events. Only the consideration of all physically plausible hazardous situations, from the occurrence of first inundations to the maximum possible water levels, makes for a comprehensive hazard and risk assessment (Fig. 5.3). This applies equally for flood situations below the design event, as required e.g. for cost-benefit-analyses of protection measures or for the assessment of residual risks due to other failures of technical or non-technical measures (e.g. late installation of mobile protection elements).

The consideration of extreme historical events can not only support flood aware-ness as realized scenarios (under historical conditions), but can also be used as a reference for the analysis of potential extreme cases under present conditions. In this regard, the intention is not to reconstruct historical hydraulic conditions or to verify historical information in terms of peak discharges. Simulation results can help to assess the probability of flood events that cause comparable water levels in the current situation. In terms of a reconstruction of historical discharges, a fur-ther investigation on historical hydraulic boundary conditions is required (Oberle, 2004). However, due to the limited historical data availability and quality, major uncertainties are expected.

5.2.5   Flood mapping – potentials and constraints

Increasingly, public flood-hazard maps are available on Internet platforms (e.g. in Germany: North Rhine-Westphalia, 2003; Rhineland-Palatinate, 2004; Saxony, 2004; Bavaria, 2005; Baden-Württemberg, 2005). However, hitherto exist-ing mapping bases for larger river areas have very little demand for accuracy and are based upon rough modeling approaches, partly without hydrodynamic-numerical simulation basis. Many state authorities have been working on the delineation of inundation zones for years with map scales of up to >1:5,000 in urban areas in order to recognise the flood hazard for discrete land-parcels and objects.

In the German federal state of Baden-Württemberg, until the year 2010 detailed and legal flood-hazard maps are being developed for all watercourses with a catch-ment area of $>10 \, \text{km}^2$. Depending on the characteristics of the water body, one- or two-dimensional models are used for flow simulation ($HQ_{10,20,100,extreme}$). An area-wide up-to-date laser scanner DTM ($1 \times 1 \, \text{m}$) with a maximum accuracy of ± 20–$30 \, \text{cm}$ and additional terrestrial surveying data (e.g. river bed, dykes) found a topographical data base.

On behalf of the Regierungspräsidium Stuttgart and in cooperation with the Uni-versity of Karlsruhe, the methodical basis for this intention was developed within the framework of the Interreg IIIB project SAFER in the Neckar catchment area and implemented in a pilot project (see case study below). The first hazard maps for the pilot section of the river Neckar were already distributed among the local authorities on a scale of 1:2,500 and are publicly available on a scale of 1:5,000 on www.hochwasser.baden-wuerttemberg.de. For areas where the potential inunda-tion area essentially depends on the amount or rather the constellation of discharge (main river, tributaries), the mapping procedure is clearly defined. It is yet uncertain how to determine effects of unpredictable failure scenarios, e.g. for dyked residen-tial areas in a standardised way as a legal basis for the precaution management. Figure 5.4 illustrates an example for the design of a scenario map with reference to

Figure 5.4. Scenario map (example) for the community Offenau (Neckar) with reference to a forecast gauge.

a forecast gauge showing the simulation result of a dyke failure scenario (Oberle, 2004).

Within the next years, in the wake of amendments in legislation, growing regional significance and new technical possibilities, flood-hazard maps with a high spatial resolution can be expected successively for many rivers in Europe. Details of procedures and mapping techniques vary from state to state and due to local concerns (e.g. data availability, vulnerability, public funds). An overview of different approaches is published by Kleeberg (2005). In contrast to hazard mapping, the assessment of damage and its visualisation in the form of risk maps is still far from being commonly practised. Risk maps help stakeholders to prioritise investments and enable authorities and citizens to prepare for disasters (e.g. Takeuchi, 2001; Merz and Thieken, 2004).

The quality of flood maps strongly depends on the quality of the DTM used. Uncertainties in DTMs are more and more overcome by an increasing availability of high-resolution digital terrain models from airborne surveys (e.g. laser scanner and aerial photographs). In spite of these technical standards and advances in practice, it can be stated that flood-risk assessment remains a quite challenging task, especially regarding the uncertainties related to extreme events exceeding the design flood or to the damage due to failures of flood control measures (Apel et al., 2004; Merz et al., 2004). However, the visualisation and communication of uncertain information in hazard maps should be optimised in a way that non-experts can understand, trust and get motivated to respond to uncertain knowledge (Kämpf et al., 2005).

As many river systems are characterized by meandering waters courses, spacious floodplains and hydraulic complex confluence of river streams, the hydraulic behaviour is changing and different for every single hydrological event. In the case of rivers with a spacious retention area, the hydraulic situation can change significantly in the event of a dyke failure or a misapplication of flood control works.

In case of a flood, due to the multitude of possible discharge and hazard scenarios, especially regarding larger river sections, forehanded simulation analysis and its visualisation in the form of hazard maps provide a sufficient basis for decision-making only to a limited extend. Hydrodynamic-numerical (HN) models should be provided for decision makers to allow short-term analysis of complex hydraulic situations in operational mode as well. However, the simulation systems available today so far don't meet these requirements.

5.2.6   Model application for decision support

In current practice, computer-aided flood simulations are commissioned to specialized experts at Universities and planning agencies by the responsible administrative authorities as a basis for their water managerial planning. Here, simulation studies with a high resolution mostly concentrate on limited river sections and a local scale. The results are documented mainly problem-specific and handed over to the decision makers as reports or analogical planning documents. However, the simulation systems are not made available. The prevalent reason is that the application of the models and the responsible interpretation of the results require a thorough understanding of the processes and principles underlying the model as well as of the software environment – an understanding that usually only experienced experts can show. An additional problem is the hitherto long calculation time when applying two-dimensional procedures.

The first case study (below) gives an example for implementing a GIS-supported 1D-HN-model in a pilot project on 200 river kilometres of the river Neckar with the water management administration of Baden-Württemberg. Due to the small retention volume as well as the primarily one-dimensional flow characteristics of the Neckar valley, the one-dimensional approach was sufficient. Recent research activities deal with the development and application of multidimensional procedures for the survey of larger river zones for an operational implementation (second case study).

Figure 5.5. 1D-HN model in developers view: connected cross-sections (discharge area) and retention zones.

5.3   HYDRODYNAMIC-NUMERICAL (HN) SIMULATION METHODS

5.3.1   One dimensional abstraction (1D-HN)

In many cases, when the flow patterns in a given river section are characterised by compact and coherent streamlines, 1-dimensional (1D) HN-models are considered as adequate for the estimation of flood-water levels and delineation of inundation zones (e.g. Baden-Württemberg, 2005). The main benefit of this model family is the computation speed.

In order to provide system geometry for the HN-model, the discharge area of the main channel and the foreshores are modeled with modified cross sections (Fig. 5.5). More complex geometries need to be divided into multiple channels. Industry standard is one for each floodplain, left and right, and another one for the main channel. Modern software supports an even higher discretisation, for example to attend groyne sections, riparian vegetation, islands or cross-section inhomogeneities in general (Fig. 5.6). In most cases these channels are calculated without interaction (there is no flux possible between these virtual channels) except for friction between the virtual channel walls, which might be calculated using the algorithms of Mertens or Pasche (DVWK, 1991).

Those areas that contribute to the retention volume of the river during a flood event are taken into account by a function of storage capacity in dependence upon the water level. This function can be determined from the digital terrain model (DTM) by means of several GIS-functionalities and can be verified by comparing

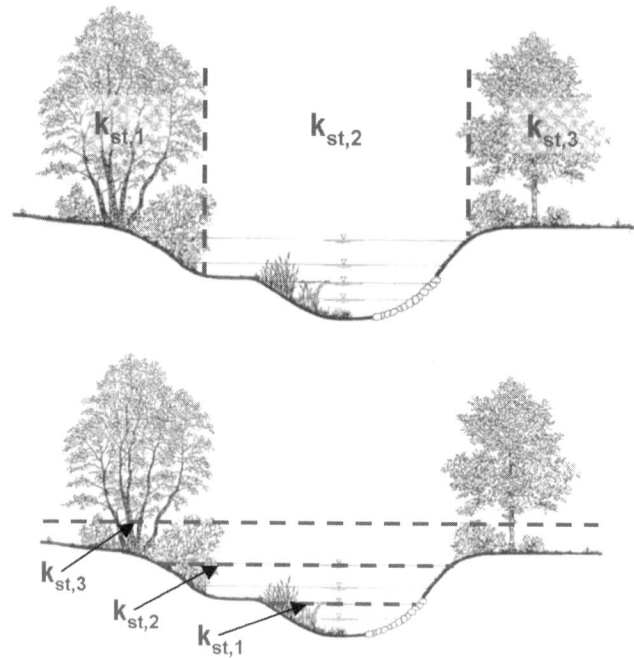

Figure 5.6. Cross-section with different friction zones for main channel and riparian vegetation (above); definition of roughness coefficient as function of water depth (below).

calculated with surveyed flood hydrographs. Modifications of the volume function of retention areas do not affect the results of a steady flow calculation. With an unsteady approach, the influence on the development of a flood wave becomes evident. A reduction of the retention volume leads to a higher peak and to an acceleration of the wave.

Within a single branch e.g. Preissmann's schema is commonly used for solving the 1D Saint-Venant equations which describe conservation of momentum and mass.

$$\frac{\partial hB}{\partial t} + \frac{\partial huB}{\partial x} = 0 \qquad \frac{\partial uhB}{\partial t} + u\frac{\partial uhB}{\partial x} = -ghB\frac{\partial z_s}{\partial x} - B\frac{\tau_{bx}}{\rho} \qquad (5.1)$$

1D Saint-Venant equations; t – time; x – flow direction; h – flow depth; u – velocity in x direction; τ – stress; z – bedlevel; ρ – density; B – branch width.

The less velocity a river can achieve, the more space it needs to enable the full discharge. Therefore, especially friction is responsible for the waterlevel.

Friction, the most important speed related energy loss that prevents the water from accelerating to infinity, is mostly modeled by the quite old empiric

Mannings-formula (also known as Strickler-formula) or the more accurate Darcy-Weisbach formula.

$$v = \frac{I^{1/2} \cdot R^{2/3}}{n} \qquad v_m = \frac{1}{\sqrt{\lambda}}\sqrt{8gRI} \tag{5.2}$$

Mannings-formula and Darcy-Weisbach formula; I – slope; R – area/outline; n, λ – friction coefficients.

Many studies deal with the nature of friction coefficients, which can be obtained from reverse engineering of historical events, or worse, from reference books. An estimation of friction development in riparian vegetation is possible, for example with the approach of Pasche (1984). Small, medium and higher vegetation can be examined separately related to barrier diameters and average distances. But there is still no general standard for ascertaining the development of friction aspects during crop rotation, vegetation periods (e.g. no plants on fields in winter), water body development or for predicting effects of construction works.

Other forms of energy losses (e.g. weirs, groynes, water power plants) can be treated with empiric formulas as well. The functionality of many 1D-software systems includes modeling schemes for meandering rivers as well as for river-regulating structures. Furthermore, additional empiric formulas can take into account minor non-one-dimensional effects like slight transversal currents or water level curvature.

5.3.2 Two-dimensional approach (2D-HN)

In cases with more complex river geometries and flow patterns (e.g. at river confluences, oxbows or other complex flow conditions), 2-dimensional (2D) models are used for a spatially differentiated hydraulic analysis. 2-dimensional means that all flow properties are considered equal over all depths at a specific point within the model.

A planar flow is possible in every previously unknown direction, not only from predefined cross-section to cross-section as in 1D-models. The flow direction might even change during flood events because of changing boundary conditions or unsteady effects. Here it is not necessary to find out the flow direction before setting up the models as in 1D modeling.

Most 2D-HN software uses unstructured mainly triangulated meshes which adapt much better to complex terrains than regular meshes (Figs. 5.7 and 5.8).

Furthermore, the possibility for spatial analysis of parameters such as water level, flow direction, flow velocity (Fig. 5.9), shear stress and others expands this model family to a highly precise and universal planning tool for disaster management, river management analysis or morphological questions.

On the other hand, the depth averaged shallow water equations contain a lot more differential terms, which multiply the computing times. It will take some more

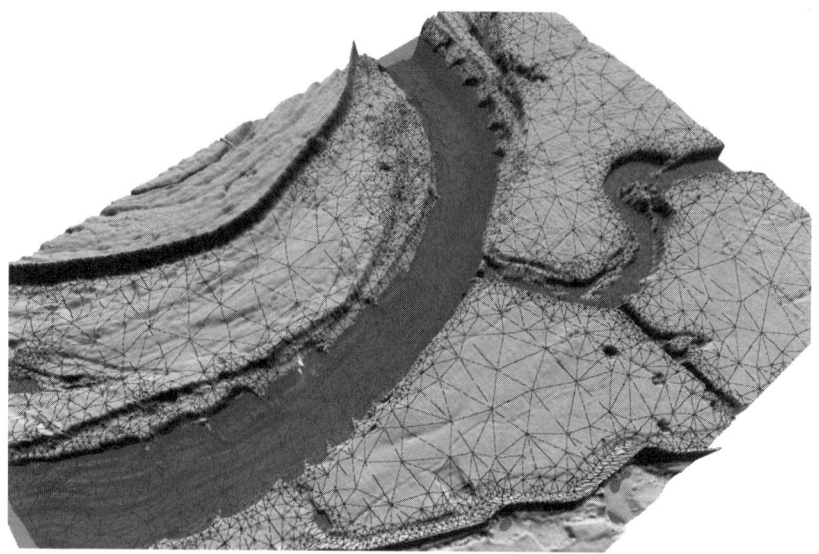

Figure 5.7. Large scale river confluence (relief view), meshed with an unstructured triangulated mesh; refined at edges with significant curvature.

years to render possible operational disaster management by faster computers and optimized calculation and modeling methods.

$$\frac{\partial h}{\partial t} + \frac{\partial hu}{\partial x} + \frac{\partial hv}{\partial y} = 0$$

$$\frac{\partial u}{\partial t} + u\frac{\partial u}{\partial x} + v\frac{\partial u}{\partial y} = -g \cdot \frac{\partial z_S}{\partial x} + \frac{1}{h\rho}\frac{\partial \sigma_x h}{\partial x} + \frac{1}{h\rho}\frac{\partial \tau_{yx} h}{\partial y} - \frac{1}{h}\frac{\tau_{Bx}}{\rho} + \frac{1}{h}\frac{\tau_{Wx}}{\rho}$$

$$\frac{\partial v}{\partial t} + u\frac{\partial v}{\partial x} + v\frac{\partial v}{\partial y} = -g \cdot \frac{\partial z_S}{\partial y} + \frac{1}{h\rho}\frac{\partial \sigma_y h}{\partial y} + \frac{1}{h\rho}\frac{\partial \tau_{xy} h}{\partial x} - \frac{1}{h}\frac{\tau_{By}}{\rho} + \frac{1}{h}\frac{\tau_{Wy}}{\rho} \quad (5.3)$$

2D – shallow water equations; t – time; x, y – co-ordinates; h – flow depth; u, v – velocity components; τ – stress; z – bedlevel; ρ – density.

Depending on the intended purposes, both types of models (1D, 2D) may be applied for stationary flow conditions (e.g. hazard assessment for a certain HQ_T) or unsteady flow conditions (e.g. for impact analyses of dyke failures).

5.3.3   Branched 1D-HN model

If more than one flood channel is well known, it is possible to combine several 1D-HN models to a single branched model (Fig. 5.10).

The flow direction is still given, but the discharge distribution enables a more accurate prognosis especially in wide-stretched floodplains. Water levels and

Figure 5.8. Small scale model for industrial zone, meshed with an unstructured triangulated mesh; refinement based on ALK-building polygons and adjusted ATKIS break lines.

velocity in outer zones, not directly covered by cross-sections, are extrapolated as in simple 1D-HN models. On the one hand, a great benefit is a much faster calculation compared to 2D models. On the other hand, the extrapolation of water levels is only valid for less widespread floodplains with minor 2D effects. Even greater streams of several hundreds kilometres were realised in branched 1D-models. Several thousand kilometres might be possible.

5.3.4   Partial-discharge model (simplified 2D-HN)

As 2D modeling based on the shallow water equations is still a time consuming process, for some applications a so called Partial Discharge (PaD) model might be a good alternative. This PaD model is based on the combination of a common one-dimensional flow equation with a digital terrain representation via a triangulated irregular network (TIN). Thereby the triangle corners are corresponding to the nodes of the one-dimensional calculation. The calculation proceeds along the TINs edges, whereas for each starting node the one having the greatest surface gradient is chosen. Thereby, idealised flux geometry is estimated based on the size of the neighbouring triangles. The time-dependency of the flow process can be approximated by controlling the total inflow volume resulting from the multiplication of discharge, typically given in cubic meters per second, by a certain period of time. This user-specified volume of water, which is the most important input parameter for calculation, is automatically divided into partial volumes, flowing through the modeled area as Partial Discharges (PaD).

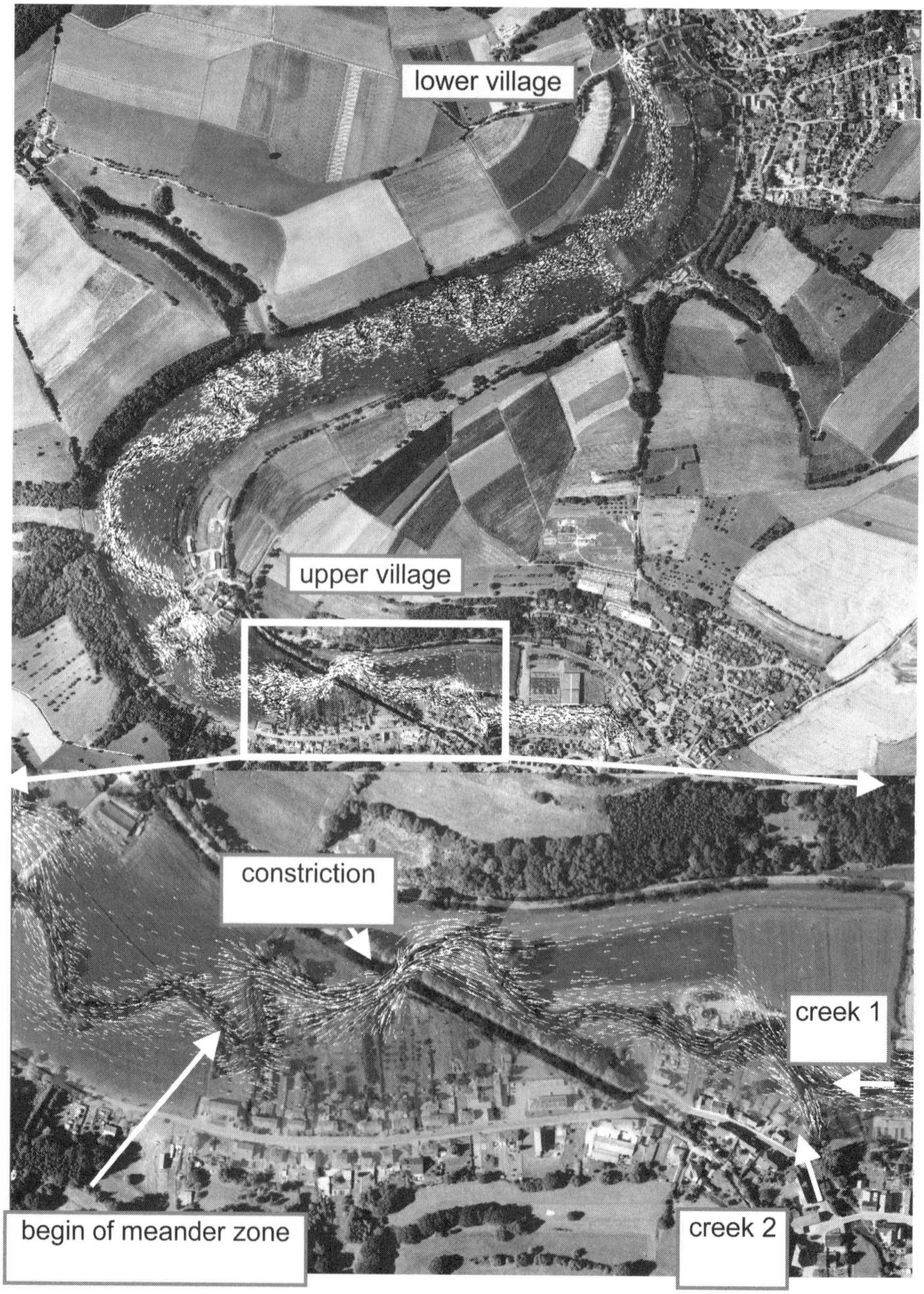

Figure 5.9. Small scale model for rural villages; velocity vectors show major preferred flow directions.

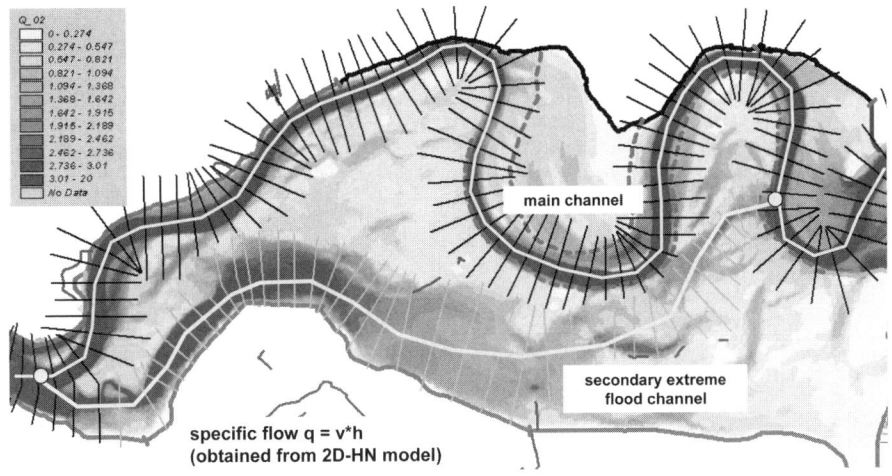

Figure 5.10. River section "Middle Elbe" with a strongly undirected flow during the 2002 extreme event. A second 1D-model branch (blue) was inserted according to specific flow (background shading) from 2D research, catching the major part of the overflow.

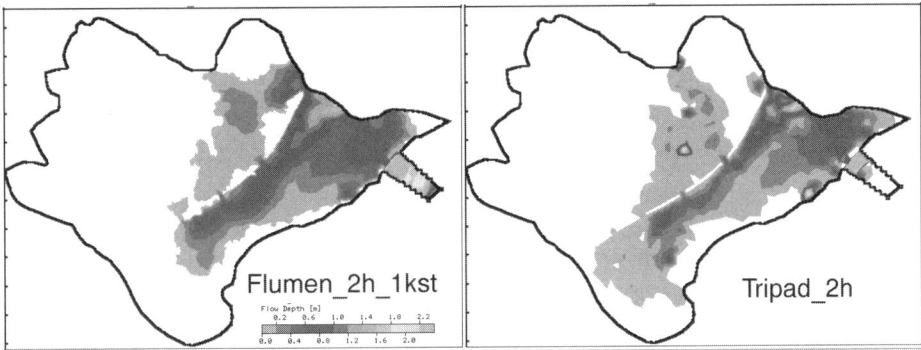

Figure 5.11. Flooding of a polder area, simulated by a two-dimensional model and a PaD model.

If all PaDs have flown in, a final state is reached which can be compared to the results of a transient two-dimensional calculation at a certain point in time. Practical tests have shown that by using this PaD model pretty fair results can be produced within only a fraction of the calculation time of a more detailed hydraulic model. Figure 5.11 shows a comparison of results using the PaD model as well as a detailed two-dimensional model by the example of the flooding of a polder area.

The biggest disadvantage of the PaD model consists in the inadequate handling of time-dependency of the flow process. So water that is once dumped cannot be removed, which makes for example the modeling of a polders draining impossible.

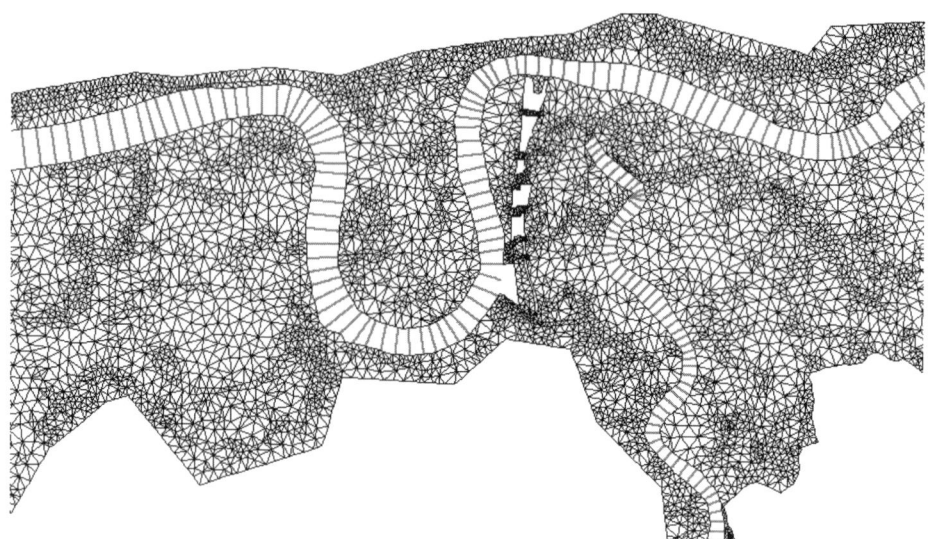

Figure 5.12. Clipping of a combined 1D/2D-model with adaptive meshing methods for the 2D simulated floodplains.

For urban areas, Beffa (1998) developed a PaD model to reproduce the main flow paths and the general inundation process well. Limitations were found for example at contractions, where the neglect of lateral forces becomes crucial.

5.3.5 Combination of one- and two-dimensional techniques (1D/2D-HN)

Another way to combine the advantages of 1D and 2D-HN models is a direct combination of fast 1D-models for a hydraulic simple main branch and connected 2D-models for areas with unknown water distribution paths as shown in Figure 5.12 (e.g. polders). The flux between both simultaneously simulated models runs over lateral weir connections between the two models (Fig. 5.13). The 2D-software package FLUMEN for example uses a Poleni type like formula but other methods are also possible (Beffa, 2002).

Different from a classic 2D model, there are at least 2 more calibration parameters. These are the virtual weir crest and the (cross-section distance/border cell size) relation. But the advantageous results are the savings of tiny cells in small but effective channels. Time steps can multiply and different flow directions are possible. Momentum transfer from one side of a 1D model to the other side is not possible, but small differences in water levels might be passed through to the other side, depending on calibration values. Therefore a 1D string should not dissect a complex two-dimensional flow area.

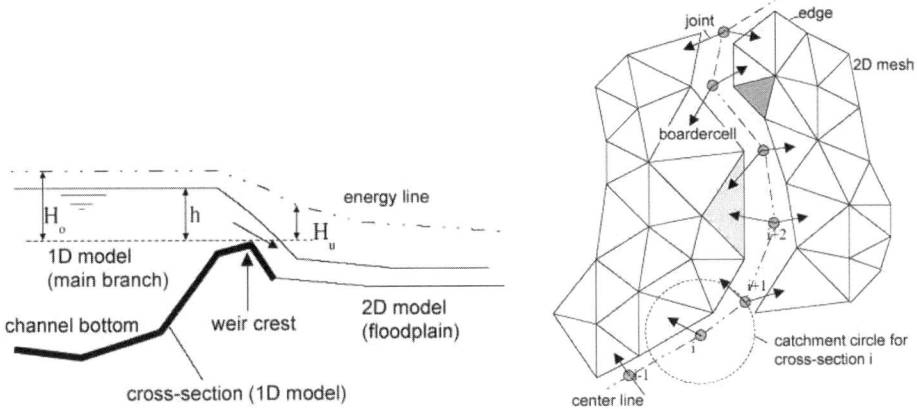

Figure 5.13. Combination of 1D and 2D models; Interaction enabled by Poleni type weir formula (graphic by Beffa, 2002).

5.3.6   Sediment transport implementation

Unfortunate for the planner, all natural flumes change their geometry by erosion and sedimentation processes. This might be a problem in urban flood management, if the channel capacity changes after artificial reshaping because the necessary discharge cross-sections are plugged by sedimentation. The result is a rise of the water level and therefore an expansion of the affected area.

There are many studies treating this problem. A group of threshold algorithms deal with the question of what impact is necessary to start or stop the material transportation. Another group tries to examine the probability of the material in motion. All of them have a strong empirical character in common. The operational range for most of them is difficult to isolate. Independent of the formula is the necessity for input of flow characterising variables, obtained by one- or multidimensional HN simulations, which might by more or less spatially discreted depending on the type of model in use.

The integration of sediment transport in HN software of all types is quite common and newer projects now allow for a sediment transport related bed adjustment during flood simulation, but results vary in a wide range. Therefore, calibration needs a higher scientific background, and validation is time consuming.

5.4   GEOGRAPHIC INFORMATION SYSTEMS (GIS)

5.4.1   Structural concepts for the abstraction of the real world

Geographic Information Systems (GIS) are software systems serving the collection, storage and analysis of spatial data. They are characterised by a combined dealing with topographic data (geometry) and factual attributes (properties). Objects (e.g. land parcels, buildings, streets of houses) are generally represented as vector models, whereas continuous information (e.g. height of the terrain, water depths, flow velocities) can be approximately handled as regular raster models ("grid", "lattice") or irregular triangulated networks ("TIN") due to spatial discretisation. In the meantime, some systems dispose of geo-data concepts that allow for a complete three-dimensional topological treatment of vector structures, which for example in the context of terrain modeling can be useful to generate breaklines or incorporate crest heights along dykes.

Orthophotos (aerial photos or satellite photos with spatial reference) and scanned topographical maps are raster datasets, in which every pixel contains a discrete value which is presented as a specific colour or grey value. If a picture file should be imported in GIS, it has to be projected to the target coordinate system using geographical coordinates of at least two fixpoints the target coordinate systems datum and the pixel resolution.

In a GIS, the geo-datasets are separately treated as layers with a fixed spatial-reference, according to their data type. For visualisation purposes a superposition with other layers (topics), analysed via various functions and combined with each other. Crucial for this is the conformance of the geo-reference system.

5.4.2   GIS-application for HN-modeling

A particularity of flood simulation (compared to other hydraulic problems) is that it requires a spatial examination of rivers and floodplains originating from a multitude of different datasets. In order to determine the exact location of input and output model data, the linear framework of river positioning generally used in the context of channel hydraulics is not sufficient. Thus, for pre- and post-processing all flood simulation methods produce large amounts of time-dependent spatial data.

In this context, a GIS turned out to be a fertile platform to assist modelers and model-users during flood related data analysis in various fields:

- Pre-analysis of the survey area (choice of simulation method, determination of model boundaries etc.)
- Collecting and processing of topographical information and its mergence (interpolation) to a digital terrain model (DTM)
- Creation of the system geometry of the HN-model (extraction of cross sectional data for 1D-models); definition of breaklines as input data for 2D-calculation meshes as well as allocation of the 2D-mesh points
- Definition of flow resistance (e.g. via analysis of digital orthophotos) and geo-coding of the model parameters
- Processing and visualisation of monitoring data as basis for model calibration (observed water levels and inundation borders, hydrological data)

Figure 5.14. GIS-supported interpolation of heterogeneous topographic data pool (left); Extraction of hydraulic relevant break-lines from high-resolution Laser-Scanner-DTM (right).

- Generation and validation of inundation zones and flood prone areas (e.g. projection of 1D-calculation results onto the study area)
- Visualisation and analysis of the calculation results (e.g. flow vectors of a 2D-flood-simulation, water level differences amongst alternative surveys)
- 3D-animation of simulated flood propagation
- Provision of a database with an interface for external data management systems

Today, linkage of one and multidimensional models with Geographic Information Systems is state of the art not only for scientific institutions. There are a number of GIS-supported hydrodynamic-numerical software products available, which can be applied effectively by experienced consultant engineers. When provided with an adequate topographical data basis and reliable calibration data, the existing tools can be used to obtain information about flood depth and flow velocity with a high temporal and spatial resolution.

Setting up a HN-model normally means to extract data with GIS methods from terrain models & land use databases and to save it in a HN-software suitable format. Digital land use databases are available in most European countries (Fig. 5.14).

High-resolution terrain data is only available if there have already been terrain related projects in these areas before. The laser scanner technique can provide terrain information of river valleys in raster data form with any resolution desired. Its accuracy depends primarily on flight path parameters, vegetation density and territorial gradients. From raw data sets, the topography and vegetation situation (Hansen and Vögtle, 1999) as well as structural facilities like buildings can be derived (Vögtle and Steinle, 2001).

Automated methodologies for reliable data interpretation will also be the subject of future research activities (Mandlburger, 2006). These will especially deal with hydraulic and hydrological requirements on algorithms to simplify the enormous information masses while keeping hydraulic-relevant topographical structures.

The enormous amount of data produced by HN modeling, especially for unsteady calculations, requires further development in data management. Latest versions of GIS programs still have problems in handling large data sources fluently, and the implementation of the height- and time variable data is still only rudimentary. GIS are the great new achievement in HN modeling. But in everyday engineering many possibilities remain unused, mainly caused by a complex and daunting usability of leading GIS products. Looking at some web based consumer orientated GIS projects like Google Earth, one can see a great slumbering potential.

In many cases the typical data ascertainment/collection at the beginning of projects is quite problematic. As data management is organized differently in every administrative region, the cross-boarder operating engineer has to search and collect data from many previously unknown sources.

Some countries open new ways of digital data distribution, using web based technologies for free public access to data like orthophotography, topographical maps, elevation models or land-use data. Others sale this normally state owned data at high prices, making it unavailable for smaller and non-profit projects.

Furthermore, a serious restraint can be the import of the tremendous amount of different types of raw data provided by authorities, geographers and other project partners. There are only few real technical problems in exchanging data from and to GIS solutions. Pre- and post-processing software provides the users with data converters. But varying co-ordinate systems throughout history and region, as well as individual data storage formats from self-made programs feed serious errors and result in a significant loss of time and money while setting up models.

5.5   CASE STUDIES

The following case studies first introduce a GIS-based flood simulation system of the river Neckar, that has been implemented by the responsible water management authorities for several years already. The model integrates a 1D-unsteady HN-model, which is acceptable due to the preponderance of 1D-current situations at the river Neckar. Furthermore, the narrow Neckar valley disposes of only small retention volumes; the weirs of the 27 barrages are completely open in case of a flood. Thus, there is no possibility of direct interventions into flood propagation.

Next, a current case study at the Middle Elbe is introduced. Due to the meandering stream course and the large retention volumes in the polder areas, as a basis of a decision support an application of a multidimensional model system is required. One solution for the operational application of a simulation model could be an interlinking of one- and multidimensional simulation procedures.

5.5.1   Case study River Neckar (south-west Germany)

On request of the Water Management Administration of Baden-Württemberg, the Institute of Water and River Basin Management (IWG) of the University of Karlsruhe has developed a flood simulation system for the Neckar River along its entire extent of more than 200 river kilometres including the adjoining flood-plain. The model was integrated into the software environment of a Geographic Information System (GIS) and enables the user via a one-dimensional hydrodynamic-numerical model, together with specially adapted GIS technologies, amongst other things to simulate any discharge constellations (Neckar and linked tributaries) and to quantify the intensity of the inundation in the potentially endangered areas.

The hydrodynamic method of the 1-dimensional procedure is based on the solution of the Saint-Venant-Equations by an implicit difference scheme (Preissmann-scheme, compare Cunge et al., 1980). The system geometry, i.e. the discharge area of the main channel and the floodplains, is represented by modified cross sections. Areas that contribute to the retention volume of the Neckar River during a flood event are taken into account by a function of storage capacity depending upon the water level. This function can be determined from the digital terrain model by means of several GIS-functionalities. The stationary model calibration was done by comparing calculated water levels with surveyed ones. In most sections, water level measurements of different recent flood events have been available, thus a calibration and validation for a spectrum of (flood) discharges was possible. The unsteady procedure can be verified by comparing calculated with surveyed flood hydrographs.

Through its GIS-interface (Fig. 5.15), the hydraulic results can be superimposed with a high-resolution DTM (grid size: 1×1 m) to determine inundation zones and respectively the boundaries. The DTM is based on elevation data from different data sources, i.e. terrestrial geodetic survey and airborne laser scanning.

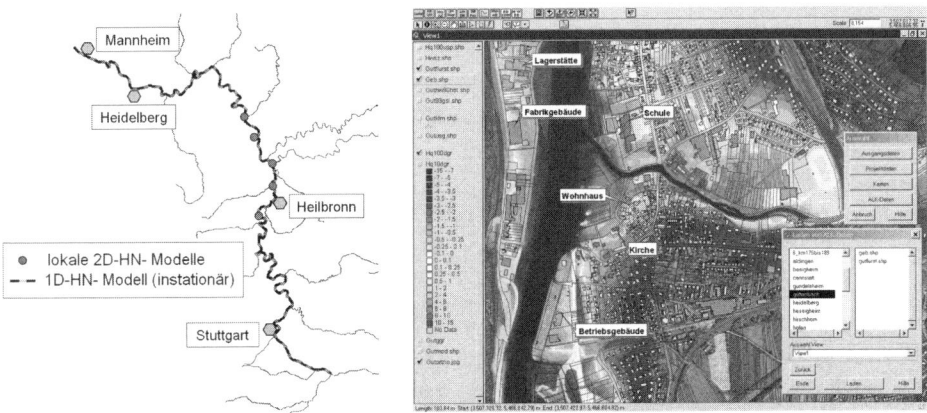

Figure 5.15. Study area "Neckar" (left); User interface for GIS-based flood analysis.

The methodology for GIS-supported flood simulation has been validated for a chosen section of the Neckar River by means of comparing calculations using a two-dimensional depth-averaged numerical model. So, the procedure-specific uncertainties as well as the accuracy of the input data were assessed and their effects on the decisive parameters (water level and inundation depth) were quantified.

The analysis draws the conclusion that it is, with respect to the required parameters, permissible to apply the one-dimensional approach to flow characteristics like those exhibited by the Neckar River. Furthermore, both the size of the study area and the goal of integrating the model into routine management by the authorities and operational application (real-time modeling and faster) supported the choice of procedure. The identification of the inundation areas on the basis of a high-resolution digital terrain model (1 × 1-m grid) only takes minutes, while a steady calculation on the basis of a two-dimensional numerical method requires several hours even when using a powerful personal computer. However, a few areas with a very complex flow situation (e.g. mouths of larger tributaries) could be recorded only insufficiently by means of the one-dimensional method. Here the additional application of local multidimensional numerical models was necessary. However, a stationary calculation based on a 2-dimensional HN procedure requires several hours even when using a powerful computer.

In order for the system to be applied by the water-management administration, a modular GIS-expert shell that is adapted to the authorities' special requirements was developed. It enables the user to carry out complex functional processes for simulation and animation (Fig. 5.16) by means of logically structured user interfaces. The system's handling and performance could be intensely tested and accordingly optimised over the course of several years during which the model was applied.

Speed and reliability of the processes employed, together with the water-level predictions of the Flood Forecasting Centre Baden-Württemberg, in principle can enable the operational use of this model in order to support decision-making during

Figure 5.16. Module for interactive 3D-real time animation and navigation.

flood events. However, in operational mode there is no necessity to modify the model (study of variations concerning the flooding of dyked areas, the control of dam levels, etc.) due to the restricted retention volume of the Neckar valley and the limited possibilities of influencing the stream-flow behaviour during a flood event. To use the model results for decision-making during a flood event, it is sufficient to calculate different discharge scenarios in advance and to make the results available to the adjoining municipalities or the disaster-control organisation in the form of flood risk maps with reference to forecast gauges or via scenario maps.

The system served as a basis for the generation of hazard maps with proto-type character in a state-wide sense. For example, hazard maps for the lower Neckar River are published on the Internet platform (Baden-Württemberg, 2005). A detailed description of the system is given by Oberle (2004).

5.5.2   Case study River Elbe (eastern Germany)

The transferability of the model techniques developed and applied for the river Neckar to other rivers depends on their individual flow characteristics in case of flood. Current research activities of the IWG concentrate on parts of the river Elbe characterised by wide inundation areas and polders. By linking 1D-systems (solving the Saint Venant Equation with the implicit differences scheme by Preissmann) and 2D-techniques (solving the complete shallow water equations based on non-structural grids) for example, complex flow processes inside polder areas as well as their impact on the unsteady flow situation in the river Elbe can be analysed. Further research focuses on the implementation of less complex 2D-HN model variants and determines if these simplifications affect the descriptive accuracy for the characterisation of the flooded regions.

The project focuses on operationally applicable models as tools for decision mak-ers in case of incoming floods. These tools will be provided for the local authorities to forecast unsteady water levels along dykes and barriers with an accuracy of one to two decimetres. Additionally, a dyke information system is set up to evaluate the current reliability of unsteady moistened dykes in relation to the calculated incom-ing hydrographs. Easy identification of bottlenecks will be possible. Reliability, alterable geometry on mouse-click and short computing time can provide different "If-Then-Else" – scenarios (e.g. polder flooding, dyke break or even emergency blasting of dykes).

Although today 2D models can reach $250 \, \text{km}^2$ and more, high-resolution models fast enough for efficient workflow on desktop computers should not exceed $70 \, \text{km}^2$. Thus the project area was divided in three high quality 2D reference models (Fig. 5.17). Additionally these models were welted after calibration and validation.

The variants of the full-dynamic models include:
- 2D-HN model with less detail accuracy – the computing time of a full 2D model can be shortened by reducing the number of cells and increasing the size of the

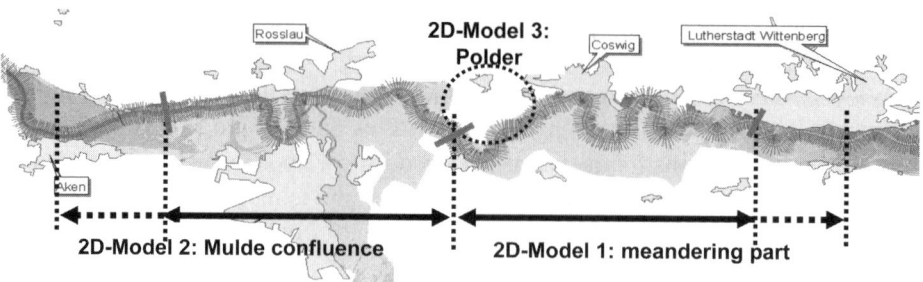

Figure 5.17.   Study area "Mittlere Elbe".

smallest computational cells while keeping the model's extension. Adaptively improved meshing methods help keeping a quality threshold with a minimum cell number (Fig. 5.18).

• Simplified 2D model – the computation speed is mainly limited by the complexity of the hydraulic equations. In a simplified 2D model terms of the hydraulic equation are assumed negligible and omitted which reduces computational time. The spatial resolution remains the same as in a fully 2D model. Implementation in process.

• Simplified 2D-HN model with lower resolution – combination of the first two simplifications in which spatial resolution and process description are both reduced.

• Branched 1D model – comparable to branched river systems in which the inundated area is discretised as a network of flow paths and retention cells and requires only a 1D description of the flow equations. An optimized branched 1D model is still more than 100 times faster than even simplified 2D models.

• Combined 1D/2D-HN models integrate the advantages of the former model types. Saving time in straight main branch sections by substituting tiny cells, but retaining spatial accuracy for 2-dimensional effects in meander and confluence zones, this hybrid type emerges as the model of choice for operational purposes. With adaptive optimized 2D models over the floodplains time steps of more than 3 seconds are possible, in spite of using a CFL number of 0.5. Simulation for far more than one hundred square kilometres is possible in operational speed units.

• Partial Discharge Model along a 2D mesh – for dyke break and polder flooding a PaD-model is tested to approximate flooding speed and maximum extent. As a dyke break produces a previously unknown change of topography, this very fast pseudo 2D model type enables an instantly generated new model with fast results.

The branched 1D model was calculated with *STREAM* (IWG). The 2D studies as well as the combined 1D/2D approach were calculated with *FLUMEN* (fluvial.ch), using the Euler method for solving the explicit system of shallow water equations. The courant number is set to 0.5 for most purposes. At present, further research work focuses amongst others on *TriPaD* (fluvial.ch), which uses simplified 2D

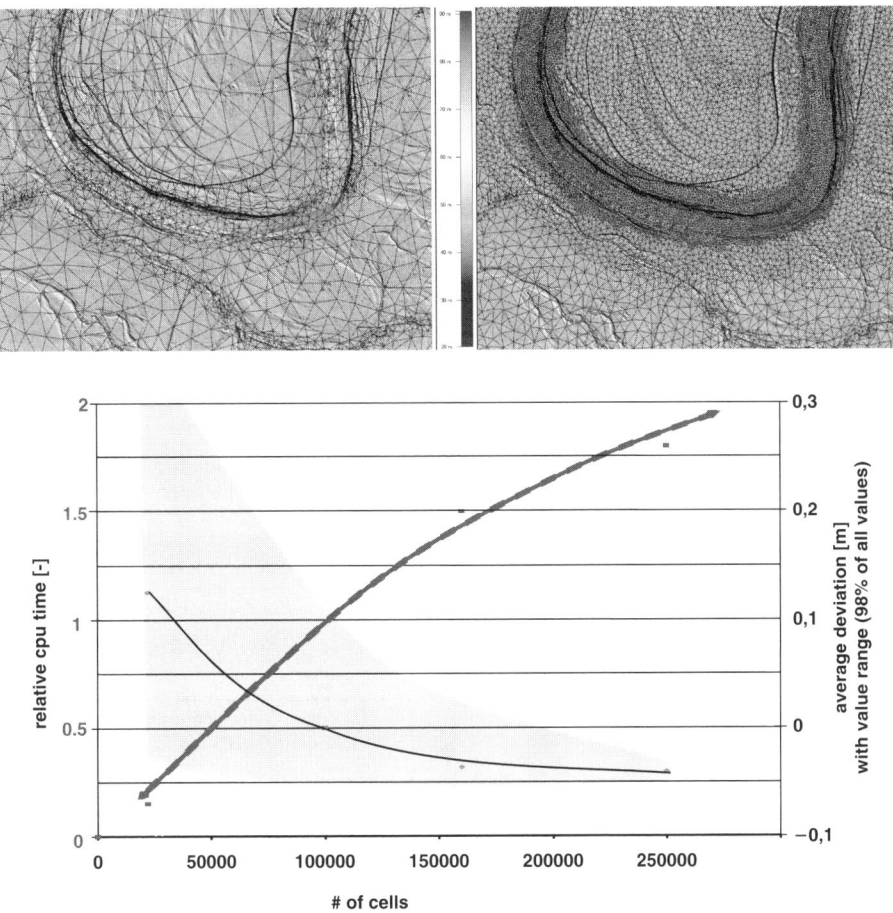

Figure 5.18. Low resolution mesh (16k cells) and 100k reference mesh of a meandering area; relation between number of cells, calculation time (dashed) and quality (average deviation (solid) & value range (shade); valid only for reference project).

modeling techniques. All examinations are based on data collected by the Wasser- und Schiffahrtsamt Dresden, Dessau fire department, ATKIS land use database, and public topographic maps. A comparison of 3 tested models is shown in Figure 5.19.

5.6   CONCLUSIONS

Nowadays, the application of one- and two-dimensional hydrodynamic-numerical methods for flood simulation and for the assessment of measures is a state-of-the-art engineering practice and a valuable means to support flood management in urban areas. Model scales from several hundred km^2 floodplains to detailed inner city alley simulations have been realised. Criteria for the selection of the modeling

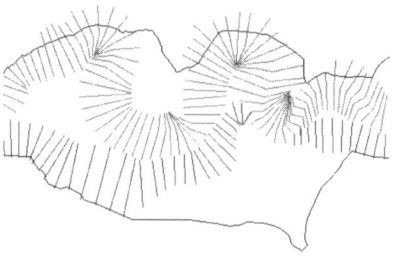

Branched 1D-HN model

– Main branch, 2 shortcuts for
  floodplains
– 1 min calculation time
– 7 cm or less average difference to a
  high-quality reference model,
  separate calibration necessary,
  calibration strongly discharge
  dependent
– Calculated with IWG STREAM

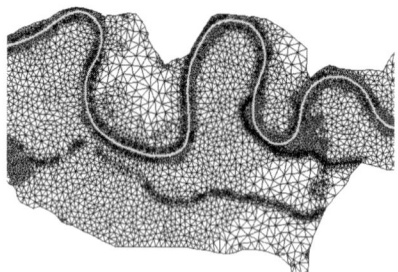

**Manually refined low resolution
2D-HN model**

– 50,000 cells
– 3 hours calculation time
– ca. 5 cm average difference to a
  high-quality reference model, high
  outlier percentage
– Calculated with FLUMEN 1.3

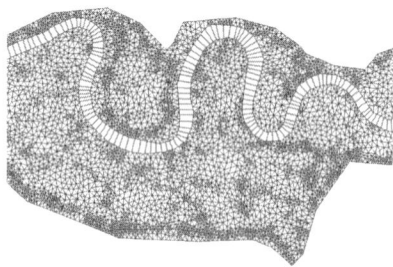

**1D model connected to two
adaptive 2D-HN models**

– 16,000 cells in floodplains
– 15 min calculation time
– ca. 3 cm or less average difference
  to a high-quality reference model,
  separate calibration necessary
– Calculated with FLUMEN 2.0

Figure 5.19. Comparison of different model types for a 47 square km floodplain. A 50 h artificial hydrograph (empty basin to medium water level plateau rising over 36 h to a stationary maximum known flood) calculated on 3.2 GHz Pentium IV single core processor.

method (1D or 2D) are the flow characteristics of the study area as well as the required target parameters.

Provided a careful model selection, the accuracy of the calculation results depends mainly on the available topographical data basis as well as on the quality and up-to-dateness of the calibration data. More and more one can resort to high-resolution laser scanner data of riversheds. However, they pose additional challenges to the modeler when it comes to data handling and filtering. The parametrisation of energy losses (e.g. induced by floodplain vegetation or river – floodplain interaction) will remain an intensive field of research in the future as well.

Ever-growing CPU power accelerates HN-models and permits the handling of greater data amounts. But as these data masses overextend human imagination, spatial organisation tools like Geographic Information Systems (GIS) enable people to manage the projects. Today GIS is an indispensable device to support pre- and post-processing functions for flood simulation. However, unnecessary complexity of established software packages and protective data philosophies retard development and burn enormous sums of time and budget.

An important future task is the provision of simulation models for decision makers. Here, more-dimensional simulation methods in connection with GIS-technologies offer an excellent potential for development. For standard purposes, calculation time mostly is no longer an issue but for operational purposes computational speed is still the main issue. Appropriate data and model simplification is the key to better and faster modeling. Many positive attempts have been made throughout the past years and only wait to be combined. Such combination of now well-known methods, like the presented 1D/2D models with adaptive meshing, will significantly extend the fields of application.

REFERENCES

Apel, H. Thieken, A.H. Merz, B. Blöschl, G.: Flood Risk Assessment and Associated Uncertainty, Natural Hazards and Earth System Sciences, 4, S. 295–308, 2004. SRef-ID: 1684-9981/nhess/2004-4-295.

Baden-Württemberg: Hochwassergefahrenkarte Baden-Württemberg, (Flood-hazard map Baden-Württemberg), Umweltministerium Baden-Württemberg, Stuttgart, 2005 (http://www.hochwasser.baden-wuerttemberg.de).

Bayern: Informationsdienst Überschwemmungsgefährdete Gebiete in Bayern, (Information service flood endangered areas in Bavaria), Bayerisches Landesamt für Wasserwirtschaft, München, 2005 (http://www.geodaten.bayern.de/bayernviewer-aqua/).

Beffa, C.: Integration ein- und zweidimensionaler Modelle zur hydrodynamischen Simulation von Gewässersystemen. In: Moderne Methoden und Konzepte im Wasserbau, ETH Zürich, 2002.

Beffa, C.: Integration von ein- und zweidimensionalen Abflussmodellen, Dresdner Wasserbauliche Mitteilungen, Heft 32, 2006.

Büchele, B. Kreibich, H. Kron, A. Ihringer, J. Theobald, S. Thieken, A. Merz, B. Nestmann, F.: Developing a Methodology for Flood Risk Mapping: Examples from Pilot Areas in Germany. In: Malzahn D., Plapp T. (eds) Disasters and Society – From Hazard Assessment to Risk Reduction. Logos-Verlag, Berlin, S. 99–106, 2004.

Büchele, B. Kreibich, H. Kron, A. Thieken, A. Ihringer, J. Oberle, P. Merz, B. Nestmann, F.: Flood-risk mapping: contributions towards an enhanced assessment of extreme events and associated risks; Natural Hazard and Earth System Sciences NHESS, European Geosciences Union [not yet published], 2006.

Cunge, J.A. Holly, F.M. Verwey, A.: Practical aspects of computational hydraulics. Institute of Hydraulic Research, Iowa, 1980.

DVWK: Hydraulische Berechnung von Fließgewässern, Heft 220, Verlag Paul Parey, Hamburg, 1991.

Hansen, W. v. Vögtle, T.: Extraktion der Geländeoberfläche aus flugzeuggetragenen Laser-Scanner-Aufnahmen. In: Photogrammetrie, Fernerkundung, Geoinformation 1999, H. 4, S. 229–236.

ICPR (International Commission for the Protection of the Rhine): Non Structural Flood Plain Management – Measures and their Effectiveness, ICPR, Koblenz, 2002.

IKoNE (Integrierende Konzeption Neckar-Einzugsgebiet): Hochwassermanagement, Partnerschaft für Hochwasserschutz und Hochwasservorsorge (flood management partnership for flood protection and precaution), Heft 4, Besigheim, 2002.

Ihringer J.: Ergebnisse von Klimaszenarien und Hochwasser-Statistik (Results from climate scenarios and flood statistics). Proc. 2nd KLIWA-Symposium, S. 153–167, Würzburg, 2004, retrieved from http://www.kliwa.de.

Kämpf, C. Kron, A. & Ihringer, I.: Design von Simulationswerkzeugen für die Praxis des Hochwassermanagements (design of simulation tools for practice of flood management). Forum für Hydrologie und Wasserbewirtschaftung, Heft 10.05, S. 157–163, 2005.

Kleeberg, H.B. (Ed.): Hochwasser-Gefahrenkarten (Flood hazard maps). Forum für Hydrologie und Wasserbewirtschaftung, Heft 08.05, 2005.

Kron A. Evdakov O. Nestmann F.: From Hazard to Risk – A GIS-based Tool for Risk Analysis in Flood Management. Van Alphen, van Beek, Tal (eds.). In: Floods, from Defence to Management, London, S. 295–301, 2005.

Kron A. Oberle P. Theobald S.: Tools for risk analysis in flood management. 4. Forum Katastrophenvorsorge, 14.–16. Juli 2003, München, Extended Abstracts, S. 82–83, 2003.

Mandelburger G.: Topographische Modelle für Anwendungen in Hydraulik und Hydrologie, Dissertationsschrift, Institut für Photogrammetrie und Fernerkundung, Technische Universität Wien, Oktober 2006.

Mignot, E.A. Paquier, S. Haider: Modeling floods in a dense urban area using 2D shallow water equations; Journal of Hydrology (2006) 327, 186–199.

Mileti, D.S.: Disasters by design. A reassessment of natural hazards in the United States. Joseph Henry Press, Washington D.C., 1999.

MUNLV (Ministerium für Umwelt und Naturschutz, Landwirtschaft und Verbraucherschutz des Landes Nordrhein-Westfalen): Leitfaden Hochwasser-Gefahrenkarten (Guideline flood-hazard maps). Düsseldorf, 2003.

Nestmann F. Kron A.: Frühwarnung und Management von Katastrophen – Beispiel Hochwasser. Veröffentlichungen zum Symposium "Naturgefahren und Kommunikation" der Umwelt- und Schadenvorsorge der SV Versicherung, 4. und 5. April 2005 in Neuhausen/Stuttgart in der Zeitschrift GAIA, S. 3–4, September 2005.

Nordrhein-Westfalen: Hochwassergefährdete Bereiche (Flood endangered areas). Landesumweltamt Nordrhein-Westfalen, Düsseldorf, (http://www.lua.nrw.de/index.htm?wasser/hwber.htm), 2003.

Oberle, P. Theobald, S. Nestmann, F.: GIS-gestützte Hochwassermodellierung am Beispiel des Neckars (GIS-based flood simulation exemplified for the Neckar River). Wasserwirtschaft, 90(7–8), S. 368–373, 2000.

Oberle, P.: Integrales Hochwasserinformationssystem Neckar – Verfahren, Werkzeuge, Anwendungen und Übertragung (Integral flood simulation system Neckar – methods, tools, applications and transfer), Dissertation, Institut für Wasser und Gewässerentwicklung, Universität Karlsruhe, Heft 226, 2004.

Oberle P. Theobald S. Evdakov O. Nestmann F.: GIS-supported flood modeling by the example of the river Neckar, Kassel Reports of Hydraulic Engineering No. 9/2000, S. 145–155 und Proceedings "European Conference on Advances in Flood Research", Potsdam Institute for Climate Impact Research, S. 652–664, No. 11/2000.

Pasche, E.: Turbulenzmechanismen in naturnahen Fliessgewässern und die Möglichkeiten ihrer mathematischen Erfassung, Mitt. Inst. Für Wasserbau und Wasserwirt. 52, Rheinisch-Westfälische Tech. Hochsch. Aachen, 1984.

Renouf, E. Paquier, A. Mignot, E.: Assessment of the exchanges between sewage network and surface water during flooding of the town of Oullins; 10th International Conference on Urban Drainage, Copenhagen/Denmark, 21–26 August 2005.

Pfeifer, N. Stadler, P. Briese, C.: Derivation of Digital Terrain Models in the SCOP++ Environment. In OEEPE workshop on Airborne Laserscanning and Interferometric SAR for Detailed Digital Terrain Models, Stockholm, 2001.

Rheinland-Pfalz: Grenzüberschreitender Atlas der Überschwemmungsgebiete im Einzugsgebiet der Mosel (Transboundary atlas of the inundation areas in the basin of the river Mosel). Struktur- und Genehmigungsdirektion Nord, Regionalstelle für Wasserwirtschaft, Abfallwirtschaft, Bodenschutz Trier, (http://www.gefahrenatlas-mosel.de), 2004.

Sachsen: Gefahrenhinweiskarte Sachsen (Hazard information map Saxony). Sächsisches Landesamt für Umwelt und Geologie, Dresden, 2004, (http://www.umwelt.sachsen.de/de/wu/umwelt/lfug/lfug-internet/interaktive_karten_10950.html).

Takeuchi, K.: Increasing vulnerability to extreme floods and societal needs of hydrological forecasting. Hydrol. Sci. J., 46(6), S. 869–881, 2001.

Theobald, S. Oberle, P. Nestmann, F.: Simulationswerkzeuge für das operationelle Hochwassermanagement, Wasserwirtschaft 94. Jg, S. 23–28, 12/2004.

Uhrich, S. Krause, J. Bormann, H. Diekkrüber, B.: Simulation von Überflutungen bei Hochwasserereignissen: Risikoeinschätzung und Unsicherheiten (Simulation of inundations due to flood events: risk assessment and uncertainties). In: Tagungsbericht 5. Workshop zur hydrologischen Modellierung: Möglichkeiten und Grenzen für den Einsatz hydrologischer Modelle in Politik, Wirtschaft und Klimafolgeforschung, ed. by Stephan, K. Bormann, H. Diekkrüger, B. Kassel University Press, Kassel, S. 59–70, 2002.

UM Baden-Württemberg (Umweltministerium Baden-Württemberg): Hochwassergefahrenkarten Baden-Württemberg, Leitfaden (Flood-hazard maps Baden-Württemberg, guideline). Stuttgart, 2005.

Vögtle, T. Steinle, E. (2001): Erfahrungen mit Laser-Scanner-Daten zur automatisierten 3D-Modellierung von Gebäuden. In: Deutscher Verein für Vermessungswesen (DVW), Landesverein Baden-Württemberg, Mitteilungen H. 2, 48 Jg., Oktober 2001.

Wind, H.G. Nieron, T.M. de Blois, C.J. de Kok, J.L.: Analysis of flood damages from the 1993 and 1995 Meuse floods. Water Resour. Res., 35(11), S. 3459–3465, 1999.

6

Flood Frequency Analysis for Extreme Events

Félix Francés & Blanca A. Botero
Technical University, Valencia, Spain

ABSTRACT: The most straightforward method for estimating flood risk at a particular point where a gauge station is located is to adjust a cumulative probability distribution function to the recorded annual maximum flows. Unfortunately this method may give rise to highly variable flood quantile estimators for high return periods. To increase reliability of the quantile estimator it can used the best statistical model and/or to increase the amount of information. The high number of distribution functions used in Hydrology is a demonstration more efforts must be done in the theoretical basis of the statistical model selection. Using historical or palaeoflood censored data can increase the information length at the gauge station. However, care must be taken with the information error and the underling stationary hypothesis. Few statistical models have been developed to introduce the non-stationarity of the long term flooding process. A very important additional output of these types of models will be their use for the future flood hazard estimation in a climate change framework.

6.1 INTRODUCTION

Flood frequency analysis consists basically of obtaining the relationship between flood quantiles and their non-exceedence probability (also referred to as return period T). It is one of the main issues in Hydrology, as it is the basis for the design of hydraulic structures (e.g., dam spillways, diversion canals, dikes and river channels), urban drainage systems, cross drainage structures (e.g., culverts, bridges and dips), flood hazard mapping, etc. Due to the social and economic implications, flood frequency analysis must be done with the highest precision. For example, if the accepted hazard for a urban flood protection system is given, the value of the design flow should be estimated as accurately as possible, since a flood quantile estimate lower than real would increase the actual flood risk at the urban area; on the other hand, an overestimation of the flood quantile would increase the cost of the protection system unnecessarily.

The most straightforward method for estimating flood risk at a particular point where a gauge station is located is to adjust a cumulative probability distribution function (cdf) to the recorded annual maximum flows. Unfortunately this method may give rise to highly variable flood quantile estimators due to:

  i) uncertainty of the statistical model,

  ii) errors in data recording (especially for the largest floods),

  iii) data series are shorter than the return period, and finally,

  iv) high variance and skewness of annual maximum flows, which cause great variability in the statistical properties of the recorded samples.

To increase reliability of the quantile estimator it can used the best statistical model and/or to increase the amount of information. The best model can be found considering either its descriptive ability or its predictive ability, the latter being specifically useful for estimating floods of large return period, as held by Cunnane (1986). Using historical or palaeoflood censored data can increase the information length at the gauge station. However, care must be taken with the information error and the underling stationary hypothesis.

6.2   DATA CLASSIFICATION

Data recorded systematically (all the time) at a flow gauge station located in a river section will be called Systematic information. On the other hand, Non-Systematic information at a river section is that information not recorded systematically, usually during a period prior to the installation of the gauge station. If there is not a gauge station, all information can be considered as Non-Systematic. The sources of this information can be historical or palaeofloods (biological or geological indicators usually, but not necessarily, previous to human registers).

The Non-Systematic information is always censored type I, in such a way we have some information concerning the flood at a given time during the Non-Systematic period because the flood was bigger than a threshold level of perception X_H (Stedinger and Cohn, 1986; Francés et al., 1994). The value of the peak flow for the floods above X_H can be known or not. Concerning the floods below X_H, it is not known their exact values, but at least it is known they were smaller than X_H. The threshold level of perception can be, for example, the position of a cave (for palaeoflood information) or the minimum discharge which produces damages in a city (for historical information). It can change with time and, in some cases, there can be two different thresholds for the same flood (following the palaeoflood example if there are two caves at different positions, the lower one with sediments and the upper one without any trace of the flood).

On the other hand, usually the Systematic data is completely known, but some times the uncertainty in the data forces to treat them also as censored.

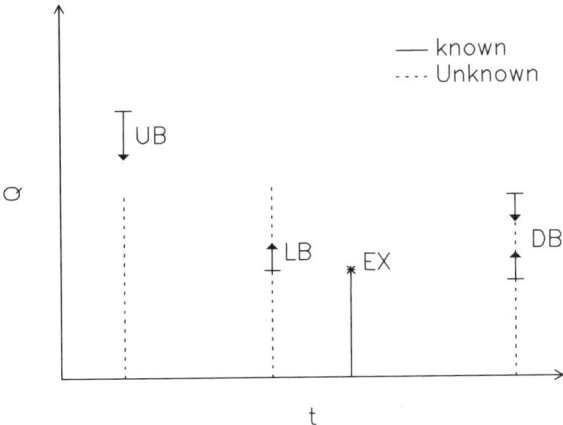

Figure 6.1. Data classification for Systematic and Non-Systematic information.

In this framework, as was presented by Naulet (2002) and represented in Figure 6.1, the Systematic and Non-Systematic flood data can be classified in:

i) EX type. The flood at time t is exactly known. It will correspond with most of the Systematic data and with some Non-Systematic floods with enough information to reconstruct the peak discharge.

ii) LB type. A flood is LB if it is only known the flood at time t was bigger than X_H. The lower bound L is equal to this threshold.

iii) UB type. In this case we only known the flood at time t was smaller than X_H, which is the information upper bound U.

iv) DB type. The floods where the known information is their lower and upper bound are considered DB.

It is important to stress that from the statistical point of view there is no difference concerning the source of the Non-Systematic information, historical or palaeoflood, and their treatment must be completely similar (Francés et al., 1994). More over, with this new classification there is not any difference also between Systematic and Non-Systematic information, with the additional advantage that always the data has a time assigned, which is crucial for non-stationary models.

6.3   DISTRIBUTION FUNCTIONS

6.3.1   Unbounded distribution functions

The list of distribution functions for flood frequency analysis that can be founded in the literature and have been applied somewhere is large. Details for the most common of them can be found for example in Kite (1988), Stedinger et al. (1993)

or Robson and Reed (1999). The classification by the number of its parameters is given by:
i) With one parameter: the exponential distribution function.
ii) With two parameters: the LogNormal (LN2) and the Extreme Value family (Gumbel or EV1, LogGumbel or EV2 and Weibull or EV3).
iii) With three parameters: the LogNormal with 3 parameters (LN3), the Pearson and LogPearson type III (LP3), the General Extreme Value (GEV), the Generalised Logistic (GL) and the Generalized Pareto (GP). Officially, the LP3 is the cdf recommended in the USA by the USWRC (1982) and in UK, the GEV was recommended by the NERC (1975), but nowadays is the GL (Robson and Reed, 1999).
iv) With four parameters: the Two Component Extreme Value or TCEV (Rossi et al., 1984). This cdf was developed for mixed population as the floods of most of Mediterranean rivers. This distribution is widely used in Italy and Spain.

Usually, distributions with one or two parameters are not able to reproduce the sample variability observed in flood records. This sample variability produces the so called "dog leg effect" (Potter, 1958) and the "separation condition" (Matalas et al., 1975). The "dog leg effect" consists in a slope change in the sample plotting positions (defined below) in a Gumbel probability paper, due to a high skewness coefficient, as it is shown in Figure 6.3 for a Mediterranean river. The "separation condition" expresses the inability of some distributions to reproduce the observed skewness and kurtosis coefficients. From these points of view, the GEV, GP3 and TCEV are recommended for high return periods compared with the size of the sample (usually T bigger than 100 years).

6.3.2   Bounded distribution functions

For a given basin there is an upper limit for its floods. This upper limit is called Probable Maximum Flood or PMF, which must depend mainly on the basin area and climate. Also, it seems clear this limit must affect the right tail of the cdf, but none of the above listed cdf has an upper limit (the GEV can have it but with low or negative skewness, which it is not the case with flood populations). Only in recent years distribution functions with an upper limit have been applied to the extreme frequency analysis. These upper bounded models can be used for PMF estimation, if this is the objective. But if the interest is in high return period quantile and because the actual existence of a flood upper bound, the predictive ability (in the sense done by Cunnane, 1986) of an upper bounded distribution function must be exploited (Takara and Loebis, 1996).
   Three models can be found in the literature:
i) The Elíasson Transformed Extreme Value type distribution (ETEV), proposed by Elíasson (1994 and 1997).

ii) The Slade-type four parameter LogNormal distribution (LN4), proposed by Slade (1936) and called in this manner by Takara and Loebis (1996).

iii) The Extreme Value with 4 parameters distribution (EV4), which was first proposed by Kanda (1981) and used first in Hydrology by Takara and Tosa (1999).

6.4   ESTIMATION METHODS

In order to estimate the distribution function parameters, several methods have been applied in Flood Frequency Analysis with Systematic and additional Non-Systematic information. The most common are:

6.4.1   The Method of Moments (MOM)

The MOM parameters are estimated solving the equations obtained matching the sample moments with the theoretical moments obtained from the distribution function. It is the simplest method, but it is the poorest from the statistical point of view. For this reason, usually is used as a first approximation for other methods.

With Non-Systematic information Tasker and Thomas (1978) proposed the so called Historically Weighted Moments, which was recommended by the US Water Resources Council Bulletin 17B (1982) and also used by Condie and Lee (1982).

Cohn et al. (1997) presented a variation of this method called the Expected Moment Algorithm (EMA), improving the estimation of the moments for censored samples.

6.4.2   The L-Moments and the Probability Weighted Moments (PWMs) methods

In this case the L-Moments or the PWMs are used to construct the equations to be solved. Because the first r L-Moments are a linear combination of the first PWMs, both will give us the same estimated parameters (Stedinger et al., 1993). These methods seem to be more robust (perform well when the population distribution is different to the fitted distribution to the sample) for short samples, because the L-Moments are always a linear combination of the ranked sample.

With additional Non-Systematic information this method was called Partial Probability Weighted Moments by Stedinger and Cohn (1986), and it have been used also by Wang (1990) and Kroll and Stedinger (1996).

6.4.3   The Maximum Likelihood Estimation method (MLE)

The MLE parameters are those which maximize the likelihood function, which is any function proportional to the joint probability density function (pdf) of the observed sample. For large samples, usually is the best estimation methodology.

The MLE method with additional Non-Systematic information has been used by many researches: with a Gumbel population by Leese (1973), Hosking and Wallis (1986a and 1986b), and Guo and Cunnane (1991); with the EV distributions family by Francés et al. (1994); with the GEV by Phien and Fang (1989); with the LN2 distribution by Condie and Lee (1982), Cohn and Stedinger (1987) and Kroll and Stedinger (1996) (in this case with censored information only); with the LP3 by Pilon and Adamowski (1993); and with the TCEV in a regional framework by Francés (1998).

6.4.4   MLE method with additional Non-Systematic information

Actually, at this moment, only with the MLE method it is possible to incorporate all kinds of censored information described earlier. For this reason and for its general better statistical properties, it is the recommended estimation method to be used.

Assuming only independence between the annual maximum floods and depending on the type of data, each year will contribute to the likelihood function in one of the next general expressions:

$$L_{EX}(\underline{\Theta}; x_i) = f_{Xi}(x_i; \underline{\Theta})$$
$$L_{LB}(\underline{\Theta}; i) = 1 - F_{Xi}(L_i; \underline{\Theta})$$
$$L_{UB}(\underline{\Theta}; i) = F_{Xi}(U_i; \underline{\Theta})$$
$$L_{DB}(\underline{\Theta}; i) = F_{Xi}(U_i; \underline{\Theta}) - F_{Xi}(L_i; \underline{\Theta})$$

where the independent but not necessarily identically distributed random variable is described each year by its pdf $f_{Xi}(.)$ or its cdf $F_{Xi}(.)$; the x_i are the known sample values (EX data in the Systematic and Non-Systematic periods); and $\underline{\Theta}$ represent the parameters to be estimated.

The final likelihood function will be computed as the product of the likelihood of the information of each year. The ML estimators are obtained by maximizing this function, or better its logarithm, over the parameters space. One additional advantage of these equations is the time assignment of each piece of information through the sub index i, which is critical for any non-stationary model. If the same functional expression for the cdf is used, the non-stationarity can be introduced changing its parameters in time as a function of others stationary parameters.

6.5   MODEL SELECTION

Once the parameters have been estimated, the next step is to select the best statistical model, because different models can give significantly different high return period quantiles. With Systematic and additional Non-Systematic information, the estimation method will be less important than the different cdf fitted to the data. Three methodologies can help this selection process.

6.5.1 High order Moments and L-Moments diagrams

From the predictive point of view, which is critical for high return period quantile, the selected cdf must be capable of reproducing the observed high order moments (skewness and kurtosis) and high order L-Moments (L-Skewness and L-Kurtosis). In order to reduce the uncertainty in the estimated high order moments and L-Moments, it is recommended to do this comparison at regional scale. One problem is that nowadays there has not been developed for all kinds of censored information the estimator algorithms for the high order moments and L-Moments.

6.5.2 Plotting Positions

The most popular way to test the model performance from the descriptive point of view is by comparison between the fitted cdf and the Plotting Positions in probability paper. For floods usually Gumbel probability paper is used, which can be obtained depicting the random variable "x" against "$-\ln[-\ln F_X(x)]$".

The Plotting Positions, also called empirical cdf, are the non exceedence probability estimate of the ordered sample. Cunnane (1978) gave the general expression for the Plotting Positions with only Systematic data:

$$F(x_{(i)}) = 1 - \frac{i - \alpha}{N + 1 - 2\alpha}$$

where $x_{(i)}$ is the descending ordered sample; N is the size of the sample; and the coefficient α depends on the population cdf in order to give a better probability estimate. For Gumbel population α must be equal to 0.44 (which is the so called Grigorten formula) and, for extreme populations as floods without assuming a priori any particular cdf, Cunnane (1978) recommends to adopt $\alpha = 0.4$.

It must be noticed for $i = 1$ or N, the coefficient of variance of the Plotting Position is close to 1, which implies a high variability in the upper and lower part of de empirical cdf.

Plotting Positions for Systematic and additional Non-Systematic data have been proposed by Benson (1950), Hirsh (1987), Hirsh and Stedinger (1987), Adamowsky and Feluch (1990), Wang (1991) and Guo and Cunnane (1991). But all of them only for the case with EX and UB data. For this case, it is recommended the E formula proposed by Hirsh and Stedinger (1987) or its generalization given by Stedinger et al. (1993):

$$F(x_{(i)}) = 1 - \frac{k'}{N + M} \frac{i - \alpha}{k' + 1 - 2\alpha} \qquad i = 1, ..., k'$$

$$F(x_{(i)}) = 1 - \frac{k'}{N + M} - \frac{N' + M - k}{N + M} \frac{i - k' - \alpha}{N - e + 1 - 2\alpha} \qquad i = k' + 1, ..., N + k$$

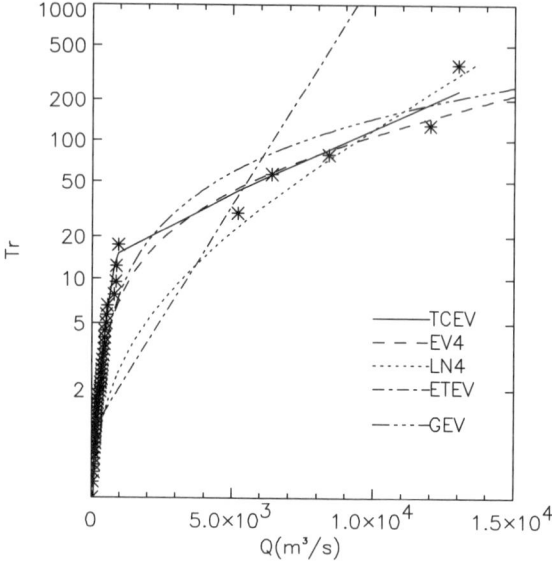

Figure 6.2. Plotting positions for the annual maximum floods (Systematic and Non-Systematic EX data) and the distribution functions fits for the Jucar River (Spain).

where $M =$ Non-Systematic period length; $k =$ number of floods above the threshold level of perception during the Non-Systematic period; $e =$ number of floods above the threshold level of perception during the Systematic period; $k' =$ total number of floods above the threshold level of perception $= k + e$. Figures 6.2 and 6.3 are examples of the comparison between the Plotting Positions using this equation and different fitted cdf.

However, nowadays there is not any plotting positions expression for the case of additional LB data, as was the case in the data of Figure 6.4. The main problem is how to order the EX floods above the threshold without knowing the exact value of the LB floods which can be also above the same threshold.

6.5.3   Goodness of fit tests

A lot of statistical tests have been developed to select a model, but most of them are not adequate to flood samples. A good recompilation can be found in the book of D'Agostino and Stephens (1986). Stedinger et al. (1993) recommend the Kolmogorov-Smirnov and the Probability Plot Correlation Coefficient tests (based on the plotting positions) and the L-Moments Ratio tests.

6.6   CONFIDENCE INTERVALS FOR THE ESTIMATED QUANTILES

It is not only important to estimate the value of a particular quantile X_T or PMF, it will be also important to have an idea about its estimation error (Figure 6.3); at

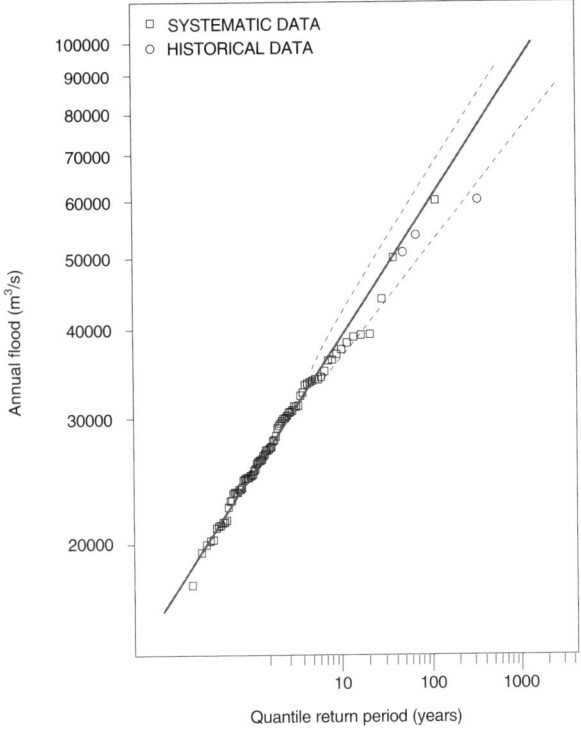

Figure 6.3. Plotting positions, Weibull fit and its 95% confident limits for the annual maximum floods in Paraná River (Argentina).

least concerning the error due to the lack of information, i.e., assuming model and data are correct.

A simple way to estimate this error is first compute the standard deviation of the quantile estimate, $S(X_T)$, and, with the hypothesis of error normality, compute the $100(1 - \alpha)\%$ confidence intervals with the expression:

$$X_T \pm S(X_T)\, z_{1-\alpha/2}$$

where z is the standard normal quantile. Theoretical expressions of $S(X_T)$ for some cdf and MOM or MLE estimation methods with only Systematic information are given by Kite (1988) and Stedinger et al. (1993). For additional Non-Systematic information fewer cases have been studied (Francés et al., 1994; Francés, 1998). Therefore, in practice, Monte Carlo simulation must be done to estimate the quantile estimate standard deviation through the sample mean square error (see for example Stedinger and Cohn, 1986 or Francés et al., 1994). The same can be applied for the PMF estimate (Figure 6.4).

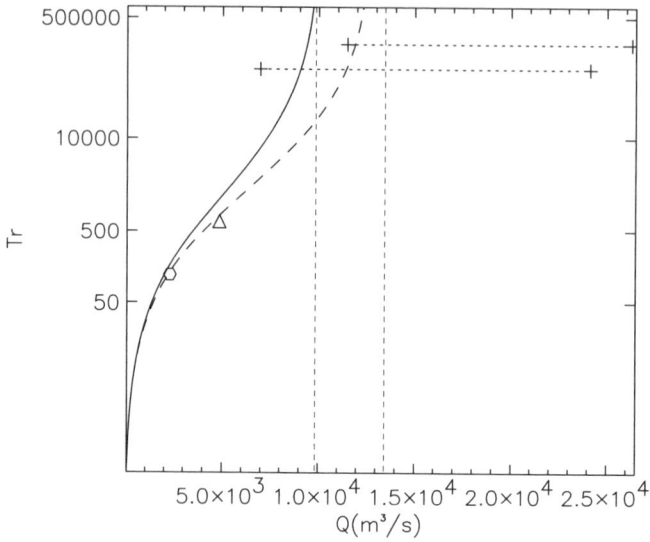

Figure 6.4. EV4 fits for the Llobregat river at Pont de Vilomara, the estimated PMF 90% confidence limits and the return period of the threshold discharge for the Contemporary climate (solid line and circle) and Holocene climate (dashed line and triangle) scenarios.

6.7 NON-STATIONARY FLOOD FREQUENCY ANALYSIS

The possible non-stationarity in hydrological series must be considered. The temporal changes in the trajectory and statistics of a state variable may correspond to natural, low-frequency variations of the climate hydrological system or to non-stationary dynamics related to anthropogenic changes in key parameters such as land use and atmospheric composition (Baldwin and Lall, 1999). Some recent studies have in fact identified a possible rising trend in the temperature and, consequently, in other hydrological variables as well (Porporato and Ridolfi, 1998). Also, in several countries, dam construction changes significantly the flood risk downstream during the last century.

As was pointed out by Porporato and Ridolfi (1998), studying the hypothetical case of a trend superimposed on a random stationary variable, with special reference to the simplified case of linear trend, even very weak non-stationarities can give rise to quite significant effects on the exceedence probability. They concluded that since the relation between exceedence probability and non-stationarity is strongly nonlinear, to ignore this fact could result in underestimation in the design and planning of hydraulics works.

Land use changes and dam effects can be eliminated using hydrological simulation. Climate changes (natural and/or human induced) must be taking into account

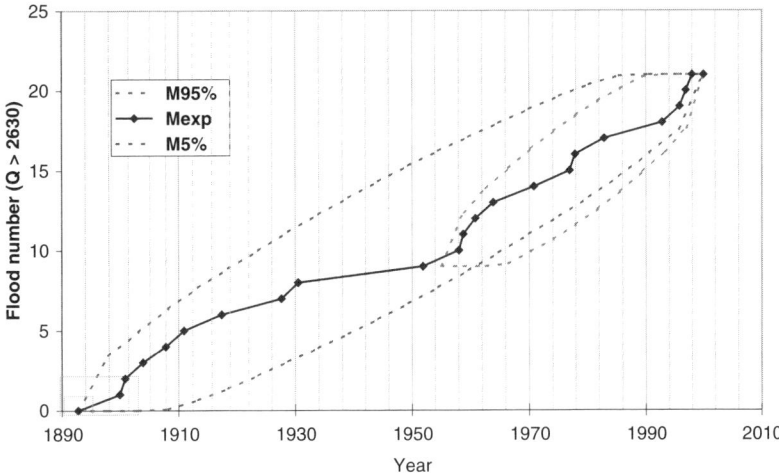

Figure 6.5. Lang's test for the number of floods above the threshold of $2630\,\mathrm{m^3/s}$ in the Ardéche River at St-Martin (France).

in the analysis in three different ways:

i)   Separate past in different climate periods and assuming for future scenario one of them, as it is done in Figure 6.4.
ii) Fit a non-stationary model, which means the past and future climate can be simulated and predicted respectively.
iii) Do nothing, if there is not any knowledge concerning the future climate.

6.7.1   Stationarity tests

Actually, the first thing to do in a Flood Frequency Analysis must be to test if the data series is or not stationary. For a Systematic record two classical tests can be done (Hirsh et al., 1993): the Pearson's Correlation Coefficient test for a linear trend and the Kendall's Correlation Coefficient test for monotonic dependence with time. Other tests can be found in Robson and Reed (1999).

To test the data stationarity for censored samples (Systematic and/or Non-Systematic), the Lang's test is proposed (Lang et al., 1999). The test null hypothesis is to suppose that the flood series can be described by a homogeneous Poisson process. The censored run can be built using all the Non-Systematic and Systematic data, and the accumulated number of floods above the threshold level of perception is computed. Figure 6.5 shows an example where the data series results to be stationary.

6.7.2   Non-stationary models

Few models and applications can be found in the literature for Non-Stationary Flood Frequency Analysis. The best of them are described below, but more efforts must be done in the future to improve especially the link between the flood model and climate variables.

Stedinger and Crainiceanu (2001) proposed two non-stationary models, called Log-Normal Trend and Log-Normal ARMA, assuming a linear trend or a low order ARMA model on the mean flood respectively. The ARMA model tries to explain the variability and persistence in flood records, but preserving the assumption that hydrology is stationary in the long run. One interesting conclusion for this type of stationary stochastic models is that even for significant persistence, there is "... *relatively little impact on the precision of the 100-year quantile describing long-term average flood risk, whereas a large impact was observed on the precision of the estimated mean flood*" (Stedinger and Crainiceanu, 2001).

Strupczewski et al. (2001) developed a Non-Stationary Flood Frequency procedure of estimation and selection between 56 models with combinations of six classical unbounded cdf, different trend functions for the first two moments and AMF and POT data. The trend functions can be linear or parabolic, and they introduce also the possibility of correlation between mean and variance. They proposed the ML method for parameter estimation and the Akaike Information Criterion for model selection.

Cunderlik and Burn (2003) proposed a second order non-stationary approach, assuming non-stationarity in the first two moments of the time series. Their model separates the regional pooled quantile function into local time-dependent component (comprising the location and scale distribution parameters) and a regional time-invariant component. Parameters of the local component are estimated from time series decomposed into a trend component and a residual time dependent random variable that represents irregular fluctuations around the trend. Non-stationarity in the location parameter is then estimated from the trend component and in the scale parameter from the variability in the residual component. They applied the non-stationary model to a group of catchments from a homogeneous region in British Columbia with regionally significant decreasing trend in the location parameter of the annual maximum flood series. Their results showed that "... *ignoring even weakly significant non-stationarity in the data series may seriously bias the quantile estimation for horizons as near as 0–20 years in the future*".

6.8   CONCLUSIONS

For high return period quantiles, the use of an improper statistical model and only the systematic flow data recorded at the closest flow gauge station will produce estimates with a high degree of uncertainty.

To increase the available information, it is recommended the use of additional Non-Systematic data from the past. If the Systematic and Non-Systematic data have similar accuracy and the proper model is used, the use of additional Non-Systematic information will reduce the quantile estimate uncertainty. In fact, all published studies dealing with the use of Non-Systematic information agree that *"historical and paleoflood information can be of tremendous value in flood frequency analyses"*, in the words of Stedinger and Cohn (1986).

Another advantage of "going to the past" is to learn about the mechanisms of catastrophic flooding in the study area. However, Non-Systematic information can violate the traditional stationary assumption in Flood Frequency Analysis. Few statistical models have been developed to introduce the non-stationarity of the long term flooding process. A very important additional output of these types of models will be their use for the future flood hazard estimation in a climate change framework.

Finally, the high number of distribution functions used in Hydrology is a demonstration more efforts must be done in the theoretical basis of the statistical model selection.

REFERENCES

Adamowski, K., and Feluch, W. 1990. Non-parametric Flood Frequency Analysis with Historical Information. *ASCE Jour. of Hydr. Eng.*, 116(e), 1035–1047.

Baldwin, C.K., and Lall, U. 1999. Seasonality of Streamflow: the Upper Mississippi River. *Water Res. Res.*, 35(4), 1143–1151.

Benson, M.A. 1950. Use of Historical Data in Flood Frequency Analysis. *Eos. Trans. AGU*, 31(3), 419–424.

Cohn, T.A., Lane, W.L., and Baier, W.G. 1997. An algorithm for computing moments-based flood quantile estimates when historical flood information is available. *Water Res. Res.*, 33(9), 2089–2096.

Cohn, T.A., and Stedinger, J.R. 1987. Use of Historical Information in a Maximum Likelihood Framework. *Jour. of Hydrol.*, 96, 215–233.

Condie, R., and Lee, K.H. 1982. Flood Frequency Analysis with Historic Information. *Jour. of Hydrol.*, 58, 47–61.

Cunderlik, J.M., and Burn, D.H. 2003. Non-stationary pooled flood frequency analysis. *Jour. of Hydrol.*, 276, 210–223.

Cunnane, C. 1978. Unbiased Plotting Positions. A Review. *Jour. of Hydrol.*, 37, 205–222.

Cunnane, C. 1986. Review of statistical models for flood estimation, Proceedings of the *Int. Symp. on Flood Frequency and Risk Analysis*, Lousiana State University, Baton Rouge, in Hydrologic Frequency Modeling, edited by V. D. Singh, D. Reidel Pub. Co.

D'Agostino, R.B., and Stephens, M.A. (eds.) 1986. *Goodness-of-fit techniques.* Marcel Dekker Inc.

Elíasson, J. 1994. Statistical Estimation of PMP Values. *Nordic Hydrology*, 25(4), 301–312.

Elíasson, J. 1997. A Statistical Model For Extreme Precipitation. *Water Res. Res.*, 33(3), 449–455.

Francés, F. 1998. Using the TCEV distribution function with systematic and non-systematic data in a regional flood frequency analysis. *Stochastic Hydrology and Hydraulics*, 12(4), 267–283.

Francés, F., Salas, J.D., and Boes, D.C. 1994. Flood frequency analysis with systematic and historical or paleoflood data based on the two-parameter general extreme value models. *Water Res. Res.*, 30(6), 1653–1664.

Guo, S.L., and Cunnane, C. 1991. Evaluation of the usefulness of historical and paleological floods in quantile estimation. *Jour. of Hydrol.*, 129, 245–262.

Hirsh, R.M. 1987. Probability Plotting Position Formulas for Flood Records with Historical Information. *Jour. of Hydrol.*, 96, 185–199.

Hirsh, R.M., Helsel, D.R., Cohn, T.A., and Gilroy, E.J. 1993. Statistical Analysis of Hydrologic Data. In D.R. Maidment (ed.), *Handbook of Hydrology*, McGraw-Hill.

Hirsh, R.M., and Stedinger, J.R. 1987. Plotting Positions for Historical Floods and their Precision. *Water Res. Res.*, 23(4), 715–727.

Hosking, J.R.M., and Wallis, J.R. 1986a. Paleoflood Hydrology and Flood Frequency Analysis. *Water Res. Res.*, 22(4), 543–550.

Hosking, J.R.M., and Wallis, J.R. 1986b. The Value of Historical Data in Flood Frequency Analysis. *Water Res. Res.*, 22(11), 1606–1612.

Kanda, J. 1981. A New Extreme Value Distribution With Lower and Upper Limits For Earthquake Motions and Wind Speeds. *Theoretical and Applied Mechanics*, 31, 351–354.

Kite, G.W. 1988. *Frequency and Risk Analysis in Hydrology.* Water Resources Publications.

Kroll, C.N., and Stedinger, J.R. 1996. Estimation of moments and quantiles using censored data. *Water Res. Res.*, 32(4), 1005–1012.

Lang, M., Ouarda, T.B.M.J., and Bobée, B. 1999. Towards operational guidelines for over-threshold modeling. *Jour. of Hydrol.*, 225, 103–117.

Leese, M.N. 1973. Use of censored data in the estimation of Gumbel distribution parameters for annual maximum flood series, *Water Res. Res.*, 9, 1534–1542.

Matalas, N.C., Slack J.R., and Wallis, J.R. 1975. Regional Skew in Search of a Parent. *Water Res. Res.*, 11, 815–826.

Naulet, R. 2002, *Utilisation de l'information des crues historiques pour une meilleure prédétermination du risque d'inondation. Application au basin de l'Ardèche à Vallon-Pon-d'Arc et St-Martin d'Ardèche*. PhD Dissertation, Université Joseph Fourier (France)/ INRS-ETE (Canada), 320 pp.

NERC 1975. *Flood Studies Report.* Natural Environment Research Council.

Phien, H.N., and Fang, T.E. 1989. Maximum likelihood estimation of the parameters and quantiles of the general extreme-value distribution from censored samples. *Jour. of Hydrol.*, 105, 139–155.

Pilon, P.J., and Adamowski, K. 1993. Asymptotic variance of flood quantile in log Pearson type III distribution with historical information. *Jour. of Hydrol.*, 143, 481–503.

Porporato, A., and Ridolfi, L. 1998. Influence of weak trends on exceedence probability. *Stochastic Hydrology and Hydraulics*, 12, 1–14.

Potter, W.D. 1958. Upper and Lower Frequency Curves for Peak Rates of Runoff. *EOS. Trans. AGU*, 39, 100–105.

Robson, A., and Reed, D. 1999. *Statistical Procedures for flood frequency estimation.* Volume 3 in Flood Estimation Handbook, Institute of Hydrology.

Rossi, F., Fiorentino, M., and Versace, P. 1984. Two-Component Extreme Value Distribution for Flood Frequency Analysis. *Water Res. Res.*, 20, 847–856.

Stedinger, J.R., and Cohn, T.A. 1986. Flood Frequency Analysis with Historical and Paleoflood Information. *Water Res. Res.*, 22(5), 785–793.

Stedinger, J.R., Vogel R.M., and Foufoula-Georgiou, E. 1993. Frequency Analysis of Extreme Events. In D.R. Maidment (ed.), *Handbook of Hydrology*, McGraw-Hill.

Stedinger, J.R. and Crainiceanu, C.M. 2001. Climate variability and flood risk assessment. Viewed at June 2003. Available in internet at http://people.cornell.edu/pages/cmc59/papers/floodriskpaper.doc

Strupczewski, W.G., Singh V.P., and Feluch, W. 2001. Non-stationary approach to at-site flood frequency modelling I. Maximum likelihood estimation. *Jour. of Hydrol.*, 248, 123–142.

Takara, K., and Loebis, J. 1996. Frequency Analysis Introducing Probable Maximum Hydrologic Events: Preliminary studies in Japan and Indonesia. *Proc. of International Symposium on Comparative Research on Hydrology and Water Resources in Southeast Asia and the Pacific.* Yogyakarta, Indonesia, 18–22 November. Indonesian National Committee for International Hydrology Programme, 67–76.

Takara, K., and Tosa, K. 1999. Storm and Flood Frequency Analysis Using PMP/PMF Estimates. *Proc. of International Symposium on Floods and Droughts.* China, 18–20 October, 7–17.

Tasker, G.D., and Thomas, W.O. (1978). Flood-Frequency Analyses with Prerecord Information. *Jour. of the Hydraul. Div. ASCE*, 104(2), 249–259.

U.S. Water Resources Council (1982). *Guidelines for Determining Flood Frequency.* Bulletin 17B, Hydrology Committee, Washington D.C.

Wang, Q.J. 1990. Unbiased Estimation of Probability Weighted Moments and Partial Probability Weighted Moments for Systematic and Historical Flood Information and Their Application to Estimating the GEV Distribution. *Jour. of Hydrol.*, 120, 115–124.

Wang, Q.J. 1991. Unbiased Plotting Positions for Historical Flood Information. *Jour. of Hydrol.*, 124, 197–206.

7

A Critical Review of Probability of Extreme Rainfall: Principles and Models

Demetris Koutsoyiannis

Department of Water Resources, Faculty of Civil Engineering, National Technical University of Athens, Heroon Polytechneiou, Zographou, Greece

ABSTRACT: Probabilistic modelling of extreme rainfall has a crucial role in flood risk estimation and consequently in the design and management of flood protection works. This is particularly the case for urban floods, where the plethora of flow control cites and the scarcity of flow measurements make the use of rainfall data indispensable. For half a century, the Gumbel distribution has been the prevailing model of extreme rainfall. Several arguments including theoretical reasons and empirical evidence are supposed to support the appropriateness of the Gumbel distribution, which corresponds to an exponential parent distribution tail. Recently, the applicability of this distribution has been criticized both on theoretical and empirical grounds. Thus, new theoretical arguments based on comparisons of actual and asymptotic extreme value distributions as well as on the principle of maximum entropy indicate that the Extreme Value Type 2 distribution should replace the Gumbel distribution. In addition, several empirical analyses using long rainfall records agree with the new theoretical findings. Furthermore, the empirical analyses show that the Gumbel distribution may significantly underestimate the largest extreme rainfall amounts (albeit its predictions for small return periods of 5–10 years are satisfactory), whereas this distribution would seem as an appropriate model if fewer years of measurements were available (i.e. parts of the long records were used).

7.1 INTRODUCTION

The design and management of flood protection works and measures requires reliable estimation of flood probability and risk. A solid empirical basis for this estimation can be offered by flow observation records with an appropriate length, sufficient to include a sample of representative floods. In practice, however, flow measurements are never enough to support flood modelling. Particularly, in urban floods the control points are numerous and the flow gauge sites scarce or non existing at all (for example in Athens, a city with a history extended over several

139

millennia, traversed by the Kephisos and Ilisos Rivers and other urban streams, no flow gauge with systematic measurements has ever operated). The obvious alternative is the use of hydrological models with rainfall input data and the substitution of rainfall for streamflow empirical information. Notably, even when flow records exist, yet rainfall probability has still a major role in hydrologic practice; for instance in major hydraulic structures, the design floods are generally estimated from appropriately synthesised design storms (e.g. U.S. Department of the Interior, Bureau of Reclamation, 1977, 1987; Sutcliffe, 1978).

However, from the birth time of science, which is typically located in the era of the Ionian philosophers (6th century BC), it is known that the empirical evidence alone never suffices to form a comprehensive and consistent picture of natural phenomena and behaviours. A theory, based on reasoning, is required to interpret empirical observations and draw such a picture. Such a theory has been sought for more than 26 centuries, since the formulation of the first logical explanations of hydrometeorological phenomena by Anaximander (c. 610–547 BC) and Anaximenes (585–525 BC) of Miletus, who studied the formation of clouds, rain and hail (Koutsoyiannis and Xanthopoulos, 1999; Koutsoyiannis et al., 2006). However, still the state of affairs regarding understanding and description of these phenomena and their behaviours may be not satisfactory.

Some of the questions in seeking a fundament for a theory are philosophical questions; for instance the concepts of infinite vs. finite and of determinism vs. indeterminism, including the notions of probability and entropy. It is necessary to briefly discuss these questions because they greatly influence our perception of hydrometeorological phenomena including rainfall and flood.

The history of infinite goes back to the 6th century BC, with Anaximander, who regarded infinite as the cosmological principle, and continues with Zeno of Elea (c. 490–430 BC) and his famous paradoxes, and later with Aristotle (384–328 BC) who introduced the notion of potential infinite, as opposed to the actual or complete infinite. The Aristotelian potential infinite "exists in no other way, but . . . potentially or by reduction" (Physics, 3.7, 206b16). It is generally claimed that the problem of mathematical infinite was tackled in the late 19th century. According to Bertrand Russell, Zeno's paradoxes "after two thousand years of continual refutation, . . . made the foundation of a mathematical renaissance" (Russell, 1903). Furthermore, "for over two thousand years the human intellect was baffled by the problem [of infinity] . . . The definite solution to the difficulties is due to Georg Cantor" (Russell, 1926; see also Crossley et al., 1990 and Priest, 2002).

In hydrometeorology, however, the concept of infinity is still not understood and this situation has led to fallacies of upper bounds in precipitation and flood, the well-known concepts of the probable maximum precipitation (PMP) and probable maximum flood (PMF) (World Meteorological Organization, 1986). These contradictory concepts are still in wide use, even though merely the Aristotelian notion of potential infinite would suffice to abandon them. To quote, for example, Dingman

(1994, p. 141) "conceptually, we can always imagine that a few more molecules of water could fall beyond any specified limit." This thinking is absolutely consistent with the Aristotelian potential infinite.

Criticisms of the PMP and PMF concepts must have started from the 1970s; among them, one of the neatest was offered by Benson (1973):

"The 'probable maximum' concept began as 'maximum possible' because it was considered that maximum limits exist for all the elements that act together to produce rainfall, and that these limits could be defined by a study of the natural processes. This was found to be impossible to accomplish – basically because nature is not constrained to limits . . . At this point, the concept should have been abandoned and admitted to be a failure. Instead, it was salvaged by the device of renaming it 'probable maximum' instead of 'maximum possible'. This was done, however, at a sacrifice of any meaning or logical consistency that may have existed originally . . . The only merit in the value arrived at is that it is a very large one. However, in some instances, maximum probable precipitation or flood values have been exceeded shortly after or before publication, whereas, in some instances, values have been considered by competent scientists to be absurdly high . . . The method is, therefore, subject to serious criticism on both technical and ethical grounds – technical because of a preponderance of subjective factors in the computation process, and because of a lack of specific or consistent meaning in the result; ethical because of the implication that the design value is virtually free from risk."

More recently, the particular hypotheses and methodologies elements of the different approaches for estimating PMP have been also criticized. The so-called statistical approach to PMP, based on the studies of Hershfield (1961a, 1965) has been revisited recently (Koutsoyiannis, 1999) and it was concluded that the data used by Hershfield do not suggest the existence of an upper limit. To formulate his method, Hershfield compiled a huge and worldwide rainfall data set (a total of 95 000 station-years of annual maximum rainfall belonging to 2645 stations, of which about 90% were in the USA), standardized each record and found the maximum over the 95 000 standardized values, which he asserted PMP. Clearly then, the PMP hypothesis is based on the incorrect interpretation that an observed maximum in precipitation is a physical upper limit; had the sample size been greater, the estimated PMP value would been greater, too.

The situation is perhaps even worse with the so-called moisture maximization approach of PMP estimation (World Meteorological Organization, 1986), which seemingly is more physically based than the statistical approach of Hershfield. This is the most representative and widely used approach to PMP, and is based on the "maximization" of the observed atmospheric moisture content (i.e. to a maximum observed value) and on the assumption that if the moisture content were maximum, then the rainfall depth would be greater than observed by a factor equal to the ratio of the maximum over the observed rainfall depth. Applying this "maximization" procedure for all observed storms, the PMP value is assumed to be the maximum over all maximized depths.

Clearly, then, the approach suffers twice by the incorrect interpretation that an observed maximum is a physical upper limit. This fallacy is used for first time to determine the maximum moisture content (formally, the maximum dew point, assuming that the observed maximum in a record of about 50 years is a physical limit; obviously, had the record length been 100 or 200 years the observed maximum dew point would most likely be higher). This logic is also used for a second time to determine the PMP as the maximum of observed maximized values. Papalexiou and Koutsoyiannis (2006) have demonstrated the arbitrariness of the approach and its enormous sensitivity to the observation records (e.g. a missing rainfall observation could result in 25% reduction of the PMP value). The arbitrary assumptions of the approach extend beyond the confusion of maximum observed quantities with physical limits. For example, the logic of moisture maximization at a particular location is unsupported given that a large storm at this location depends on the convergence of atmospheric moisture from much greater areas.

In conclusion, it is surprising that the contradictory PMP and PMF concepts are regarded by many as concepts more physically based than a probabilistic approach to extreme rainfall and flood. This is particularly the case because the PMP and PMF concepts are greatly based on probabilistic or statistical assumptions, which in addition are rather misrepresentations of physical phenomena and indicate confused interpretation of probability. In turn, as will be discussed in the next section, these very concepts may have also affected probabilistic approaches of hydrological processes, in an attempt to make them consistent with the unsupported assumption of an upper bound.

This situation harmonizes with a dominant logic in hydrometeorology that probability does not offer a physical insight and is not related to understanding of physical phenomena, but rather it is only an unavoidable modelling tool. In contrast, understanding and insights are regarded as pertinent to deterministic thinking and to mechanistic explanations of phenomena. This logic, however, ought to have been abandoned at the end of the 19th century, after the development of statistical thermophysics and later of the quantum physics which rely upon the concepts of probability and statistics and depart from mechanistic physics. More recently, the study of chaotic dynamical systems and the astonishing results that the evolution of even the simplest nonlinear systems is unpredictable after a short lead time, have demonstrated the ineffectiveness of deterministic thinking. In this respect, even a faithful follower of determinism is inevitably forced to accept probabilistic description of phenomena for practical problems. However, when using probabilistic descriptions the gain may be greater if these descriptions are not regarded as incomprehensible mathematical models but rather as insightful physical descriptions.

The notion of indeterminism is at least as old as Heraclitus (c. 535–475 BC) and the notion of probability is the extension (quantifying transformation) of the

Aristotelian idea of "potential" (Popper, 1982, p. 133). The mathematical formalism of probability is much older than the recent notion of chaotic systems albeit its concrete fundament was offered in the mid 20th century by Kolmogorov (1933). The notion of probability may imply indeterminism from the outset (all events are possible, usually with different probabilities, but eventually one occurs) and may differ from the deterministic thinking (only one event is possible but it may be difficult to predict which one).

The notion of probability in synergy with the notion of infinite can remove paradoxical impressions related to upper bounds of physical quantities such as rainfall: The probability that rainfall exceeds any positive number x decreases toward zero as x decreases, becomes inconceivably small for very high x and becomes precisely zero for $x = \infty$. So, there is no need to assume such controversial concepts as PMP. This was explained half a century ago by the famous statistician Feller (1950), using another example, the age of a person:

"The question then arises as to which numbers can actually represent the life span of a person. Is there a maximal age beyond which life is impossible, or is any age conceivable? We hesitate to admit that man can grow 1000 years old, and yet current actuarial practice admits no bounds to the possible duration of life. According to formulas on which modern mortality tables are based, the proportion of men surviving 1000 years is of the order of magnitude of one in $10^{10^{36}}$ a number with 10^{27} billions of zeros. This statement does not make sense from a biological or sociological point of view, but considered exclusively from a statistical standpoint it certainly does not contradict any experience. There are fewer than 10^{10} people born in a century. To test the contention statistically, more than $10^{10^{35}}$ centuries would be required, which is considerably more than $10^{10^{34/}}$ lifetimes of the earth. Obviously, such extremely small probabilities are compatible with our notion of impossibility. Their use may appear utterly absurd, but it does no harm and is convenient in simplifying many formulas. Moreover, if we were seriously to discard the possibility of living 1000 years, we should have to accept the existence of a maximum age, and the assumption that it should be possible to live x years and impossible to live x years and two seconds is as unappealing as the idea of unlimited life."

In hydrometeorology, the introduction and development of the concepts of probability and statistics have been closely related to the study of extreme rainfall and flood and were greatly determined by the design needs of flood protection works. Empirical ideas similar to the modern probability concepts had been formulated in hydrology about a century ago (for instance, the hydrological frequency curves known as "duration curves"; Hazen, 1914). At about the same time, great mathematicians were developing the theoretical foundation of probability of extreme values (von Bortkiewicz, 1922a, b; von Mises, 1923; Fréchet, 1927; Fisher and Tippet, 1928; Gnedenco, 1941). Around the 1950s the empirical and theoretical approaches converged to form the branch of hydrology now called hydrologic statistics, whose founders were Jenkinson (1955), Gumbel (1958) and later Chow (1964). However, as already stated above, based on the PMP example, the current state of knowledge is not satisfactory and several important questions still wait for answers. For instance, Klemeš (2000) argues that "The distribution models used now, though disguised in rigorous mathematical garb, are no more, and quite likely less, valid for estimating the probabilities of rare events than were the extensions 'by eye' of duration curves employed 50 years ago." Obviously, however, the probabilistic approach to extreme values of hydrological processes signifies a major progress in hydrological science and engineering as it quantifies risk and disputes arbitrary and rather irrational concepts and approaches.

The most important questions that have not received definite answers yet are related in one or another manner to the notion of infinite. These questions concern the asymptotic distribution of maxima, a distribution that assumes a number of events tending to infinity, and are focused on the distribution tails, i.e. the behaviour of the distribution function as the hydrological quantity of interest tends to infinity.

Thus, if one is exempted from the concept of an upper limit to a hydrological quantity and adopts a probabilistic approach, one will accept that the quantity may grow to infinity with decreasing probability of exceedence. In this case, as probability of exceedence tends to zero, there exists a lower limit to the rate of growth which is mathematically proven. This lower limit is represented by the Gumbel distribution, which has the "lightest" possible tail. So, abandoning the PMP concept and adopting the Gumbel distribution can be thought of as a step from a finite upper limit to infinity, but with the slowest possible growth rate towards infinity. Does nature follow the slowest path to infinity? This question is not a philosophical one but has strong engineering implications. If the answer is positive, the design values for flood protection structures or measures will be the smallest possible ones (among those obtained by the probabilistic approach), otherwise they will be higher. These questions are studied in this article with the help of some recent works.

7.2   BASIC CONCEPTS OF EXTREME VALUE DISTRIBUTIONS

It is recalled from probability theory that, given a number n of independent identically distributed random variables, the largest (in the sense of a specific realization) of them (more precisely, the largest order statistic), i.e.:

$$X := \max\{Y_1, Y_2, \ldots, Y_n\} \tag{7.1}$$

has probability distribution function

$$H_n(x) = [F(x)]^n \tag{7.2}$$

where $F(x) := P\{Y_i \leq x\}$ is the common probability distribution function of each of Y_i. Herein, $F(x)$ will be referred to as parent distribution. If n is not constant but rather can be regarded as a realisation of a random variable with Poisson distribution with mean v, then the distribution of X becomes (e.g. Todorovic and Zelenhasic, 1970; Rossi et al., 1984),

$$H'_v(x) = \exp\{-v[1 - F(x)]\} \tag{7.3}$$

Since $\ln[F(x)]^n = n\ln\{1 - [1 - F(x)]\} = n\{-[1 - F(x)] - [1 - F(x)]^2 - \cdots\} \approx -n[1 - F(x)]$, it turns out that for large n or large $F(x)$, $H_n(x) \approx H'_n(x)$. Numerical investigation shows that even for relatively small n, the difference between $H'_n(x)$ and $H'_v(x)$ is small (e.g. for $n = 10$, the relative error in estimating the exceedence probability $1 - H_n(x)$ from (7.3) rather than from (7.2) is about 3% at most).

In hydrological applications concerning the distribution of annual maximum rainfall or flood, it may be assumed that the number of values of Y_i (e.g. the number of storms or floods per year), whose maximum is the variable of interest X (e.g. the maximum rainfall intensity or flood discharge), is not constant. Besides, the Poisson model can be regarded as an acceptable approximation for such applications. Given also the small difference between (7.3) and (7.2), it can be concluded that (7.3) should be regarded as an appropriate model for the practical hydrological applications discussed in this article.

The exact distributions (7.2) or (7.3), whose evaluation requires the parent distribution to be known, have rarely been used in hydrological statistics. Instead, hydrological applications have made wide use of asymptotes or limiting extreme value distributions, which are obtained from the exact distributions when n tends to infinity. Gumbel (1958) developed a comprehensive theory of extreme value distributions. According to this, as n tends to infinity $H_n(x)$ converges to one of three possible asymptotes, depending on the mathematical form of $F(x)$. Obviously, the same limiting distributions may also result from $H'_v(x)$ as v tends to infinity. All three asymptotes can be described by a single mathematical expression introduced by Jenkinson (1955, 1969) and become known as the general extreme value (GEV) distribution. This expression is

$$H(x) = \exp\left\{-\left[1 + \kappa\left(\frac{x}{\lambda} - \psi\right)\right]^{-1/\kappa}\right\}, \quad \kappa x \geq \kappa\lambda(\psi - 1/\kappa) \tag{7.4}$$

where $\psi, \lambda > 0$ and κ are location, scale and shape parameters, respectively; ψ and κ are dimensionless whereas λ has same units as x. (Note that the sign convention of κ in (7.4) may differ in some hydrological texts.) Leadbetter (1974) showed that this holds not only for maxima of independent random variables but for dependent random variables, as well, provided that there is no long-range dependence of high-level exceedences.

When $\kappa = 0$, the type I distribution of maxima (EV1 or Gumbel distribution) is obtained. Using simple calculus it is found that in this case, (7.4) takes the form

$$H(x) = \exp[-\exp(-x/\lambda + \psi)] \tag{7.5}$$

which is unbounded from both from above and below $(-\infty < x < +\infty)$.

When $\kappa > 0$, $H(x)$ represents the extreme value distribution of maxima of type II (EV2). In this case the variable is bounded from below and unbounded from above $(\lambda\psi - \lambda/\kappa \leq x < +\infty)$. A special case is obtained when the left bound becomes zero $(\psi = 1/\kappa)$. This special two-parameter distribution is

$$H(x) = \exp\left\{-\left(\frac{\lambda}{\kappa x}\right)^{1/\kappa}\right\}, \quad x \geq 0 \tag{7.6}$$

In some texts, (7.6) is referred to as the EV2 distribution. Here, as in Gumbel (1958), the name EV2 distribution is used for the complete three-parameter form (equation (7.4)) with $\kappa > 0$. Distribution (7.6) is referred to as the Fréchet distribution.

When $\kappa < 0$, $H(x)$ represents the type III (EV3) distribution of maxima. This, however, is of no practical interest in hydrology as it refers to random variables bounded from above $(-\infty < x \leq \lambda\psi - \lambda/\kappa)$. As discussed in the introduction, many regard an upper bound in hydrological quantities as reasonable. Even Jenkinson (1955) regards the EV3 distribution as "the most frequently found in nature, since it is reasonable to expect the maximum values to have an upper bound". However, he leaves out rainfall from this conjecture saying "to a considerable extent rainfall amounts are 'uncontrolled' and high falls may be recorded". In fact, he proposes the EV2 distribution for rainfall (note that he uses a different convention, referring to EV2 as type I). In a recent study, Sisson et al. (2006), even though detecting EV2 behaviour of rainfall maxima, attempt to incorporate the idea of a PMP upper bound within an EV2 modelling framework (see also Francés, this volume).

The simplicity of the above mathematical expressions is remarkable. This extends to the inverse function $x(H) \equiv x_H$ that is used to estimate a distribution quantile for a given non-exceedence probability H. This is

$$x_H = (\lambda/\kappa)[\exp(\kappa z_H) - 1] + \lambda\psi \tag{7.7}$$

where z_H is the so-called Gumbel reduced variate, defined as

$$z_H := -\ln(-\ln H) \tag{7.8}$$

For the Gumbel distribution, (7.7) takes the special form

$$x_H = \lambda(z_H + \psi) \tag{7.9}$$

which implies a linear plot of x_H versus z_H (a plot known as the Gumbel probability plot). For the Fréchet distribution, (7.7) takes the form

$$x_H = \lambda \psi \exp(\kappa z_H) \tag{7.10}$$

which implies a linear plot of $\ln x_H$ versus z_H (a plot referred to as the Fréchet probability plot).

The close relationship between the distribution of maxima $H(x)$ and the tail of the parent distribution $F(x)$ allows for the determination of the latter if the former is known. The tail of $F(x)$ can be represented by the distribution of x conditional on being greater than a certain threshold ξ, i.e. $G_\xi(x) := F(x|x > \xi)$, for which:

$$1 - G_\xi(x) = \frac{1 - F(X)}{1 - F(\xi)}, \quad x \geq \xi \tag{7.11}$$

If one chooses ξ so that the exceedence probability $1 - F(\xi)$ equals $1/v$, the reciprocal of the mean number of events in a year (this is implied when the partial duration series is formed from a time series of measurements, by choosing a number of events equal to the number of years of record), and denote $G(x)$ the conditional distribution for this specific value, then:

$$1 - G(x) = v[1 - F(x)] \tag{7.12}$$

Combining equation (7.12) with equation (7.3) it is obtained that:

$$G(x) = 1 + \ln H'_v(x) \tag{7.13}$$

If $H'_v(x)$ is given by the limit distribution $H(x)$ in equation (7.4), then it is concluded that for $\kappa > 0$:

$$G(x) = 1 - \left[1 + \kappa \left(\frac{x}{\lambda} - \psi \right) \right]^{-1/\kappa}, \quad x \geq \lambda \psi \tag{7.14}$$

which is the generalized Pareto distribution. Similarly, for $\kappa = 0$:

$$G(x) = 1 - \exp(-x/\lambda + \psi), \quad x \geq \lambda \psi \tag{7.15}$$

which is the exponential distribution. For the special case $\psi = 1/\kappa$:

$$G(x) = 1 - \left(\frac{\lambda}{\kappa x} \right)^{1/\kappa}, \quad x \geq \lambda/\kappa \tag{7.16}$$

In this way, a one to one correspondence between the type of the extreme value distribution and the type of the tail of the parent distribution is established. The EV1 distribution ($\kappa = 0$, equation (7.5)) corresponds to an *exponential* parent distribution *tail* (equation (7.15)), else known as *short tail*, or *light tail*. The EV2 distribution ($\kappa > 0$, equation (7.4) including the special case (7.6)) corresponds to an *over-exponential* parent distribution *tail* (equation (7.14) including the special case (7.16)), else known as *hyper-exponential tail, Pareto tail, power-law tail, algebraic tail, long tail, heavy tail* and *fat tail*.

From the distribution functions $H(x)$ and $G(x)$, two return periods can be defined as follows:

$$T := \delta/[1 - G(x)], \quad T' := \delta/[1 - H(x)] \tag{7.17}$$

where δ is the mean interarrival time of an event that is represented by the variable X. In both cases X represents annual values, so $\delta = 1$ year; δ is most commonly omitted but here we kept it for dimensional consistency, given that the return period has units of time, typically expressed in years.

Equation (7.16) is precisely a power law relationship between the distribution quantile x and the return period T:

$$x = (\lambda/\kappa)(T/\delta)^\kappa \tag{7.18}$$

In the generalized Pareto case (equation (7.14)), the corresponding relationship is

$$x = (\lambda/\kappa)[(T/\delta)^\kappa - 1 + \kappa\psi] \tag{7.19}$$

whereas in the exponential case the corresponding relationship is

$$x = \lambda[\ln(T/\delta) + \psi] \tag{7.20}$$

7.3   THE DOMINANCE OF THE GUMBEL DISTRIBUTION

Due to their simplicity and generality, the limiting extreme value distributions have become very widespread in hydrology. In particular, EV1 has been by far the most popular model. In hydrological education is so prevailing that most textbooks contain the EV1 distribution only, omitting EV2. In hydrological engineering studies, especially those analysing rainfall maxima, the use of EV1 has become so common that its adoption is almost automatic, without any reasoning or comparison with other possible models. Sometimes, it is also suggested, or even required, by the guidelines or regulations of several organizations, institutes and country services. Historically, several reasons have been contributed to the prevailing of the Gumbel distribution:

- *Theoretical reasons.*   Most types of parent distributions functions that are used in hydrology, such as exponential, gamma, Weibull, normal, lognormal, and the EV1 itself (e.g. Kottegoda and Rosso, 1997) belong to the domain of attraction of the Gumbel distribution. In contrast, the domain of attraction of the EV2 distribution includes parent distributions such as Pareto, Cauchy, log-gamma (also called log-Pearson type 3), and the EV2, which traditionally are not in very common use in hydrology, particularly in rainfall modelling.

- *Simplicity.*   The mathematical handling of the two-parameter EV1 is simpler than that of the three-parameter EV2.
- *Accuracy of estimated parameters.*   Obviously, two parameters are more accurately estimated than three. For the former case, mean and standard deviation (or second L-moment) suffice, whereas in the latter case the skewness is also required and its estimation is extremely uncertain for typical small-size hydrological samples.
- *Practical reasons.*   Probability plots are the most common tools used by practitioners, engineers and hydrologists, to choose an appropriate distribution function. As explained earlier, EV1 offers a linear Gumbel probability plot of observed x_H versus observed z_H (which is estimated in terms of plotting positions, i.e. sample estimates of probability of non-exceedence). In contrast, a linear probability plot for the three-parameter EV2 is not possible to construct (unless the shape parameter κ is fixed). This may be regarded as a primary reason of choosing EV1 against the three-parameter EV2 in practice. For the two parameter EV2 (Fréchet) distribution, a linear plot ($\ln x_H$ versus z_H) is possible as discussed earlier. However, empirical evidence shows that, in most cases, plots of x_H versus z_H give more straight-line arrangements than plots of $\ln x_H$ versus z_H.

From a practical point of view, the choice of an EV1 over an EV2 distribution may be immaterial if small return periods T are considered. For instance, in typical storm sewer networks, designed on the basis on return periods of about 5–10 years, the difference of the two distributions is negligible; besides, in such return periods even interpolation from the empirical distribution would suffice. However, for large T (>50 years), for which extrapolation is required, EV1 results in probability of exceedence of a certain value significantly lower than EV2. That is, for large rainfall depths, EV1 yields the lowest possible probability of exceedence (the highest possible T) in comparison to those of EV2 for any value of κ. For $T > 1000$, the return period estimated by EV1 could be orders of magnitude higher than that of EV2 (see Figure 7.3 and its discussion in section 5).

This should be regarded as a strong disadvantage of EV1 from the engineering point of view. Normally, this would be a sufficient reason to avoid the use of EV1 in engineering studies. Obviously, this disadvantage of EV1 would be counterbalanced only by strong empirical evidence and theoretical reasoning. In practice, the small size of common hydrological records (e.g. a few tens of years) cannot provide sufficient empirical evidence for preferring EV1 over EV2. This will be discussed further in section 5. In addition, the theoretical reasons, exhibited above, are not strong enough to justify the adoption of the Gumbel distribution. This will be discussed in section 4.

7.4 THEORETICAL JUSTIFICATION OF THE DISTRIBUTION TYPE

As discussed above, the rainfall process at fine time scales (hourly, daily) has been modelled by distributions belonging to the domain of attraction of EV1 such as gamma or Weibull. However, the adoption of these distributions is rather empirical, not based on theoretical reasoning. Thus, the above theoretical justification of the EV1 distribution is inconsistent. In contrast, recently three arguments have been formulated that favour the EV2 over the EV1 distribution, which are summarized below.

7.4.1 Argument 1: Asymptotic versus actual distribution

What matters in hydrological applications, is the actual distribution of maxima, i.e. $H'_n(x)$ or $H_n(x)$ as given in (7.2) or (7.3), respectively. The asymptotic distribution $H(x)$ for $n \to \infty$ provides a useful indication of the behaviour in the tails but not necessarily a model for practical use. It has been observed (Koutsoyiannis, 2004a) that the convergence of $H_n(x)$ to $H(x)$ may be enormously slow. This is demonstrated in Figure 7.1, which depicts Gumbel probability plots of the exact distribution functions of maxima $H_n(x)$ for $n = 10^3$ and 10^6 for a parent distribution function that is Weibull ($F(y) = 1 - \exp(-y^k)$) with shape parameter $k = 0.5$. The parent distribution belongs to the domain of attraction of the Gumbel limiting distribution, so the Gumbel probability plot tends to a straight line as $n \to \infty$. However, even for n as high as 10^6 the curvature of the distribution function is apparent. Obviously, in hydrological applications, such a high number of events within, say, a year, is not realistic (it can be expected that the number of storms or floods in a location will not exceed the order of 10–10^2). Thus, the limiting distribution for $n \to \infty$ may be not useful. The slow convergence in this case should be contrasted with fast convergence in other limiting situations; for example the distribution of the sum of a number of variables to the normal distribution, according to the central limit theorem, is very fast, so that about 10–30 events suffice to obtain an almost perfect approximation to the normal distribution.

Let us assume that the Weibull distribution (which belongs to the domain of attraction of EV1) with shape parameter smaller than 1 (e.g. $k = 0.5$ as in the example of Figure 7.1) can be a plausible parent distribution of storms and floods at a fine time scale, which is known to be positively skewed and with J-shaped density function. Accordingly, as observed in Figure 7.1, the probability plot of the exact distribution of maxima should be a convex curve, rather than a straight line, which indicates that, for a relatively small n, a three-parameter EV2 distribution may approximate sufficiently the exact distribution. Thus, even if the parent distribution belongs to the domain of attraction of the Gumbel distribution, an EV2 distribution can be a choice better than EV1.

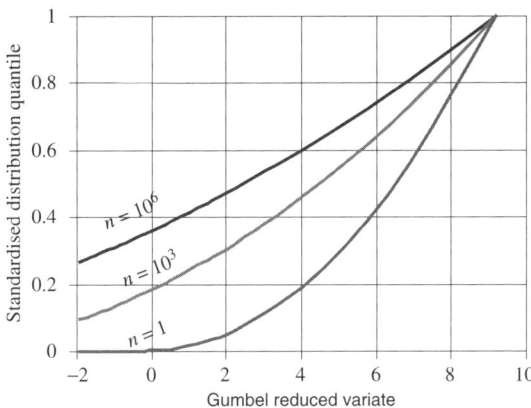

Figure 7.1. Gumbel probability plots of exact distribution function of maxima $H_n(x)$ for $n = 10^3$ and 10^6, also in comparison with the parent distribution function $F(y) \equiv H_1(y)$, which is Weibull with shape parameter $k = 0.5$. The distribution quantile has been standardised by $x_{0.9999}$ corresponding to $z_H = 9.21$ (from Koutsoyiannis, 2004a).

7.4.2   Argument 2: Change of domain of attraction due to parameter changes

In argument 1 it was assumed that the random variables Y_i whose maximum values are studied are independent and identically distributed ones. However, it is more plausible to assume that different Y_i have the same type of distribution function $F_i(y)$ but with different parameters. The statistical characteristics (e.g. averages, standard deviations, etc.) and, consequently, the parameters of distribution functions exhibit seasonal variation. In addition, evidence from long geophysical records shows that there exist random fluctuations of the statistical properties on multiple large time scales (e.g. tens of years, hundreds of years, etc.).

   In this respect, it has been shown theoretically that a gamma parent distribution, which belongs to the domain of attraction of EV1, switches to the EV2 domain of attraction if its scale parameter varies randomly following another gamma distribution function (Koutsoyiannis, 2004a). This point was also made by Katz et al. (2005) for an exponential parent distribution, which is a special case of the gamma distribution function. In addition, it was demonstrated using Monte Carlo simulations (Koutsoyiannis, 2004a) that a gamma parent distribution function with constant shape parameter and scale parameter shifting between two values, which are sampled at random with specified probabilities, results in an actual (for $n = 5$) extreme value distribution which is closely approximated by an EV2 distribution, whereas the EV1 distribution departs significantly from the simulated actual distribution.

7.4.3   Argument 3: Principle of maximum entropy

The principle of maximum entropy is a well established mathematical and physical principle, defined on grounds of probability theory, that can infer the detailed structure or behaviour of a system from rough (macroscopical) information of the system. For a stochastic system, the principle can determine the distribution function of the system states, from assumed macroscopical constraints (e.g. moments) of the system. The classical definition of entropy φ, known as the Boltzmann-Gibbs-Shannon entropy, is

$$\varphi := E[-\ln f(Y)] = -\int_{-\infty}^{\infty} f(y) \ln f(y) \, dy \qquad (7.21)$$

where $f(y) := dF(y)/dy$ denotes the probability density function of the parent variable and $E[\cdot]$ denotes expectation.

In a recent study, Koutsoyiannis (2005a) has shown that the principle of maximum entropy can predict and explain the distribution functions of hydrological variables using only two "macroscopic" statistical properties of observed time series (equality constraints), the mean μ and the standard deviation σ, as well as the inequality constraint that the variables under study are non-negative quantities. For variables with high variation ($\sigma/\mu > 1$) the classical entropy φ fails to apply with these constraints. In this case, a generalized definition of entropy, due to Tsallis (1988, 2004) should be used instead. This is

$$\varphi_q = \frac{1 - \int_0^{\infty} [f(x)]^q \, dx}{q - 1} \qquad (7.22)$$

and precisely reproduces φ when $q = 0$. Maximization of φ_q with the aforementioned constraints results in Pareto tail of the parent distribution with shape parameter $\kappa = (1 - q)/q$. Now, there is sufficient empirical evidence that at small time scales rainfall exhibits high variation ($\sigma/\mu > 1$). In this case, maximization of Tsallis entropy yields power-type (Pareto) distribution.

7.5   EMPIRICAL JUSTIFICATION OF THE DISTRIBUTION TYPE OF EXTREME RAINFALL

In seeking empirical evidence to justify the distribution type, one must be aware of bias in statistical estimations and error probability in statistical tests that emerge from typical hydrological samples. In fact, estimation bias and error probability are very large and this explains why the inappropriateness the EV1 distribution was not understood for so many years. Specifically, typical annual maximum rainfall series with record lengths 20–50 years completely hide the EV2 distribution and display EV1 behaviour. This was initially demonstrated by Koutsoyiannis and Baloutsos (2000) using an annual series of maximum daily rainfall in Athens,

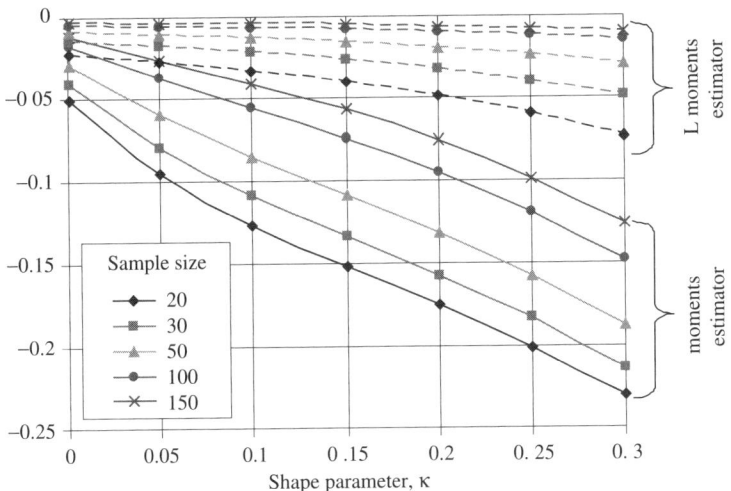

Figure 7.2. Bias in estimating the shape parameter κ of the GEV distribution using the methods of moments and L-moments (from Koutsoyiannis, 2004a).

Greece, extending through 1860–1995 (136 years). This series was found to follow EV2 distribution, but if smaller parts of the series were analysed, the EV1 distribution seemed to be an appropriate model.

A systematic Monte Carlo simulation study to address this problem has been done in Koutsoyiannis (2004a). Some of the results, concerning the estimation bias, are depicted in Figure 7.2. A negative bias, defined as estimated κ minus true κ, is apparent, for both the moments and L-moments estimators. It can be observed that for true $\kappa = 0.15$ (a value that is typical for extreme rainfall, as will be discussed later) and for a record length of 20 years the bias of the method of moments is -0.15, which means that the estimated κ will be zero! Even for a record length of 50 years the negative bias is high ($b = -0.12$), so that κ will be estimated at 0.03, a value that will not give good reason for preferring EV2 to EV1.

The situation is improved if L-moments estimators are used as the resulting bias is much lower. However the method of L-moments is relatively new (Hosking et al., 1985; Hosking, 1990) and its use has not been very common so far. In addition, even the method of L-moments is susceptible to type II error (no rejection of the null false hypothesis of an EV1 distribution against the true alternative hypothesis of EV2 distribution) with a high probability. As demonstrated in Koutsoyiannis (2004a) for $\kappa = 0.15$ and record length 20 years the frequency of not rejecting the EV1 distribution is 80%. Even for record length 50 years this frequency is high: 62%.

The results of this analysis show that (a) only long records (e.g. 100 years or more) could provide evidence of the distribution type of extreme rainfall, and (b) even with these records, the estimation of the shape parameter κ of the GEV distribution is highly uncertain, and an ensemble of many records should be used to obtain a reliable estimate.

In this respect, Koutsoyiannis (2004b) compiled an ensemble of annual maximum daily rainfall series from 169 stations of the Northern Hemisphere (28 from Europe and 141 from the USA) roughly belonging to six major climatic zones. All series had lengths from 100 to 154 years, the top three (in terms of length) being Florence, Genoa and Athens, with record lengths 154, 148 and 143 years respectively. The empirical distribution of one of the stations (Athens, Greece) is shown in Figure 7.3, on Gumbel probability plot, along with the theoretical EV2 and EV1 distributions fitted by several methods. The plot clearly shows that (a) the EV2 distribution fits the empirical one better than the EV1 distribution; for the highest observed daily rainfall (~150 mm), EV2 and EV1 assign return periods of ~200 and ~1000 years (differing by a factor of 5), respectively; for a rainfall depth of ~220 mm, EV2 and EV1 assign return periods of ~1000 and ~100 000 years (differing by two orders of magnitude), respectively. These observations demonstrate how important the correct choice of the theoretical model is and how much the EV1 distribution underestimates the return period of extreme rainfall.

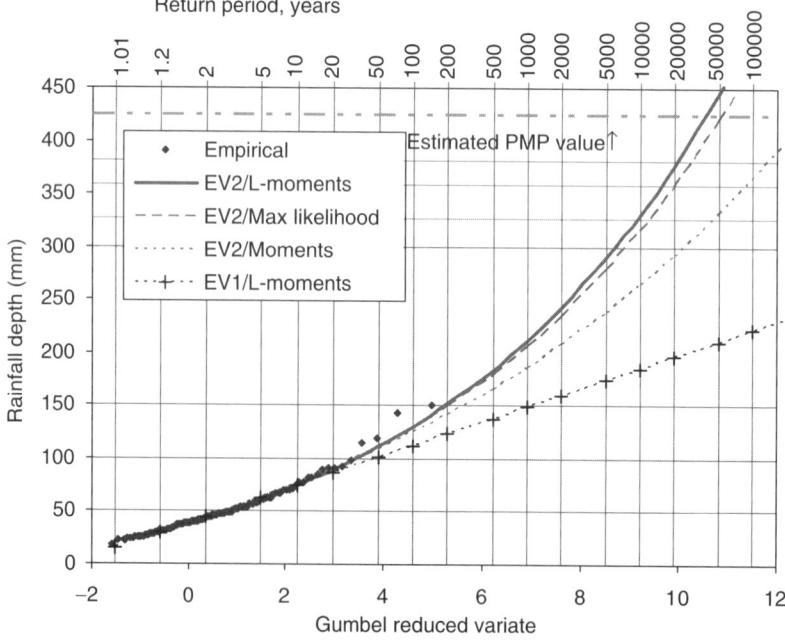

Figure 7.3. Empirical distribution and theoretical EV2 and EV1 distributions fitted by several methods for the annual maximum daily rainfall series of Athens, National Observatory, Greece (Gumbel probability plot; from Koutsoyiannis, 2004b). The PMP value (424.1 mm) was estimated by Koutsoyiannis and Baloutsos (2000).

In addition, a PMP value, estimated by Hershfield's method is also plotted in Figure 7.3. As discussed above, this value should not be regarded as an upper bound of rainfall but just as a value with high return period. It turns out from Figure 7.3 that the return period of this PMP values is around 50 000 years. It may be useful to mention that the aforementioned critical revisit (Koutsoyiannis, 1999) of Hershfield's data set, on which his method was based, revealed that Hershfield's PMP should be regarded as a rainfall value with return period of about 65 000 years.

These findings are representative of a general behaviour of all 169 rainfall records. In fact, in more than 90% of the records the estimated κ by the methods of maximum likelihood and L-moments were positive. The small percentage of non-positive κ in the remaining records is fully explained as a statistical sampling effect. This provides sufficient support for a general applicability of the EV2 distribution worldwide. Furthermore, the ensemble of all samples were analysed in combination and it was found that several dimensionless statistics, including the coefficient of variation of the annual maximum series, are virtually constant worldwide, except for an error that can be attributed to a pure statistical sampling effect. This enabled the formation of a compound series of annual maxima, after standardization by mean, for all 169 stations. The empirical distribution of the compound series is shown in Figure 7.4, on Gumbel probability plot, along with the theoretical

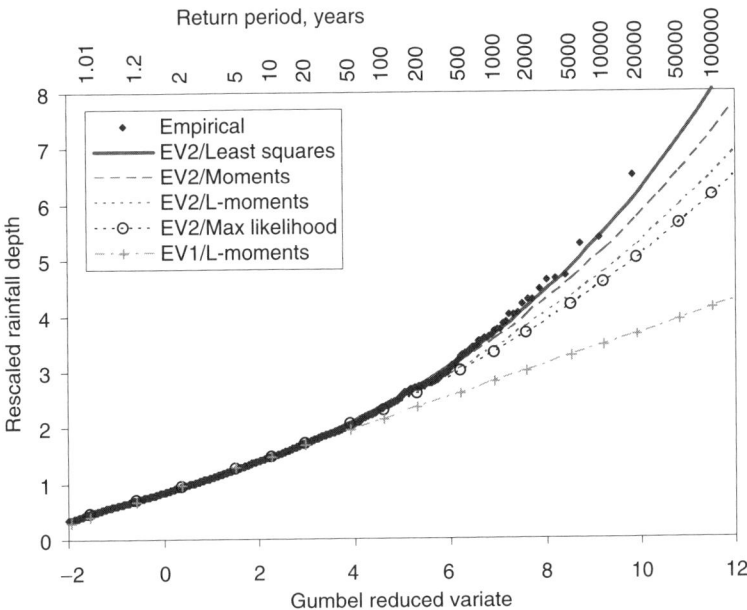

Figure 7.4. Empirical distribution and theoretical EV2 and EV1 distributions fitted by several methods for the unified record of all 169 annual maximum rescaled daily rainfall series (18 065 station-years; from Koutsoyiannis, 2004b).

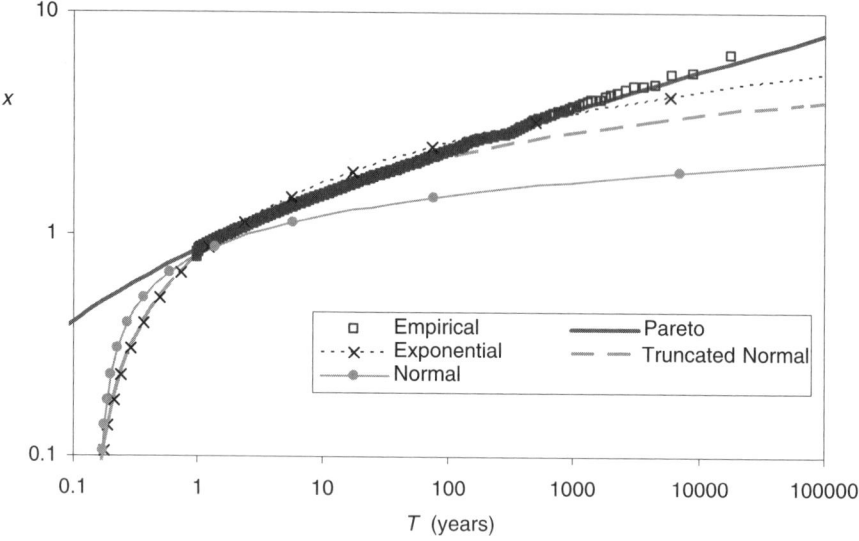

Figure 7.5. Plot of daily rainfall depth from the unified standardized sample above thresh-old, formed from data of 168 stations worldwide, vs return period, in comparison to Pareto, exponential, truncated normal and normal distributions (adapted from Koutsoyiannis, 2005a).

EV2 and EV1 distributions fitted by several methods. The plot clearly shows that the EV2 distribution fits the empirical one whereas the EV1 distribution is totally inappropriate. The compound series also supported the estimation of a unique κ for all stations, which was found to be 0.15.

The same data set was revisited in Koutsoyiannis (2005a) in a framework investi-gating the applicability of the maximum entropy principle in hydrology. In this case, instead of series of annual maxima, the series-above-threshold were constructed for 168 out of 169 records (in the Athens case only the annual maximum values were available, and thus the construction of a series-above-threshold was not pos-sible). All series were standardized by their mean and merged in one sample with length 17 922 station-years. The empirical distribution of this sample is depicted in Figure 7.5 (double logarithmic plot), where values lower than 0.79 are not shown, as this number is the lowest value of the merged series-above-threshold. In addi-tion, several theoretical distribution functions are also plotted. Among these, the Pareto distribution is obtained by the maximum entropy principle for coefficient of variation $\sigma/\mu = 1.19$. The agreement of the Pareto distribution with the empirical one is remarkable. The Pareto distribution is precisely consistent with the EV2 dis-tribution of the annual maximum, as justified in section 2. The shape parameter of the Pareto distribution, as obtained by the maximum entropy principle, is 0.15, the same value with the one obtained by fitting the EV2 distribution in the compound series of annual maximum rainfall.

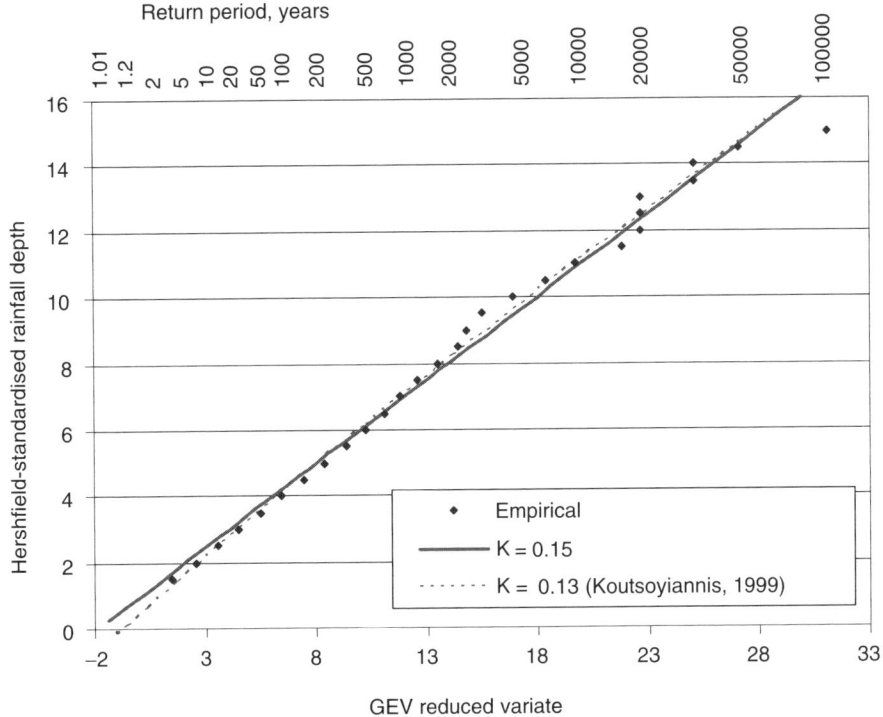

Figure 7.6. Empirical distribution of standardized rainfall depth $k = (X - \mu)/\sigma$ for Hershfield's (1961a) data set (95 000 station years from 2645 stations), as determined by Koutsoyiannis (1999), and fitted EV2 distributions with $\kappa = 0.13$ (Koutsoyiannis, 1999) and $\kappa = 0.15$ (Koutsoyiannis, 2004b) (EV2 probability plots with fixed $\kappa = 0.15$).

Additional empirical evidence with same conclusions is provided by the afore-mentioned Hershfield's (1961a) data set. Koutsoyiannis (1999) showed that this is consistent with the EV2 distribution with $\kappa = 0.13$. The plot of Figure 7.6 (EV2 probability plot with fixed $\kappa = 0.15$, which is further explained in section 7) indicates that the value $\kappa = 0.15$ can be acceptable for that data set too. This enhances the trust that an EV2 distribution with $\kappa = 0.15$ can be thought of as a generalized model appropriate for mid latitude areas of the north hemisphere.

Additional empirical evidence with same orientation was provided by Chaouche (2001) and Chaouche et al. (2002). Chaouche (2001) exploited a data base of 200 rainfall series of various time steps (month, day, hour, minute) from the five continents, each including more than 100 years of data. Using multifractal analyses he showed that (a) an EV2/Pareto type law describes the rainfall amounts for large return periods; (b) the exponent of this law is scale invariant over scales greater than an hour; and (c) this exponent is almost space invariant.

Other studies have also expressed scepticism for the appropriateness of the Gumbel distribution for the case of rainfall extremes and suggested hyper-exponential tail behaviour. Thus, Wilks (1993), who investigated empirically several distributions which are potentially suitable for describing extreme rainfall, using rainfall records of 13 stations in the USA with lengths ranging from 39 to 91 years, noted that EV1 often underestimates the largest extreme rainfall amounts and suggested an update and revision of the Technical Paper 40 (Hershfield, 1961b), a widely used climatological atlas of United States that was compiled fitting EV1 distributions to annual extreme rainfall data. Coles et al. (2003) and Coles and Pericchi (2003) concluded that inference based on the Gumbel model to annual maxima may result in unrealistically high return periods for certain observed events and suggested a number of modifications to standard methods, among which is the replacement of the Gumbel model with the GEV model. Mora et al. (2005) confirmed that rainfall in Marseille (a raingauge included in the study by Koutsoyiannis, 2004b) shows hyper-exponential tail behaviour. They also provided two regional studies in the Languedoc-Roussillon region (south of France) with 15 and 23 gauges, for which they found that a similar distribution with hyper-exponential tail could be fitted; this, when compared with previous estimations, leads to a significant increase in the depth of rare rainfall. On the same lines, Bacro and Chaouche (2006) showed that the distribution of extreme daily rainfall at Marseille is not in the Gumbel law domain. Sisson et al. (2006) highlighted the fact that standard Gumbel analyses routinely assign near-zero probability to subsequently observed disasters, and that for San Juan, Puerto Rico, standard 100-year predicted rainfall estimates may be routinely underestimated by a factor of two. Schaefer et al. (2006) using the methodology by Hosking and Wallis (1997) for regional precipitation-frequency analysis and spatial mapping for 24-hour and 2-hour durations for the Washington State, USA, found that the distribution of rainfall maxima in this State generally follows the EV2 distribution type.

7.6   THE DISTRIBUTION TAILS IN OTHER HYDROLOGICAL PROCESSES

The theoretical arguments presented in section 4 that support the EV2 over the EV1 distribution are not related merely to rainfall but rather to any process with high variability. Thus, it could be expected that other processes should also exhibit a similar behaviour.

This is the case for flood runoff. In fact, as demonstrated by Koutsoyiannis (2005b, c) and Gaume (2006), there are theoretical reasons by which we can conclude that the type of extreme value distribution in rainfall and runoff will be the same. If rainfall follows the EV1 distribution, then it can be shown that runoff will also follow the EV2 distribution. Conversely, if runoff follows the EV2 distribution, then rainfall should necessarily follow the EV2 distribution. Perhaps, the EV2 distribution in flood is easier to verify empirically (due to magnification of variability of extremes) and thus, the EV1 distribution has not been as standard in flood modelling as is in rainfall modelling. Thus, the log-gamma model, which belongs to the domain of attraction of EV2 has more frequently used in flood modelling. For instance, this model is the federally adopted approach to flood frequency in the USA (US Water Resources Council, 1982). But when flood frequency is estimated from rainfall, which is modelled using the EV1 model, then the flood frequency becomes necessarily consistent to the EV1, as explained above. Several more recent studies have also supported a three-parameter GEV over an EV1 distribution for floods (Farquharson et al., 1992; Madsen et al., 1997).

Similar results have been provided by fractal/multifractal analyses. Thus, Turcotte (1994) studied flood peaks over threshold in 1200 stations in the United States and concluded that they follow a fractal law, which essentially is described by equation (7.18). Pandey et al. (1998) established power-law distributions for daily mean streamflows in 19 river basins in the USA. Similarly, Malamud and Turcotte (2006) examined six river basins from different climatic regions and hydrologic conditions in the USA and concluded in power law distributions using either flood peaks over threshold or all daily mean streamflows, also considering in some cases paleoflood data.

Naturally, other hydrological processes driven by runoff are anticipated to follow long-tail distributions, too. However, it may be more difficult to verify empirically the type of distribution tail in such cases, because instrumental records are typically much shorter. Nevertheless, reconstructions of time series are possible in some other cases, for instance, in sediment yield time series from sediment deposits. Thus, Katz et al. (2005) were able to detect long tail behaviour in the annual sediment yield time series Nicolay Lake on Cornwall Island, Canada. In addition, Katz et al. (2005) provide an excellent review of the tail behaviours of several ecological variables.

7.7   PRACTICAL ISSUES FOR THE APPLICATION OF THE EV2 DISTRIBUTION

As discussed in section 3, the simplicity and the two-parameter form of the EV1 distribution are strong points that made it prevail in hydrology. However, if the shape parameter of the EV2 distribution is fixed (in extreme rainfall $\kappa = 0.15$, as discussed in section 5) the general handling of the distribution becomes as simple as that of the EV1 distribution. For example, the estimation of the remaining two parameters becomes similar to that of the EV1 distribution. That is, the scale parameter can be estimated by the method of moments from:

$$\lambda = c_1\sigma \tag{7.23}$$

where $c_1 = \kappa/\sqrt{\Gamma(1-2\kappa) - \Gamma^2(1-\kappa)}$ or $c_1 = 0.61$ for $\kappa = 0.15$, while in the EV1 case $c_1 = 0.78$. The relevant estimate for the method of L-moments is:

$$\lambda = c_2\lambda_2 \tag{7.24}$$

where λ_2 is the second L-moment and $c_2 = \kappa/[\Gamma(1-\kappa)(2^\kappa - 1)]$ or $c_2 = 1.23$ for $\kappa = 0.15$, while in the EV1 case $c_2 = 1.443$. The estimate of the location parameter for both the method of moments and L-moments is:

$$\psi = \mu/\lambda - c_3 \tag{7.25}$$

where $c_3 = [\Gamma(1-\kappa) - 1]/\kappa$ or $c_3 = 0.75$ for $\kappa = 0.15$, while in the EV1 case $c_3 = 0.577$.

If, in addition to λ and ψ, the shape parameter is to be estimated directly from the sample (which is not advisable but it may be useful for comparisons) the following approximate equations can be used (Koutsoyiannis, 2004b):

$$\kappa = \frac{1}{3} - \frac{1}{0.31 + 0.91C_s + \sqrt{(0.91C_s)^2 + 1.8}} \tag{7.26}$$

$$\kappa = 8c - 3c^2, \quad c := \frac{\ln 2}{\ln 3} - \frac{2}{3 + \tau_3} \tag{7.27}$$

where C_s and τ_3 are the regular and L skewness coefficients, respectively. The former corresponds to the method of moments and the resulting error is smaller than ±0.01 for $-1 < \kappa < 1/3$ $(-2 < C_s < \infty)$. The latter corresponds to the method of L-moments and the resulting error is smaller than ±0.008 for $-1 < \kappa < 1$ $(-1/3 < \tau_3 < 1)$.

The construction of linear probability plots is also easy if κ is fixed. It suffices to replace in the horizontal axis the Gumbel reduced variate $z_H = -\ln(-\ln H)$ (equation (7.8)) with the GEV reduced variate $z_H = [(-\ln H)^{-\kappa} - 1]/\kappa$. An example of such a plot is depicted in Figure 7.6.

7.8   RESULTING INTENSITY-DURATION-FREQUENCY CURVES

The construction of rainfall intensity-duration-frequency (IDF) relationships or curves is one of the most common practical tasks related to the probabilistic description of extreme rainfall. Unfortunately, however, the construction is typically performed by empirical procedures (e.g. Chow et al., 1988). Even the terms "duration" and "frequency" in IDF are misnomers; in fact, "duration" should read "timescale" (in order not to be confused with the duration of a rainfall event) and "frequency" should read "return period". Thus, the IDF relationships are mathematical expressions of the rainfall intensity $i(d, T)$ averaged over timescale d and exceeded on a return period T.

The recent theoretical advances in the probabilistic description can support a more theoretically based, mathematically consistent, and physically sound approach. A few assumptions are needed to support such an approach, namely:

1. The separability assumption, according to which the influences of return period and timescale are separable (Koutsoyiannis et al., 1998), i.e.,

$$i(d, T) = a(T)/b(d) \qquad (7.28)$$

where $a(T)$ and $b(d)$ are mathematical expressions to be determined.

2. The similarity assumption, according to which the distribution of average rainfall intensity conditional on being wet is statistically similar for all time scales (Koutsoyiannis, 2006).

3. A stochastic description of rainfall intermittency, which, as suggested by Koutsoyiannis (2006) should be a generalization of a Markov chain process that results applying the maximum entropy principle to the rainfall occurrence process.

4. A probabilistic distribution of the rainfall depth at any scale, which as discussed above should be of Pareto/EV2 type.

Based on assumptions 1-3, Koutsoyiannis (2006) showed that the function $b(d)$ can be approximated for relatively short timescales by the expression (here written in slightly different form)

$$b(d) = (1 + d/\theta)^{\eta} \qquad (7.29)$$

where $\theta >$ is a parameter with units same as the timescale d and η is a dimensionless parameter with values in the interval (0, 1). This resembles an expression historically established with empirical considerations. The approximate character of (46) as well as that of assumptions 1 and 2 should be underlined. At the same time, it should be noted that (46) is more accurate than a pure power law of $b(d)$, which has been suggested by modern fractal approaches. Particularly, (46) implies a decrease of rainfall intensity on small timescales, as compared to what is predicted by a power law. This is very important for the design of urban drainage networks that have small concentration times.

Furthermore, assumption 3 combined with (7.19) results in

$$i(d, T) = (\lambda/\kappa)[(T/\delta)^{\kappa} - 1 + \kappa\psi] \qquad (7.30)$$

By comparison of (7.28) with (7.26), we conclude that only the scale parameter λ should be a function of timescale d and particularly that $\lambda \sim (1 + d/\theta)^{-\eta}$. We easily then deduce that the final form of the IDF will be

$$i(d, T) = \lambda' \frac{(T/\delta)^\kappa - \psi'}{(1 + d/\theta)^\eta} \tag{7.31}$$

where $\psi' := 1 - \kappa\psi$ and $\lambda' := (\lambda/\kappa)(1 + d/\theta)^\eta$, which should be constant, independent of d, Notice that (7.29) is dimensionally consistent and that the return period T refers to the parent distribution (and thus it can take values smaller than $\delta = 1$ year, but necessarily greater than $\delta\psi^{1/\kappa}$). Also, notice that the numerator of (7.29) differs from a pure power law that has been commonly used in engineering practice. By virtue of (7.13) and (7.17), (7.29) can be easily converted in terms of the return period of the distribution of maxima and takes the form

$$i(d, T) = \lambda' \frac{[-\ln(1 - \delta/T')]^{-\kappa} - \psi'}{(1 + d/\theta)^\eta} \tag{7.32}$$

In the latter case, obviously T' should be greater than $\delta = 1$ year. All parameters are precisely the same in both (7.31) and (7.32). Consistent parameter estimation techniques for these relationships have been discussed in Koutsoyiannis et al. (1998).

7.9   CONCLUSIONS

Historically, the modelling of rainfall has suffered from several fallacies, such as the existence of an upper bound (PMP), and empirical practices that do not have theoretical support. Rational thinking and fundamental scientific principles, formulated since the birth of science in ancient Greece, can help combat such fallacies.

Probability, statistics and stochastic processes have offered a better alternative in perceiving and modelling of the rainfall process. However, even the probabilistic approaches have suffered from misconceptions and bad practices that have resulted in underestimation of rainfall variability and uncertainty. Among them is the wide application of the Gumbel or EV1 distribution, which has been the prevailing model for rainfall extremes despite the fact that it yields unsafe (the smallest possible) design rainfall values.

More recent studies have provided theoretical arguments and general empirical evidence from many rainfall records worldwide, which suggest a long distribution tail and favour the EV2 distribution of maxima. Simultaneously, they explain that the broad use of the EV1 distribution worldwide is in fact related to statistical biases and errors due to small sample sizes, rather than to the real behaviour of rainfall maxima, which should be better described by the EV2 distribution. Similar behaviours have been also detected in other hydrological processes such as streamflow and sediment transport.

The new methodological framework is more theoretically consistent, and more mathematically and physically sound (justified by the physico-mathematical principle of maximum entropy). Simultaneously, it is very simple so as to allow its easy implementation in typical engineering tasks such as estimation and prediction of design parameters, including the construction of IDF curves. The new framework imposes also some requirements for stochastic models of rainfall, many of which are currently not consistent with the long tail behaviour of the rainfall distribution.

REFERENCES

Bacro, J.-N. and A. Chaouche (2006), Incertitude d'estimation des pluies extrêmes du pourtour méditerranéen: illustration par les données de Marseille, *Hydrol. Sci. J.*, 51(3), 389–405.

Benson, M.A. (1973), Thoughts on the design of design floods, in *Floods and Droughts, Proc. 2nd Intern. Symp. in Hydrology*, pp. 27–33, Water Resources Publications, Fort Collins, Colorado.

Chaouche K. (2001), Approche multifractal de la modélisation stochastique en hydrologie. Thèse, Ecole Nationale du Génie Rural, des Eaux et des Forêts, Centre de Paris, France (http://www.engref.fr/thesechaouche.htm).

Chaouche, K., P. Hubert and G. Lang (2002), Graphical characterisation of probability distribution tails, *Stoch. Environ. Res. Risk Assess* 16(5), 342–357.

Chow, V.T. (1964), Statistical and probability analysis of hydrology data, Part I, Frequency analysis, In: Chow, V.T. (Ed.), *Handbook of Applied Hydrology*, McGraw-Hill, New York, pp. 8.1–8.42 (Section 8-I).

Chow, V.T., D.R. Maidment and L.W. Mays (1988), *Applied Hydrology*, McGraw-Hill.

Coles, S. and L. Pericchi (2003) Anticipating catastrophes through extreme value modelling. *Appl. Statist.*, 52, 405–416.

Coles, S., L.R. Pericchi and S. Sisson (2003), A fully probabilistic approach to extreme rainfall modeling, *J. Hydrol.*, 273(1–4), 35–50.

Crossley, J.N., C.J. Ash, C.J. Brickhill, J.C. Stillwell and N.H. Williams (1990), *What Is Mathematical Logic?* Dover, New York.

Dingman, S.L. (1994), *Physical Hydrology*, Prentice Hall, Englewood Cliffs, New Jersey.

Farquharson, F.A.K., J.R. Meigh and J.V. Sutcliffe (1992), Regional flood frequency analysis in arid and semi-arid areas, *J. Hydrol.*, 138, 487–501.

Feller, W. (1950), *An introduction to Probability Theory and its Applications*, Wiley, New York.

Fisher, R.A. and L.H.C. Tippet (1928), Limiting forms of the frequency distribution of the largest or smallest member of a sample, *Proc. Cambridge Phil. Soc.*, 24, 180–190.

Fréchet, M. (1927), Sur la loi de probabilité de l'écart maximum, *Ann. de la Soc. Polonaise de Math.*, Cracow, 6, 93–117.

Gaume, E. (2006), On the asymptotic behavior of flood peak distributions, *Hydrol. Earth Syst. Sci.*, 10, 233–243.

Gnedenco, B.V. (1941), Limit theorems for the maximal term of a variational series, *Doklady Akad. Nauk SSSR*, Moscow, 32, 37 (in Russian).

Gumbel, E.J. (1958), *Statistics of Extremes*, Columbia University Press, New York.

Hazen, A. (1914), Storage to be provided in impounding reservoirs for municipal water supply, *Trans. Am. Soc. Civil Engrs*, 77, 1539–1640.

Hershfield, D.M. (1961a), Estimating the probable maximum precipitation, *Proc. ASCE, J. Hydraul. Div.*, 87(HY5), 99–106.

Hershfield, D.M. (1961b), Rainfall Frequency Atlas of the United States, U.S. Weather Bur. *Tech. Pap. TP-40*, Washington, DC.

Hershfield, D.M. (1965), Method for estimating probable maximum precipitation, *J. American Waterworks Assoc.*, 57, 965–972.

Hosking, J.R.M. (1990), L-moments: analysis and estimation of distributions using linear combinations of order statistics, *J. Roy. Statist. Soc. Ser. B* 52, 105–124.

Hosking, J.R.M., J.R. Wallis and E.F. Wood (1985), Estimation of the generalized extreme value distribution by the method of probability weighted moments. *Technometrics* 27(3), 251–261.

Hosking, J.R.M. and J.R. Wallis (1997), *Regional Frequency Analysis – An Approach Based on L-Moments*, Cambridge Univ. Press, New York.

Jenkinson, A.F. (1955), The frequency distribution of the annual maximum (or minimum) value of meteorological elements, *Q. J. Royal Meteorol. Soc.*, 81, 158–171.

Jenkinson, A.F. (1969), Estimation of maximum floods, *World Meteorological Organization, Technical Note No 98*, ch. 5, 183–257.

Katz, R.W., G.S. Brush and M.B. Parlange (2005), Statistics of extremes: modeling ecological disturbances, *Ecology*, 86(5), 1124–1134.

Klemeš, V. (2000), Tall tales about tails of hydrological distributions, *J. Hydrol. Engineering*, 5(3), 227–231 and 232–239.

Kolmogorov, A.N. (1933), Grundbegriffe der Wahrscheinlichkeitrechnung, Springer, Berlin. Published in English in 1950 as Foundations of the Theory of Probability, Chelsea, New York.

Kottegoda, N.T. and R. Rosso (1997), *Statistics, Probability, and Reliability for Civil and Environmental Engineers*, McGraw-Hill, New York.

Koutsoyiannis, D. (1999), A probabilistic view of Hershfield's method for estimating probable maximum precipitation, *Water Resour. Res.*, 35(4), 1313–1322.

Koutsoyiannis, D. (2004a), Statistics of extremes and estimation of extreme rainfall, 1, Theoretical investigation, *Hydrol. Sci. J.*, 49(4), 575–590.

Koutsoyiannis, D. (2004b), Statistics of extremes and estimation of extreme rainfall, 2, Empirical investigation of long rainfall records, *Hydrol. Sci. J.*, 49(4), 591–610.

Koutsoyiannis, D. (2005a), Uncertainty, entropy, scaling and hydrological stochastics, 1, Marginal distributional properties of hydrological processes and state scaling, *Hydrol. Sci. J.*, 50(3), 381–404.

Koutsoyiannis, D. (2005b), Interactive comment on "On the asymptotic behavior of flood peak distributions – theoretical results" by E. Gaume, *Hydrol. Earth Syst. Sci. Discuss.*, 2, S792–S796 (www.copernicus.org/EGU/hess/hessd/2/S792/).

Koutsoyiannis, D. (2005c), Interactive comment on "On the asymptotic behavior of flood peak distributions – theoretical results" by E. Gaume, *Hydrol. Earth Syst. Sci. Discuss.*, 2, S838–S840 (www.copernicus.org/EGU/hess/hessd/2/S838/).

Koutsoyiannis, D. (2006), An entropic-stochastic representation of rainfall intermittency: The origin of clustering and persistence, *Water Resour. Res.*, 42(1), W01401.

Koutsoyiannis, D. and G. Baloutsos (2000), Analysis of a long record of annual maximum rainfall in Athens, Greece, and design rainfall inferences, *Natural Hazards*, 22(1), 31–51.

Koutsoyiannis, D., D. Kozonis and A. Manetas (1998), A mathematical framework for studying rainfall intensity-duration-frequency relationships, *J. Hydrol.*, 206(1–2), 118–135.

Koutsoyiannis, D., N. Mamassis and A. Tegos (2006), Logical and illogical exegeses of hydrometeorological phenomena in ancient Greece, *Proceedings of the 1st IWA International Symposium on Water and Wastewater Technologies in Ancient Civilizations*, 135–143, International Water Association, Iraklio, 2006.

Koutsoyiannis, D. and T. Xanthopoulos (1999), Τεχνική Υδρολογία (*Engineering Hydrology*), 3rd Ed., National Technical University of Athens, Athens (in Greek).

Leadbetter, M.R. (1974), On extreme values in stationary sequences, *Z. Wahrscheinlichkeitstheorie u. Verwandte Gebiete*, 28, 289–303.

Madsen, H., C.P. Pearson and D. Rosbjerg (1997), Comparison of annual maximum series and partial duration series methods for modeling extreme hydrologic events, 2, Regional modeling, *Water Resour. Res.*, 33(4), 759–769.

Malamud, B.D. and D.L. Turcotte (2006), The applicability of power-law frequency statistics to floods, *J. Hydrol.*, 322, 168–180.

Mora, R.D., C. Bouvier, L. Neppel and H. Niel (2005), Approche régionale pour l'estimation des distributions ponctuelles des pluies journalières dans le Languedoc-Roussillon (France), *Hydrol. Sci. J.*, 50(1), 17–29.

Papalexiou, S. and D. Koutsoyiannis (2006), A probabilistic approach to the concept of probable maximum precipitation, *Advances in Geosciences*, 7, 51–54.

Pandey, G., S. Lovejoy and D. Schertzer (1998), Multifractal analysis of daily river flows including extremes for basins of five to two million square kilometres, one day to 75 years, *J. Hydrol.*, 208, 62–81.

Popper, K. (1982), *Quantum Physics and the Schism in Physics*, Unwin Hyman, London.

Priest, G. (2002), *Beyond the Limits of Thought*, Oxford.

Rossi, F., M. Fiorentino and P. Versace (1984), Two-component extreme value distribution for flood frequency analysis, *Water Resour. Res.*, 20(7), 847–856.

Russell, B. (1903), *Principles of Mathematics*, Allen and Unwin.

Russell, B. (1926), *Our Knowledge of the External World*, revised edition, Allen and Unwin.

Schaefer, M.G., B.L. Barker, G.H. Taylor and J.R.Wallis (2006), Regional precipitation-frequency analysis and spatial mapping for 24-hour and 2-hour durations for Washington State, *Geophysical Research Abstracts*, 8, 10899.

Sisson, S.A., L.R. Pericchi and S.G. Coles (2006), A case for a reassessment of the risks of extreme hydrological hazards in the Caribbean, *Stoch. Environ. Res. Risk Assess.*, 20, 296–306.

Sutcliffe, J.V. (1978), *Methods of Flood Estimation, A Guide to Flood Studies Report*, Report 49, Institute of Hydrology, Wallingford, UK.

Todorovic, P. and E. Zelenhasic (1970), A stochastic model for flood analysis, *Water Resour. Res.*, 6(6), 1641–1648.

Tsallis, C. (1988), Possible generalization of Boltzmann-Gibbs statistics, *J. Statist. Phys.* 52, 479–487.

Tsallis, C. (2004), Nonextensive statistical mechanics: construction and physical interpretation, in *Nonextensive Entropy, Interdisciplinary Applications* (ed. by M. Gell-Mann and C. Tsallis), Oxford University Press, New York, NY.

Turcotte, D.L. (1994), Fractal theory and the estimation of extreme floods, *J. Res. Nat. Inst. Stand. Technol.*, 99(4), 377–389.

US Department of the Interior, Bureau of Reclamation (1977), *Design of Arch Dams*. US Government Printing Office, Denver, Colorado, USA.

US Department of the Interior, Bureau of Reclamation (1987), *Design of Small Dams*, third edn. US Government Printing Office, Denver, Colorado, USA.

US Water Resources Council (1982), *Guidelines for Determining Flood Flow Frequency*, Bull. 17B, Hydrol. Subcom., Office of Water Data Coordination, US Geological Survey, Reston, VA.

von Bortkiewicz, L. (1922a), Variationsbreite und mittlerer Fehler, *Sitzungsberichte d. Berliner Math. Ges.*, 21, 3.

von Bortkiewicz, L. (1922b), Die Variationsbreite bein Gauss'schen Fehlergesetz, *Nordisk Statistik Tidskrift*, 1(1), 11 & 1(2), 13.

von Mises, R. (1923), Über die Variationsbreite einer Beobachtungsreihe, *Sitzungsber, d. Berliner Math. Ges.*, 22, 3.

Wilks, D.S. (1993), Comparison of three-parameter probability distributions for representing annual extreme and partial duration precipitation series, *Water Resour. Res.* 29(10), 3543–3549.

World Meteorological Organization (1986), *Manual for Estimation of Probable Maximum Precipitation*, Operational Hydrology Report 1, 2nd edition, Publication 332, World Meteorological Organization, Geneva.

8

Role of Detention and Retention Basins in Stormwater Management and Environmental Protection

Miodrag Jovanović

University of Belgrade, Faculty of Civil Engineering, Serbia

ABSTRACT: The key role of detention and retention basins in stormwater management, sedimentation/pollution control, and environmental protection, is evaluated through best management practice design recommendations. An overview of the engineering approach to stormwater routing and analysis of basins' sediment trapping efficiency is given. Design topics such as: stormwater treatment efficiency, control structures, release impacts, vegetation and landscape, maintenance, legal and social implications of basin construction, are discussed. Finally, two illustrative case studies are presented – one pertaining to urban stormwater management by a system of detention basins, and the other, pertaining to a retention basin with the primary role to provide environmental, ambient stream corridor preservation.

8.1 INTRODUCTION

Detention and retention basins are most efficient means for stormwater management. Traditionally, such basins were designed as flood control reservoirs. However, presently they acquire an additional role, that of water quality treatment and environmental protection facilities. As this new role is becoming more important with rapid urbanization, which demands higher drainage capacities and cause unfavourable environmental impacts, the traditional design techniques for detention/retention basins need to be reassessed in order to meet both the old and the new objectives.

For the purpose of easier communication, a terminological distinction is made between "detention basins" and "retention basins". (In literature, term "basins" is often substituted by "ponds", implying small urban facilities.)

By definition, detention basins are facilities which temporarily store stormwater runoff and subsequently release water at a slower rate than it is collected by the drainage system. There is very little infiltration of the stored stormwater. Such facilities are designed with the goals of controlling peak discharge rates and providing settling of suspended sediments and pollutants. Detention basins are designed for flood events of various probability of occurrence, typically from 0.1% to 1.0% (10–100 year return periods). Between flood events no water storage is envisaged, thus they act as "dry" basins. A typical detention basin is schematically depicted in Fig. 8.1.

Enlarged detention basins, designed to extend detention beyond that required for stormwater storage and to provide better water quality effects, are called "extended detention basins". Such basins offer better possibilities for reduction of the total sediment load and of constituents associated with these sediments. The amount of this reduction depends on a number of factors such as:

- size of the basin;
- peak outflow rate;
- size distribution and mineral composition of sediments;
- amount of unsettleable active clay and colloidal content;
- type of associated pollutants and their concentration.

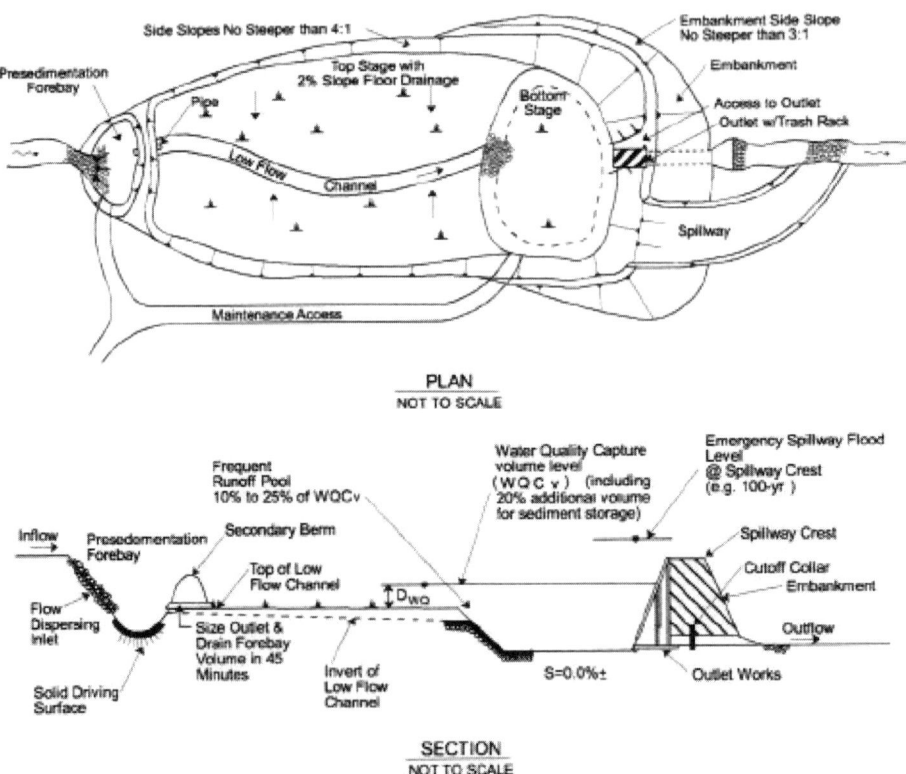

Figure 8.1. Schematic depiction of a typical detention basin (this page) and a typical retention basin (next page) (*Source*: Clar et al., 2004c).

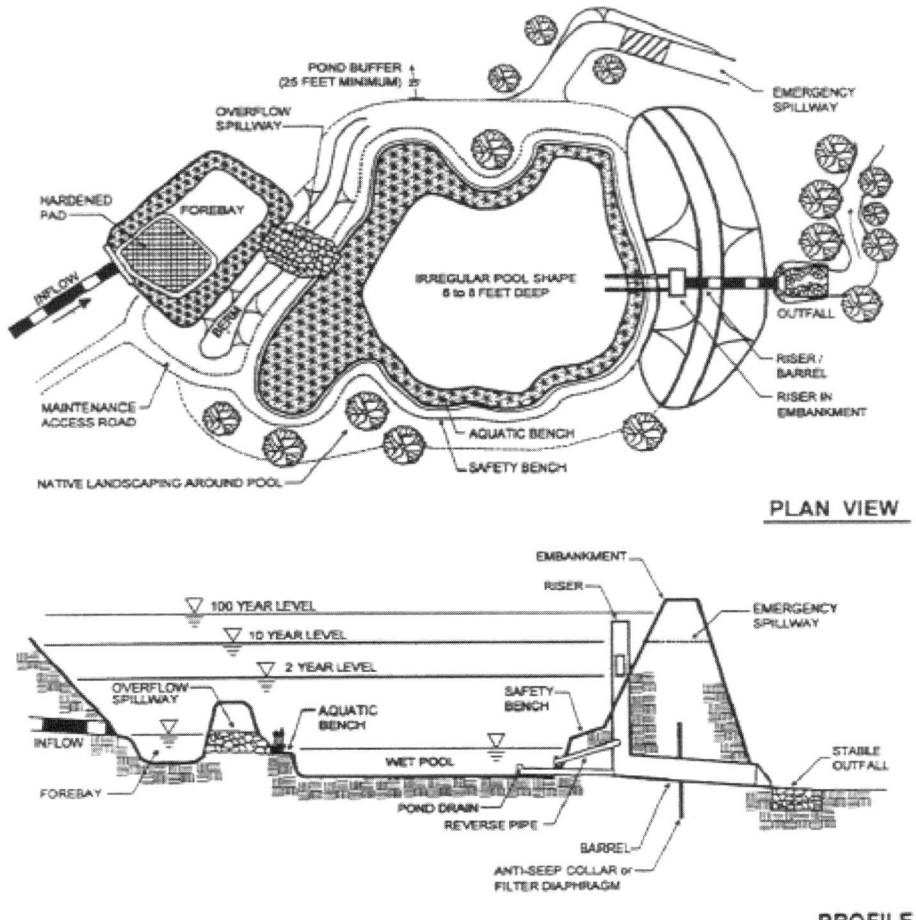

Figure 8.1. Continued.

By definition, retention basins are permanent "wet" basins, which act as small artificial lakes. They are designed to control flood flows, and to remove pollutants from stormwater. Retention basins in most cases have better removal rates of solids than temporary detention basins, due to the fact that retention basins are permanent storage facilities, allowing further water treatment in time intervals between storms. In this respect, a retention basin can be sized to remove nutrients and dissolved constituents, which is not the case with small detention basins. A typical retention basin is schematically depicted in Fig. 8.1.

The third type of basins – the "infiltration basins", designed to infiltrate and reduce a significant fraction of the runoff volume, and control groundwater recharge and base flow impacts, will not be considered in this article.

8.2   DESIGN CONSIDERATIONS

The flood management projects have been traditionally evaluated primarily by using economic benefit-cost criteria. However, there are cases when not-easily quantifiable requirements, such as preservation of biodiversity and ambient quality, are equally important. As problems of environmental preservation become more and more acute, the relative importance of such requirements increases. Thus, technical criteria for evaluation of flood management schemes and water quality performance need to be integrated with criteria of non-technical nature, and as a result, an optimal solution may not always be the least costly one.

There are a number of topics which need to be addressed when designing a detention or retention basin:
- physical factors;
- storage volume;
- stormwater treatment efficiency;
- dam embankments, inlet and outlet structures;
- impact control;
- vegetation and landscape;
- operation, inspection, and maintenance;
- legal and social implications.

8.2.1   Physical factors

There are a number of physical factors involved in detention/retention basin design, the principal two being topography and soil characteristics.

Topography. Selection of the type and size of basins may be strongly influenced by local topography. If the terrain is flat, the basins may be limited by depth to water table. This means that in flat regions, it is difficult to build a basin with an embankment as would be possible in steep regions, and therefore it may be necessary to construct the basin by excavation. In the case of mountainous terrain, detention or retention basin embankment heights are restricted. The type of the basin outlet is also controlled by topographical conditions. Weirs and open channel outlets are suitable for flat terrain, while drop structures are preferable for steep slopes.

Soil characteristics. Highly permeable soils are not acceptable for retention basins because of excessive drawdown during dry periods. Exfiltration rates can be reduced by different measures, such as compacting of the bed, incorporating clay to the soil, or providing a synthetic liner. Excavating the pool into the groundwater table can ensure its permanency, but prior to design, the groundwater regime needs to be thoroughly studied.

In karstic regions where the groundwater regime is of random nature, geotechnical field investigations and testing is prerequisite. In this case, detention and retention basins may require liners, and their ponding depth may be limited.

In general, geotechnical site investigations and tests are always recommended for design purposes. Measurements of bulk density and infiltration rates are to be carried out both in situ and laboratory. Hydrologic soil classifications and infiltration rates can be found in best management practices publications (ASCE, 1998; Clar et al., 2004c; Urbonas et al., 1993).

8.2.2   Storage volume

Detention basins. Dry detention basins have been in use since the late sixties of the twentieth century. As intensity of urbanization increases, building detention basins becomes almost inevitable in order to reduce the peak flow to match the pre-development peak flow. The design storms have typical return periods of 2, 10, 25, 50, and 100 years.

Experience shows that there are limits upon effectiveness of detention practices for providing downstream flood control. This is due to the fact that design of detention facilities is often confined to the limits of the property for which the facility is being designed without regard for potential downstream impacts. In such cases, releases from detention basins may result in situations where the flood is simply transferred or even intensified downstream, with an increased river channel erosion and flood damage.

After the basin is sized for peak flow reduction, it should be checked for water quality control, and if the desired removal percentage, according to effluent criteria, is not achieved, the designed volume needs to be modified in order to satisfy both flood reduction and water quality control requirements. The storage volume should detain the flow long enough to capture the pollutants, and provide space for deposited solids. To estimate this volume, long-term analysis of the sediment yield should be made. Detention basins are especially useful for sedimentation of clay sized particles, which account for approximately 80% of the suspended sediment mass found in stormwater (Clar et al., 2004c). Extended detention basins offer good possibilities for removing suspended constituents, while possibility of removing solubles is limited. Generally however, detention basins are inferior to retention basins when removal rates of solids are concerned.

Retention basins. Storage volume of a retention basin, as a permanent dual purpose stormwater management facility providing flood peak attenuation and water quality enhancement, is designed considering the drainage area, hydrologic conditions (runoff peak and volume), and criteria for treatment of nutrients and various pollutants. The storage volumes are variable, depending on site-specific conditions and the drainage area, which can be small-scale (typically 20–100 acres), or large, regional scale (several hundreds of acres). The minimum drainage area should sustain the retention pond during summers, which means that the drainage area should have a sufficient base flow to prevent excessive drawdown of the permanent water body during dry seasons.

The primary removal mechanism for pollutants in retention basins is by settling of the solid materials; therefore such basins should be sized to maximize sedimentation. This means that after a certain storage volume is defined for flood control purposes, a check should be made of the basin's ability to remove pollutants. The principal parameter in this respect is the hydraulic detention (residence) time:

$$T_d = \frac{\forall_b}{n_e \cdot \forall_r} \text{ [years]}, \tag{8.1}$$

where n_e is the average number of runoff events per year, and $\forall_b, \forall_r$ is the basin storage volume and the volume of runoff for an average storm, respectively.

Field investigations and analyses of settling indicate that retention basins sized for nutrient removal with a minimum detention time $T_d = 2$ weeks and a minimum ratio $\forall_b/\forall_r = 4$ achieve total suspended sediment removal rates of 80–90% (Clar et al., 2004c).

In addition to the storage volume, depth of the retention basin is an important design parameter because it affects solids settling. The basin should be shallow enough to ensure aerobic conditions and avoid thermal stratification, yet deep enough to prevent algal bloom or resuspension of deposited sediments by storm inflow and wind generated currents. The average depths of 2–4 m generally satisfy the adverse requirements. Such depths are higher than the depth of sunlight penetration and thus prevent emergent plant growth, reduce the risk of thermal stratification, and provide favourable conditions for survival of fish and other living organisms.

These general considerations must be followed by more detailed analysis concerning the designed detention/retention basin performance in respect to peak-flow attenuation, and sediment and pollutant removal.

8.2.2.1 Stormwater routing

Stormwater routing is an important part of basin design project in order to determine the necessary storage volume, as well as to check that the basin outflow structures provide a peak discharge low enough to meet a pollutant loading criteria.

For dry detentions basins, volume is assumed to be constant – the inflow volume must be equal to the outflow volume, provided that the infiltration can be neglected. Discharge routing in the basin must satisfy the continuity equation:

$$\frac{d\forall(t)}{dt} = Q_{in}(t) - Q_{out}(t), \tag{8.2}$$

where t is time, $\forall$ is the storage volume, and Q_{in}, Q_{out} are the inflow and the outflow discharge rates, respectively.

If the storage volume in Eq. 2 is expressed in terms of the basin water surface area (Ω) and the water surface elevation (Z), the continuity Eq. 2 takes the following form:

$$\Omega(Z)\frac{dZ}{dt} = Q_{in}(t) - Q_{out}(Z), \tag{8.3}$$

The initial problem, defined by the ordinary differential equation 3, requires an initial condition: $Z(0) = Z_o$, where Z_o is a prescribed water surface value at time $t = 0$, and a pre-defined function $\Omega(Z)$, obtained from survey data. Mathematical model based on Eq. 3 assumes that inertial effects can be neglected, and that the water level in the basin is always horizontal.

The inflow discharge hydrograph $Q_{in}(t)$ is defined by hydrological analysis of the catchment runoff (Haan et al., 1994) and the flow in the stream entering the basin, for the design flood event.

The outflow discharge is calculated by the energy conservation (Bernoulli) equation expressed as a discharge relation. If the outflow control structure takes the form of a rectangular broad-crested weir, the following expression holds for free and submerged weir flows:

$$Q_{out}(Z) = \begin{cases} C_{Q,1}b\sqrt{2g}(Z - Z_w)^{3/2} \\ C_{Q,2}b\sqrt{2g}(Z - Z_w)(Z - Z_{tw})^{1/2} \end{cases} \tag{8.4}$$

where b and Z_w are the weir effective breadth and crest elevation, respectively, Z_{tw} is the tail water elevation, $C_{Q,1}$ and $C_{Q,2}$ are discharge coefficients for the free and submerged weir flow, respectively (which take into account the effects of the approach velocity, friction, and the shape of the weir upstream contour), and g is the gravitational acceleration.

The submergence criterion is:

$$
\begin{aligned}
Z_{tw} - Z_w &\le \frac{2}{3}(Z - Z_w) \quad \text{--free flow} \\
Z_{tw} - Z_w &> \frac{2}{3}(Z - Z_w) \quad \text{--submerged flow}
\end{aligned} \tag{8.5}
$$

In the critical case when: $Z_{tw} - Z_W = 2/3(Z - Z_W)$, $C_{Q,1} = 0.385C_{Q,2}$.

An essentially similar expression to Eq. 4, can be used in the case when the outflow control structure takes the form of a spillway, but an appropriate discharge coefficient must be used.

When water is released from the basin through a submerged outflow pipe, discharge is calculated as:

$$Q_{out}(Z) = \begin{cases} C_{Q,3}A_o\sqrt{2g}(Z - Z_{tw})^{1/2} & \text{--for } (Z - Z_{tw}) > 0 \\ 0 & \text{--for } (Z - Z_{tw}) \le 0 \end{cases} \tag{8.6}$$

where A_o is the cross-sectional area of the outflow pipe, and $C_{Q,3}$ is the stage dependent discharge coefficient.

With defined discharges on the right hand side, Eq. 3 can be solved for water surface elevation in the basin (Z). Due to its non-linear nature, this equation must be solved numerically by an appropriate method, such as the improved Euler method, or one of the higher order predictor-corrector or Runge-Kutta methods (Ferziger, 1981; Rubin et al., 2001). A schematic diagram in Fig. 8.2 displays typical results of calculation.

It may be remarked that in preliminary analyzes, a simplified routing calculations can be done, based on the assumption that the inflow and outflow hydrograph are triangular in shape. In this case, the peak outflow discharge can be easily calculated using algebraic relationships. Although such an approach may be justified by the level of analysis, shortage of time or resources, the present computer software availability is such (Newman et al., 1999) that simplified approaches of this kind are obsolete.

One important result of the stormwater routing is the "detention time". This term refers to an average residence time for the given inflow of water and suspended solids. For the steady flow when inflow equals the outflow, the detention time T_d is the time required for the rate of flow Q to displace the basin volume $\forall$: $T_d = \forall/Q$. For variable flow, storm episodes all have different detention times. Detention time is the key factor for water quality assessment (Tapp et al., 1982).

8.2.2.2   *Sediment routing*
Sediments originate in the catchment through erosion processes, and are transported in the river system in the direction of the coast. Amount and timing of sediment production in the catchment depend on many factors, such as: geology, topography,

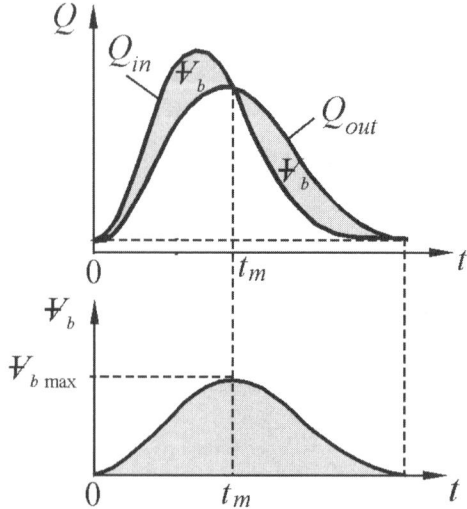

Figure 8.2. Results of stormwater routing – hydrographs and storage volume of detention/ retention basin.

land use, climate, hydrology, and human activities. Sediment management is by nature a long-term process, which has to be carried out at the catchment scale. Fig. 8.3 shows that many existing and planned regulations and policies will have impacts on the quantity and quality of sediments in a given catchment. These regulatory activities deal with discharge of pollutants and hence affect the quality of sediments. They will have to be translated into inputs for a suite of coupled and GIS-based, integrated models which will have to be used not only to predict changes in erosion, sediment transport, and deposition, but also to identify and quantify impacts of climate, land use, and socio-economic changes. All this is said to point out that design of detention/retention basins fits into a much broader context of sediment management.

Suspended sediment routing in a basin can be accomplished with different level of complexity. On one side, there are very sophisticated mathematical models based on fluid dynamics equations with turbulence closure (Chapra, 1997; Rubin et al., 2001), and on the other side, there are simple overflow rate methods convenient for engineering practice (ASCE, 1998; Clar et al., 2004c; Driscoll et al., 1986; Urbonas et al., 1993). Basic principles of the latter approach are briefly reviewed in sequel.

Sediment trapping performance of a basin can be evaluated considering the overflow rate calculated by using the peak discharge of the design storm, the basin

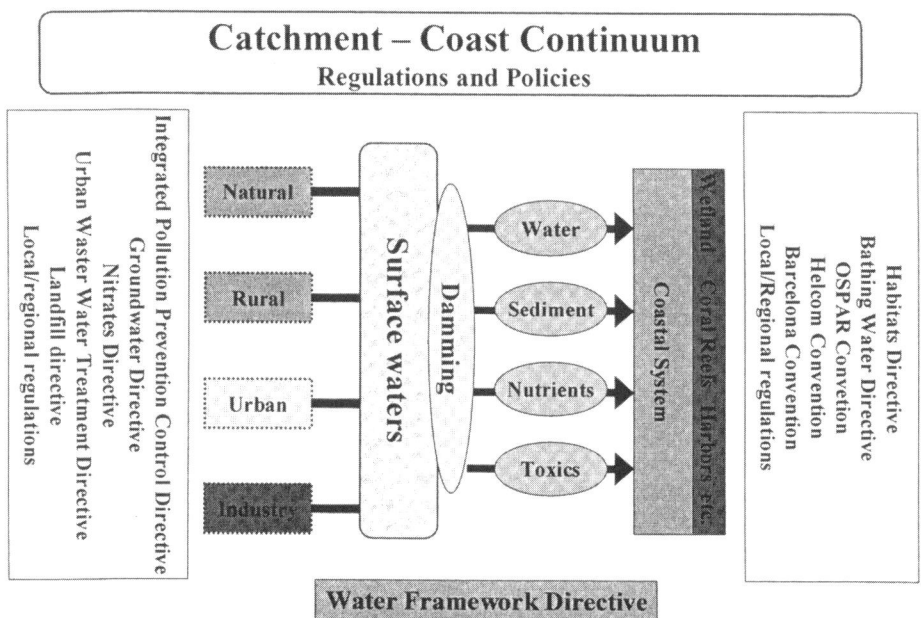

Figure 8.3. Upstream and downstream regulations and policies relevant for sediment quantity and quality and impact assessment downstream (*Source*: SedNet, 2004).

size, and the settling velocity of particles. Assuming an ideal settling basin of mean depth H and surface area Ω, the critical settling velocity of particles is: $V_c = H/T_d = Q/\Omega$, where: T_d is detention time (available time for particles to settle), and Q is the peak overflow discharge.

The settling velocity depends on the size and mineral characteristics of particles. A number of formulas or nomograms are available for determining settling velocity of individual spherical particle under ideal conditions (see for instance Chang, 1988; Rubin et al., 2001). According to the Stokes law, the quiescent settling velocity (W) of small particles is proportional to the square of their diameter (d):

$$W = \frac{g(\rho_s - \rho)d^2}{18\mu}, \tag{8.7}$$

where ρ_s and ρ are sediment and water densities, respectively, and μ is the temperature dependent dynamic viscosity of water.

Sediment trapping in the basin is analyzed for a several particle size classes, considering the "trapping efficiency factor" – the ratio of velocities: $TE_i = W_i/V_c$, where $i = 1, 2, \ldots, N$ is the particle class number. Particles for which $TE \geq 1$ will all settle (100%), while particles for which $TE < 1$ will settle with fraction W_i/V_c of all particles.

The total percentage of settled (removed) particles will be (Fig. 8.4):

$$P_t = \int_0^1 \left(\frac{W}{V_c}\right) dp \approx \sum_{i=1}^{i=N} \left(\frac{W_i}{V_c}\right) \Delta p_i = \sum_{i=1}^{i=N} TE_i \cdot \Delta p_i. \tag{8.8}$$

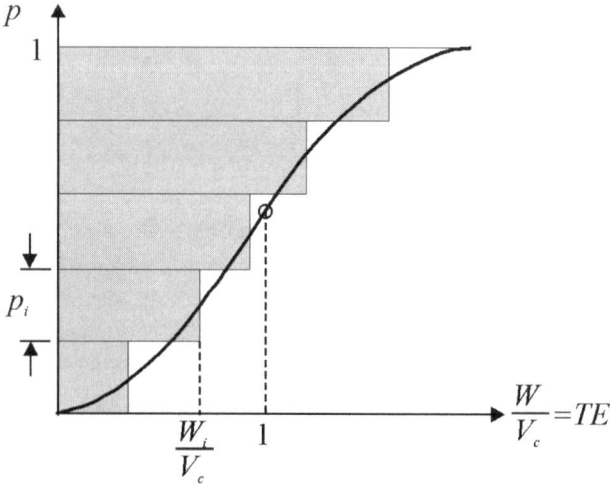

Figure 8.4. Trapping efficiency for different particle size classes.

(In summation it is clear that for $TE_i > 1$, one should take $TE_i = 1$.)

Size distribution of trapped sediments. For the total estimated sediment yield from the catchment (M_s) (for calculation method see for instance Clar et al., 2004b), the individual class masses of trapped and discharged sediments – $m_{t,i}$, and $m_{d,i}$, respectively, can be calculated by simple expressions:

$$m_{t,i} = \Delta p_i \cdot M_s \cdot TE_i \tag{8.9}$$

$$m_{d,i} = \Delta p_i \cdot M_s \cdot (1 - TE_i), \tag{8.10}$$

giving the total mass of the trapped sediment:

$$M_t = \sum_{i=1}^{i=N} m_{t,i} = M_s \sum_{i=1}^{i=N} \Delta p_i \cdot TE_i \tag{8.11}$$

$$M_d = \sum_{i=1}^{i=N} m_{d,i} = M_s \sum_{i=1}^{i=N} \Delta p_i (1 - TE_i). \tag{8.12}$$

Thus, fractions of one particular size class in the total mass of trapped and removed sediment are respectively:

$$f_{t,i} = \frac{m_{t,i}}{M_t} = \frac{\Delta p_i \cdot TE_i}{\sum_{i=1}^{i=N} \Delta p_i \cdot TE_i} \tag{8.13}$$

$$f_{d,i} = \frac{m_{d,i}}{M_t} = \frac{\Delta p_i (1 - TE_i)}{\sum_{i=1}^{i=N} \Delta p_i (1 - TE_i)} \tag{8.14}$$

Clay sized particles. Clay sized particles are smaller than 0.062 mm. It can be assumed that a certain fraction ($p_{c,i}$) of the mass of each particle class can be attributed to clay sized particles: $(\Delta p_i M_s) \cdot p_{c,i}$. Therefore, the total mass of trapped and discharged clay sized particles is respectively:

$$M_{t,c} = M_s \sum_{i=1}^{i=N} \Delta p_i \cdot p_{c,i} \cdot TE_i \tag{8.15}$$

$$M_{d,c} = M_s \sum_{i=1}^{i=N} \Delta p_i \cdot p_{c,i} (1 - TE_i). \tag{8.16}$$

Active clay. Assessment of the active clay fraction is very important for evaluation of impacts detention/retention basins exert on flow of pollutants and nutrients. As is known, pollutants and nutrients have a certain mass of settleable particulates, which are sorbed on the active clay (Rubin et al., 2001).

If a number of such pollutants or nutrients is denoted by $k = 1, 2, \ldots, N_s$, and if each of them has a mass of trapped and discharged settleable particulates $m_{ds,k}$ and $m_{ts,k}$ respectively, the total masses of trapped and discharged material will be:

$$M_{t,s} = \sum_{k=1}^{k=N_s} m_{ts,k} \qquad (8.17)$$

$$M_{d,s} = \sum_{k=1}^{k=N_s} m_{ds,k}, \qquad (8.18)$$

giving the total masses of active clay, trapped and discharged:

$$M_{t,ac} = M_{t,c} - M_{t,s} \qquad (8.19)$$

$$M_{d,ac} = M_{d,c} - M_{d,s}. \qquad (8.20)$$

For a total clay yield from the catchment: $M_c = \sum_{i=1}^{i=N} M_s \cdot \Delta p_i \cdot p_{c,i}$, the basin's trapping efficiency for the active clay (pollutants and nutrients) is (Clar et al., 2004c):

$$TE_{ac} = \frac{M_{t,ac}}{M_c - M_{s,in}}, \qquad (8.21)$$

where $M_{s,in} = \sum_{k=1}^{k=N_s} m_{s,in}$ is the total mass of N_s settleable particulates present in the inflow to the basin.

Dynamic trapping efficiency. The previously defined trapping efficiency TE can be used for predicting sediment removal rate for a single storm event. For permanent retention basins, one should include impact of settling between storms, taking into account the quiescent settling and the average time interval between storms. The quiescent removal rate for a single particle class can be estimated as:

$$Q_{r,i} = W_i \cdot \Omega_b \quad [\text{m}^3/\text{s, m}^3/\text{day}] \qquad (8.22)$$

where Ω_b is the surface area of the basin [m^2].

If the mean runoff volume and the mean time interval between storms are denoted $\forall_r$ and T_{st}, respectively, the removal ratio for a particular retention basin will be:

$$r_i = \frac{T_{st} \cdot Q_{r,i}}{\forall_r}. \qquad (8.23)$$

For dynamic settling in a retention basin with variable stormwater inflows (runoff volumes), the trapping efficiency TE_i needs to be statistically averaged over all storms. There are some models for prediction of long-term trapping efficiency, and the one most commonly used is the model proposed by EPA (Driscoll et al., 1986).

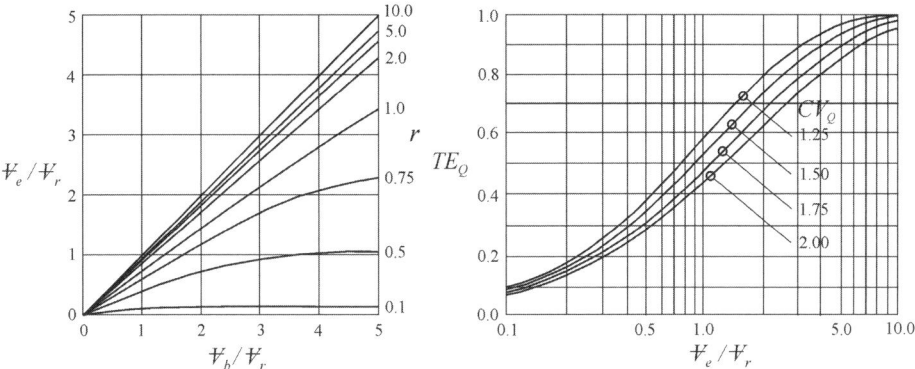

Figure 8.5. Diagrams for determining trapping efficiency of retention basins under quiescent conditions (after Driscoll et al., 1986).

According to this model, which is based on gamma distributed flows with a given mean flow and a coefficient of variation CV_Q, the dynamic trapping efficiency is:

$$DTE_i = \left[\frac{1/CV_Q^2}{1/CV_Q^2 - \ln(TE_i)} \right]^{1/CV_Q+1}, \qquad (8.24)$$

where TE_i is the trapping efficiency for particles in class "i", for the mean storm flow.

The coefficient of variation CV_Q in Eq. 8.24 is to be approximated by using the coefficient of variation of rainfall intensity.

The total, long-term removal efficiency of a retention basin is determined by combined effects of stormwater trapping and trapping during quiescent intervals between storms. An approximate method for evaluating the total long-term trapping efficiency is based on the equation (Driscoll et al., 1986):

$$TTE_i = 1 - (1 - DTE_i)(1 - TE_Q), \qquad (8.25)$$

where TE_Q – the trapping efficiency of retention basins under quiescent conditions (intervals between storms), is obtained by empirical diagrams in Fig. 8.5.

8.2.2.3   *Routing of chemical pollutants*
Chemical pollutants are brought into a detention or retention basin by the watershed runoff. Since most of dissolved chemicals are conservative, trapping is a result of settling of the settleable component of the chemicals – "particulate chemicals", and settling of active clay particles with sorbed chemicals.

The settleable fraction is assumed to be part of the clay sized fraction, and trapping of these particles is calculated the same way as for clay particles, as previously

Table 8.1. Pollutant removal percentages for basins designed for water quality control (*Source*: ASCE 2001; Clar et al., 2004c).

Type of basin	Total suspended solids	Nitrogen	Phosphorus	Lead	Zinc	BOD
Detention (dry) basins	50–80	0 (dissolved) 10–30 (total)	0 (dissolved) 10–50 (total)	35–80	35–70	20–40
Retention (wet) basins	70–85	50–70 (dissolved) 30–40 (total)	50–70 (dissolved) 50–65 (total)	25–85	25–85	20–40

shown. The only difference is that fraction of clay sized particles that are chemical particulates (settleable particles) is defined from pollution concentration for an average storm ("event mean concentration"). The mass of sorbed and dissolved pollutant is defined in terms of concentrations of the solid and liquid phases based on testing of local soils. Details can be found in literature (Clar et al., 2004b; Whipple et al., 1981).

8.2.3   Stormwater treatment efficiency

In addition to treatment of suspended solids, detention/retention basins have an important role in removal of nutrients. Retention basins are superior to detention basins for the treatment of nutrients (nitrogen and phosphorus) in urban stormwater. While detention basins rely on solids-settling processes, permanent retention basins remove dissolved nutrients through several physical, chemical and biological processes. A comparison of removal efficiencies, based on several sources (ASCE, 2001; Clar et al., 2004c), is given in Table 8.1.

As can be seen, permanent retention basins are most suitable for reducing nutrient loadings. In this respect, they can significantly alleviate eutrophication problems in downstream lakes, reservoirs, and estuaries. However, according to EPA design guides (Clar et al., 2004c), a permanent retention basin requires 2–7 times more volume than an extended detention basin, resulting in 50–150% higher costs. This cost increase is justifiable only when flood reduction is the primary objective, as is the case when benefits from flood control exceed costs of water quality treatment.

Retention basins need a reliable source of runoff or ground water to maintain the volume of the permanent pool. Thus retention basins require larger drainage areas, typically in the order of tens of hectares. From previous considerations it is obvious that retention ponds and basins are advantageous solutions in urban areas where nutrient loadings are expected to be high.

8.2.4   Dam embankment, inlet and outlet structures

Detention basins. The dam embankment forming the basin is to be designed in such a way to withstand overtopping in case of storms larger than the design storm. Alternatively, an emergency spillway should be constructed. Analysis of potential hydraulic consequences of sudden embankment failure is to be a part of design documentation. Side slopes of the dam embankment must remain stable under saturated soil conditions, and must be very mild (less than 1:4) for safety reasons, recreational use, and prevention of erosion.

It is recommendable to design a forebay with a volume of approximately 10% of the basin design volume in order to increase sediment deposition at the inflow, thus reducing cleaning costs of the basin itself. For the forebay to be relatively frequently cleaned, it must have a solid bottom.

The basin inlets can be of different types, such as rundown shutes with an energy dissipator, baffle shutes, manholes, inflow pipes with diffusing basins, etc. (ASCE, 1985; DeGroot, 1982). Such structures should prevent erosion of the basin bottom and banks at the inflow point, and prevent resuspension of the already deposited sediments.

The basin outlet should ensure a slow release of the design capture volume over the design emptying time. Such outlets may have multiple orifices, controlling the release of both large and small storms. Design of outlets needs to fulfill requirements concerning the release rate. If the outlet works are oversized, the emptying rate would be too high for small storms, and residence times too short for fine particles to settle, resulting in limited water quality benefit. The problem of ensuring the right release rates for a range of inflows is solved by multiple stage outlets, as will be latter shown by a case study.

Retention basins. As with detention basins, the inlet design should dissipate flow energy. It should also diffuse the inflow plume as it enters the forebay. Inlet designs include energy dissipators in the form of paved rundowns, manholes, lateral benches with wetland vegetation, and large rocks. To reduce the frequency of major cleanout of the basin, a sediment forebay with a solid bottom should be constructed near the inlet to trap coarse sediment particles. As in the case of detention basins, the storage capacity of the forebay may be about 10% of the basin's storage. The forebay is usually separated from the main basin by a lateral sill with wetland vegetation (ASCE, 1992).

A typical outlet facility consists of a riser with a trash rack and an antivortex device. The outlet capacity should be designed for the desired flood control performance. Means for preventing seepage are to be provided along the outlet conduits passing through or under the dam embankment. The riser is typically sized to drain the basin in two days so that sediments may be removed mechanically when necessary. A locking gate valve must be provided at the outlet of the drain pipe (Clar et al., 2004c).

An emergency spillway must be designed to protect the basin's dam embankment from overtopping. The return period of the design storm for the emergency spillway depends on the hazard classification which can vary from country to country.

8.2.5 Impact control

Dry detention basins and wet – retention basins, have different ability to control impacts. They can be efficient in control of physical impacts such as flooding and channel erosion. A detention or retention basin is designed in such a way to reduce a designed flood peak to the capacity of an existing sewer system, which has become undersized under growing urbanization. A hydraulic analysis must be carried out in order to check that super positioning of peaks will not result in aggravated downstream flooding conditions. This is especially important when several basins act in succession. If detention or retention basins are to be used for channel erosion control, their effectiveness is optimized by the choice of the design storm, and volume sizing.

In order to evaluate potential channel erosion due to high release flows from the basin, unsteady flow calculations are to be performed. The flow is usually treated as one-dimensional (1D) in space, assuming that the water surface in any cross-section is horizontal, and that the velocity is averaged over the cross-section. The mathematical model of 1D unsteady flow is based on the equations of the mass and momentum conservation (Cunge et al., 1980):

$$\frac{\partial A}{\partial t} + \frac{\partial Q}{\partial x} - q = 0 \tag{8.26}$$

$$\frac{\partial Q}{\partial t} + \frac{\partial}{\partial x}\left(\frac{Q^2}{A}\right) + gA\left(\frac{\partial Z}{\partial x} + S_f\right) = 0, \tag{8.27}$$

where t, x, Q, Z are time, distance along the channel, discharge, and water surface elevation, respectively. Variable A is the cross-sectional area, q is the lateral inflow per unit length, and S_f is the energy grade due to friction, which may be defined in terms of a roughness coefficient (n) and the hydraulic radius (R):

$$S_f = \frac{n^2 Q|Q|}{R^{4/3}A^2}. \tag{8.28}$$

The system of partial differential equations (8.26–8.27), with appropriate initial and boundary conditions, is numerically solved by either a finite difference or a finite element method. The Preissmann four point implicit finite difference numerical scheme is the standard approach for this class of 1D unsteady flows (Cunge et al., 1980).

By calculating the temporal and spatial distributions of discharge and flow area, distribution of the mean cross-sectional velocity $V(x,t) = Q(x,t)/A(x,t)$ can be evaluated, and compared with the critical velocity for the incipient motion of sediment particles in the river bed, or soil particles in the channel banks. Alternatively, channel's resistance to fluvial erosion may be assessed applying the critical shear stress criterion for incipient motion. The topic of stable channels is extensively treated in river engineering literature (see for instance Chang, 1988).

A general design principle is to ensure that the channel which receives water from the basin's outlet should be protected from erosive impact of high velocities. This can be accomplished by various measures, such as provision of riprap lining, stilling basins, check dams, and other structures which can reduce velocities to non erosive levels.

Other impacts include pollution impacts, thermal impact, and various biological impacts. As far as pollution impact is concerned, dry detention basins generally do provide chemical pollutant removal, but their removal efficiency performance is highly variable. Retention basins, as permanent water storage facilities, have much better pollutant removal efficiency. Extended detention basins and retention basins tend to increase thermal impacts to receiving streams, but in most cases it is not always clear to what extent a moderate increase of temperature in receiving streams presents a real threat to the environment.

In general for habitat and biological impacts, effectiveness of detention and retention basins is subject to the design storm, the holding period for the volumes captured, and the biological characteristics of the receiving water.

8.2.6   Vegetation and landscape

Detention basins. Vegetation in basins ensures erosion control and sediment entrapment. The choice of vegetation is locally dependent, as basin slopes are usually planted with native grasses. The bottom is subject to sediment deposition and frequent inundation, thus could be lined with marshy vegetation, riparian shrub, low weeds, or gravel.

Detention basins should fit into the landscape, and be an integral part of the environment, without impairing the ambient quality. Moreover, building a detention basin offers an opportunity to improve esthetics of the landscape and upgrade local residents' quality of life through recreation. It is clear however, that each basin is a case of its own, and that the designer needs to consider a number of local conditions and constraints, various aspects of safe operation, inspection and maintenance, and to identify community needs and concerns. Landscape architects should be consulted as well (IRMA et al., 2001).

Retention basins. Similarly to detention basins, the side slopes of retention basins should be very mild (1:4 or less) to facilitate maintenance. A littoral zone should be established around the perimeter of the basin to promote emergent vegetation along the shoreline. This vegetation reduces erosion, helps removal of dissolved nutrients, reduces the formation of floating algal mats, and provides habitat for aquatic and wetland wildlife. The strip of emergent wetland vegetation should be at least 3 m wide, with water depths 0.2–0.5 m. The total wetland area should be of an order of 25–50% of the basin's water surface area (Clar et al., 2004c). Aquatic and wetland vegetation is obtained from transplant nursery stocks. Agricultural specialist should be consulted for guidelines pertaining to the performance of different types of aquatic and wetland vegetation.

In order to establish and maintain wetland coverage, a landscaping plan is needed, including delineation of landscaping zones and planting configuration. A landscaping plan should be based on plant species commonly found in the region.

8.2.7   Operation, inspection and maintenance

Although detention and retention basins require practically no operational attention, their performance and functional reliability depend on good maintenance. Failure to maintain basins leads to the following unfavourable consequences:

- increased risk of flooding downstream areas; in an extreme case, loss of life and property due to failure of dam embankment;
- downstream channel instability with increased sediment deposition;
- degraded landscape and basin appearance.

These consequences can be avoided by regular inspection and maintenance. It must be stressed that retention basins and wetlands require systematic (often subtle) maintenance measures in order to stay healthy aquatic ecosystems.

Regular annual inspections and additional inspections following extreme storm events are prerequisite. Objectives of inspection is to check for accumulation of sediment and debris, state of the inlet and outlet structures, embankment (settlement, signs of slope erosion, etc.), level of vegetative growth, and possible damages of the system. Responsibility for inspection is shared between state and local regulatory agencies and the basin owner. Inspection reports which are produced on a regular basis ensure that the maintenance responsibility is fully respected by the involved parties.

Basin maintenance can be divided into functional and esthetic maintenance. The functional maintenance has two components – preventive and corrective maintenance (Clar et al., 2004c). Regular preventive maintenance is performed to ensure full functionality of the system. It includes grass maintenance, debris removal, maintenance and repair of mechanical equipment, mosquito elimination, etc. The corrective maintenance is performed periodically and upon emergency. It consists of measures intended to restore functionality or safe operation of the system, such as: sediment removal, dam embankment slope repairs, removal of vegetation (whose root system undermines the dam embankment), structural repairs, erosion repairs, removal of snow and ice accumulations, etc.

It may be noted that sediment removal consists of relatively frequent minor interventions for unblocking control inlet and outlet structures, and major basin life-cycle maintenance. In the latter case, time schedule depends on the annual total suspended sediment load and the basin area. A rough estimate of the average annual sediment accumulation in the basin (δ_s) can be made using a simple relation (ASCE and WEF, 1998):

$$\delta_s = P_e \cdot C \cdot \left(\frac{\Omega_c}{\Omega_b} \right) \cdot TE \ \text{[mm]}, \tag{8.29}$$

where P_e – effective precipitation (annual runoff-producing precipitation in mm), C – average annual volumetric concentration of suspended sediments in runoff, TE – trapping efficiency of the basin, and Ω_c, Ω_b – surface area of the catchment and the basin, respectively.

The sediment accumulation calculation can be used for estimation of sediment removal frequency and maintenance costs. Caution is necessary, because due to many uncertainties, chances are that the actual accumulation rate will be higher and that more frequent cleanout will be needed. This should be kept in mind since sediment removal projects are costly, requiring survey works, mechanized equipment, transport and disposal of the removed material, restoration of the original basin bottom and slopes, reestablishing vegetation, and in the case of retention basins, draining the basin before works. The removed sediments can only be temporarily stored on the site, until a permanent disposal site is available.

More frequent removal of sediments is necessary when hazardous waste is present, such as heavy metals. Sampling should reveal if buildup of heavy metals or other constituents – pollutants is present, in which case site rehabilitation and total cleanout may be necessary. Similar approach holds for sampling and removal of toxic wastes, with procedures defined by legal regulations imposed by each country.

8.2.8 Legal and social implications

Dry detentions basins can be constructed at most urban and suburban sites, and are designed to fit into the urban and suburban landscape. When dry, they can be used for recreational purposes, as playgrounds. Thus, in detention basin design, factors such as multiple uses, esthetics, and maintenance, should be equally considered. Permanent retention basins offer even higher esthetic quality. Sediment deposits and debris are under water and thus, out of sight. A relatively large water surface area can be used for more diversified recreational and tourist activities. However, potential heating of water during summer months can be a limiting factor.

A design project for retention basins may be coped with various legal and social restrictive implications (Affeltranger, 2001; ASCE, 2001; NSDW, 2004). Basin location may be in some regions prohibited by law, or exceptionally allowed, only upon application and provision of additional technical documentation. The legal requirements in this respect may vary from country to country. For instance, one potential constraint on the use of retention basins may be that filling of wetland areas may be restricted by regulations or by initiative of various environmentalist groups or non-governmental organizations. Unexpected problems may also emerge form conflicting interests of a number of involved parties, such as local authorities, urban planners, farmers, civil construction companies, etc.

8.3   CASE STUDIES

Two illustrative case studies are presented from author's native country – Serbia. The first presents a system of three detention basins in Belgrade designed for stormwater management and water quality control. The second pertains to a retention basin in a semi-urban region, designed to provide flood protection for several villages while fully providing stream corridor environmental preservation.

8.3.1   The Kumodraz brook in Belgrade

The $7.8 \, km^2$ catchment of the Kumodraz brook in Belgrade is of varying land use, ranging from rural areas in the upstream part to the densely populated downstream parts. The Kumodraz brook is introduced into the sewer system at the central point of the catchment (Fig. 8.6). This system presently consists of about 1000 pipes. It is a separate stormwater and wastewater system in the upper catchment region, and a combined sewer system in the downstream region.

The design objectives include provision of (Faculty of Civil Engineering – Belgrade, 1999; Despotović et al., 1999):

- efficient stormwater drainage without surcharging for return period of 2 years;
- flood protection of the urbanized part of the catchment for return period of 10 years;
- separation of stormwater and wastewater systems in areas where the combined system exists;
- new separate system for presently non-drained areas;
- mitigation of the first-flush effect in the stormwater drainage system;
- preservation of the Kumodraz brook in its natural state at the upstream part of the catchment.

A complex design solution includes both reconstruction of the existing system and addition of new components – new pipe network and detention facilities (Fig. 8.7). In order to prevent flooding of the downstream highly urbanized part of the catchment, a significant attenuation of stormwater runoff is proposed by means of three detention basins, located 0.5, 1.0, and 1.5 km upstream from the point where the brook is introduced into the sewer system (Fig. 8.6). In this way, high runoff flows are to be reduced to an acceptable measure for the downstream sewage system, and detained long enough to improve the water quality by settling fine sediments and pollutants.

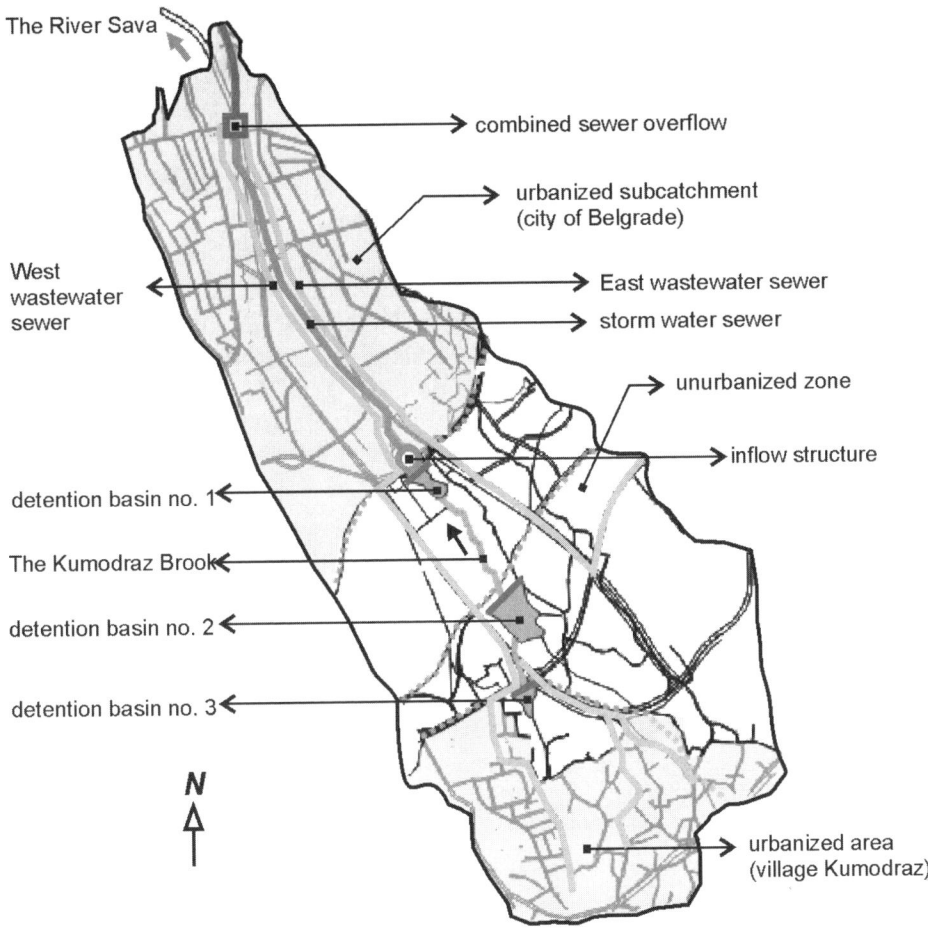

Figure 8.6. The watershed of the Kumodraz brook in Belgrade and the proposed flood management system by three detention basins (*Source*: Faculty of Civil engineering – Belgrade, 1999; Despotović et al., 1999).

The three detention basins are formed by constructing dam embankments of variable heights (between 6 and 9 m), at best available locations. For instance, it is proposed that the dam embankment of detention basin no. 3 take the advantage of an existing roadway earth structure, with an impervious outer layer added (Fig. 8.8). This basin's outflow control structure, consisting of a shaft spillway with two outlet pipes, offers a flexible flow control for various storm events (Fig. 8.8).

Detention basins have a total storage capacity of about 170,000 m^3. According to the designed stormwater management strategy, in the upstream part of the catchment, (which covers approximately 60% of the total area, or 400 hectares), surface runoff should be retained for all storm events up to 10-year return period,

Figure 8.7. Designed sewage system for the existing and planned urban areas with detention basins 2 and 3, shown at their maximum storage capacity (*Source*: Faculty of Civil engineering – Belgrade, 1999).

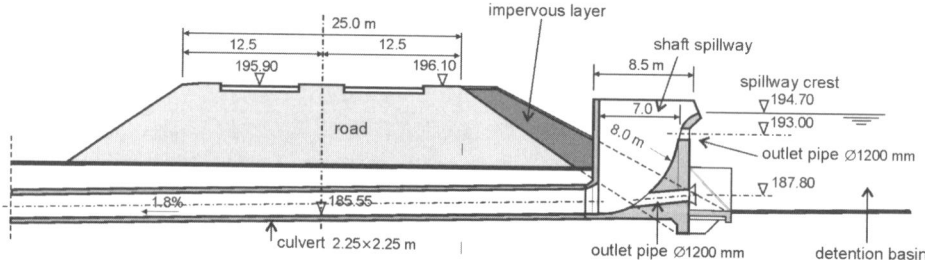

Figure 8.8. Dam embankment and outlet works of detention basin no. 3 (*Source*: Faculty of Civil engineering – Belgrade, 1999).

and significantly retained for floods up to 100-year return period. The capacities of the outlet control structures (shaft spillways and risers), are designed in such a way that the 10-year flood is fully retained, ensuring the normal functioning of the stormwater trunk sewer, whose maximum capacity is about $3 \text{ m}^3/\text{s}$. Results of stormwater routing are presented in Fig. 8.9.

The detention times of the basins for the 10-year flood vary from 1.2 to 19 hours. The basins are expected to trap 92% of coarse sediment load averaged over all storms, and the total basins' trapping efficiency for the clay sized particles is about 56%.

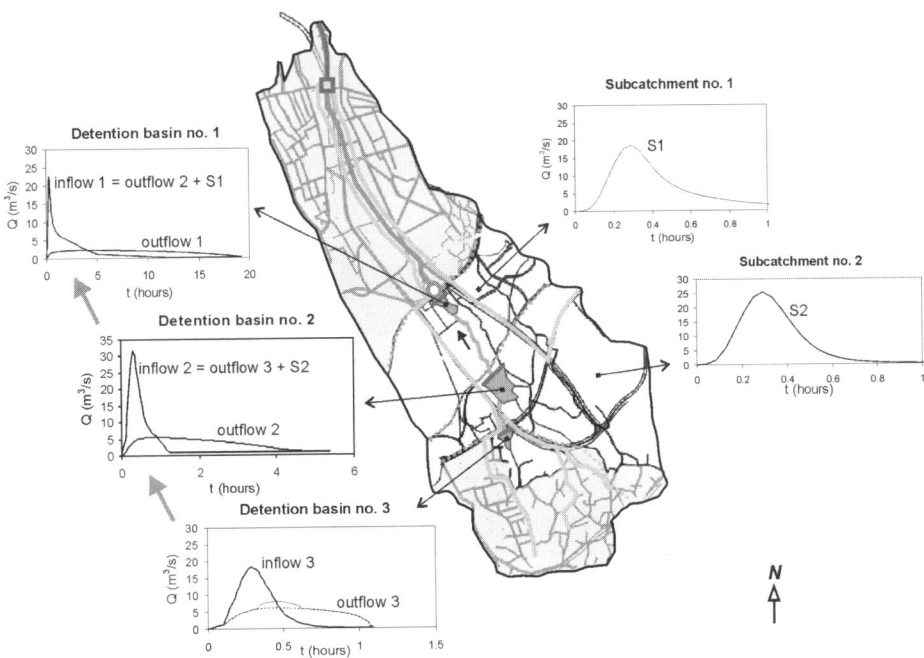

Figure 8.9. Attenuation of the 10-year flood by the system of cascading detention basins (*Source*: Faculty of Civil engineering – Belgrade, 1999; Despotović et al., 1999).

The Kumodraz brook channel is designed to accommodate flows up to the 10-year return period. The described stormwater management scheme for this urban environment gives an opportunity for river training to be based on the "natural concepts", promoting usage of stone and biological materials for protection works, thus preserving esthetic values of the environment.

8.3.2   The Karas River

The Karas River is a trans-boundary river, about 20% of its catchment area being in Serbia. The 1975 major flood endangered the 32 km long river valley, about 3000 hectares of farmland, 7 villages with about 20,000 people, and important infrastructure (Fig. 8.10).

The integrated flood management concept is applied for this semi-urban region, with the following objectives (Mladenović Babić and Jovanović, 2002, 2004):

• protection of farmland, villages, and infrastructure;
• preservation of the surface water and groundwater quality;
• preservation of the biodiversity of the river ecosystem;
• preservation of the ambient quality of the region, upgrading recreation and tourism.

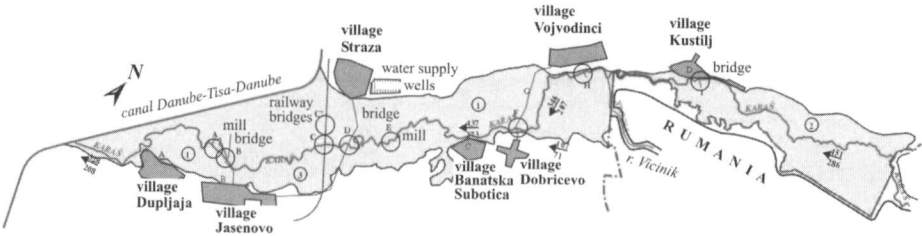

Figure 8.10. The flood plain of the Karas River with the endangered villages and infrastructure (*Source*: Mladenović Babić and Jovanović, 2002).

Figure 8.11. The Karas River scenery.

The last two items were of particular importance, since the Karas River is the last river in the region with the preserved ecosystem, of a high aesthetic value (Fig. 8.11).

Several flood management strategies are considered. A conventional approach is to build levees along the main channel, protecting the endangered region against the chosen 100-year design flood. The capacity of the main channel is $70\,\mathrm{m^3/s}$, and duration of excessive flow rates that can be expected is to 3–4 days in an average year. However, this solution, favoured by the local construction companies, has a number of disadvantages:

- limited expansion space for floods will increase velocities in the main channel;
- erosion potential in the main channel will be increased;
- negative effects of increased velocities on ecosystem are to be expected;
- new drainage systems will be required;
- reconstruction (elevation) of bridges will be necessary;
- extensive earth works will degrade the stream corridor.

In order to eliminate the listed shortcomings, a better flood management strategy is proposed, based on a combination of a retention basin and a flood relief canal (Fig. 8.12). The main objective is to preserve the river channel in its natural state, as much as possible. No levees are to be built along the downstream reach, which

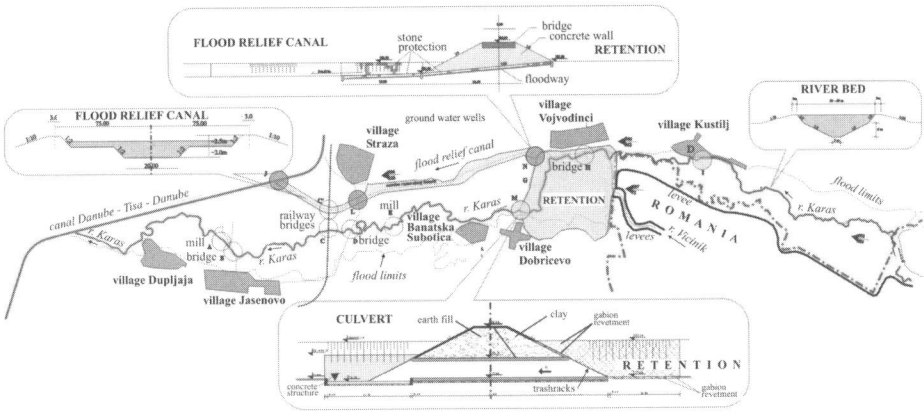

Figure 8.12. The flood management scheme for the Karas River (*Source*: Mladenović Babić et al., 2004).

has the highest esthetic and ecological value. For this reason, a flood relief canal is designed to accept all excess water (above $70\,\mathrm{m^3/s}$, expected several times per year), and a retention basin, to increase the safety margin for extreme flood events. Upstream from the retention basin, the conventional solution with levees along the river channel is to be applied.

The flood relief canal, formed as a grass-covered canal with mild slopes, would be in operation once in 4 years, on the average. During periods between floods, it is to be used as pasture land.

The location of the retention basin, which has an area of 415 hectares, is chosen applying the criterion of the least valuable farming land. Downstream from the retention basin, water is to be distributed between the flood relief canal and the river channel, for which two special structures are required – a culvert, to direct water from the retention basin into the main river channel, and an inlet structure, to divert water into the flood relief canal (Fig. 8.12). The retention basin's hydraulic residence time varies between 30 and 180 days, depending on the flood event.

The proposed flood management scheme is justified through a feasibility study and detailed technical documentation. Although somewhat more expensive than the alternative with the levees, the proposed scheme is judged to be more advantageous for two reasons:

- it offers a higher degree of safety for this trans-boundary region, with no danger for Serbia and Romania even if 100-year floods on rivers Karas and its tributary, the Vicinik river (Fig. 8.10), were to coincide, and
- it provides efficient means for environmental preservation, developing possibilities for recreation and tourism.

Figure 8.13. Weir and ramp drop structures.

Following the intention that the river channel should be preserved in its present state, a special attention is paid on the impact the river training may have on the flora and fauna. Only sporadic interventions were considered, applying the following two basic principles:

(i) All meanders are to be preserved, and no channel enlargement is permitted. Cut-offs are likely to increase velocities above the tolerable limits for some species, degrading biodiversity and stability of the ecosystem.

(ii) All river training measures along the river banks are to be extremely restrictive. Removal of the riparian vegetation reduces bank resistance to erosion, and can be harmful to the ecosystem (increases water temperature, reduces the amount of biologic material in the food-chain, etc.). Extensive excavation or removal of soil also degrades the scenery.

In accordance with the above principles, the designed river training measures consist only of bank protection and the longitudinal bed slope control. The bank protection by vegetation and stone is to be applied only to the sharpest river bends. Ramp structures for longitudinal slope control are designed. Experience with this kind of drop structures shows excellent impact on water quality through increase of the oxygen content (Fig. 8.13).

The social implications of the Karas river flood management project are twofold. Firstly, considerable efforts are needed in order to:

• convince local authorities that the proposed solution is the optimal one;

• resolve conflict of interests (local construction operative would prefer to build levees);

• make amends with farmers whose land will be used for the retention basin.

Secondly, for this project to be approved, the public must become fully aware not only of all direct (economic) benefits, but also of many indirect benefits, connected with the healthy environment, esthetics, quality of life, etc.

8.4   CONCLUSIONS

There is no general pattern for designing urban flood management and pollution protection schemes. In dealing with each particular case, an integrated approach should be taken, adjusting a technical solution to the local, often not easily quantifiable requirements, such as preservation of biodiversity, or esthetic value of the stream corridor. In coping with rapid urbanization, individual detention basins, or systems of detention basins, provide efficient means for flow control – stormwater peak-reduction to the pre-development level. This role of detention basins is very important when the primary objective is maintaining proper operation of an existing sewer system in a rapidly developing urban region. In addition to flood management and water quality treatment, retention basins can play the key role in stream corridor preservation. Various pro-active flood control strategies can be designed using retention basins and flood-relief canals, while preserving streams in their original state. This promotes the concept of "natural" river training. Only an integrated approach to stormwater management can lead to design solutions of highest environmental and social value.

REFERENCES

Affeltranger, B., 2001. Public participation in the design of local strategies for flood mitigation and control. IHP-V, Technical Documents in Hydrology, No. 48, UNESCO, Paris

ASCE, 2001. A Guide for Best Management Practice (BMP) Selection in Urban Developed Areas, ASCE, Reston, VA

ASCE, 1985. Final Report of the Task Committee on Stormwater Detention Outlet Control Structures, ASCE, New York

ASCE and WEF, 1992. Design and Construction of Urban Stormwater Management Systems, ASCE Manuals and Reports of Engineering Practice No. 77, New York, WEF Manual of Practice FD-20, Alexandria, VA

ASCE and WEF, 1998. Urban Runoff Quality Management, ASCE Manuals and Reports of Engineering Practice No. 87, Reston, VA, WEF Manual of Practice FD-23, Alexandria, VA

Chang, H.H., Fluvial Processes in River Engineering, John Wiley & Sons, New York

Chapra, S.C., 1997. Surface Water Quality Modeling, McGraw Hill, New York

Clar, M.L., Barfield, B.J., O'Connor, T.P., 2004. Stormwater Best Management Practice Design Guide – Volume 1: General Considerations, National Risk Management Research Laboratory, EPA/600/R-04/121

Clar, M.L., Barfield, B.J., O'Connor, T.P., 2004. Stormwater Best Management Practice Design Guide – Volume 2: Vegetative Biofilters, National Risk Management Research Laboratory, EPA/600/R-04/121A

Clar, M.L., Barfield, B.J., O'Connor, T.P., 2004. Stormwater Best Management Practice Design Guide – Volume 3: Basin Best Management Practices, National Risk Management Research Laboratory, EPA/600/R-04/121B

Cunge, J., Holly, F.M., Verwey, A., 1980. Practical Aspects of Computational River Hydraulics, Pitman, Boston, London, Melbourne

DeGroot, W.G., 1982. Stormwater Detention Facilities, ASCE, New York

Despotović, J., Petrović, J., Jaćimović, N., Djordjević, S., Jovanović, M., Djukić, A., Babić, B., Prodanović, D., 1999. Preliminary Design for Reconstruction of Stormwater and Wastewater System in a Developed Urban Area – A Case Study. Proceedings of the 8th ICUSD, Sydney, Australia

Driscoll, E.D., DiToro, D., Gaboury, D., Shelly, P., 1986. Methodology for Analysis of Detention Basins for Control of Urban Runoff Quality, EPA, Washington CD, Report No. EPA 440/5-87-01 (NTIS No. PB87-116562)

Faculty of Civil Engineering Belgrade, 1999. Design of Stormwater and Wastewater System in the Kumodraz Catchment, Institute for Hydraulic and Environmental Engineering, (in Serbian)

Ferziger, J.H., 1981. Numerical Methods for Engineering Application, John Wiley & Sons, New York

Haan, C.T., Barfield, B.J., Hayes, J.C., 1994. Design Hydrology and Sedimentology for Small Catchments, Academic Press, San Diego, CA

IRMA, NCL, Delft hydraulics, 2001. Interactive Flood Management and Landscape Planning in River Systems, http://ncrweb.org/downloads/pub13.pdf

Mladenović Babić, M., Jovanović, M., 2002. Alternative solutions for the Karas river flood management, Institute for Water Resources Development "Jaroslav Černi", Belgrade (in Serbian)

Mladenović Babić, M., Jovanović, M., Knežević, Z., 2004. The Integrated Flood Management – The Karas River Case Study, 3rd European Conference on River Restoration, Zagreb, Croatia

NSDW – National SUDS Working Group, 2004. Interim Code of Practice for Sustainable Drainage Systems, ISBN 0-86017-904, http://www.ciria.org/suds/

Newman, T.L., Omer, T.A., Driscoll, E.D., 1999. SWMM Storage Treatment for Analysis/ Design of Extended-Detention Ponds, Conference on Stormwater and Urban Water Systems Modeling, Toronto, Ontario

Rubin, H., Atkinson, J., 2001. Environmental Fluid Mechanics, Marcel Dekker, New York, Basel

SedNet – European Sediment Research Network, 2004. Contaminated Sediments in European River Basins, http://www.SedNet.org

Tapp, J.S., Ward, A.D., Barfield, B.J., 1982. Designing Sediment Ponds for Theoretical Detention Time, Proceedings of the ASCE, 108(HY1)

Urbonas, B.R., Stahre, P., 1993. Stormwater – Best Management Practices Including Detention, Prentice Hall, Englewood Cliffs, N.J.

USDA Natural Resources Conservation Service, 2000. Stream Corridor Restoration – Principles, Processes, and Practices, http://www.usda.gov/stream restoration

Whipple, W., Hunter, J.V., 1981. Settleability of Urban Runoff Pollution, Journal of the Water Pollution Control Federation, 53, 1726

9

Flood Induced Indirect Hazard Loss Estimation Models

William Veerbeek

Dura Vermeer Business Development, Hoofddorp, The Netherlands

ABSTRACT: Indirect flood losses can cause substantial economic and social disruptions that may last over long periods of time. Due to differentiations in land use and different flood conditions, simple ratios between direct and indirect flood losses are inadequate to reach accurate estimates. Current models for estimating indirect hazard impact are based on economic flow where economies are treated as a series of interconnected activities. The purpose of this article is to give an overview of the state-of-the-art in indirect hazard loss estimation models. These include unit-loss models, Input-Output models, econometric models and increasingly popular Computable General Equilibrium models. Most of these models have been primarily used to estimate effects in policy changes. Yet, since they are strongly founded micro-economics, derivates of these models have been successfully applied in estimating indirect hazard impact, including indirect flood losses. The focus in these models is on the estimation of higher-order effects, by relating local disruption to economic ripple effects on a higher level. While experience in application of these models in actual events is still limited, the first results are promising. Improved accuracy can be reached by adapting the models further to incorporate specific local geographic conditions as well as adjusting models to smaller timeframes.

9.1 INTRODUCTION

Quantification of economic losses as a result of floods and other natural hazards is vital in determining potential economic impact, vulnerability and flood management policies. Increased risk as a result of climate change combined with rapid urbanization urge for further insight into the volume and distribution of economic losses. Since modern economies are characterized by a widespread set of interdependent activities, the estimation of indirect hazard impact, e.g. flood impact, becomes more urgent than ever. An illustration is the interruption of lifeline utilities and infrastructure, which could potentially lead to widespread economic and social repercussions due to the damage to critical links within their networks. The same can be said for business interruption of production facilities that provide vital goods or services. Insight on the higher-order effects caused by local disruptions is essential for estimating impact and adjusting policies to decrease vulnerability. Additionally, indirect losses account for transfer costs of resources during post-disaster conditions as well as substitution in demand. Current studies on resilience engineering (Hollnagel et al., 2006) and vulnerability analysis confirm a "network approach" in estimating impact.

Current methods and models on flood loss estimation, providing retrospective (i.e. backward analysis based on empirical data) as well as prospective analysis (i.e. scenario analysis and prediction), has almost primarily been focused on the impact of direct losses (e.g. Penning-Rowsell & Chatterton, 1977; Parker et al., 1987; Parker, 2000). Due to the perceived complexity, much less attempts have been made to design and implement models that estimate indirect hazard losses. Current approaches to estimate indirect damage are mainly based on two perspectives: Unit-Loss methods (Penning-Rowsell et al., 1977) are focusing on individual micro-economic losses, while (derivates of) Input-Output models (Cochrane, 1974) focus on macro-economic effects by expressing relations between local disruptions and the economy as a whole. While especially (derivates of) Input-Output models have undergone many alterations and extensions in order to increase accuracy and flexibility, practical application in indirect loss estimation is still limited.

The purpose of this article is to give a general overview on the state-of-the-art on contemporary methods and models for estimating indirect hazard loss including a general qualification on their pitfalls and potentials. Furthermore, general issues surrounding the conceptual framework as well as the empirical aspects of indirect flood damage estimation are identified. In addition, further research topics are suggested that could increase accuracy and application in general, and topics that extend these models indirect flood loss assessment.

9.2   CONSTRAINTS

9.2.1   General constraints

Indirect flood loss is conceptualized by Parker (1987) as "those caused through interruption and disruption of economic and social activities as a consequence of direct flood damages." This separates them from direct losses that focus primarily on losses resulting from the replacement, repair and restoration of damaged objects caused by a flood event. This includes losses resulting from damaged stocks/supplies, infrastructure and lifeline utility networks. An important issue in the estimation of primary flood losses, which can be extrapolated to indirect losses, is the method used for valuing recourses. Damage losses should be estimated at prices that represent their efficient allocation instead of current market prices (Rose, 2004). Since market prices often do not account for inefficiencies, or don't even exist in the case of environmental damage, they do not reflect the real costs resulting from flood damage or other catastrophes. Losses resulting from business interruption, which is a subcomponent of total indirect losses, are difficult to estimate because businesses and other parties (e.g. insurance companies) use heuristics (i.e. biased estimates based on rules of thumb) in determining the loss of sales. These are not necessarily reflecting actual losses (Rose, 2004). Furthermore, a clear distinction should be made between gross and net output values. The former reflecting the total value of products including its intermediate stages, while the later reflects the value of the final product. The final value of the product doesn't necessarily replicates the total value, since market fluctuations (e.g. depreciation) might present different price levels. Therefore, in the case of estimating flood damage, the gross output values should be used.

Special caution should be taken not to "double-count" stock losses when estimating primary and indirect damage. While stock losses are a function of primary damage, they are sometimes incorrectly included in the costs resulting from business interruption, which are in themselves are a partial function of reduced stocks. The same constraints exist for counting the costs of damaged production facilities (e.g. equipment). Double-counting also appears when objects perform several functions simultaneously (e.g. a combined road-levee structure); the costs of restoration or replacement should only be included once.

9.2.2   Constraints related to indirect loss estimation

The primary aim when assessing indirect flood losses is the estimation of higher-order effects. Higher-order effects are the upscale economical effects resulting from local economic disruption. The extrapolation of regional impacts to a national or supra-national level is not always straightforward. While business interruption might have an impact at a regional level, the implications on a national level might be minimal because of compensation by businesses outside the calamity area. This means that the net damage on a national level will be minimal. Consequently, with the exception of a temporal instability in which markets adapt, the higher-order effects can be dismissed. Hence, when estimating higher-order effects, it is fundamental to have good view on the scale level the impact is estimated on since these can differ dramatically. When compared to general hazard loss estimation, the flood context doesn't pose any special constraints on the estimation of higher-order effects; no additional model specifications are required for assessing higher-order effects.

Special attention when estimating indirect hazard losses has to be put on the recovery period of the affected region. In many cases, the disruption of businesses isn't maximized over the recovery period; through time, production will recover gradually instead of step-wise. Therefore, indirect flood loss estimation due to business interruption cannot be estimated over a single point in time but has to be regressed over the recovery period. Regaining production level, or more generally regaining the original operational level, gives an indication of the resiliency of the affected activity. Resilience is in general defined as the capacity of a system to absorb perturbations and reorganize (Walker, 2004). It is a system property that can be accounted for on both the individual level (e.g. an individual business) or on regional level and provides a vital strategy to reduce losses during the recovery period (Rose et al., 2005). Note that resilience on a regional or even national level is highly associated to higher-order effects; when impact because of business interruption is minimized because of compensation by other firms, or substitution of demand outside the affected area, the system as a whole proves to be resilient. Thus, when higher-order effects are minimal, the aggregate resilience level is high. Of course, full resiliency would only occur in perfect market conditions, which in reality don't exist.

Finally, there are general difficulties related to the available data concerning indirect flood losses. First of all the amount of available data is limited, since most effort in data acquisition is put into estimating primary hazard losses. Secondly, the existing data is often not centrally acquired but scattered into fragments distributed through various agents (e.g. individual businesses, insurance companies). Furthermore, loss data is often incomplete since no standard methods are available in data collection. Limited data availability combined with incompleteness make indirect loss estimation based on simple fractions of primary losses highly improbable. In combination with the particularity of conditions that surround individual flood events, fixed ratio methods based on goodness-of-fit approximations can only be used for gaining initial approximations.

9.3   MODELS ON INDIRECT FLOOD LOSS ESTIMATION

9.3.1   Unit-loss methods

Originating in the estimation of direct flood losses (Penning-Rowsell, 1977), the unit-loss (UL) model is an aggregate method in which indirect flood losses are estimated through the combined losses for a wide variety of properties. The aim of the method is to provide standard average or site survey loss data for residences, businesses, lifeline utility failures and infrastructure. The method provides a standardized method in which Data acquisition is performed by providing representative economic agents with a set of detailed questionnaires. The purpose of these questionnaires in estimating indirect flood losses is to acquire loss data on costs resulting from business interruption as a function of various flood damage properties. The acquired data is then matched and averaged to estimate some general figures on the aggregate losses. The individual loss factors comprise a wide set of damage components. These include relatively specific factors (e.g. delay costs produced by road capacity decline), for which specific calculation methods have been developed based on empirical survey data. Categories comprise calculation methods for losses in:

   i) retail, distribution, office and leisure service businesses
   ii) manufacturing activities
   iii) traffic delay
   iv) utility disruption
   v) public services
   vi) emergency services
   vii) households

The averaged aggregate loss-data produce a set of tables that estimate higher-order effects by using general indirect flood loss figures. These figures do not include regional elasticity coefficients, or multiplier effects (i.e. changes in spending causing a disproportionate change in aggregate demand).

   The main advantage of the UL model over other models is that it has been specifically developed for flood loss estimation. Most of the calculation methods describe individual flood losses as a function of flood specific properties (e.g. depth). This makes the UL model fit to give a detailed account on local indirect losses for almost every economic sector. The estimates are to a large extend based on precise survey data, which makes the model vulnerable in actual application in prospective analysis since no adequate structure exists that allows for robust generalizations. Yet, the availability of sound survey data is a general problem in flood loss estimation.

   The main setback when using UL models for estimating indirect flood losses is the absence of an underlying economic model that accounts for higher-order effects; the interdependencies between economic agents and sectors are not accounted for. While higher-order effects can be extensive, they are many steps removed from indirect losses that are accounted for in UL models. Furthermore, the model doesn't acknowledge specific behavioral strategies and phenomena that occur during disruptions resulting from hazard impact. These include effects on prices and resulting substitution in demands, resiliency improvement strategies like substitution of production resources and multiplier effects that amplify or dampen indirect losses. Additionally, UL models only implicitly acknowledge locally dependent factors. While the acquired survey data is solidly based on local conditions, aggregate higher-order effects are based on general figures therefore decreasing accuracy and ignoring the quality of the acquired data.

   In conclusion, the compiled indirect loss data makes UL models primarily useful for estimating indirect damage on a local perspective (i.e. for the affected area), with the limitation that no substitution of production recourses takes place. Accurate estimations for actual flood events can only be made in case of low-impact floods with a short duration. Since UL models emphasize on data acquisition, they are still used frequently to provide input figures for models that specialize in the estimation of higher-order effects.

9.3.2 Input-Output models

Input-Output (I-O) models are used widely in the field of regional economics. Application in natural hazard loss estimation dates back relatively long (Cochrane, 1974), and has been refined over the last decades into a widely accepted method. The popularity of I-O models is mainly due to its relatively wide scope and the moderate data requirements when compared to other loss estimation methods. The model rooted in the classic Leontief Input-Output model which consists of a sectoral linear production equation satisfying a demand vector (Leontief, 1986).

Basically, an I-O is a static, linear model that combines all purchases and sales between sectors of an economy, based on the technical relations of production (Miller et al., 1985). This means that I-O models consider economies as a series of interdependent activities in which capital flow occurs, making them well suited to examine the rippling effects throughout an economy resulting from some local disruption (Rose, 2006). Note that I-O models are not necessarily developed for predicting indirect hazard losses, but are suitable for being used at varying occasions. Depending on the used I-O model and the scale on which effects are estimated, I-O models separate flows into internal and external flows. Internal flows comprise of capital flow within the area in focus, while external flow signifies the flow to adjacent areas. This distinction is essential in estimating indirect flood loss damage including higher-order effects.

I-O models consist of a matrix containing input (demand) and output (production) figures for a series of economical sectors. The figures are distributed over the several sectors for both within and outside of the region in question. The output values within the I-O model are demand driven. The basic equation representing the total demand of industry i is conceptualized as:

$$y_i = x_i - \sum_{j=1}^{m} x_{ij} \tag{9.1}$$

where x_{ij} is the amount of production by i absorbed by j as input, and y_i is the total demand from the outside industry i. The aggregate output from a region is the sum of all sectors. Outputs are affected by a set of coefficients that represent various properties. These include induced effects (income or employment effects triggered by household consumption expenditures), indirect effects (backward inter-industry supply linkages) and production effects (the amount of products needed from other companies to produces a product). The aggregate gross output satisfying the final demand is conceptualized as:

$$Y = (I - A)X \tag{9.2}$$

where Y is the final demand matrix, I is the identity matrix of the gross output matrix X, and A is the set of coefficients. Additionally a set of regional coefficients are used to further adjust the model to local conditions.

The optimization within the model is to maximize the all total sector outputs (eq. 9.1) to address the total demand (eq. 9.2), resulting in a Walrasian general equilibrium. Changes in production of business j will lead to excess demand, which has to be compensated by another business j'. The model assumes that full market saturation will always occur. Since it is not feasible that economies are closed, an outside demand vector y_i is introduced which represents external agents (e.g. government agencies). Sectoral differentiation is mostly based on a subdivision of agriculture, industry, services extended by households, government expenditure, import, investments and taxes for import tables. Yet, categories they can be expanded to express any economic relation.

The use of I-O model within a retrospective context of flood loss estimation is straightforward. Since business interruption will affect the output figures for various sectors within the I-O table, it is relatively simple to calculate both the internal and external supply changes, thus accounting for higher-order effects of local disruptions within the economical network. Higher-order effects express excess demands which cannot be compensated by substitution. An example of a successful application of this method is the Indirect Economic Loss Module of the HAZUS Loss Estimation Methodology (FEMA, 2001), in which ripple effects for a series of flood events on various locations can be estimated. Other currently used models are RIMS (Bureau of Economic Analysis, 2006) and IMPLAN (Minnesota IMPLAN group, 2006). For prospective estimations, the level of business interruption is based on a comparison of similar flood events, combined with I-O tables reflecting local conditions.

Although further refinements have been suggested in applying I-O models for hazards loss estimation, the underlying I-O model has remained unaltered. Boisvert (1992) and Cochrane (1997) extended the I-O model with flexible treatment of imports, which compensate for shortages of regionally produced inputs in the aftermath of an earthquake. Cole (1988) improved the method for estimating the temporal impact (recovery time) on higher-order effects. Further refinements include transportation impacts (Gordon et al., 1998; Cho et al., 2001), lifeline impacts (Rose et al., 1997), resilience and time-phasing impacts (Okuyama et al., 2000; Rose et al., 1997), ripple effects on adjacent and non-adjacent regions (Cole, 1998; Okuyama, 1999).

The main advantage of using I-O models for estimating indirect flood losses is the clear and straightforward insight it provides in higher-order effects as a result of a local disruption. Furthermore, the focus on sector interdependencies and flow within and between regions produces a transparent analysis of the degree of autonomy and economic dependencies within an economy. This extends its function to regional vulnerability analysis in as a function of hazard impact. The relative simplicity of the I-O model generates also weak points in application. First of all there is a lack of behavioral content; I-O models don't provide means to integrate assumptions on the behavior of interacting economic agents. Agents (e.g. consumers, producers, government) base their strategies and behavior on different utility functions which determine their activities. I-O models incorporate the implicit assumption that some central planner has a goal of maximization of growing needs. In reality this assumption is not necessarily satisfied (Shvyrkov et al., 1980). A second, more general problem is that a formal relation between output and price is omitted. In case of catastrophes, prizes are potentially influenced by a limited production volumes resulting in a higher pressure on demand. Omitting this mechanism results in an underestimation of higher-order effects. Related to this setback is the exclusion of resource constraints which could potentially result from hazard impact. I-O models assume maximum supply saturation, which in reality is dynamically satisfied instead of instantly. This is further extended by the fact that I-O models don't incorporate information on substitution which in turn compensates higher-order effects. This addresses the linearity of I-O models and implies ignoring economies of scale. In general though, these shortcomings account for all described hazard loss models; the sheer complexity of the relations between micro and macro economy domains, models and estimates will always result in suboptimal assessment and neglect of factors. Inclusion of these factors into the I-O model, though improving their accuracy, could result non-operability since every extension is also dependent on additional datasets which might not be available or infeasible to acquire. Yet most inaccuracies within the use of I-O models results from incorrect use including double counting, mixing gross en net figures, etc. (Rose, 2004).

Since I-O models are not specifically designed for assessing flood impacts, many flood specific attributes in their estimations are absent. This implies that data on loss of production volumes, changes in demands within the affected region, local effects on utility lifeline disruption should be acquired from other sources (e.g. UL models). The acquired data serves as adjusted Input-Output figures on which elasticities are based.

The methods for using I-O models in prospective context to perform scenario analysis (e.g. impact analysis, resource availability, resiliency, recovery rate) are somewhat ad hoc. Currently, no standards have been defined for application. As already stated, indirect flood loss data is often sparse and incomplete. Therefore, estimation based on a goodness-of-fit towards retrospective time-series data is a difficult process which influences the confidence interval of the estimations dramatically. To estimate the robustness of the predictions several methods can be applied. These include sensitivity tests on the model's major parameters, the incorporation of a probability distribution on the major parameter values and a Monte Carlo simulation on the relation of the parameter's distribution of values (West, 1986). What is remarkable is that there is only a very limited amount of studies performed on the validation and verification of I-O models. Although the application of I-O models for prospective estimates always incorporates a limited accuracy, further research on calibration to existing data (including constraints) would improve their accuracy.

9.3.3   Computable General Equilibrium models

The current state-of-the-art in hazard loss estimation models is set by Computable General Equilibrium (CGE) models. CGE models share many common properties with I-O models. Like for I-O models, the conceptual starting point for a CGE model is the circular flow of commodities in a closed economy. Likewise, CGE-models incorporate multi-sectoral interdependencies by using an Input-Output table as the core of all calculations. Yet, in addition to I-O models, CGE-models incorporate price models and individual behavioral components as a function of resource constraints. This separates them from I-O models which are founded on a general assumption of global supply-demand saturation. A CGE-model is "a multi-market simulation model based on the simultaneous optimizing behavior of individual consumers and firms in response to price signals, subject to economic account balances and resource constraints" (Rose, 2005).

The main agents in a CGE model are households, firms and the government which exchange capital through the provision of goods & services and taxes. The circular flow, which is a resulting system property emerging from the agent's behavior accounts for the price level of goods & services. For example producers optimize their behavior by minimizing costs, average cost pricing and adjusting to household's demands. The general equilibrium (GE) paradigm is based on a set of theorems based on neoclassical economics. These include Pareto efficiency of equilibriums, efficiency of equilibriums and the existence, (local) uniqueness and stability of an equilibrium (an in dept explanation of these theorems is outside the scope of this paper).

The general equilibrium paradigm is based on the following set of assumptions (Wing, 2004):

i) Commodity market clearance. The value of gross output of an industry must equal the sum of the values of the intermediate uses of a good and the final demand that absorbs a commodity. This is similar to factor market clearance in which firms fully employ the representative agent's endowment of a particular factor. Market clearance is conceptualized as:

$$y_i = \sum_{j=1}^{N} x_{ij} + c_i + s_i \qquad (9.3)$$

where y_i is the produced commodity, x_{ij} is the demand by producer j for commodity x_i and c_i, s_i are the demands by the representative agents as an input to consumption and saving activities.

ii) Factor market clearance. This implies that a firm fully employs the representative agent's endowment of a primary factor, conceptualized as:

$$V_f = \sum_{j=1}^{N} v_{fj} \tag{9.4}$$

where v_{fj} is the representative agent's endowment for a primary factor f.

iii) Zero profit. The value of gross output of a sector must equal the sum of the values of inputs of intermediate goods and primary factors that the industry employs in it production. This is conceptualized as:

$$p_j y_j = \sum_{i=1}^{N} p_i x_{ij} - \sum_{f=1}^{F} w_f v_{fj} \tag{9.5}$$

where $p_j y_j$ is the value of output generated by producer j, $p_i x_{ij}$ is the values of the input of the i intermediate good and $w_f v_{fj}$ are the primary factors employed in production.

iv) Income balance. The representative agent's income is made up of the receipts from the rental of primary factors that balance the agent's gross expenditure on satisfaction of commodity demands which is conceptualized as:

$$m = \sum_{f=1}^{F} w_f V_f \tag{9.6}$$

where $w_f V_f$ is the value of producers' payments for the use of primary factors.

General equilibrium is reached through combining equation 9.3 to 9.6. Note that equilibrium is always reached for a specific instance in time, therefore providing an expressive model in which the effects of changing conditions can be traced back, including the effects of resilient strategies during the recovery period. Yet, in practice these assumptions are not necessarily followed within CGE models, making them actually more fit to represent actual conditions. Alternatives include non-market clearing (e.g. unemployment in the labor market), imperfect competition (e.g. monopoly pricing) or stable demands regardless of price fluctuations (e.g. government demand).

The conversion from GE to a CGE model is applied by setting up an algebraic framework that account for the described elasticities, operating on a set of matrices consisting of an Input-Output matrix and a Social Accounting matrix. The Input-Output matrix (similar to the one used I-O models) expressing flow consists of three matrices combining an $N \times N$ input-output matrix of industries' uses of commodities, an $F \times N$ matrix of primary factor inputs to industries, and an $N \times D$ matrix of commodity uses by final demand activities. The resulting figures form a social accounting matrix, which is a snapshot of the inter-industry and inter-activity flows of value within an economy that is in equilibrium in a particular benchmark period. While I-O models are defined by static relationships, production and consumption maximization within CGE-models for real applications in an imperfect market imply a dynamic nonlinear behavior.

Calibration of the CGE-model is done through optimizing quantities of the variables by using actual data. This optimization process poses a nonlinear complementary problem (Ferris and Pang, 1997). Important questions during this process are the existence and uniqueness of the equilibrium in the CGE-model. This means that the outcomes must converge and be robust towards perturbations during convergence. The robustness of the outcomes during calibration is defined by the variability within the aggregate data. In real world problems, specific agents are often modeled separately, resulting in complex nested utility or production functions. This makes mathematical proof of existence and uniqueness of the solution impossible (Rose et al., 2005).

CGE-models are widely used to analyze aggregate welfare and distributional impacts of policies like tax alteration (e.g. Weyant 1999). Only recently, their potential is acknowledged in hazard loss estimation by applying constraints on demand or supply resulting from local disruption affecting produced commodities y_i (eq. 3). Production functions in CGE models are applied to aggregate categories of major inputs of capital, labor, energy and materials. Constraints in these functions make the CGE-model converge to a new equilibrium. By comparing the pre- and post-change equilibrium vectors of prices, activity levels, demands and income levels, the impact may be evaluated. As in IO-models, CGE-models estimate the impact disruptions (e.g. flood impact) cause on the aggregate welfare level in a theoretical consistent way by quantifying the change in the income and consumption of the representative agents that result from the interactions and feedbacks among all the markets in the economy.

Due to their relative novelty and perceived complexity, application of CGE-models estimating hazard loss for actual events is limited. Exceptions are the impacts of utility lifeline disruptions (Rose et al., 2004) and the impacts of earthquakes (Rose et al., 2002). Most applications have been experiments with synthetic models (Boisvert, 1992; Brookshire et al., 1992).

Using CGE models in hazard loss estimation undermines some of the basic assumptions CGE models are founded on. The main assumption is that the economy is always in equilibrium. This is no problem when the period of analysis is relatively long, but does pose problems in case of hazards loss estimation since these typically involve shorter periods characterized by non-equilibrium conditions. These temporal constraints have to be accounted for by extending the model with adaptive production functions (Rose et al., 2005). Furthermore, the overly flexible nature of CGE-models results in underestimation of actual losses resulting from hazard impacts (Boisvert, 1992). Typically, the process of input substitution (e.g. the costs for changing from tap water to bottled water during the recovery period after flood impact) is overestimated which results in an underestimation of the actual costs made for substitution. Another illustration of underestimation due to substitution is the impact of interruption of lifelines utility networks. While I-O models indicate that a twenty percent curtailment in lifeline service would lead to at least a twenty percept decrease of sectoral output, CGE models indicate a drop in sectoral output of only a couple of percentage points owing to the inherent ease of substitution of inputs (Rose et al., 2004). This type of underestimation is mainly due to incorrect application. Instead of using elasticities applicable to the relatively recovery period, often elasticities are used for long periods in time in which naturally a high level of substitution will take place. A more fundamental reason why CGE models underestimate impact is the inherent interdependence between supply, price and demand: When input is curtailed (e.g. because of flood impact), prices will be raised resulting in a lower demand. Yet, when demand decreases, prices will in turn drop again resulting in a demand stimulus. Standard rules o thumb on multiplier effects therefore do not apply.

The accuracy of CGE-models can be improved by including survey data on resiliency and direct impact (Rose et al., 2002), which can be obtained by applying data acquisition methods described for UN models. This data can be used to revise the individual production function parameters per sector or agent to better predict region wide impacts of direct flow or stock damages. When CGE models are used for prospective cases, the same methods can be used as when applying I-O models: scenario analysis, sensitivity tests on parameters, probability distributions for parameters and stochastic simulations. Since the complexity of CGE models is far greater than within I-O models, special care should be taken into sensitivity tests on parameters; in general, higher complexity results in more difficult control of model behavior. The non-linear behavior of the model also accounts for some perceived hesitation in application of CGE models due to their "black-box" properties; the optimization process within the model is difficult to validate and verify.

Within the context of flood loss estimation, CGE models should be extended with a set of functions that acknowledge the relatively short recovery period dampening the inherent overestimation of substitution of inputs. Rose and Liao (2004) implemented a method to improve accuracy by optimizing recalibration of the production function parameters to reflect actual resiliency. This extension exploits locally bounded individual differences in survey data instead of using general values and is tested in a case study on the economic impacts of a disruption to the Portland Metropolitan Water System as a result of a hypothetical major earthquake. Furthermore, specific flood loss data that represents changes production volumes, changes in demand, behavioral changes, distribution, etc. should be acquired. As in the case of I-O models, data provided by UL models is a necessity to perform assessment by CGE models. This implies that CGE models can only operate in actual hazard loss estimation when ample knowledge on the local disruption exists.

Implicitly, CGE models used for hazard loss estimation measure the economic resilience of the affected area. Whenever economic disruption on a regional level, or even local disruptions in production occur only in a very limited extend, this accounts for a high level of resilience. Since both the actual impact as well as the recovery due to for instance substitution is absorbed within the model, CGE models can identify both the robustness as well as the adaptive capacity therefore identifying areas which account for the aggregate vulnerability level towards hazard impact.

9.3.4   Econometric models

Econometric models comprise of a series of statistically estimated equations representing the aggregate functioning of an economy (Glickman, 1971). The relations between an arbitrarily chosen set of variables and their effects on others (e.g. the effect of production losses on the demand on a specific product) are regressed over a set of time series or cross-section data. Regression methods include various forms of linear regression functions, Bayesian networks, neural networks or even genetic algorithms. Thus in principle, econometric models are nothing more than extended general statistical estimation models combined into a single framework.

In principle, econometric models show great potential in especially prospective scenario analysis because their ability to generalize implicit relations into a set of robust equations. The predictions made by the models are show no major bias is addressed before in the underestimation and overestimation of impact produced by I-O models and CGE models. Furthermore in comparison to I-O models or CGE models, their foundations, validation and verification methods are well defined. Yet, all assumptions made by a model are based on the function regression of data sets. Therefore econometric models are fundamentally incapable making sound estimations on events outside the confidence interval they are trained on during regression. To reach a high confidence interval, time series data has to span a period of often 10 years or more, making them impractical in the domain of regional economics since such amounts of data are often not available. This problem extends to using econometric models in estimating the effects of floods, or disasters in general. Because of the low frequency, relatively sparse data is often available for coefficient estimation during function regression.

At the moment enough data is available to ensure accuracy of the regressed functions, econometric models can substitute some of the explicit equations expressing elasticities in I-O or CGE models. This puts constraints on the applied regression method since some regression methods pose "black-box" problems (e.g. neural networks), making validation and verification within especially CGE models difficult.

9.4 FUTURE DEVELOPMENTS

The weaknesses in currently used hazard loss estimation models seem related to the perspective from which economic disruptions are estimated. Although all described models are firmly embedded into micro-economic theory, there is a general difference within the approach between (derivates of) unit-loss models and (derivates of) Input-Output based models. While the former focus on the acquisition of specific locally based hazard loss data, the later focus on the relation between local disruption and higher-order effects. In fact, one can argue that many of the described models are complementary. A logical future development would therefore be to integrate the local specificity of unit-loss models into the flow-based framework of Input-Output based models. Furthermore, the temporal aspects of hazard loss estimation, which cover often relatively short time periods, should be explicitly acknowledged within the modeling framework. In the effort to integrate specific local conditions, further research has to be performed on the integration of spatial models. Since data availability is extending the use of Geographical Information Systems (GIS), the spatial implications of local disruptions can be analyzed with higher accuracy when hazard loss estimation models are connected to GIS (e.g. utility lifeline or infrastructural disruptions within a distribution network). Furthermore, new insights on the theory of spatial mechanisms in urban development (Batty, 2005) can improve this framework and gain further insight in the dynamics of spatial economy.

On the other hand, further knowledge has to be gained on the driver's side of hazard loss estimation: the effects of natural disasters on specific sectors within industry, trade and services. Current quality and availability of survey data hardly justifies generalizations within loss estimation. The increase of flood events in recent years (Dartmouth Flood Observatory, 2006) should compensate this lack to some extend. Note that the estimation of indirect flood losses is dependent on estimation models for primary flood losses; e.g. the collapse of a bearing wall due to lateral pressure or water velocity has a substantial effect on the estimated indirect losses.

9.5 CONCLUSION

Compared to direct hazard loss estimation methods, the modeling of indirect hazard loss estimation is still in a preliminary phase; no standards have been defined and many of the models used are still open to substantial improvements. This is mainly due to the complex environment they are subject too. Since models estimating indirect losses do not only account for the losses suffered by the affected region, but are particularly focused on the estimation of higher-order effects, the complexity of economic relations has to be absorbed by the model.

Current models used in practice for estimating indirect hazard loss can be subdivided into two main categories: (derivates of) Unit-Loss models and (derivates of) Input-Output based models. Indirect hazards loss within Unit-Loss models is based on aggregate loss data acquired by large scale survey of individual businesses. Higher-order effects are only accounted for in a limited fashion since no underlying model is included that accounts for economic flow. This makes Unit-Loss models only applicable in a limited set of cases in which a complete and accurate set of empirically acquired loss data exists. Unit-Loss models are therefore hardly fit to be used in prospective analysis, since generalization and implementation of economic flow and behavior cannot be expressed. The second branch of models is based on economic flow by using an input-output data. These models, which originate in the Leontief model for closed economies, are widely used for estimating policy effects. Due to their ability to estimate higher-order effects, they have become an increasingly popular method for estimating the effects of local economic disruptions, including hazard loss estimation. A typical problem in Input-Output models is their rigid response which is practically devoid of behavioral content of economic agents. Input-Output models overestimate indirect flood losses, which in practice means that they are useful for calculating the upper bound of the potential indirect damage resulting from disruption.

The current state-of-the-art in estimating higher-order effects are Computable General Equilibrium Models which are derivates of Input-Output models. These incorporate many of the best features of other models, but without many of their limitations. Within the context of hazard loss estimation, let alone floods impact, their use is a relatively novelty. Yet they provide an excellent framework for analyzing vulnerability, impact, resilient strategies and policy responses. Especially when calculating the impact of lifeline disruption (e.g. telecommunications, energy and infrastructure), they are superior to alternative approaches.

Within hazards loss estimation, I-O models extended with an adjustment for inventories are probably better suited to recovery periods of less than one week (Rose, 2006). For all other cases, CGE models are better suited to estimate higher-order effects resulting from disruptions. Special notice should be taken to the implementation of resiliency within economic agents. While I-O models are not able to implement substitution and other resilient strategies, CGE models are. Thus, acknowledging some of the dynamics within responses to flood impacts. CGE models typically contain a large number of variables and parameters and are structurally complex. This makes them far more difficult in application than I-O models. Due to the inherent flexibility within standard CGE models, they underestimate impacts by overly maximizing substitution and conservation. This means that CGE models define the lower bound of the potential indirect damage resulting from disruption. Finally, econometric models show great potential in the application of hazards loss estimation. Unfortunately, these models depend on extensive time series data of more than 10 years or extensive cross-sectional data which are generally unavailable. As a result, their use is limited.

All models are dependent to some degree on the availability and quality of flood loss data. Due to limited availability, initial improvements in accuracy should result from further calibration of the models to actual events. Secondly, especially Input-Output and CGE models can be further extended by incorporating methods that address spatial and temporal factors; estimations on hazard loss are defined by local characteristics and relatively short periods of disruption.

Although still in their infancy, indirect hazard loss models offer a great potential to get more insight into the potentially large disruptions resulting from hazard impact. Future developments and outcomes will have an increasing effect on flood management policies and urban planning. The integration of spatial data within GIS combined with economic models may drive many strategic decisions on the perception of contemporary development policies.

REFERENCES

Batty, M. 2005, Cities and Complexity, Understanding Cities with Cellular Automata, Agent-Based Models and Fractals, MIT Press, Cambridge, USA

Booyens, H.J., Viljoen, M.F. and De Villiers, G. 1999, Methodology for the calculation of industrial flood damage and its application to an industry in Vereeniging, Water SA 25, pp 41–46

Boisvert, R. 1992, Direct and Indirect Economic Losses from Lifeline Damage, Indirect Economic consequences of a Catastrophic Earthquake, Final Report by Development technologies to the Federal Emergency Management Agency

Bureau of Economic Analysis 1997, Regional Multipliers, A User Handbook for the Regional Input-Output Modeling System (RIMSII), U.S. Department of Commerce, Bureau of Economic Analysis, Washington D.C., USA

Cho, S., Gordon, P., Moore, H., Richardson, M., Shinozuka, M. and Chang, S.E. 2001, Integrating Transportation Network and Regional Economic Models to Estimate the Cost of a Large Urban Earthquake, Journal of Regional Science, Vol. 41, pp 39–65

Cho, S., Fan, Y.Y. and Moore, J.E. 2003, Modeling Transportation Network Flows as a Simultaneous Function of Travel Demand, Network Damage, and Network Level of Service, the Proceedings of the ASCE Technical Council on Lifeline Earthquake Engineering (TCLEE) 6th U.S. Conference and Workshop on Lifeline Earthquake Engineering, August 10–13, 2003, Long Beach, CA, pp. 868–877

Cochrane, H.C. 1974, Predicting the economic impact of earthquakes, Working Paper No. 15, Institute of Behavioral Science, University of Colorado, Boulder, USA

Cochrane, H.C. 1997, Forecasting the Economic Impact of a Mid-West Earthquake, in Economic Consequences of Earthquakes: Preparing for the Unexpected (editor Jones, B.), MCEER, Buffalo New York, USA

Cole, S. 1988, The Delayed Impacts of Plant Closures in a Reformulated Leontief Model, Papers of the Regional Science Associations, Vol. 65, pp 135–149

Cole, S. 1998, Decision Support for Calamity Preparedness: Socioeconomic and Interregional Impacts, in Engineering and Socioeconomic Impacts of Earthquakes: An Analysis of Electricity Lifeline Disruptions in the New Madrid Area (editor Shinozuka, M.), MCEER, Buffalo, USA

Darthmouth Flood Observatory 2006, World Atlas of Flood Hazard. [url] http://www.dartmouth.edu/~floods/Atlas.html

De Bruijn, K.M. 2005, Resilience and Flood Risk Management, A Systems Approach applied to Lowland Rivers, DUP Science, Delft, Netherlands

Dutta, D. Herath, S. and Musiake, K. 2001, Direct Flood Damage Modeling towards Urban Flood Risk Management, Urban Safety Engineering proceedings, pp 127–143

Federal Emergency Management Agency 2001, Earthquake Loss Estimation Methodology (HAZUS), National Institute of Building Sciences, Washington DC, USA

Ferris, M.C. and Pang, J.S. 1997, Engineering and Economic Applications of Complementary Problems, SIAM Review Vol. 39(9.4), pp 669–713

Glickman, N. 1971, An Economic Forecasting Model for the Philadelphia Region, Journal of Regional Science, Vol. 11, pp 15–32

Gordon, P., Richardson, H. and Davis, B. 1998, Transport-Related Impacts of the Northridge Earthquake, Journal of Transportation and Statistics, Vol. 1, pp 22–36

Hale, A. and Heijer, T. 2006, Defining Resilience, Chapter 3 in Resilience Engineering, Concepts and Precepts (editors Hollnagel, E., Woods, D.D., and Leveson, N.), Ashgate, Hampshire, UK

Minnesota IMPLAN group, Inc. 2006, IMPLAN Professional. [online] URL: http://www.implan.com/products.html

Leontief, W. 1986, Input-Output Economics, Oxford University Press, Oxford, UK

Meyer, V. and Messner, F. 2005, National Flood Damage Evaluation Methods, A Review of Applied Methods in England, the Netherlands, the Czech Republic and Germany, UFZ-Discussion Papers, 21/2005

Okuyama, Y., Hewings, G.J.D., Kim, T., Boyce, D., Ham, H. and Sohn, J. 1999, Economic Impacts of an Earthquake in the New Madrid Seismic Zone: A Multiregional Analysis, in Optimizing Post-Earthquake Lifeline System Reliability (editors Elliot, W. and McDonough, P.), Proceedings of the 5th U.S. Conference on Lifeline Earthquake Engineering, Seattle, USA

Okuyama, Y., Hewings, G.J.D. and Sonis, M. 2000, Sequential Interindustry Model (SIMJ) and Impact Analysis: Applications for Measuring Economic Impact of Unscheduled Events, Regional Economics Application Laboratory, University of Illinois, Urbana, USA

Parker, D.J., Green, C.H. and Thompson, P.M. 1987, Urban Flood protection Benefits, a project appraisal guide "The Red Book", Gower Publishing Company, Brookfield, USA

Parker, D.J. 2000, Floods, Volume I, Routledge, London, UK

Penning-Rowsell, E.C. and Chatterton, J.B. 1977, The benefits of flood alleviation: a manual of assessment techniques, Saxon House/Gower, Farnborough, UK

Rose, A.Z., Cao, Y. and Oladosu, G. 2000, Simulating the economic impacts of climate change in the Mid-Atlantic Region, Climate Research, Vol. 14, pp 175–183

Rose, A.Z. and Lim, D. 2002, Business Interruption Losses from Natural Hazards: Conceptual and Methodological Issues in the Case of the Northridge Earthquake, Environmental Hazards: Human and Social Dimensions, Vol. 4, pp 1–14

Rose, A. Z. 2004, Economic Principles, Issues, and Research Priorities in Natural Hazard Loss Estimation, in Modeling the Spatial Economic Impacts of Natural Hazards (editors Okuyama, Y. and Chang, S.), Springer, Heidelberg, Germany, pp 13–36

Rose, A.Z. and Liao, S.Y. 2004, Modeling Regional Economic Resilience to Disasters: A Computable General Equilibrium Analysis of Water Service Disruptions, Journal of Regional Science, Vol. 45, pp 75–112

Shvyrkov, Y.M. 1980, Centralized Planning of the Economy, Progress Publishers, Moscow, Russian Federation

Smith, D.I., Handmer, J.W., Greenaway, M.A. and Lustig, T.L. 1990, Losses and Lessons from the Sydney Floods of August 1986, Volume 1 and Volume 2, Australian National University, Centre for Resource and Environmental Studies, Canberra, Australia

Walker, B., Holling, C.S., Carpenter, S.R. and Kinzig, A. 2004, Resilience, adaptability and transformability in social-ecological systems, Ecology and Society Vol. 9(9.2): [online] URL: http://www.ecologyandsociety.org/vol9/iss2/art5

Webb, G.R., Tierney, K.J. and Dahlhamer, J.M. 2002, Predicting Long-term business recovery from Disaster: A comparison of the Loma Prieta Earthquake and Hurricane Andrew, Environmental Hazards, Vol. 4, pp 45–58

West, G.R. 1986, A Stochastic Analysis of an Input-Output Model, Econometrica, Vol. 54, pp 363–374

Weyant, J. 1999, The Cost of the Kyoto Protocol: a Multi-Model Evaluation, Energy Journal special issue, pp 1–24

Wing, I.S. 2004, Computable General Equilibrium Models and Their Use in Economy-Wide Policy Analysis,Technical Note No. 6. MIT Joint Program on the Science and Policy of Global Change, Cambridge, USA

10

Flood Damage Estimation and Flood Risk Mapping

Andreas Kron

Institute of Water and River Basin Management, University Karlsruhe (TH), Germany

ABSTRACT: The damages in Germany due to severe flood disasters in the last decades amount to billions of Euro. For example, the Rhine floods in 1993 with 530 M €, and in 1995 with 280 M €, the Odra flood in 1997 with 330 M €, the Danube flood in 1999 with 412 M €, the Elbe and Danube flood in 2002 with 11800 M € (Kron, 2004). The need of specific research efforts and spatial data, particularly hazard or risk maps, for an improved risk assessment and prevention on regional and local level is evident.

The paper deals with requirements on flood and flood risk mapping in the context of the shift from classical flood protection to flood risk management. Both for hydrological and hydraulic modelling the existing methods for flood hazard estimation have to be enhanced for extreme and catastrophic events. As the flood risk is depending on flood parameters and the vulnerability of affected objects in the second part a GIS-Based software-tool is presented which enables estimations of the direct flood damage based on different types of stage-damage functions. With damage estimation for the whole range of occurring flood events from first damage-causing floods up to extreme events, it is possible to provide information for an integrated flood risk management.

10.1   INTRODUCTION

"Hazard" is defined as the occurrence of a flood event with a defined exceedance probability and "risk" as the potential damages associated with such an event, normally expressed as monetary losses. It becomes clear that hazard and risk quantification depend on spatial specifications (e.g. area of interest, spatial resolution of data). With regard to flood risk, the local water level is decisive for the occurrence of damage (e.g. Smith, 1994; see below). Therefore, a high level of detail, i.e. an appropriate scale of flood maps is a fundamental precondition for a reliable flood risk assessment. Detailed spatial information on flood hazard and vulnerability is necessary for the development of regional flood-management concepts, planning and cost-benefit analysis of flood-protection measures and, extremely important,

213

for the preparedness and prevention strategies of individual stakeholders (communities, companies, house owners, etc.). Moreover, the mapped information that an area or object is potentially endangered due to a given flood scenario directly implies legal and economical consequences such as competencies of public authorities for flood control and spatial planning, owner interests, insurance polices, etc.

To determine the hazards, first the relevant natural processes and parameters have to be analyzed and quantified. For the assessment of flood potential, usually events with a defined maximum recurrence interval are taken into account. This information is available for most of the bigger rivers. But there is still a lack of knowledge for events greater than the design discharge. These Floods may be rare, but can cause severe damages and fatalities in case of overtopping or even destruction of protection structures. They can also cause extraordinary hydraulic situations like clogging of bridges due to debris, which need to be analyzed separately. But even for floods below the design discharge, there is a residual risk due to factors such as soaked dikes in long duration flood events or delay in the operation of mobile protection elements.

In Germany, the federal states (Bundesländer) are responsible for flood management and also, for the generation of flood maps. Many state authorities have been working for years on the delineation of inundation zones with map scales of up to ≥1:5000 in urban areas in order to recognise the flood hazard for discrete land-parcels and objects. Increasingly, public flood-hazard maps are available on internet platforms (e.g. Nordrhein-Westfalen, 2003; Rheinland-Pfalz, 2004; Sachsen, 2004; Bayern, 2005; Baden-Württemberg, 2005). In the next years, with respect to amendments in legislation, regional significance and technical possibilities, flood-hazard maps with high spatial resolutions can be expected successively for all rivers in Germany. Details of procedures and mapping techniques vary from state to state and due to local concerns (e.g. data availability, vulnerability, public funds). An overview of different approaches is published by Kleeberg (2005). For example in Baden-Württemberg, two types of flood-hazard maps will be provided for all rivers with catchment areas $>10 \, km^2$. The first map will show the extent of inundation zones of the 10-, 50- and 100-year event, supplemented by an "extreme event" being in the order of magnitude of a 1000-year event and, as documented, information on historical events. The second map will provide the water depths of the 100-year event.

Apart from the determination of flood hazard, the vulnerability in terms of potential economic losses must be estimated. On the basis of the accumulated values at risk in the areas and the functional relationship between the parameters of the flood events and the resulting damage, risk potentials can be identified, quantified and the expected damage can be estimated not only for single events but in the whole range of damaging flood events. It has to be ensured, that the used analytical tools enables the consideration of these factors in order to support the stakeholders.

In Germany the assessment of damage and its visualisation as risk maps is still far from being commonly practised. Risk maps, however, help stakeholders to prioritise investments and they enable authorities and people to prepare for disasters (e.g. Takeuchi, 2001; Merz and Thieken, 2004). One reason may be that the risk (i.e. the damage in a specific area due to a certain flood event) varies more in time than the principal hazard. This means, a hazard map can be interpreted free from local changes in vulnerability and needs therefore not to be updated so frequently as it may be expected from a risk map. Another reason may be that risk maps may hamper economical interests (e.g. land prices).

Good examples for risk assessments and maps are – among others – the ICPR Rhine-Atlas (ICPR, 2001), the programme of flood-hazard mapping in Baden-Württemberg (UM Baden-Württemberg, 2005), the integrated flood management conception in the Neckar river basin (IkoNE, 2002), the DFNK approach for the city of Cologne (Apel et al., 2004; Grünthal et al., in press), and the risk assessment in England and Wales (Hall et al., 2003). Since flood risk encompasses the flood hazard and the consequences of flooding (Mileti, 1999), such analyses require an estimation of flood impacts, which is normally restricted to detrimental effects, i.e. flood losses. In contrast to the above discussed hydrological and hydraulic investigations, flood damage modelling is a field which has not received much research attention and the theoretical foundations of damage models need to be further improved (Wind et al., 1999; Thieken et al., 2005).

10.2   HAZARD MAPPING

10.2.1   Estimation of flood probability

The estimation of flood frequencies is well-known as a key task in flood hazard assessment. The first regionalization methods for flood estimates in Baden-Württemberg were developed in the 1980's (Lutz, 1984). In 1999, the following regionalization approach for mean annual peak discharges (MHQ) and peak discharges (HQT) for recurrence intervals T from 2 to 100 years, partially 200 years was published (LfU, 1999), followed by an updated version on CD in 2001 (LfU, 2001). The approach is based on flood-frequency analyses at 335 gauges which cover catchment areas from less than $10\,km^2$ ($\sim$7% of all gauges) to more than $1000\,km^2$ ($\sim$7%) and periods of records varying from a minimum of 10 to more than 100 years (average 45 years). By this means, regionalized MHQ and HQT are provided for approximately 3400 locations of the river network in Baden-Württemberg, completed by analogous information at 375 gauges and as longitudinal profiles for 163 major rivers.

Actually, the availability of reliable and spatially distributed event parameters for extreme floods is a fundamental prerequisite for any comprehensive

Table 10.1. Calculated "Climate Change Factor" for Baden-Württemberg. The area of Baden-Württemberg is disposed in five regions 1–5 each with same factors (from Baden-Württemberg, 2005).

Recurrence interval [a]	Climate Change Factor $f_{T,K}$				
	1	2	3	4	5
2	1.25	1.50	1.75	1.50	1.75
5	1.24	1.45	1.65	1.45	1.67
10	1.23	1.40	1.55	1.43	1.60
20	1.21	1.33	1.42	1.40	1.50
50	1.18	1.23	1.25	1.31	1.35
100	1.15	1.15	1.15	1.25	1.25
200	1.12	1.08	1.07	1.18	1.15
500	1.06	1.03	1.00	1.08	1.05
1000	1.00	1.00	1.00	1.00	1.00

flood-risk management. For instance, peak discharges for recurrence intervals up to $T = 100$ years (corresponding to an exceedance probability of one percent per year) are commonly accounted in flood mapping and flood-protection planning. Peak discharges for larger events with recurrence intervals up to 1000 or even 10,000 years are required for dam safety analyses (cf. DIN 19700), hazard mapping for extreme cases, related risk analyses and emergency planning purposes. For example in Baden-Württemberg, a guideline gives specific technical recommendations for the dimensioning of flood-protection measures (LfU, 2005a). These recommendations already include the preventative consideration of potential impacts of future climate change on peak discharges by a so-called "Climate Change Factor" proposed by Ihringer (2004) based on statistical analyses of downscaled regional climate-model outputs. The results shows a plausible trend, that for forecast conditions 2050 an increase of flood peak discharge can be expected for small and mean flood water flows (Table 10.1). The effects on climate change on peak discharges can be calculated:

$$HQ_{T,K} = f_{T,K} \cdot HQ_T \qquad (10.1)$$

with $HQ_{T,K}$ = peak discharge with climate change, $f_{T,K}$ = Climate Change Factor, HQ_T = peak discharge without Climate Change.

The effects on the Flood hazard and Damage situation are presented below and shown in Table 10.3.

10.2.2   Flood hazard modelling

To quantify flood hazard and risk in urban areas or at individual locations, flood discharges (e.g. HQ_{100}) have to be transformed into hydraulic parameters like water levels, inundation depths or flow velocities by means of hydrodynamic models. In many cases when the flow patterns in a given river section are characterised by compact and coherent streamlines, 1-dimensional (1D) models are considered as adequate for the estimation of flood-water levels and delineation of inundation zones (e.g. Baden-Württemberg, 2005). In cases with more complex river geometries and flow patterns (e.g. at river confluences or other complex flow conditions), 2-dimensional (2D) models are used for a spatially differentiated hydraulic analysis, especially when local parameters like flow direction, flow velocity, shear stress, etc. are requested. Depending on the intended purposes, both types of models (1D, 2D) may be applied for stationary flow conditions (e.g. hazard assessment for a certain HQ_T) or unsteady flow conditions (e.g. for impact analyses of dike failures).

At the Neckar river, the pilot area of this part of the study, a complex flood-information system was set up since the late 1990 (Oberle et al., 2000), consisting of a series of 1D- and 2D-HN-models which are interactively connected with a geographical information system (GIS). This system enables the simulation of different flood scenarios in order to evaluate, for example, effects of river engineering measures on flood waves. Through its GIS-interface, the hydraulic results can be superposed with a high resolution DTM (grid size: 1×1 m) to determine inundation zones and respectively the boundaries. The DTM is based on elevation data from different data sources, i.e. terrestrial and airborne surveys. Apart from topographical information, flood-relevant spatial data like flood marks, flood impact area, retention zones and legally defined flood areas, are integrated in the GIS. Linkups to aerial photographs of recent flood events complete the volume of spatial data sets.

With respect to the main target parameters of flood-risk analysis and mapping (water levels, inundation zones/depths) and to the flow characteristics along the Neckar river, a generally 1-dimensional HN procedure was chosen. The choice of this procedure was supported by the fact that the handling of the system and the computing time should match with the size of the study area (approx. 220 river kilometres) and the goal, to install the system as operational tool for daily working practice in the water management authorities. Finally, the calculation of a flood event and the visualisation of inundation depths in the GIS only takes minutes with this system, so that analyses can be realised also based on actual flood forecasts. Some river sections with more complex flow conditions (e.g. tributary mouths) could be assessed only insufficiently by means of an 1-dimensional approach. Here, local 2D-HN-models were additionally applied. However, a stationary calculation on the base of a 2-dimensional HN procedure requires several hours even using a powerful computer.

The hydrodynamic method of the above mentioned 1-dimensional procedure is based on the solution of the Saint-Venant-Equations by an implicit difference scheme (Preissmann-scheme, compare Cunge et al., 1980). Under the normal flow conditions of the Neckar river, this approach is valid and very efficient even for large river sections with respect to data handling, model build-up, model calibration and validation as well as sensitivity analyses and, finally, studies of variants. The functionality of the system includes modelling schemes for looped and meshed river systems as well as for river-regulating structures (e.g. weirs, groins, water power plants). The system geometry of the HN-model, i.e. the discharge area of the main channel and the floodplains, is represented by modified cross sections. The model calibration is done by comparing calculated water levels with surveyed ones. In most sections, water level measurements of different recent flood events are available, thus a calibration and validation for a spectrum of (flood) discharges is possible. A detailed description of the system is given by Oberle (2005).

In most cases, HN-models are applied to documented flood events from the last decades (in order to assess the present hydraulic conditions) or to statistical flood events with recurrence intervals up to 100 or 200 years (e.g. for delineation of inundation zones or as design events for flood protection measures). With regard to the increasing process complexity, for instant in the case of overtopping or even destruction of a flood protection structure, and to the lack of measurements for the calibration and validation of model parameters for such cases, the application of HN-models to larger floods is rarely practised. However, despite of the uncertainties, it is necessary to apply HN-models for floods that exceed the 100- or 200-year level, as they are the most relevant situations in terms of residual risks, causing severe damages and fatalities. In particular with respect to residual risk, it is obvious that model parameters should apply for extreme events. Only the reflection of all physically plausible hazardous situations from the occurrence of first inundations to the maximum possible water levels yields to a comprehensive hazard and risk assessment. This is applies equally for flood situations below the design event, as required e.g. for cost-benefit-analyses of protection measures or for the assessment of residual risks due to other failures of technical or non-technical measures (e.g. late installation of mobile protection elements).

Often, historical flood marks indicate much higher water levels than current flood protection level and thus, should serve as realistic reference scenarios for extreme events. In the upper part of the Neckar, the flood with the highest ever recorded water levels occurred in 1824. The water marks can be found at several buildings in flooded communities giving impressions of the severity of historic floods. They can be taken into account for all flood related planning. With adapted HN-models (discharge-relevant areas, roughness coefficients, etc.) it is possible to assess if similar flood water levels could appear in the present situation.

The present HN-calculations at the Neckar river has confirmed that the historical event 1824 was much higher than today's design flood. Fig. 10.1 shows the calculated maximum water levels of the Neckar river for the 100-year flood (HQ_{100}) and the historical flood of 1824 (under actual hydraulic conditions). For example, around the community of Offenau (Neckar-km 98.0), the historical water level

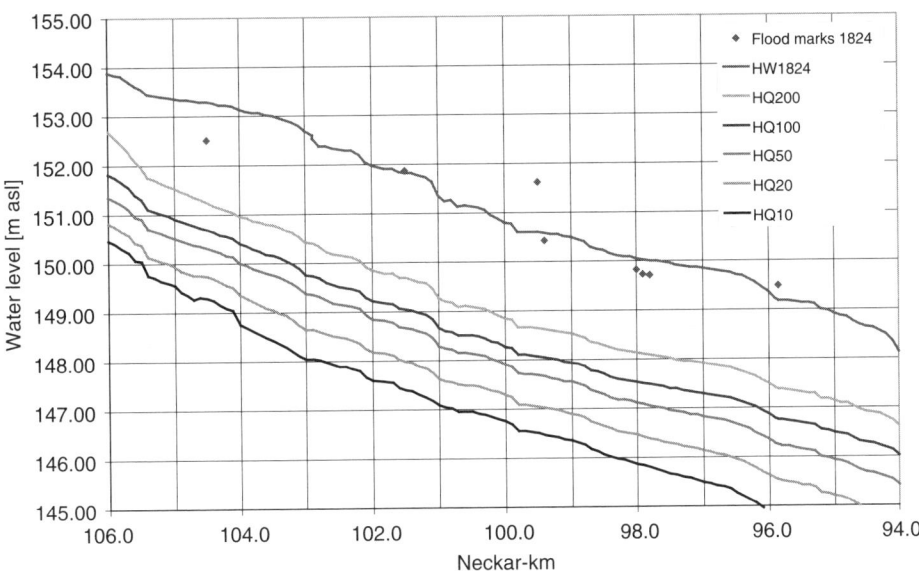

Figure 10.1. Maximum water levels along the Neckar river for statistical flood events (HQ_T with $T = 10, 20, \ldots, 200$ years) and the historical event 1824 (HW1824), the latter as reconstructed from water-level marks. Note location of the community Offenau at the river Neckar (98.0 km upstream of the confluence with the river Rhine).

of 1824 was approximately 2.5 m higher than the dikes that have been built for a 100-year flood.

It has to be emphasised that the consideration of extreme historical events can not only support flood awareness as realised scenarios (under historical conditions), but also used as reference for the analysis of potential extreme cases under present conditions. In this regard, the intention here was not to reconstruct historical hydraulic conditions or to verify historical information in terms of peak discharges. The results shown can help for example to assess the probability of flood events that cause comparable water levels in the actual situation. In terms of a reconstruction of historical discharges, a further investigation on historical hydraulic boundary conditions is required (Oberle, 2005). However, due to the limited historical data availability and quality, major uncertainties are expected.

10.2.3   Hazard mapping

The above presented hydrological and hydraulic models, i.e. the regionalization approach for the estimation of extreme events (HQ_T) as well as the GIS-based flood information system for the Neckar river served as basis for the generation of hazard maps with prototype character in a state-wide sense. Coordinated by the water management authorities in Baden-Württemberg, until 2010 flood hazard maps will be drawn up for all rivers with a catchment area $> 10 \, \text{km}^2$.

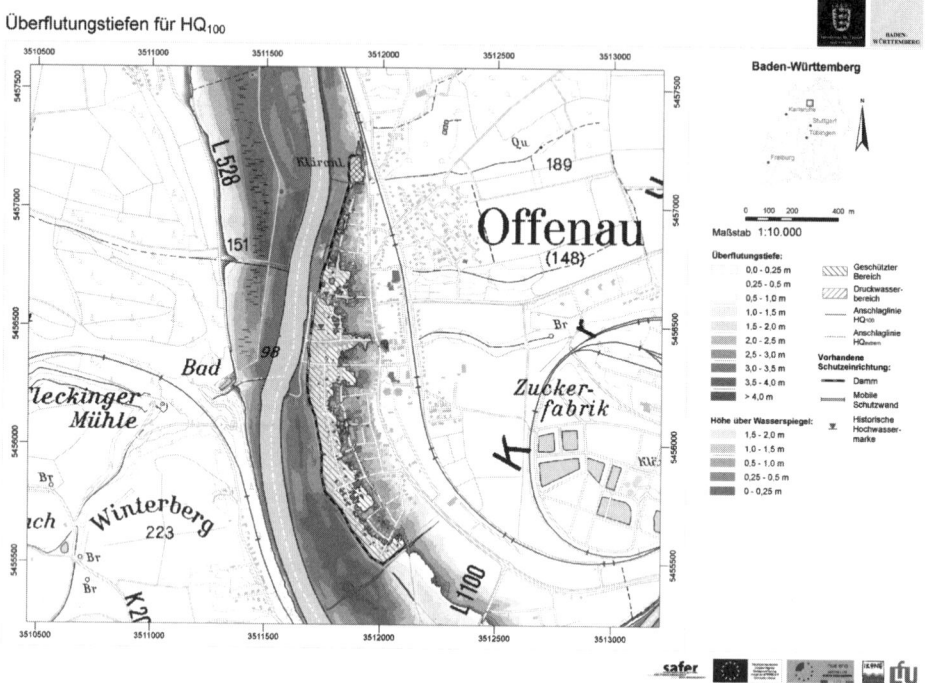

Figure 10.2. Standard Map Design Type 1 – Depth of flood for a 100-year flood (Baden-Württemberg, 2005).

For example, hazard maps for the lower Neckar river (Fig. 10.2) are published on the mentioned internet platform (Baden-Württemberg, 2005). The flood hazard maps for the Neckar river basin are developed within the framework of the INTERREG III-project SAFER and serve as a basis to establish flood partnerships.

10.3   FLOOD RISK MAPPING

10.3.1   Flood-damage estimation

Based on the knowledge on accumulated values in the areas at risk and relationships between event parameters and resulting damage, flood risks can be identified and quantified, i.e. expected damages for a given flood scenario can be calculated. This information about flood risk for individual buildings, settlement areas and river basins is indispensable to inform the population and stakeholders about the local flood risk, for planning of flood control measures and for benefit-cost analyses of these measures.

The comprehensive determination of flood damage involves both, direct and indirect damage. Direct damage is a damage which occurs due to the physical contact of flood water with human beings, properties or any objects. Indirect damage is a damage which is induced by the direct impact, but occurs – in space or time – outside the flood event. Examples are disruption of traffic, trade and public services. Usually, both types are further classified into tangible and intangible damage, depending on whether or not these losses can be assessed in monetary values (Smith and Ward, 1998). Although it is acknowledged that direct intangible damage or indirect damage play an important or even dominating role in evaluating flood impacts (FEMA, 1998; Penning-Rowsell and Green, 2000), the largest part of the literature on flood damage concerns direct tangible damage (Merz and Thieken, 2004).

A central idea in flood damage estimation is the concept of damage functions. Most functions have in common that the direct monetary damage is related to the inundation depth and the type or use of the building (e.g. Smith, 1981; Krzysztofowicz and Davis, 1983; Wind et al., 1999; NRC, 2000; Green, 2003). This concept is supported by the observation of Grigg and Helweg (1975) "that houses of one type had similar depth-damage curves regardless of actual value". Such depth-damage functions, also well-known as stage-damage-functions, are seen as the essential building blocks upon which flood damage assessments are based and they are internationally accepted as the standard approach to assess urban flood damage (Smith, 1994).

Generally, there are two main approaches in estimating the direct damages due to flooding. The first (ex post-) method is based on the collection of historical flood damage information. Relevant damage determining information (e.g. building type, water depth in cellar and ground floor, damage for building fabric, fixed and mobile inventory, etc.) can be included in databases out of which average stage-damage functions can be derived for different classified building types.

In Germany, most stage-damage curves are based on the most comprehensive German flood damage data base HOWAS that was arranged by the Working Committee of the German Federal States' Water Resources Administration (LAWA) (Buck and Merkel, 1999; Merz et al., 2004). But recent studies have shown that stage-damage functions may have a large uncertainty (e.g. Merz et al., 2004). These uncertainties may be caused by various reasons. Beneath the building specific parameters there is a multiplicity of factors with impact on damages. Damages in regions that are frequently affected and "experienced" in flooding are less than in regions that are unprepared. For flash floods induced by local heavy precipitation the warning time is very short so the affected people are not able to safeguard their valuables. If damage information is collected by questioning of affected people the time between flooding and questioning is also relevant. If people are asked immediately after flooding the damages may be overestimated under the impression of loosing life standard. On the other hand, long term damages like mildewing of cellar walls are recognised much later. It is still not possible to survey all determining factors, so large uncertainties cannot be avoided.

The second method is based on predefined relationships between water depth and monetary damages. Probably the most comprehensive approach has been the Blue Manual of Penning-Rowsell and Chatterton (1977) which contains stage-damage curves for both residential and commercial property in the UK. In this study synthetic depth-damage-function for different types of residential buildings was derived on base of typical examples of house types, age, inventory and social class of the inhabitants for the UK. The hypothetic damages for building fabric and contents are given for both flood durations, less and more than 12 hours. Other ex ante approaches bases on What-If-Surveys where experts are appraising the potential damages on building structure, inventory etc. for different water levels. With these expertises, stage-damage-functions can be derived. Depending on the extent of the project area and required level of accuracy the What-If-Surveys surveys can be carried out with different levels of detail:

• For spacious areas resp. approximately identification of areas with relevant damage potential stage-damage-functions can be derived from sampling surveys of specific building types for example one-family houses or apartment buildings. The functions for the surveyed buildings are assigned for all buildings of the same type or age.

• If a higher level of accuracy is required or if high valued objects are placed within the project area, these objects has to be identified and individual stage-damage-functions has to be determined. These approaches can be used e.g. for cost-benefit analyses of regional or local flood protection measures.

• If the damages have to be calculated very precisely for every single object, there is no other way than to examine every possibly affected building or object.

As outlined above, stage-damage-functions for different building types or building uses are an internationally accepted standard approach for flood damage estimation. While the outcome of most stage-damage-functions is the absolute monetary loss to a building, some approaches provide relative depth-damage-functions, determining the damage e.g. in percentage of the building value (e.g. Dutta et al., 2003). If these functions are used to estimate the loss due to a given flood scenario property values have to be predetermined (Kleist et al., 2004). However, using these functions, one has to be aware that the damage estimation generally is associated with large uncertainties, as recent studies asserted (Merz et al., 2004). One approach to reduce the uncertainty connected with stage-damage-functions is their specific adjustment to the area of interest (Buck and Merkel, 1999).

Recent flood events have shown, that during slowly rising river floods the maximum water level during the flood event is responsible for the resulting damage. In these cases, the gradient of the flood wave is small and for this reason there are no damaging effects due to flow velocity impacts. Major damages are caused by wetting of inventory and building structure in the cellar and the ground floor.

This does not apply for flash floods e.g. in mountainous areas where, due to high flow velocity, buildings may collapse partly or totally. Therefore, it is obvious that flood damage depends, in addition to building type and water depth, on many factors which are not considered using stage-damage-functions. One factor is the flow velocity, but there are also others like the duration of inundation, sediment concentration, availability and information content of flood warning, and the quality of external response in a flood situation (Smith, 1994; Penning-Rowsell et al., 1994; USACE, 1996). Although a few studies give some quantitative hints about the influence of some of the factors (McBean et al., 1988; Smith, 1994; Wind et al., 1999; Penning-Rowsell and Green, 2000; ICPR, 2002; Kreibich et al., 2005), there is no comprehensive approach to consider these factors in a loss-estimation model. Using actual flood damage data from the 2002 flood in Germany, we followed this idea here and developed a multifactorial approach for damage estimation.

The flood-damage estimation can be undertaken on different levels of spatial differentiation:

- On local scale, the damages can be estimated based on spatial data and stage-damage-functions for individual buildings or land parcels.
- On a more aggregated level, the approach can be based on statistical information about population, added values, business statistics or capital assets for land-use units. These values are published yearly by responsible state authorities (statistical offices).
- Large-scale analyses may be carried out for larger land-use units, like communities or ZIP-code areas, considering that they may be only partially flooded.

During the last years, the computational power increased in a way that today flood damage analyses even for larger river courses can be undertaken with a high level of detail. In this context, the question of spatial scale of damage analysis is moving from limitations concerning the area size to limitations concerning the quality respectively the level of detail of available spatial data sets of hazard and vulnerability.

10.3.2   GIS-based flood damage estimation on local scale based on stage-damage curves

On the basis of the accumulated values in the areas at risk and the knowledge about the functional interrelation between the event parameters and the resulting damage, risk potentials can be identified and quantified and the damage expectancy values can be calculated. This information about flood risk for individual buildings, settlement areas and river basins is indispensable to inform the population and stakeholders about the local flood risk, for flood control planning measures and for benefit-cost analyses of projected hydraulic measures.

As discussed above, it is commonly required in flood-risk assessment to locate accessible information about hazard and vulnerability at a high spatial resolution (e.g. for cost-benefit analyses, for local protection measures, rating of risks for insurance purposes). In view of these practical requirements, a GIS-based tool for damage estimation was developed in the "Risk-Map Germany" – Project of CEDIM (Center for Disaster Management and Risk Reduction Technology, www.cedim.de). This tool supplements the above mentioned flood information system at the Neckar river, i.e. it builds directly on the water level information for individual endangered objects based on hydrodynamic calculations.

The GIS-based tool for damage estimation on local scale uses the following procedure.

- Selection of the project area (spatial, postal- codes or areas of communities).
- Identification and categorization of each building in the project area (based on ALK-Data).
- Estimation of the flood-sill for each structure (lowest damaging water level).
- Estimation of the ground-floor elevation (above cellar).
- Estimation of the values for building-structure and contents (fixed/mobile inventory).
- Estimation of the stage-damage-functions, differentiated for different types of buildings, cellar/floor, building structure/inventory.
- Calculation of the water-level for each object in the area.
- Estimation of the damages to buildings and contents for different water-levels based upon the type and use of each building.

The tool provides the selection of the project area on the base of different spatial or administrative areas: barrages, communities or ZIP-code areas. The area of interest or spatial objects can be selected from tables or as graphical selection in the GIS.

For the damage-estimation, the water depth close to or inside the object is the determining factor. With the HN-modelling in connection with the digital terrain model the water depths above the terrain is calculated. The assumption that the damaging water depth inside the object is the same as the depth over terrain is correct if the ground floor elevation has the same elevation as the surrounding territory and if there are no protection measures. In this case, the relevant elevation of the object basis can be calculated on the base of the DTM as the mean value of the terrain altitude on the buildings base. A second option in the tool is to enter the first floor elevation and the height of the flood-sill for each single object. Thus, local object features and protection measures can be considered.

Due to the fact that the absolute damage depends on a variety of factors being specific for every single building or land parcel, a meaningful damage estimation can be expected from the application of such stage-damage-functions with adaptation for individual objects or – in terms of exposure and vulnerability – uniform spatial units. For that reason, the possibility to apply different functions was

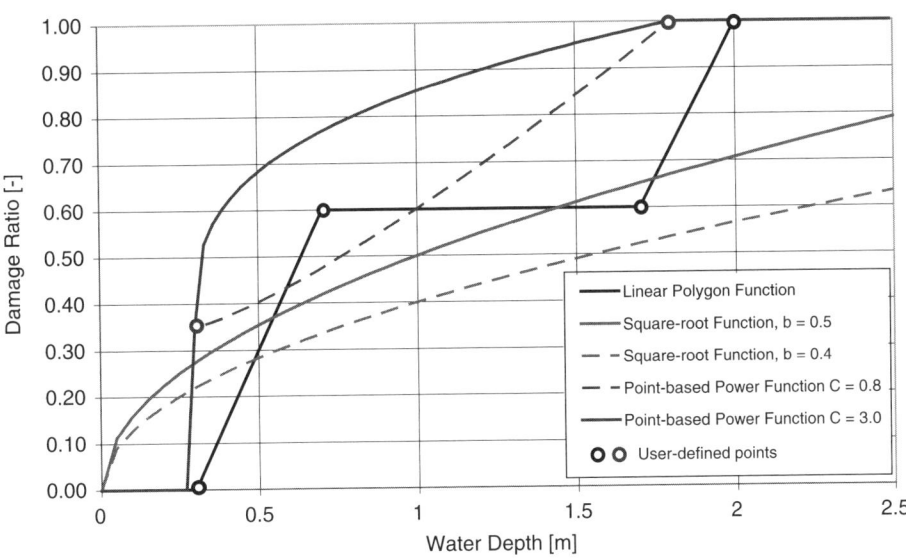

Figure 10.3. Examples of selectable function types in the GIS-based damage analysis tool. The Linear Polygon function is given by the user defines points. The shape of the Square-Root is determined by the multiplying parameter *b*, the shape of the Point-based Power Function is defined by start and end point as well as the exponent C.

implemented in this software module, where the user can choose at least one of the three following function types: 1. Linear Polygon Function, 2. Square-Root Function, or 3. Point-based Power Function. Examples for the different Functions types are shown in Fig. 10.3.

10.3.2.1 *Linear polygon function*

The user interface allows to enter 5 pairs of variates (h_i/S_i) of water-depth and damage, which are interpolated sectional with linear functions. Between the minimum $(i = 1)$ and maximum $(i = 5)$ pair, the function can be noted:

$$S = S_i + \left(\frac{S_{i+1} - S_i}{h_{i+1} - h_i} \right) \cdot (h - h_i), \quad h_i \leq h < h_{i+1} \tag{10.2}$$

where $S =$ estimated damage, S_i, $h_i =$ user-defined nodes of the function, $h =$ water depth.

The first pair of variates (h_1/S_1) defines the minimum water depth below which the damage is zero. The last point (h_5/S_5) sets the possible maximum damage; for water depths above the damage stays constant. The Polygon function (10.3) allows a simple adaptation to individual damage symptoms of different types of objects.

10.3.2.2   Square-root function

In practical view, square-root stage-damage functions provide good results for damage estimation (Buck and Merkel, 1999). Therefore, a square-root function is implemented in the damage estimation tool as second function type, where the parameter b is user-defined:

$$S = b \cdot \sqrt{h}$$

(10.3)

with S = estimated damage, h = water depth, b = user-defined parameter.

The parameter b characterises the damage for $h = 1$ m. Hence, using Eq. (10.4), the damage progression can be described with only one parameter. For the sample damage estimation in this paper (see below), the damage functions for different building types in the project area were chosen based on the flood-damage database HOWAS (Buck and Merkel, 1999).

10.3.2.3   Point-based power function

In some cases, damage does not occur until the stage rises a threshold height in the building. For example in rooms or storeys, where floors and walls are tiled, damage can be negligible until the water level affects the electrical installation (power sockets). On the other hand, the maximum damage is often obtained, when the inventory is submerged; a further rise of the water level does not increase the damage in a relevant manner. For these cases, a power function can be chosen in the tool, where the points of first and maximum damage as well as the exponent B determining the gradient of the function can be individually defined.

$$S = S_{max} \cdot \left[\frac{S_0}{S_{max}} + \left(1 - \frac{S_0}{S_{max}}\right) \cdot \left(\frac{h - h_0}{h_{max} - h_0}\right)^{1/c} \right]$$

(10.4)

with S = estimated damage, h = water depth, (h_0/S_0) = point of first damage, (S_{max}/h_{max}) = point of maximum damage, C = user-defined exponent.

The creation, editing, and choice of these three functions are realised by different masks, that allow the user to conveniently handle the input. For the Polygon and the Power function types, the damage can be calculated in absolute monetary units (EUR) or percents of damage.

Before starting the calculation, the relevant flood event must be selected. According to the coupling of the damage estimation tool to the flood information system for the Neckar river, the outcomes of the hydraulic calculations, i.e. water surfaces, can be directly used as input for damage estimation. The implementation of the damage estimation tool in a GIS-software environment is realised in four dialogue modules shown in Fig. 10.4.

Hence, the GIS-based damage estimation tool enables the user to assess the flood damage to single buildings in flood-prone areas and the spatial aggregation of the event-specific damage for a defined group of buildings or areas. Fig. 10.5 shows

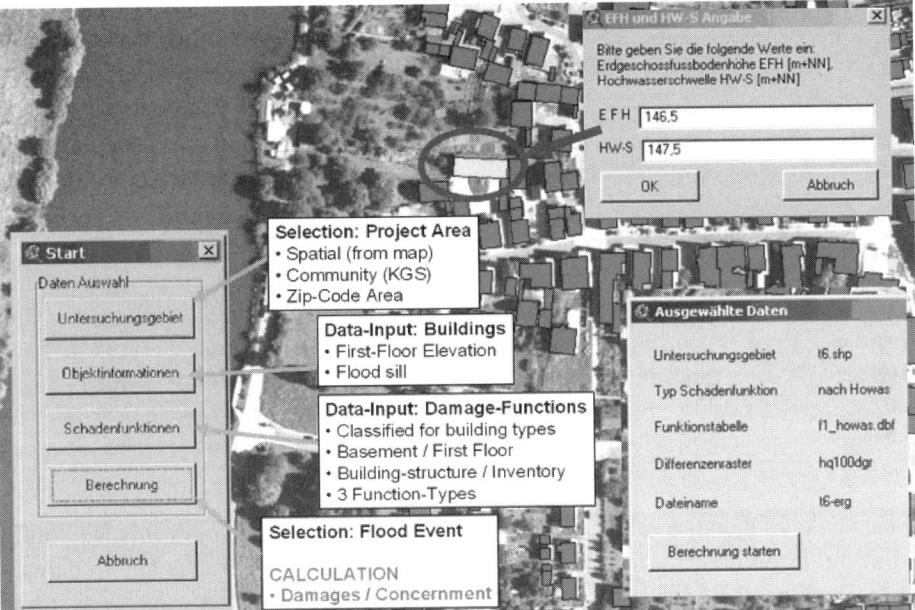

Figure 10.4. GIS-based damage analysis tool (screenshot of graphical user interface). The damage or the general involvement can be calculated by selecting the area of interest, land-use and event information, and damage-relevant factors.

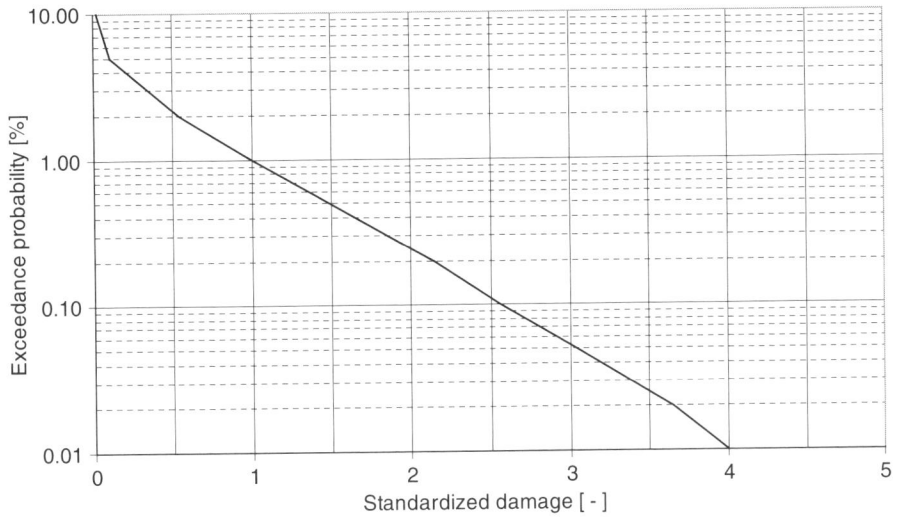

Figure 10.5. Standardized damage to residential buildings in a test community exceedance probabilities from 0.1 to 0.0001 (i.e. damage to a 100-year flood = 1, situation without flood protection measure).

Table 10.2. Standardized damage to buildings and inundated areas for the test community.

Exceedance probability [%]	Affected buildings [-]		Inundated area [m²]		Standardized Damages [%]
	Residential	Total	Residential	Settlement	
10	4	8	4,000	6,000	0.002
5	14	40	15,000	29,000	0.01
2	71	126	41,000	71,000	53
1	127	242	81,000	129,000	100
0.50	223	378	143,000	220,000	194
0.20	237	403	156,000	237,000	215
0.10	266	447	174,000	267,000	255
0.05	296	496	192,000	298,000	304
0.02	328	540	203,000	326,000	364
0.01	342	571	207,000	341,000	400

the calculated damage values for a test community for a range of events, beginning from the flood causing the first damage up to the 1000-year event. The damage values in Fig. 10.5 are standardized to the 100-year event. That means for example, that the damage caused by the 1000-year flood is approximately 2.6 times higher than the one caused by the 100-year event.

Furthermore, the tool includes functionalities to cope with cases where detailed land-use data e.g. from the Automated Real Estate Map (ALK) are not available or where the assessment could be simplified. As revealed in Table 10.2, it is possible to make assumptions e.g. about the number of affected houses in flood-prone areas, in order to give an overview on flood risk without explicitly calculating monetary damage. For damage calculations on a more aggregated spatial level, the values at risk can be derived from statistical data for administrative districts and related to their spatial unit (EUR/m²). In this case, the damage estimation can be delivered by spatial intersection of flood-hazard information (inundation zone) with land-use data e.g. data from the Authoritative Topographic-Cartographic Information System in Germany (ATKIS) in order to calculate the extension the inundated settlement area (see columns 4 and 5 in Table 10.2).

Usually, the flood-damage calculation is provided for cost-benefit analyses of flood protection measures. For this purpose, the costs of a flood-protection measure can be compared to its benefit, i.e. the expected damage up to the design event respectively the residual risk after the implementation of the measure, normally expressed as mean annual damage (MAD). The MAD is defined as the average or mean of all harmed values. It is calculated as the integral of the damage-probability function.

$$MAD_{i,k} = \int_{i}^{k} S_i(P_i) \cdot dP_i \qquad (10.5)$$

Table 10.3. Effects on climate change on the water levels at Neckar km 98.0 and the damages to residential buildings in the community of Offenau (without protection measures).

Recurrence interval [a]	Climate change factor $f_{T,K}$ [-]	Increase of peak discharge [m^3/s]	Increase of water level [m]	Damage increasing factor [-]
10	1.40	672	1.13	25.3
20	1.33	647	1.04	8.9
50	1.23	536	0.81	3.6
100	1.15	392	0.58	2.0
200	1.08	231	0.34	1.1
500	1.03	96	0.13	1.1
1000	1.00	0	0	1.0

where MAD = mean annual damage, $S_i(P_i)$ = expected damages S_i for a given probability P_i, I = probability of flood event, that causes first damages, k = prob. of flood event with maximum damages.

For the above mentioned sample community, a dike designed for a 100-year event provides a significant reduction of MAD, but the residual risk due to a larger flood event still accounts for approximately 40 percent of the original value.

The main advantage of the presented damage estimation on local scale is, that the damage-determining factors are given both on hazard side (being in general the water depth) as well as on side of vulnerability (stage-damage-functions for individual objects). With the presented tool one can easily calculate potential damages and effects on any changes in the flood hazard situations in short time. As an example Fig. 10.3 shows the impact of future climate change to the damages on residential buildings. With the above-mentioned Climate Change Factors especially for floods with high recurrence intervals there is a significant increase of peak discharge and water levels. As the damages are a function of water level it can be seen, that damages for frequent events are increasing significant. The maximum can be identified for a 10-year flood with resulting damages that are app. 25-times higher than in the today situation. And even for the 100-years flood the damages are twice as high as today.

For estimations on an aggregated spatial level, where areas of the same building type may be defined (e.g. ATKIS-data), but no information on individual objects is available, one has to make assumptions on the spatial distribution of buildings and building types. Furthermore, as the water depth in an inundated area varies in space, the definition of the damage-determining water level gets more uncertain with increasing spatial units. Thus, using stage-damage-functions, one has to define the damage-relevant depth or use a statistical approach to estimate the spatial distribution.

10.4 CONCLUSIONS

Flood management and consequently flood mapping is a key task and ongoing development in the sphere of competency of state authorities in Germany. To support this demanding target from the scientific side, especially aiming at a more reliable flood-risk assessment, these studies were focussed on the improvement of the following methods: 1. The estimation of extreme events which exceed the design flood of flood protection measures, 2. The assessment of flood hazard and risk over the whole spectrum of possible damage-relevant flood events, 3. The damage estimation via the consideration of various building- and event-specific influences on the resulting damage. In detail, a regionalization approach for flood peak discharges was presented, especially regarding recurrence intervals of 200 to 10,000 years and a large number of small ungauged catchments. In addition, the newly developed multifactorial approach for damage estimation considers more damage influencing factors besides the water depth, like building quality, contamination and precautionary measures.

Flood damage and risk assessment is actually practised with an increasing level of detail and accuracy for more and more areas, especially in regions which are frequently or were recently affected by floods, like Baden-Württemberg. In this regard, it is on the one hand self-evident that flood hazard and risk, in particular the residual risk of flood protection measures, must be assessed on a level of detail which supports local planning and precaution. This includes the assessment for direct and indirect damages for the whole range of relevant flood events from the occurrence of first damages up to catastrophic events with low recurrence probability (10^{-3} or lower). With the coupling of hydraulic (hazard) simulation models and Flood Damage Estimation tools Risk assessment can be derived with high level of detail.

Furthermore, as hazard and vulnerability are not constant in time, corresponding analyses and maps must be updated after significant changes with a minimum of additional expenses. The presented information systems and analysis tools are a base for these purposes, not only for our specific study areas. On the other hand, as discussed, flood hazard and risk assessment is still associated with large uncertainties, even in areas where a rather good data and model base is available (as for example given at the Neckar river by means of the flood information system with relatively well-known hydraulic conditions and substantial spatial data). Considering that integrated flood-risk management implies decisions under uncertainty, the effectiveness of detailed risk analyses has to be critically reflected.

ACKNOWLEDGEMENT

Most results presented in this paper have been worked out in the framework of the "Risk-Map Germany"-Project of CEDIM (Center for Disaster Management and Risk Reduction Technology, www.cedim.de) and are based upon the following publications:

Büchele, B., Kreibich, H., Kron, A., Ihringer, J., Oberle, P., Thieken, A., Merz, B., Nestmann (2006). Flood-risk mapping: contributions towards an enhanced assessment of extreme events and associated risks, Nat. Hazards Earth Syst. Sci., 6, 621–628, 2006.

Kron, A., Evdakov, O., Nestmann, F. (2005). From hazard to risk – a GIS-based tool for risk analysis in flood management. In: Van Alphen, van Beek, Tal (eds) Floods, from Defence to management, London, 2005, 295–301.

Büchele, B., Kreibich, H, Kron A., Ihringer, J., Theobald, S., Thieken, A., Merz, B., Nestmann, F. (2004). Developing a Methodology for Flood Risk Mapping: Examples from Pilot Areas in Germany. In: Malzahn, D., Plapp, T. (eds) Disasters and Society – From Hazard Assessment to Risk Reduction. Logos-Verlag, Berlin, 99–106.

REFERENCES

Apel, H., Thieken, A.H., Merz, B., Blöschl, G.: Flood Risk Assessment and Associated Uncertainty, Natural Hazards and Earth System Sciences, 4, 295–308, 2004. SRef-ID: 1684-9981/nhess/2004-4-295.

Baden-Württemberg: Hochwassergefahrenkarte Baden-Württemberg (Flood-hazard map Baden-Württemberg). Umweltministerium Baden-Württemberg, Stuttgart, 2005 (http://www.hochwasser.baden-wuerttemberg.de).

Baden-Württemberg: Festlegung des Bemessungshochwassers für Anlagen des technischen Hochwasserschutzes, Leitfaden. Landesanstalt für Umweltschutz Baden-Württemberg, Oberirdische Gewässer, Gewässerökologie, Heft 92, Karlsruhe, 2005.

Bayern: Informationsdienst Überschwemmungsgefährdete Gebiete in Bayern (Information service flood endangered areas in Bavaria). Bayerisches Landesamt für Wasserwirtschaft, München, 2005 (http://www.geodaten.bayern.de/bayernviewer-aqua/)

BMI (Bundesministerium des Innern): Bundesregierung zieht vorläufige Schadensbilanz der Hochwasserkatastrophe: bisher finanzielle Hilfe im Umfang von über 700 Millionen Euro geleistet (Federal government draws provisional damage balance of the flood disaster: hitherto financial assistance provided to the extent of over 700 million euros). Press release (6.11.2002, http://www.bmi.bund.de/dokumente/ Pressemitteilung/ix_90912.htm), 2002.

Bobée, B., Cavadias, G., Ashkar, F., Bernier, J. Rasmussen, P. F.: Towards a systematic approach to comparing distributions used in flood frequency analysis, J. Hydrol., 142, 121–136, 1993.

Buck, W., Merkel, U.: Auswertung der HOWAS-Schadendatenbank. Institut für Wasserwirtschaft und Kulturtechnik der Universität Karlsruhe. HY98/15, 1999.

Buck, W., Monetary Evaluation of Flood Damage (2004). In: Malzahn, D., Plapp, T. (eds) Disasters and Society – From Hazard Assessment to Risk Reduction. Logos-Verlag, Berlin, 123–128.

Büchele, B., Kreibich, H., Kron, A., Ihringer, J., Theobald, S., Thieken, A., Merz, B., Nestmann, F. (2004). Developing a Methodology for Flood Risk Mapping: Examples from Pilot Areas in Germany.

Cunge, J.A., Holly, F.M., Verwey, A.: Practical aspects of computational hydraulics. Institute of Hydraulic Research, Iowa, 1980.

Deutsche Rück (Deutsche Rückversicherung AG): Das Pfingsthochwasser im Mai 1999 (The flood at Whitsun in May 1999). Deutsche Rück, Düsseldorf, 1999.

DKKV (Deutsches Komitee für Katastrophenvorsorge e.V.): Hochwasservorsorge in Deutschland – Lernen aus der Katastrophe 2002 im Elbegebiet (Flood risk prevention in Germany – Lessons Learned from the 2002 disaster in the Elbe region), Schriftenreihe des DKKV 29, DKKV, Bonn, 2003.

Dutta, D., Herath, S. Musiake, K.: A mathematical model for flood loss estimation, J. Hydrology, 277, 24–49, 2003.

DWD: Starkniederschlagshöhen für die Bundesrepublik Deutschland (Heavy precipitation depths for the Federal Republic of Germany). Deutscher Wetterdienst, Offenbach, 1997.

Engel, H.: The flood event 2002 in the Elbe river basin: causes of the flood, its course, statistical assessment and flood damages, Houille Blanche, 2004(6), 33–36, 2004.

FEMA (Federal Emergency Management Agency) (1998). Costs and benefits of natural hazard mitigation, Federal Emergency Management Agency, Washington DC.

Green, C. (2003). Handbook of Water Economics: Principles and Practice, 443 pp., John Wiley and Sons, Chichester.

Grigg, N.S. Helweg, O.J.: State-of-the-art of estimating flood damage in urban areas, Water Resour. Bull., 11(10.2), 379–390, 1975.

Grünthal, G., Thieken, A.H., Schwarz, J., Radtke, K., Smolka, A., Gocht, M., Merz, B., Comparative risk assessment for the city of Cologne, Germany – storms, floods, earthquakes, Natural Hazards (in press).

Hall, J.W., Dawson, R.J., Sayers, P.B., Rosu, C., Chatterton, J.B., Deakin, R.: A methodology for national-scale flood risk assessment, Water & Maritime Engineering, 156, 235–247, 2003.

ICPR (International Commission for the Protection of the Rhine) (2001). Rhine-Atlas, ICPR, Koblenz, retrieved from http://www.rheinatlas.de.

ICPR (International Commission for the Protection of the Rhine) (2002). Non Structural Flood Plain Management – Measures and their Effectiveness, ICPR, Koblenz.

Ihringer, J. Softwarepaket für Hydrologie und Wasserwirtschaft. Anwenderhandbuch (Software package for hydrology and water resources management. User handbook). Universität Karlsruhe, 1999.

Ihringer, J.: Ergebnisse von Klimaszenarien und Hochwasser-Statistik (Results from climate scenarios and flood statistics). Proc. 2nd KLIWA-Symposium, 153–167, Würzburg, 2004, retrieved from http://www.kliwa.de.

IKoNE (Integrierende Konzeption Neckar-Einzugsgebiet): Hochwassermanagement, Partnerschaft für Hochwasserschutz und Hochwasservorsorge, Heft 4, Besigheim 2002.

IKSE (International Commission for the Protection of the Elbe): Dokumentation des Hochwassers vom August 2002 im Einzugsgebiet der Elbe (Documentation of the flood in August 2002 in the Elbe catchment), Magdeburg, 2004 (in German).

Kleeberg, H.B. (ed.): Hochwasser-Gefahrenkarten (Flood hazard maps). Forum für Hydrologie und Wasserbewirtschaftung, Heft 08.05, 2005.

Kleist, L., Thieken, A., Köhler, P., Müller, M., Seifert, I., Werner, U.: Estimation of Building Values as a Basis for a Comparative Risk Assessment. In: Malzahn, D., Plapp, T. (eds) Disasters and Society – From Hazard Assessment to Risk Reduction. Logos-Verlag, Berlin, 115–122, 2004.

Kreibich, H., Thieken, A.H., Petrow, T., Müller, M., Merz, B. (2005) Flood loss reduction of private households due to building precautionary measures – Lessons Learned from the Elbe flood in August 2002, NHESS – Natural Hazards and Earth System Sciences, 5: 117–126.

Kron, W.: Zunehmende Überschwemmungsschäden: Eine Gefahr für die Versicherungswirtschaft? (Increasing flood damage: a hazard for the insurance industry?). ATV-DVWK: Bundestagung 15.–16.09.2004 in Würzburg, DCM, Meckenheim, 47–63, 2004.

Kron, A., Oberle, P., Theobald S.: Werkzeuge zur Risikoanalyse im Hochwassermanagement (Tools for risk analysis in flood management). 4. Forum Katastrophenvorsorge, 14.–16. Juli 2003, München, 2003.

Krzysztofowicz, R., Davis D.R. (1983). Category-unit loss functions for flood forecast-response system evaluation, Water Resources Research, 19(6), 1476–1480.

LfU (Landesanstalt für Umweltschutz Baden-Württemberg): Hochwasserabfluss-Wahrscheinlichkeiten in Baden-Württemberg (Flood discharge probabilities in Baden-Württemberg). Oberirdische Gewässer/Gewässerökologie 54, Karlsruhe, 1999.

LfU (Landesanstalt für Umweltschutz Baden-Württemberg): Hochwasserabfluss-Wahrscheinlichkeiten in Baden-Württemberg – CD (Flood discharge probabilities in Baden-Württemberg – CD). Oberirdische Gewässer/Gewässerökologie 69, Karlsruhe, 2001.

LfU (Landesanstalt für Umweltschutz Baden-Württemberg): Festlegung des Bemessungshochwassers für Anlagen des technischen Hochwasserschutzes (Definition of the design flood for technical flood protection measures). Oberirdische Gewässer/ Gewässerökologie 92, Karlsruhe, 2005a.

LfU (Landesanstalt für Umweltschutz Baden-Württemberg): Abflusskennwerte in Baden-Württemberg. Teil 1: Hochwasserabflüsse, Teil 2: Mittlere Abflüsse und Mittlere Niedrigwasserabflüsse (Flow parameters in Baden-Württemberg: Part 1: Flood discharges, Part 2: Mean flows and low flows). Oberirdische Gewässer/ Gewässerökologie 94, Karlsruhe, 2005b.

Lutz, W.: Berechnung von Hochwasserabflüssen unter Anwendung von Gebietskenngrößen (Calculation of flood discharges with application of spatial parameters). Mitteilungen des Instituts für Hydrologie und Wasserwirtschaft, Universität Karlsruhe, Heft 24, 1984.

McBean, E.A., Gorrie, J., Fortin, M., Ding, J., Moulton, R.: Adjustment Factors for Flood Damage Curves, J. Wat. Res. Plan. Man., 114(6), 635–645, 1988.

Merz, B., Thieken, A.H.: Flood risk analysis: Concepts and challenges, Österreichische Wasser- und Abfallwirtschaft, 56(3–4), 27–34, 2004.

Merz, B., Kreibich, H., Thieken, A.H., Schmidtke, R.: Estimation uncertainty of direct monetary flood damage to buildings, NHESS – Natural Hazards and Earth System Sciences, 4, 153–163, 2004.

Merz, B. Thieken, A.H.: Separating Aleatory and Epistemic Uncertainty in Flood Frequency Analysis, J. Hydrol., 309, 114–132, 2005.

Mileti, D.S.: Disasters by design. A reassessment of natural hazards in the United States. Joseph Henry Press, Washington DC, 1999.

MUNLV (Ministerium für Umwelt und Naturschutz, Landwirtschaft und Verbraucherschutz des Landes Nordrhein-Westfalen): Leitfaden Hochwasser-Gefahrenkarten (Guideline flood-hazard maps). Düsseldorf, 2003.

MURL (Ministerium für Umwelt, Raumordnung und Landwirtschaft des Landes Nordrhein-Westfalen): Potentielle Hochwasserschäden am Rhein in NRW (Potential flood damage at the Rhine in North-Rhine Westfalia). Düsseldorf, 2000.

Nordrhein-Westfalen: Hochwassergefährdete Bereiche (Flood endangered areas). Landesumweltamt Nordrhein-Westfalen, Düsseldorf, 2003 (http://www.lua.nrw.de/index.htm?wasser/hwber.htm).

NRC (National Research Council): Risk analysis and uncertainty in flood damage reduction studies. National Academy Press, Washington DC, 2000.

Oberle, P., Theobald, S. Nestmann, F.: GIS-gestützte Hochwassermodellierung am Beispiel des Neckars. Wasserwirtschaft, 90 (7–8), 368–373, 2000.

Parker, D.J., Green, C. H., Thompson, P. M.: Urban Flood Protection Benefits: A project appraisal guide. Gower Technical Press, Aldershot, 1987.

Penning-Rowsell, E.C. Chatterton, J.B.: The Benefits of Flood Alleviation: A Manual of Assessment techniques. Gower Technical Press, Aldershot, 1977.

Penning-Rowsell, E.C. Green, C.: New Insights into the appraisal of flood-alleviation benefits: (1) Flood damage and flood loss information, J. Chart. Inst. Water E., 14, 347–353, 2000.

Penning-Rowsell, E., Fordham, M., Correia, F.N., Gardiner, J., Green, C., Hubert, G., Ketteridge, A.-M., Klaus, J., Parker, D., Peerbolte, B., Pflügner, W., Reitano, B., Rocha, J., Sanchez-Arcilla, A., Saraiva, M.d.G., Schmidtke, R., Torterotot, J.-P., van der Veen, A., Wierstra, E., Wind, H.: Flood hazard assessment, modelling and management: Results from the EUROflood project, in Floods across Europe: Flood hazard assessment, modelling and management, edited by E.C. Penning-Rowsell and M. Fordham, 37–72, Middlesex University Press, London, 1994.

Rheinland-Pfalz: Grenzüberschreitender Atlas der Überschwemmungsgebiete im Einzugsgebiet der Mosel (Transboundary atlas of the inundation areas in the basin of the river Mosel). Struktur- und Genehmigungsdirektion Nord, Regionalstelle für Wasserwirtschaft, Abfallwirtschaft, Bodenschutz Trier, 2004 (http://www.gefahrenatlas-mosel.de).

Sachsen: Gefahrenhinweiskarte Sachsen (Hazard information map Saxony). Sächsisches Landesamt für Umwelt und Geologie, Dresden, 2004 (http://www.umwelt.sachsen.de/de/wu/umwelt/lfug/lfug-internet/interaktive_karten_10950.html).

Smith, D.I.: Actual and potential flood damage: a case study for urban Lismore, NSW, Australia, Appl. Geography, 1, 31–39, 1981.

Smith, D.I.: Flood damage estimation – A review of urban stage-damage curves and loss functions, Water SA, 20(3), 231–238, 1994.

Smith, K. Ward, R.: Floods – Physical processes and Human Impacts. Chichester: Wiley, 1998.

SSK (Sächsische Staatskanzlei): Augusthochwasser 2002 – Der Wiederaufbau in Sachsen ein Jahr nach der Flut (Flood event August 2002 – the reconstruction in Saxony one year after the flood). Dresden, 2003.

Takeuchi, K.: Increasing vulnerability to extreme floods and societal needs of hydrological forecasting, Hydrol. Sci. J., 46(6), 869–881, 2001.

Thieken, A.H., Müller, M., Kreibich, H., Merz, B.: Flood damage and influencing factors: New insights from the August 2002 flood in Germany, Water Resour. Res. (in press).

Uhrich, S., Krause, J., Bormann, H., Diekkrüber, B.: Simulation von Überflutungen bei Hochwasserereignissen: Risikoeinschätzung und Unsicherheiten (Simulation of inundations due to flood events: risk assessment and uncertainties). In: Tagungsbericht 5. Workshop zur hydrologischen Modellierung: Möglichkeiten und Grenzen für den Einsatz hydrologischer Modelle in Politik, Wirtschaft und Klimafolgeforschung, ed. by Stephan, K., Bormann, H., Diekkrüger, B., Kassel University Press, Kassel, 59–70, 2002.

UM Baden-Württemberg (Umweltministerium Baden-Württemberg): Hochwassergefahrenkarten Baden-Württemberg, Leitfaden. Stuttgart, 2005.

USACE (U.S. Army Corps of Engineers). Risk-based analysis for flood damage reduction studies, Engineering Manual 1110-2-1619, Washington DC, 1996.

Wind, H.G., Nieron, T.M., de Blois, C.J., de Kok, J.L.: Analysis of flood damages from the 1993 and 1995 Meuse floods, Water Resour. Res., 35(11), 3459–3465, 1999.

11

Flood Risk Modelling in Urban Watercourses – Results of the European FLOWS Project

Erik Pasche & Nicole von Lieberman
Technical University Hamburg-Harburg, Hamburg, Germany

11.1   INTRODUCTION

In the past the main objective of flood risk mitigation in the City of Hamburg has been concentrated on storm surges coming from the river Elbe. They have caused extreme damage and many losses of life in 1962 that this flood risk has been given priority. In 2002 extreme precipitation over the City of Hamburg caused great damage along the smaller urban rivers demonstrating the City's vulnerability against flash floods of urban catchments. The City became aware that the flood risk management concept for these urban rivers needed to be revised and modern technology and strategy of flood management to be included. The legal framework came from the German Government who has released a couple of new guidelines and water acts for flood risk management, e.g. the five-point program, "Das Gesetz zur Verbesserung des vorsorgenden Hochwasserschutzes" and the "LAWA-guidelines" for a future oriented flood risk management. They require an integrative flood management strategy which includes the whole catchment of a river and gives priority to more space for the rivers. This new policy means expanding the relationship between the city, space and water, and setting a new, changed benchmark for building, planning and expansion of the city. However, considering the central theme of a "Growing city" which should have water implanted in the centre, this is not an easy undertaking – especially since the space in Hamburg is generally limited and expensive. It is necessary to create a basic understanding and awareness of the water (most notably the inland water), and also, to control runoff of the surface water. In the future, the water will be included as an important component in the existing and current city planning. The citizens should be informed about the possible individual and collective flood preparedness measures regarding their houses and be prepared for "living with floods". This is even more important since the climate changes favour extreme weather situations, and heavy rain and flooding become increasingly likely.

Because of this, Hamburg is planning to improve the requirements for a forward-looking flood management of inland water. In agreement with the new water act the first step is the determination of the 100-year inundation areas along its urban rivers and the evaluation of the flood risk on these areas. However no clear concept existed about the appropriate method of flood mapping and evaluation of flood risk. The special hydrological situation of the North German Lowland requires extensions of existing hydrological and hydraulic models to simulate the interaction between groundwater and surface runoff which can have a substantial influence on the flood flow in lowlands. Additionally a method had to be developed through which the flood risk can be better understood by stakeholders and urban planners.

11.2   OBJECTIVES OF THE PROJECT

The main objective of this project was the development and introduction of appropriate modelling and mapping tools at the water authorities of Hamburg to determine inundation and flood risk along the urban rivers of Hamburg. Making use of innovative information technologies, e.g. Web-technology, Open Geographical Systems and Graphical User Interfaces the access and application of these models have to be simplified that urban planners can use these sophisticated tools in a cross-sectoral way to study by themselves the flood risk along existing and planned urban developments and to select and prove the efficiency of appropriate compensation measures such as SUDS (sustainable urban drainage systems), dike reallocation, flood polders and river restoration. The quality and applicability of these new instruments need to be tested at a sample river of Hamburg.

11.3   STUDY AREA

During the storm event of the summer 2002 the watercourse Kollau was most heavily affected by the flood. At several locations the flow exceeded the capacity of that channel bed inundating residential areas, industrial facilities and an important regional road, the freeway A7. The observed problems and exposure of this flooding are representative for many urban watercourses of Hamburg. Additionally many data had been already available reducing the effort of data acquisition considerably. Therefore this brook has been chosen as pilot watercourse for the control and verification of the new modelling instruments which have been mainly developed by the Institute of River & Coastal Engineering at the Hamburg University of Technology. In close cooperation with the district office Eimsbüttel, the Authority for City Development and Environment of the Free and Hanseatic City of Hamburg (BSU) the researchers of the university applied this new technology, evaluated the data quality and developed a strategy to introduce these new modelling and mapping tools in the administration and at local consulting companies.

11.4   THEORETICAL STUDIES

As basis for the modelling instruments two existing models of the Institute of River & Coastal Engineering have been chosen which represented the state-of-the-art in the deterministic rainfall-runoff modelling on the basis of a semi-distributive lumped approach and in one dimensional hydrodynamic water surface modelling (Pasche et al., 2005) based on the Darcy Weisbach equation. The hydrological model has its strengths in the modelling of hydrological systems in flat regions with mild or no surface contour which are representative for the City of Hamburg. In these regions the flood flow in rivers is substantially influenced by runoff from groundwater. The hydraulic model has its strengths in modelling the effect of wooden vegetation which dominates many river banks and flood-plains on restored rivers of Hamburg. For the assessment of the risk a flood damage model has been developed which estimates the damage by a mezzo-scale approach. On the basis of the land use units and the water depth the spatial distribution of the flood damage is evaluated and classified in 3 levels of exposure.

11.4.1   Methodology for rainfall-runoff modelling

In the hydrological model the whole catchment is subdivided into smaller units, the sub catchments. They cover only parts of a whole water shed area but represent still drainage units in which the hillslope runoff (surface and subsurface) confluents to one point, the drainage point of the sub catchment. All sub catchments are linked by channel sections. Their ends match with the drainage point of the sub catchment and drain into the adjacent channel section (fig. 11.1). Still in each sub catchment the vegetation cover, land use and the hydrogeology might vary. Therefore the subcatchments are further subdivided in hydrotopes. They are areas in which homogeneous conditions with respect to the vertical runoff processes occur. In general this is given for areas with homogeneous land use, pedology and geology. They can easily determined within a GIS by overlaying and intersecting these three thematic maps.

In each hydrotop the vertical flow process is subdivided into a chain of reservoirs. The central element of this vertical chain is the unsaturated soil zone which comprises the root zone and further underlying zones of unsaturated soil (Fig. 11.2). It can be modelled by any number of reservoir layers to get the vertical variation of the soil structure. For each layer the mass conservation is balanced by the

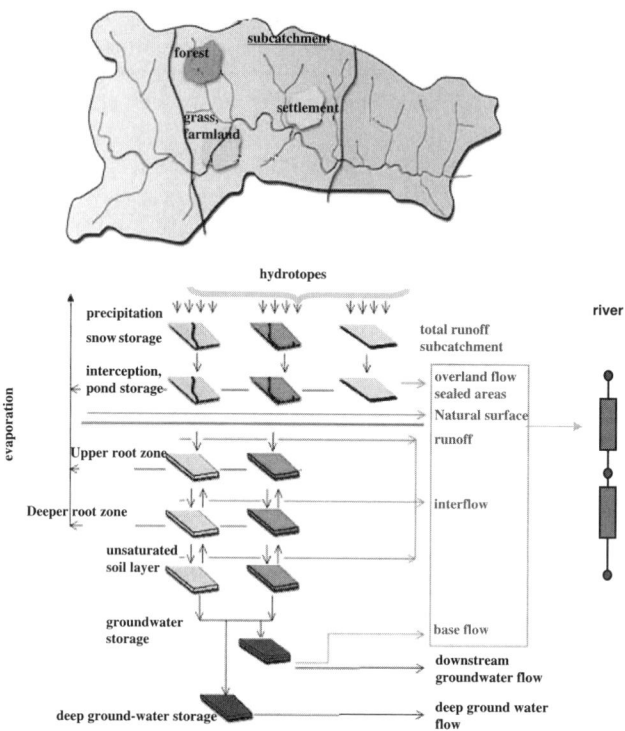

Figure 11.1. Structure of the rainfall-runoff model.

1-dimensional continuity equation of soil water

$$\frac{d(m \cdot sw_i(t))}{dt} = inf_{micro,i}(t) - (perc_i(t) + intf_i(t)) - evt_i(t) + ca(t) + q_B(t) \quad (11.1)$$

The change of water content $sw(t)$ in each soil layer is the sum of direct infiltration of the precipitation water into the soil matrix, the micropores ($inf_{micro}(t)$ [mm/h]), the capillary uprise $ca(t)$, the interflow $intf(t)$ [mm/h], the evapotranspiration $evt(t)$ [mm/h] and the exfiltration of soil water out the layer called percolation $perc(t)$. In the last layer of the unsaturated soil zone the percolation rate is replaced by the rate of groundwater recharge $q_B(t)$ [mm/s]. Due to rising and falling of the groundwater table the size m [mm] of the last layer of the unsaturated zone has to be adjusted. In case of high groundwater table soil layers need to be totally eliminated.

The challenge lies in the determination of the infiltration and exfiltration rates. As they are mainly a function of the soil water content $sw(t)$. Based on an approach of Ostrowski (1982) and the infiltration theory of Horton and Green-Ampt the potential infiltration rate into the soil matrix can be described by the following

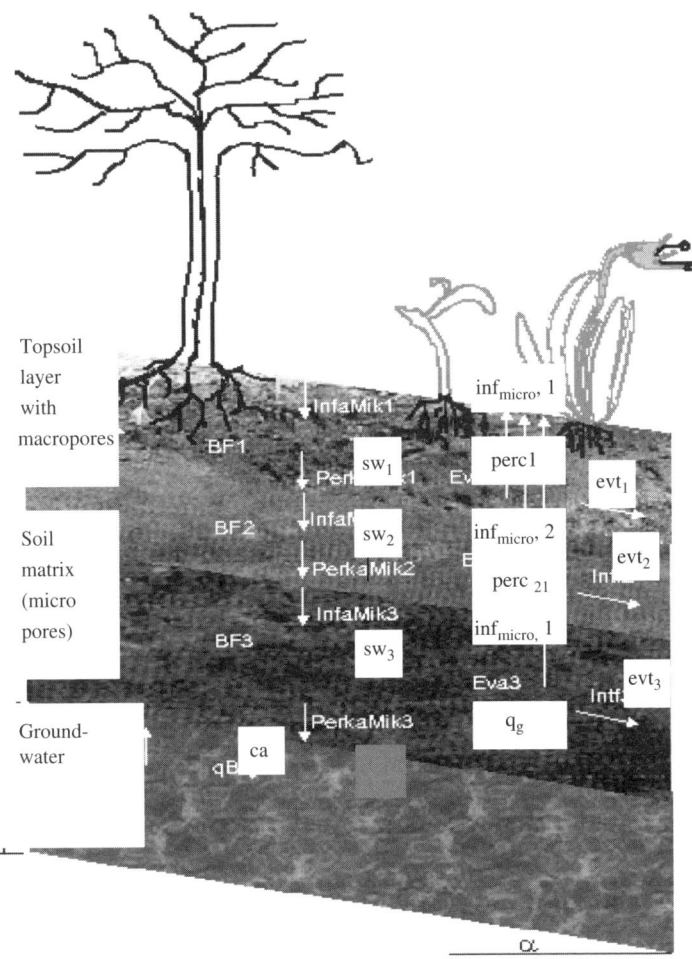

Figure 11.2. Soil water dynamics.

equation:

$$inf_{micro,pot}(t) = k_Z(\psi(t))_i \cdot \frac{sw_{max} - sw(t)}{sw_{max} - wp} \qquad (11.2)$$

with
$k_Z(\psi(t))_i$ = hydraulic conductivity in dependence on the soil water
         tension [mm/h]
$\psi(\overline{sw(t)})$ = mean soil water tension within the micropores of the root layer [mm]
wp = wilting point in [mm]
fc = field capacity [mm]
sw_{max} = maximum soil water content at saturation in [mm].

The real infiltration $inf_{micro,i}(t)$ into the micropores is the minimum of the potential infiltration rate and the maximum available water on the surface which corresponds to the intensity of the precipitation above surface $inf_{max}(t)$ in [mm/h].

For the percolation $perc(t)$ of the soil water into the deeper soil layer it is assumed that only the micropores contribute to this exfiltration of water into deeper layers.

The potential percolation $perc_{pot}(t)$ is calculated in dependence on the soil water tension and the amount of free soil water $(sw(t) - fc)$:

$$perc_{pot}(t) = k_Z(\psi(t)) \cdot \frac{sw(t) - fc}{sw_{max} - fc} \quad \text{for } sw(t) \geq fc \quad (11.3)$$

If the potential percolation $perc(t)$ of a soil layer i exceeds the infiltration rate of the following soil layer $i + 1$, interflow is generated within the layer i:

$$intf_i(t) = perc_i(t) - inf_{micro,i+1}(t) \quad (11.4)$$

As mentioned above the storm water hydrograph at the outlet of a sub catchment is mainly a function of the infiltration-excess overland flow, the interflow (saturation-excess overland flow) and the groundwater base flow.

In semi-distributed models these runoff components are determined by a lump model approach. In the simplest way this is a linear reservoir which has been extended to linear reservoir cascades or parallel reservoir cascades. However the empirical parameters (retention coefficient k and the number n of the reservoirs) are empirical parameters with little physical background and have to be determined by calibration. In ungauged catchments these methods can not be applied reliably as the validity of the empirical parameters determined at other catchments is in general not given. A better physical basis is given for the method of isochrones. In this approach the runoff process is divided in a translation and retention process. The first one is described on the basis of the isochrones which represents lines of the same transport time of the runoff within the catchment. The area between two lines is arranged in a time-area-diagram corresponding to the transport time of the downstream and upstream isochrone. Assuming over the whole sub catchment a unit infiltration-excess runoff of 1 mm the time-area-diagram represents the unit translation hydrograph u_t which can be calculated for discrete time steps $n\Delta t$:

$$u_t(n\Delta t) = \frac{\sum_{i=1}^{n} A(i\Delta t) - \sum_{i=1}^{n-1} A[(i-1)\Delta t]}{\Delta t \cdot A_e} \quad (11.5)$$

with $A(i\Delta t)$ = area of the sub catchments that has already drained at the moment, $i\Delta t$ and A_e = total area of the catchment.

For the determination of the isochrones Geographical Information Systems have been applied very successfully. Based on a Digital Terrain Model (DTM) the flow paths on the overland can be determined. The total transport time t_c (concentration time) along this flow path can be derived by segmenting and determining the flow velocity v_i within each segment i of length s_i.

$$t_c = \sum_{i=1}^{n} \frac{s_i}{v_i} \tag{11.6}$$

In most publications the kinematic wave equation and the Manning's formula are used for the determination of the flow velocity. Pasche/Schröder (1994) showed that the sheet flow around vegetation is better described on the basis of the Darcy-Weisbach law and a resistance law derived from flow around cylindrical bodies.

$$v = \sqrt{2 \cdot g \cdot I_o \cdot \frac{a^2}{d_P \cdot c_{WR}}} \tag{11.7}$$

with I_o = hillside slope, d_P = diameter of the vegetation element [m], a = distance of the vegetation elements [m].

The drag coefficient c_{WR} can be assumed according to DVWK (1991) to $c_{WR} = 1.5$.

When the sheet flow reaches the micro channels (rills and ephemeral gullies) a boundary layer flow is generated, in which the mean velocity is dependent on the water depth. This shows that the transport time in catchments is not a constant but varies in dependence on the intensity of the available infiltration-excess runoff. Consequently the time-area diagram can not be a Unit Hydrograph valid for all precipitation events but has to be adjusted. Due to the uncertainties in hillslope hydraulics and the lack of detailed information about the interrill areas, structure of rills and ephemeral gullies the Unit Hydrograph of the time-area diagram is sent through a linear reservoir u_R. It also considered other retention effects which are not included in the calculation of isochrones, e.g. trapping in ponds and hollows.

$$u_{TR}(n\Delta t) = \sum_{i=1}^{n} u_t(i\Delta t) \cdot u_R[n\Delta t - (i-1)\Delta t] \tag{11.8}$$

with

$$u_R[n\Delta t - (i-1)\Delta t] = \frac{1}{k} \cdot e^{-[n\Delta t - (i-1)\Delta t]/k} \tag{11.9}$$

and k = retention coefficient of the surface runoff. This parameter is compensating all the deficiencies in your time-area-diagram and thus is less a physically based parameter but more a calibration parameter. It was found to vary between $k = 4.0$ to $20\,h$ for surface runoff in natural environments.

On the basis of this translation-retention Unit Hydrograph $u_{tr}(t)$ the hydrograph of the surface runoff at the exit of the sub catchment can be calculated by the lump

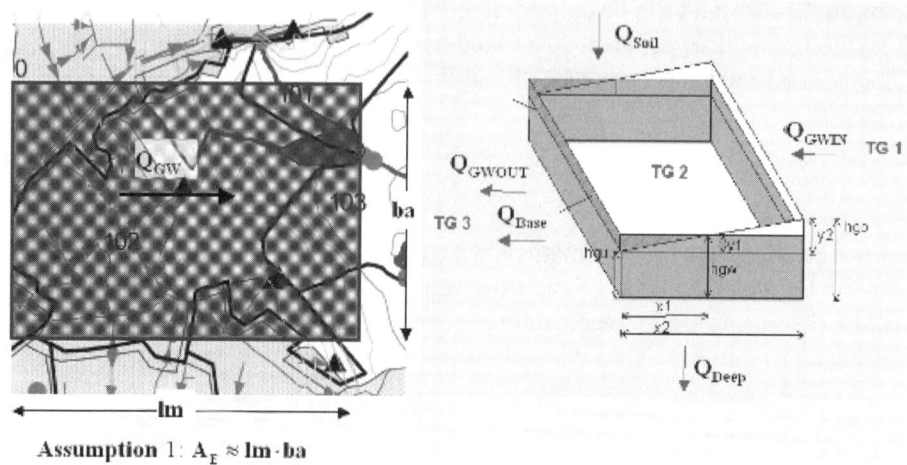

Figure 11.3. Concept and definition sketch of the groundwater reservoir.

model approach:

$$Q_a(n\Delta t) = \frac{1}{3.6} A_e \sum_{i=1}^{n} i_{eff}(i\Delta t) \cdot u_{TR}[n\Delta t - (i-1) \cdot \Delta t]\Delta t \qquad (11.10)$$

with $i_{eff}(i\Delta t)$ = generation rate of infiltration-excess runoff in [mm/h] at moment $i\Delta t$.

This isochrone model can be also applied for modelling the subsurface run-off (interflow). As the flow velocity within the macro pores was assumed to be dependent on the slope of the hillside the flow paths within the macro pores can be assumed to be parallel to the ones of the surface runoff. Then the transport time of the macro pore flow along each flow path can be calculated on the basis of equation 11.6. Also the retention coefficient of the linear reservoir has to be adjusted to the subsurface flow conditions and needs to be found by calibration.

For the determination of the groundwater induced base flow into the river channels the lumped model approach has been successfully applied in semi-distributed rainfall-runoff models (Pasche et al., 2004). In this approach the groundwater aquifer is approximated by a reservoir with horizontal groundwater table. The mass balance of this reservoir is given by the 1-dimensional continuity equation of the groundwater body (Fig. 11.3):

$$\frac{dh_{GW}(t)}{dt} = \frac{1}{3.6}qg(t) + [Q_{GWIN}(t) - Q_{Base}(t) - Q_{GWOUT}(t)$$

$$- Q_{Deep}(t)]/(A_E \cdot 10^6) \qquad (11.11)$$

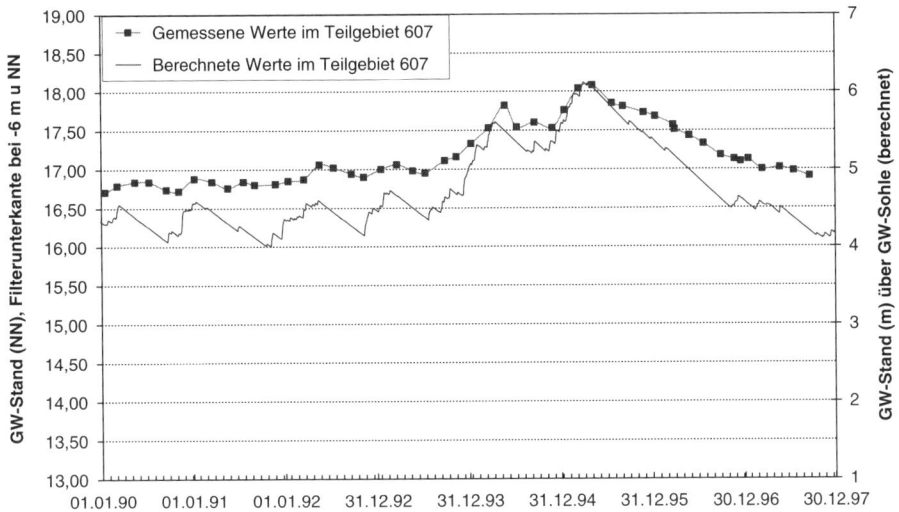

Figure 11.4. Mean groundwater stages within a subcatchment.

The changing rate of the ground water stage $h_{GW}(t)$ is balanced with the groundwater recharge $qg(t)$, the groundwater inflow $Q_{GWIN}(t)$ [m³/s], the baseflow $Q_{Base}(t)$, the outflow $Q_{GWOUT}(t)$ of the aquifer to the downstream side and the outflow $Q_{Deep}(t)$ into a second deeper groundwater aquifer. Based on the linear reservoir theory the groundwater outflow $Q_{GWOUT}(t)$ and the base flow $Q_{Base}(t)$ are calculated in dependence on the stored volume $S(t)$ in the groundwater aquifer. As an example the resulting equations are given for groundwater stages between the bottom height h_{GU} [in m above bottom of aquifer] and the highest elevation h_{GO} [in m above bottom of aquifer] of the channel:

$$Q_{Base}(t) = \frac{0.5}{k_{Base}} \cdot \frac{h_{GW}(t) - h_{GU}}{h_{GO} - h_{GU}} \cdot (h_{GW}(t) - h_{GU}) \cdot n_e \cdot A_E \qquad (11.12)$$

$$Q_{GWIN}(t) = \frac{1}{k_{GW}} (h_{GW}(t) - h_{GU}) \left[1 - 0.5 \frac{h_{GW}(t) - h_{GU}}{h_{GO} - h_{GU}} \right] \cdot n_e \cdot A_E \qquad (11.13)$$

with A_e = area of the sub catchment in [m²], n_e = effective porosity of the groundwater aquifer [-].

The retention coefficients k_{Base} [h for base flow] and k_{GW} for the groundwater outflow are empirical coefficients which need to be determined by calibration. The elevations h_{GO} and h_{GU} stand for the thickness of the aquifer layer. The difference between these two parameters stands for the part of the aquifer which participates at the generation of the base flow. Although Pasche et al. (2004) could show that these parameters can correspond with the physical parameters (Fig. 11.4), they are not identical and have to be verified by calibration. Especially for low lands with mild topographic gradients and strong backwater effect within the groundwater aquifer they need to be reduced.

11.4.2   Methodology of hydraulic modelling

Good consensus exists between engineers in the formulation of the basic flow equations for rivers. For one-dimensional flood-routing the theoretical basis is the well-known Saint Venant equation (Pasche et al., 2004)

$$\frac{1}{gA}\frac{\partial Q}{\partial t} + \frac{1}{gA}\frac{\partial(\alpha'Q^2/A)}{\partial x} + \cos\Theta\frac{\partial z_{WSP}}{dx} + I_R - \cos\varphi\frac{v_e q_e}{gA} = 0 \qquad (11.14)$$

with Q = total discharge [m^3/s], A = cross section [m^2], q_e = inflow/outflow [m^3/sm], z_{Wsp} = water surface elevation [m a.O.], I_R = friction slope [-], α' = energy coefficient [-], Θ = longitudinal slope of river bottom in [°], φ = angle between inflow/outflow and main flow [°], $i, i+1$ = downstream, upstream profile [-], x = length of river along thalweg [m].

   The most relevant parameter for the one dimensional equation is the friction slope I_R and the energy coefficient α'. The simplest approach use only one parameter to quantify the friction slope I_R. It includes all flow losses and represents the variable roughness along the wetted parameter by a mean value. In dependence of the flow formula this parameter is named Manning's n (GMS-formula) or Darcy Weisbach coefficient λ (DW-formula).

$$I_R = \frac{1}{8g}\frac{\lambda}{r_{hy}}\frac{Q^2}{A^2} \qquad (11.15) \qquad\qquad I_R = \frac{n^2}{r_{hy}^{4/3}}\frac{Q^2}{A^2} \qquad (11.17)$$

$$\tau_{S0,i} = \frac{\lambda}{8}\rho u_i\sqrt{u_i^2+u_j^2} \qquad (11.16) \qquad\qquad \tau_{S0,i} = \frac{\rho g n^2}{h^{1/3}}u_i\sqrt{u_i^2+u_j^2} \qquad (11.18)$$

Darcy-Weisbach-formula   Gauckler-Manning-Strickler-formula

   Only the DW-formula is physically based which can be seen by the units of the empirical parameters. While the Darcy-Weisbach-coefficient λ is dimensionless the Manning's n parameter has the physically senseless units of [s^2/m$^{1/3}$]. Making use of the boundary layer theory and the theory of flow around bodies physically based equations can be derived in which the Darcy-Weisbach-coefficient λ is expressed in terms of directly determinable geometric parameters or equivalent parameters, like the equivalent sand roughness k_s in the Colebrook-White formula.

$$\frac{1}{\sqrt{\lambda_{So}}} = -2.03 \cdot \log\left(\frac{2.51}{f\,Re\sqrt{\lambda_{So}}} + \frac{k_s}{f\,14.84R}\right) \qquad (11.19)$$

with $Re = \frac{UR}{v}$ = Reynolds Number, R = hydraulic radius, v = kinematic viscosity and f = form factor.

BWK (1999) has shown that on this basis nearly all relevant friction losses can be well quantified. Pasche (1984) has shown that especially for natural rivers the DW-formula has advantages as the flow losses caused by non-submerged wooden vegetation and by momentum transfer at the interface between river and flood plain can be well described by the following formulas:

$$\lambda_P = \frac{4 \cdot h_P \cdot d_P}{a_x \cdot a_y} \cdot c_{WR} \cdot \cos(\alpha_{lat}) \qquad (11.20)$$

$$\frac{1}{\sqrt{\lambda_T}} = -2 \cdot \log\left[0.07 \cdot \left(\frac{c \cdot b_m}{b_{III}}\right)^{1.07} \cdot \Omega\right] \qquad (11.21)$$

$$\Omega = \left[0.07 \cdot \frac{a_{NL}}{a_x}\right]^{3.3} + \left[\frac{a_{NB}}{a_y}\right]^{0.95} \qquad (11.22)$$

with a_x, a_y = distance of vegetation elements in both horizontal directions [m]; h_P = water depth in front of the vegetation element [m]; d_P = diameter of vegetation element [m]; α_{lat} = lateral inclination of bottom [°]; c_{WR} = drag coefficient of wooden vegetation [-] according to Pasche (1984); a_{NL}, a_{NB} = wake length and wake width [m] according to BWK (1/1999); c = form factor [-]; b_m, b_{III} = contributing width of vegetation zone and river [m] according to BWK (1999).

Pasche et al. (2006) showed that on the basis of the GMS-formula the quality of the flood modelling could be substantially improved. All parameters of the DW-formula have been estimated on the basis of a field survey in which the corn fraction of the bed material and the geometry of the wooden vegetation has been determined and transferred into equivalent sand roughness and average vegetation parameters a_x, a_y, and d_P according to the recommendations of BWK (1999). Only little adjustments were necessary in the calibration procedure. With one parameter set the whole range of monitored flow events could be well reproduced (Fig. 11.5). In contrary one parameter set of the GMS-formula did not give a good fit over the whole range of observed flow events. Especially in rivers with wooden vegetation on the bank the GMS-formula underestimated the water stage with increasing deviation for rising water stages.

The momentum coefficient α' results from the integration of the square of the velocity $v(y,z)$ over the cross-section A divided by the square of the mean velocity v times the total area A:

$$\alpha' = \frac{1}{v^2 A} \iint_A v(y,z)^2 dA \qquad (11.23)$$

In compact cross-sections its value is nearly 1.0. But in natural rivers with wooden vegetation on the bank and adjacent flood-plains the momentum coefficient can not be neglected. Separating the whole channel into the three independent stream tubes of left flood-plain (fl, l), main channel (ch) and right flood-plain (fl, r) the

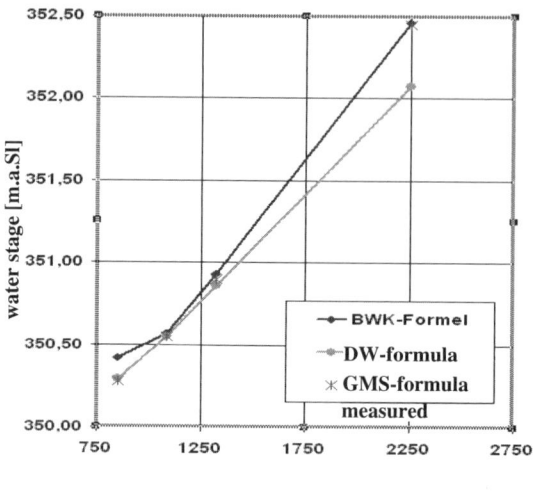

Figure 11.5. Water stages of the river Danube at gauge Neustadt.

momentum coefficient can be calculated by:

$$\alpha' = \frac{A[A_{fl,l}[R_{fl,l}/\lambda_{fl,l}] + A_{ch}[R_{ch}/\lambda_{ch}] + A_{fl,r}[R_{fl,r}/\lambda_{fl,r}]]}{[A_{fl,l}[R_{fl,l}/\lambda_{fl,l}]^{1/2} + A_{ch}[r_{ch}/\lambda_{ch}]^{1/2} + A_{fl,r}[R_{fl,r}/\lambda_{fl,r}]^{1/2}]^2} \qquad (11.24)$$

This approach underlies the assumption that the flow velocity within each stream tube (left and right flood-plain, stream channel) can be approximated by a constant velocity. On flood-plains with extreme changes of land use cover or water depth this assumption is no longer valid and equation 11.24 needs to be extended by dividing the stream tubes in smaller units. Additionally in compound channels the friction slope I_R has to adjusted in a similar way as the energy coefficient:

$$I_R = \frac{1}{[A_{fl,l}(R_{fl,l}/\lambda_{fl,l})^{0.5} + A_{fl,r}(R_{fl,r}/\lambda_{fl,r})^{0.5} + A_{ch}(R_{ch}/\lambda_{ch})^{0.5}]^2} \cdot \frac{Q^2}{8g} \qquad (11.25)$$

11.4.3   Methodology of risk modelling

Risk is a result of inundation probability and exposure of the objects and population on the flood-plain. While the inundation probability results from the application of the hydrological and hydraulic modelling tools the exposure needs the modelling of the flood damage.

11.4.3.1   *Damage assessment*
Within this project a method have been applied which uses standard asset values and damage functions as basis for the damage assessment. The asset can be

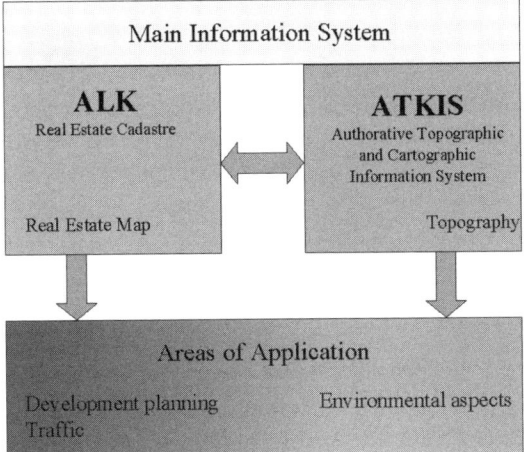

Figure 11.6. Survey data, ALK and ATKIS.

either recorded for each individual object (micro-scale approach) or derived from statistical records in which the total asset and total area is given for the main land use categories and administrative regions (macro-scale approach). This last method, also called regionalisation method, has been applied within this project. It determines the flood damage in the 5 steps:

Step 1: Categorization of the land use on the flood-plain in standard land use units
In macro-scale approaches the land use is taken from geodetic maps which are in general provided by surveying agencies. These maps can vary in spatial resolution and in the refinement of the land use units. In Germany two types of land use maps are available. The Official Real Estate Map (ALK, scale 1:1000) and the Authoritative Topographic and Cartographic Information System (ATKIS) with the Digital Landscape Model (DLM, scale 1:5000). These data represent the basis for urban and landscape planning (Fig. 11.6). The Automated Real Estate Map includes geometric data and coordinates of the boundaries and outer building contours of the Real Estate parcels. It is scaled in 1:1000. In a second data bank, the ALB (Official Real Estate Book) the attributes of the parcels are stored. The information of the real estate parcels are referenced in these two data sources by the same parcel identification number. At present this data model is transferred into a new one which merges these two databanks to one single databank, called ALTKIS.

The land use provided by ATKIS is based on a map scale of 1:25,000. Thus the land use in these maps is less refined and does not reference the real estate parcels. ATKIS was initiated by the German Federal State and developed by the AdV (WORKING COMMITTEE OF THE SURVEYING AUTHORITIES OF THE STATES OF THE FEDERAL REPUBLIC OF GERMANY). It provides digital models of the earth's

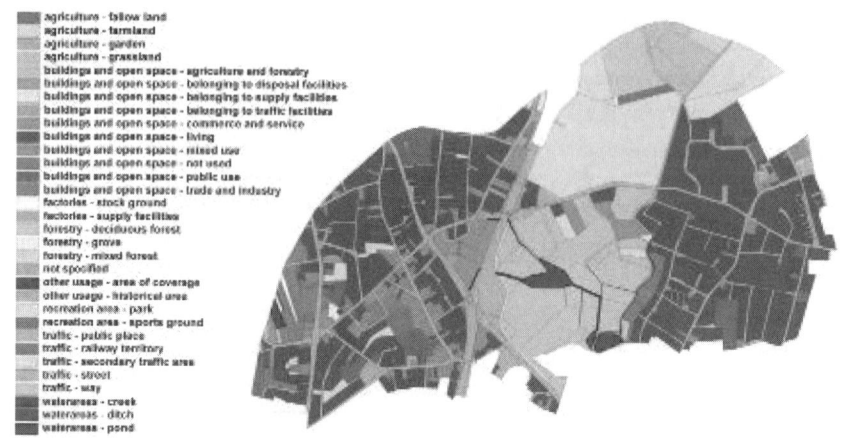

Figure 11.7. Land use units of the ALK, structured due to the decade of the ALB-numbers (example from the northern part of the river Kollau).

Figure 11.8. Land use categories derived from the ALK (example from the northern part of the river Kollau).

surface and represents the first official geobased information system of Germany. It is provided by the FEDERAL AGENCY FOR CARTOGRAPHY AND GEODESY AND THE STATE SURVEY OFFICES.

Within this project the land use was taken from the ALK and ALB. In a first step the different land use units of the ALK have been reduced to nine categories and a mapping tool have been developed which maps the ALK land use into these land use categories (Fig. 11.7, Fig. 11.8 and Tab. 11.1).

Table 11.1. Merged land use groups.

Land use categories	Object type from ALK-layer 21	Object specification as per ALK	
Settlement	1100–1400	Buildings and open space	Public use, living, commerce and service
	2100–2900	Buildings and open space	Mixed use, supply fac., disposal fac., agriculture and forestry fac., recreation area
Industry and commerce	1700	Buildings and open space	Trade and industry
	3100–3600	Factories	Mining land, heap, stock ground, supply fac., disposal fac.
Traffic and public places	4100–4900	Recreation area	Sports ground, park, campsite
	5100–5900	Traffic	Street, way, public place, railway territory, airport, shipping traffic, secondary traffic area
Agriculture	9100; 9300–9400	Other usage	Training area, historical area, graveyard
	6100; 6300	Agriculture	farmland, garden
Meadows	6200	Agriculture	Grassland
Forest	7100–7600	Forestry	Deciduous forest, coniferous forest, mixed forest, grove
Water	8100–8900	Waterarea	River, channel, harbour, creek, ditch, lake, coastal waters, pond, swamp
	Parts of 9200	Other usage	Area of coverage
Nature	Residual objects according to definitions	Areas which don't include asset values and which won't be refurbished after a flooding event	

Step 2: Derivation of assets and damage functions for the land use categories on the basis of statistical material

The monetary value of each land use category is determined on the basis of the net asset. This includes the cost of construction and of the inventory. It can be expressed for purchase prices or alternatively for actual prices. The first asset value represents the costs at the moment of purchase and can be calculated either as gross value which assumes a constant value during the whole lifecycle of the asset or as net value which includes depreciation and amortization. If the assets are changing this concept can not be applied. In this case preference has to be given to the asset concept of actual prices. It includes all costs to reconstruct and restore the asset. Also for this price concept the alternatives of net and gross costs exist. As buildings damaged by flood have to be restored in its original state the net asset value for actual prices seems to represent the damage potential in the most realistic way. Therefore this price concept has been taken for the evaluation of the asset within this project.

In general the statistical agencies release their material in an aggregated form (e.g. capital stock for the whole country, for each "Bundesländer" or for each county). To get the asset for each property this material has to be disaggregated. The assumption of a constant asset for each property would be too unrealistic. Thus it is assumed that the asset is proportional to the size of the property (area of the territory). For the damage assessment the grid cells have been taken as the smallest units. Consequently the asset is also needed per grid cell. To ensure that the asset has not to be adapted to the size of the grid cell the total asset has to be broken down to an asset value per square meter (in the following referenced as specific asset). For the different land use categories of Tab. 11.1 the applied method will be explained to determine the specific asset from the aggregated statistical material.

Specific asset for the land use category "settlement". The total asset of the land use category settlement is composed of three capital stocks: buildings, its inventory and the vehicles of the land users.

The capital stock for residential buildings can be gained from the German Federal State Agency of Statistics. It publishes every year the net asset of this capital stock and the total area of the corresponding land use category for the whole of Germany and each German "Bundesland". Thus this information would be sufficient to determine the specific asset for residential buildings. But the density of settlements vary extremely within the different regions. Especially in large urban areas apartments and small properties with high building density dominate while in rural areas single family homes are dominating. The density of settlement effects the specific asset and thus need to be considered. It is well reflected through the ratio of total area of settlements and the number of residents within a region. This influence of the building density on the specific asset can be considered if the total capital stock is transformed into a residential stock per capita.. Then for the smallest administrative unit (e.g. neighbourhood, district) this per capita stock is aggregated to a total capital stock and finally divided by the total residential area of this administrative units giving a specific asset of buildings.

The same variability can be observed for the capital stock of motor vehicles as the density of vehicles is closely related to the density of residents. Thus the same method for the determination of the specific asset should be applied for motor vehicles as for the buildings. Unfortunately the capital stock of motor vehicles (on the basis of actual price) is not published by the statistical agencies in Germany. Thus the total capital stock can only determined by assuming an average value for motor vehicles multiplied with the number of motor vehicles given for the smallest administrative unit. Of course this unit has to be the same as taken for the capital stock of building. The third component, the capital stock of the inventory is also not recorded by the statistical agencies. Similar to the capital stock of vehicles it needs to be derived on the basis of rough assumptions. A good source of information can be the insurance industry. They in general release average values for the inventory of their clients. Multiplying this average value with the number of households within the administrative unit will finally result in a specific asset of inventory which is comparable to the one of the vehicles and the buildings.

Capital assets of the economic sectors. In general the national statistical agencies record the capital stock of the economy on a yearly basis. For Germany the regional statistical offices (Statistische Landesämter) determine the capital stock on different levels. For the present purpose the level A6 with the classification into the following six economic sectors is most appropriate:
1. Agriculture, forestry and fishing
2. Production industry (without Building industry)
3. Building industry
4. Commerce, service and traffic
5. Financial business, renting services
6. Public and private services (without pecuniary character (foundations))

The statistical records give the net capital asset at replacement value for each sector. Unfortunately the economic sectors do not correspond fully with the land use categories of Tab. 11.1. Thus the total area for each sector can not exactly be derived from the geographical land use databank (ALK, ATKIS). Sectors 2, 3 and 4 need to be summed up and projected on the land use category "industry". The land use category "traffic and public places" is best approximated by taking the sector 6 as basis.

Additionally to the capital asset, the stock value needs to be calculated. Taking an approach of Pasche/Kräßig (2005) the stock value is determined as a percentage of the capital asset for each economic sectors.

Similar to the category "settlement" the specific asset of each economic sector results by dividing the total of capital asset and stock through the total area of each sector within the administrative unit which is for the economic sectors the "Bundesland".

To some extend the sector trading and services are included in the land use category settlement. This is due to mixed urban developments. These areas are normally specified within the land use maps, but no definite projection on the economic sector can be made as private households and business by trading and services are to be found in one building (Tab. 11.1). Thus it lies within the freedom of the expert to what extend this area of mixed urban development is included in the total area of the land use category of industry.

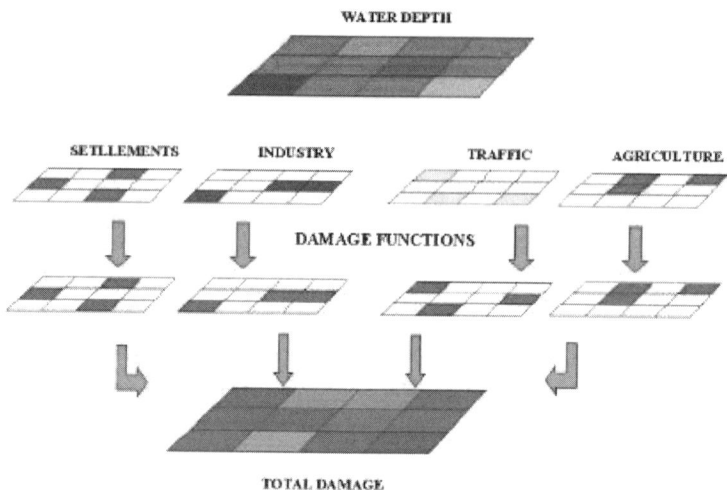

Figure 11.9. Applied scheme of map overlay for damage assessment.

Step 3: Map overlay of land use and inundation probability to get units of equal land use and inundation probability

Damage is a result of inundation and vulnerability of the affected objects. Thus to derive areas of equal damage the inundation maps need to be overlaid and intersected with the land use maps. On the basis of Geographical Information System (GIS) different themes of graphical map objects can be easily overlaid and their polygons intersected and formed to new objects which are sharing the attributes of the overlaid themes.

But damage is strongly dependent by a third geographical parameter, the water depth during flood. Thus the areas of equal land use and inundation probability need to be further divided into areas with constant water depth. Thus the whole area of damage assessment is subdivided in a grid. The resolution of the grid cells needs to be adjusted to the variability of the topography in which the water depth can be approximated as a constant value. Each grid cell is overlaid with the inundation map and the land use map to project the attributes of inundation probability and land use category on each grid cell (Fig. 11.9).

Step 4: Calculation of the damage on the basis of the capital stock/asset and the damage functions

In case of flooding the damage will not be total but a percentage of the value of the property. Several studies in Germany (IKSR, 2001; IKSE, 2003) have determined functions which give this percentage of damage in dependence of the main land use categories defined in step 1 and the maximum local water depth at inundation (Fig. 11.10, Fig. 11.11).

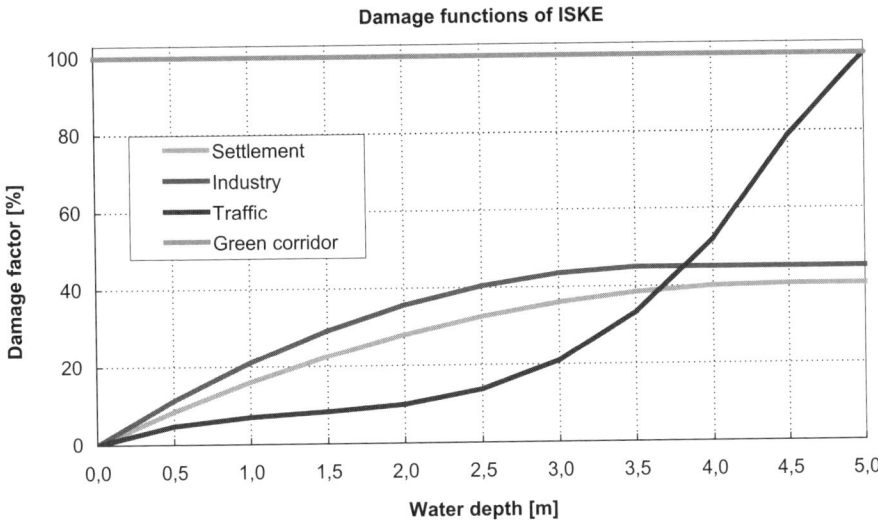

Figure 11.10. Damage function, given by the International Commission for the protection of the river Elbe (functions taken from IKSE [2003]).

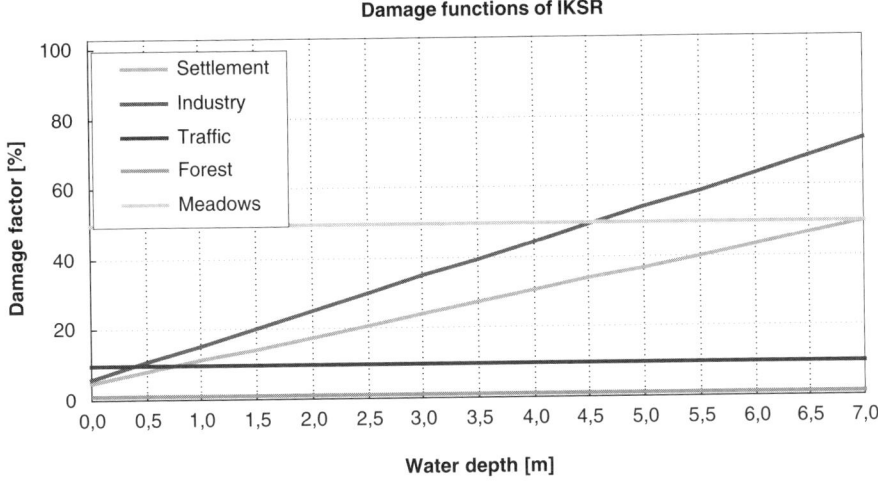

Figure 11.11. Damage function of the International Commission for the protection of the river Rhine (functions taken from IKSR [2001]).

These damage functions vary considerably indicating that more parameters influence the damage. As Pasche/Geissler (2005) showed one of the most important factors influencing the damage is the resilience performance which expresses the capability to resist the flood by wet- and dry-proofing of the affected buildings. Up to 80% the flood damage can be reduced through such measures. Still more research is needed to consider this effect in the damage assessment. Therefore the resilience performance could not be considered in this project and the existing damage functions given in Fig. 11.10 have been applied without modifications or extensions.

On the basis of these damage functions the damage within each grid cell determined in step 4 can be calculated by the following equation:

$$D_{i,j}^{Ln} = C_{i,j}^{Ln} \cdot V^{Ln} \qquad (11.26)$$

with $D_{i,j}$ = specific damage for grid cell $[i,j]$ in ($€/m^2$), $C_{i,j}^{Ln}$ = percentage of damage in dependence of land use category Ln and the water depth $h_{i,j}$ in grid cell $[i,j]$ in [%] and V^{Ln} = specific asset/capital stock for each land use type Ln in [$€/m^2$]. The result of such a grid cell oriented damage calculation is demonstrated in Fig. 11.12. It shows the spatial distribution of specific damage for a certain flood event and a given region affected by flood. Multiplying this specific damage with the area of the grid cell and adding up this value for all grid cells gives the total damage for this area and flood event.

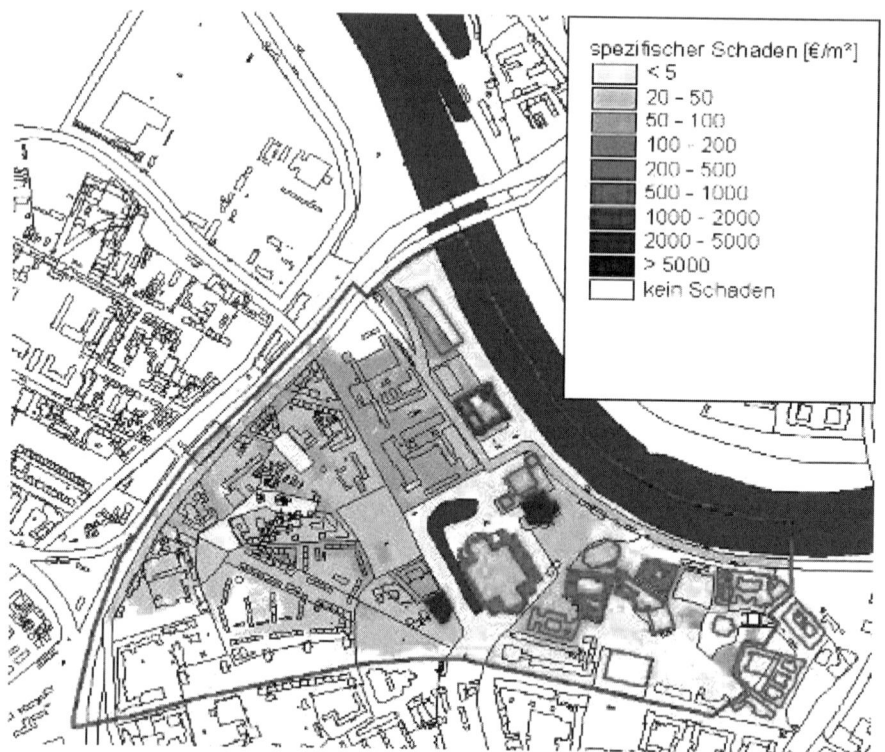

Figure 11.12. Spatial distribution of specific damage for a given flood event.

Step 5: Calculation of the probable annual damage
One extreme flood event (e.g. a 100-year flood) is not sufficient to assess the flood risk of an urban region. Especially the frequent floods have a strong effect on the overall exposure of flood and need to be included in the flood risk assessment although the inundated area and damage at each event is less than at extreme events. Thus risk assessment methods for flood risk need to include the damage of all probable flood events. But the occurrence of a flood is uncertain and consequently the total flood damage is not a deterministic but statistical value. In Germany the method of the DVWK (1985) is applied to include the probability in the damage assessment. The method regards all probable flood events which can occur within a certain period of time. Its duration is set equivalent to a return period of a flood which is in general regarded as a risk and taken for the design of flood defence measures. In Germany this "design flood" is for urban areas in general a one hundred year flood. But in some cases of high damage potential even higher probabilities are taken, e.g. a 200-year flood along the river Rhine for the City of Cologne.

With these assumptions and statistical approaches the method gives a probable annual flood damage $\bar{S}$:

$$\bar{S} = \sum_{i=1}^{max} \frac{S(p_{i-1}) + S(p_i)}{2} \cdot \Delta P_i \qquad (11.27)$$

with S = damage at a flood event of a certain probability p, Δp_i = the difference of probability between the flood event $i - 1$ and i, and max = the maximum number of flood events to be considered for damage assessment which is due to the design flood chosen for the area under consideration. A sample application of this formula is given in Tab. 11.2.

11.4.3.2   *Risk classification*

The regionalisation method of damage assessment calculates the probable annual flood damage on statistical grounds and thus shows only the average for the different land use categories. In reality the assets vary for the different land use units. Thus the damage determined by this method might be misunderstood by the land users. Also the damage does not imply directly the consequences for the land users. Thus a method has been developed which classifies the probable annual flood damage in the three categories high, medium and moderate. By referring necessary measures of flood risk management to each category politicians, water authorities and the land users directly recognize their risk and the necessary actions to take.

The difficulty lies in the definition of the threshold between the 3 categories. In this project the idea has been taken that the severity of flood damage is very much a function of the feasibility to insure the exposed object against flood. In Germany insurance companies have introduced the risk assessment system ZÜRS which also

Table 11.2. Sample calculation of the probable annual damage.

	Return period T_i [a]	P_i [1/a]	ΔP_i [1/a]	Flood level [NN + m]	$S(P_i)$ [1000 €]	$S(i) = \dfrac{S(i-1) + S(i)}{2}$ [1000 €]	$S(i) \cdot \Delta P_i$ [1000 €/a]	Summation of $S(i) \cdot \Delta P_i$ [1000 €/a]	Probable annual damage [1000 €/a]
1	1	1	0.8	166.22	45.0	0.00	0.00		
2	5	0.2	0.1	167.00	90.0	0.00	0.00	54.00	
3	10	0.1	0.06	167.49	120.0	0.00	0.00	226.50	
4	25	0.04	0.02	168.07	270.0	0.00	0.00	755.70	
5	50	0.02	0.01	168.37	455.0	0.00	0.00	2,242.55	
6	100	0.01	0.005	168.69	630.0	0.00	0.00	6,201.53	
7	200	0.005		169.02	770.0	0.00	0.00	16,344.07	25,824.35

defines 3 categories of flood risk which takes a 10 year flood as threshold between the category *high* and *medium* and a 50 year flood for the threshold between the categories medium and *moderate*. The risk category *moderate* ends at a 100 year flood. No insurance will be given for objects within the inundated area of 10 year flood probability. Between the 10 and 50 year flood insurance coverage is limited and above 50 year flood full insurance coverage will be given in general. The deficiency of this method is that the vulnerability of the exposed objects is not taken into account which means that this zoning system is not really indicating the risk. Therefore this approach has been extended by addressing specific damages to the thresholds which are characteristic for the inundation area of 10 and 50 year flood probability. In the end the following probable annual damages have been defined as thresholds:

- $0.1 \, €/m^2/a$ for the threshold between moderate and medium risk
- $1.0 \, €/m^2/a$ for the threshold between medium and high

Assuming an average size of $800 \, m^2$ for a property of domestic housings these specific damages are equivalent to $80 €$ and $800 €$ of total annual damage per property. This amount will be the minimum charge for the yearly insurance rate if you want to insure the property against this flood risk. The lower threshold comes close to the yearly rate of the general insurance of a property and thus should be acceptable by property owner. The second one is far beyond an acceptable rate and confirms the assumption that properties beyond this threshold are facing an unacceptable risk.

11.4.4   Software Concept for an Integrative Flood Risk Modelling

The three methodologies to be applied in flood risk assessment have to be portrayed into a computer code leading to three independent software modules, a rainfall-runoff-model, a water profile model and a risk model. All three modules need extensive input data (further referenced as model data). They have to be derived from the physiographic, land use and hydro-meteorological conditions of the catchment and its watercourses. These spatial and time-dependant data represent the real world data (further referenced as reality data). During the software run further data are produced, the result data. The acquisition and preparation of the reality and model data are extremely time-consuming and error-sensitive. The result data need to be further processed for visualization and analysis. Thus the availability of efficient data management tools for pre- and post-processing is a crucial prerequisite for the applicability and acceptance of these three models by engineers and planners.

In cooperation with the FLOWS-project WP3cviii innovative information system, called KALYPSO Enterprise, has been developed which consists of a client based user interface, a server-based databank and a processing service which

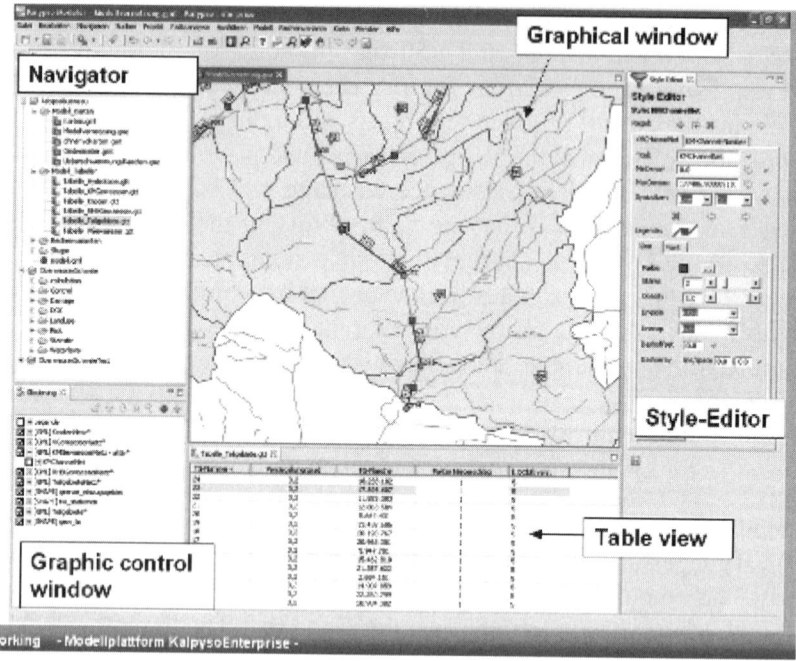

Figure 11.13.   User-Interface of KALYPSO-Enterprise.

controls the operation of the models and the data exchange between them (Fig. 11.13). The software has been developed on the basis of the Eclipse framework and makes extensively use of international standards according to the recommendations of the Open Geospatial Consortium (OGC). GML3 standard has been applied for the input and result data accomplishing an efficient and error-minimizing data exchange between the different processing units of KALYPSO Enterprise. Additionally the following OWS standards for the exchange of geographical data are supported: Web Map Service (WMS) by an implementation of the Open Source software *degree* (www.deegree.org), Web Feature Service (WFS) and Styled Layer Descriptor (SLD). A graphical window for map visualization and editing, and a control window for the structuring and selection of map themes provides GIS-functionality. The time-dependant hydro meteorological data (process data) can be accessed via the new OGC-standard Sensor Observation Service (SOS).

The main components of KALYPSO Enterprise and their connectivity are given in Fig. 11.14. It has been produced as an Open Source product and can be downloaded via www.kalypso.wb.tu-harburg.de.

Describe the outcomes of the project. Include a text description as well as photos (high-resolution digital images), videos (preferably digitised), GIS maps, Internet links to web pages produced by the project and permission for publication.

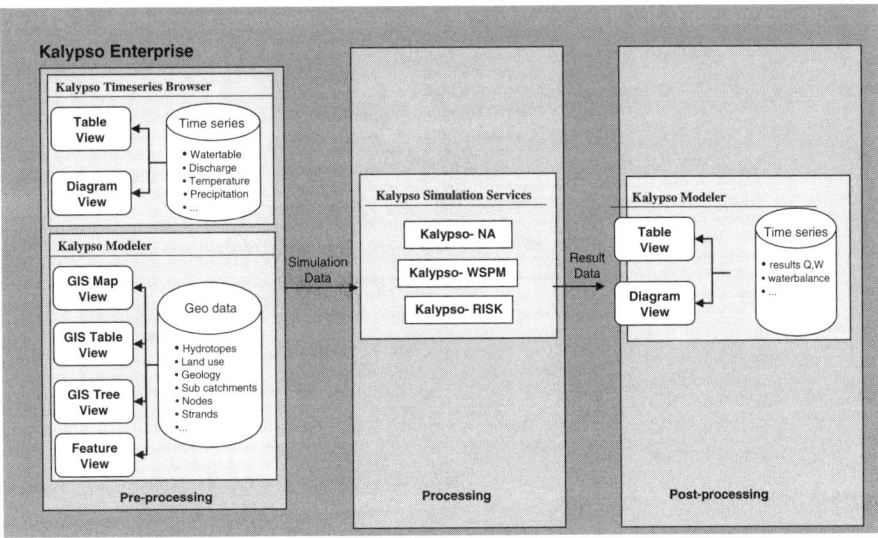

Figure 11.14. Overview KALYPSO Enterprise.

11.5   RESULTS

11.5.1   Establishment of KALYPSO Enterprise within the city of Hamburg

One of the key outcomes of the project has been the development of the software tool KALYPSO Enterprise with the three modelling components "KALYPSO-Storm", "KALYPSO-WSPM" and "KALYPSO-Risk".

They cover the full methodology explained in chapter 4 and integrates the most relevant functions to build the model data (pre-processing component), to define, organize and execute the simulation cases (processing component) and to manage and to process the result data (post-processing component). An overview of the main functions and modules of KALYPSO Enterprise is given in figure (Fig. 11.14).

After intensive testing and verification of the software system it has been implemented at the Authority for City Development and Environment of the Free and Hanseatic City of Hamburg. Its employees responsible for the watercourses have been instructed and trained in its application. Prior to the installation a new server had been installed as "FLOWS Server" at the computer centre of the administration and Web Map Services and Sensor Observation Services had been attached to some of their main geographical and hydrological databases (Fig. 11.15).

11.5.2   Development of flood risk maps for the catchment Kollau

The second very important outcome of the project was the development of flood risk maps for the urban catchment Kollau.

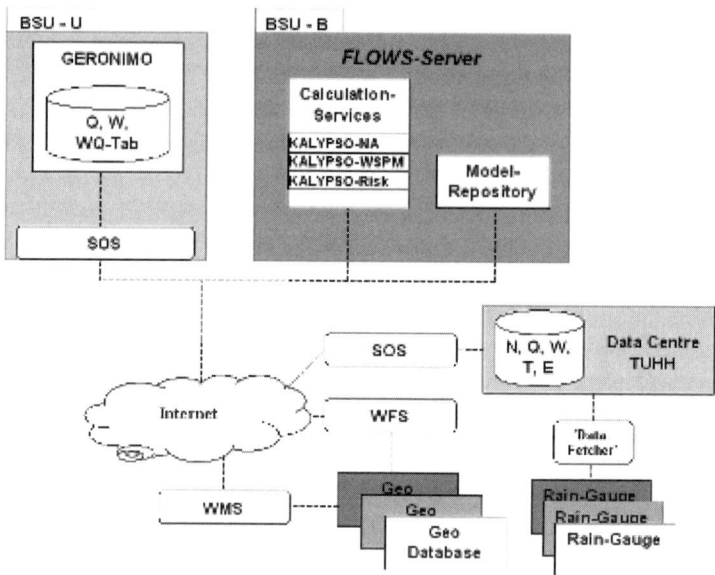

Figure 11.15. Implementation scheme of KALYPSO-Enterprise at the BSU.

11.5.2.1 *Update of reality and model data*

In the first step new available data about the study area has been invented, further processed and integrated into the databases (Tab. 11.3).

The geographical data have been provided by the Hamburg Agency of Geoinformation and Surveying (LGV) and the Department of Environmental Protection (BSU-U), the hydrological data by BSU-U and the meteorological data by the German Weather Service (DWD), Hamburg Agency for Waste and Storm Water Management (HSE). Additional meteorological data came from 4 rain-gauges in the catchment Kollau which are in the responsibility of the Department of River Management (BSU-B), but operated by the Institute of River & Coastal Engineering of the TUHH. Nearly no information was available about the geometry of the watercourses and the 16 flood retention basin. Thus they had to be surveyed and documented in profiles (watercourses) and a digital terrain model (retention basin).

Most of the reality data and model data have been already determined in a preliminary research study which the Institute of River & Coastal Engineering of the TUHH has been carried out for the BSU-B (research report of the River & Coastal Engineering: Hydrologische Modellierung des Gewässernetzes im Einzugsgebiet der Kollau in der Freien und Hansestadt Hamburg, BSU, 2003). But they needed to be updated, improved with respect to the quality and gaps of the database had to be closed. The greatest improvement resulted from LIDAR-data reducing the grid cell to 2 m and the elevation error to ±10 cm of the digital terrain model on the Kollau

Table 11.3. Data overview.

No.	Data	Where from?	Format/Location
	Geobasic Data		
9	Digital Map (DK5) 1:5000	LGV	TIFF
10	Digital City Map DSGK 1:1000	LGV	TIFF/Shape
11	Digital air photograph 1:5000	LGV	TIFF
12	Digital elevation data (2*2 m-raster)	LGV	asc-ascii
	Meteorological time series		
1	precipitation (daily, 5 minutes)	DWD/BSU/HSE	DataCenter
2	Wind speed and direction (daily)	DWD/BSU/HSE	DataCenter
3	Temperature (daily)	DWD/BSU/HSE	DataCenter
4	Sun duration	DWD/BSU/HSE	DataCenter
5	Humidity (daily)	DWD/BSU/HSE	DataCenter
	Hydrology		
6	Measured watertable	BSU-U	GERONIMUS
7	Watertable-Discharge-Relationship	BSU-U	GERONIMUS
8	Groundwater isolines	BSU-U	GERONIMUS-Visor, VOB, Oracle-SIDE
	Stadtplanung		
13	Land use map	BSU/ST	geodatabase, shape-file
	Pedological/Geological data		
14	Geological Map	BSU-U/Geologisches Landesamt	Geronimus
	Drainage		
15	Stormwater pipe net	HSE	oracle, shape-file
	Reservoirs		
16	Digital terrain model	Bezirke	AutoCad DWG-Format
17	Watertable-Discharge-Relationship	BSU und Bezirke	
	River data		
18	Cross sections	BSU und Bezirke	D66
19	Construction catalog	BSU und Bezirke	Catalog

flood plains. This data source has been a perfect basis for an exact mapping of the inundation areas. Inconsistencies in the water surface model could be removed by a new topographic measurement of some channel profiles and updating the model data. Further updates were done for the land use, the hydro meteorological data and storm water pipe network. Unfortunately a long gap in rain records at the DWD-operated rain-gauge Fuhlsbüttel could not be closed within this project. Although this gauge is not located within the catchment it is most important as it provides the longest rain record within the City of Hamburg (from 1967 on). However the recording gap between 1980 and 1996 reduces the useable time-series

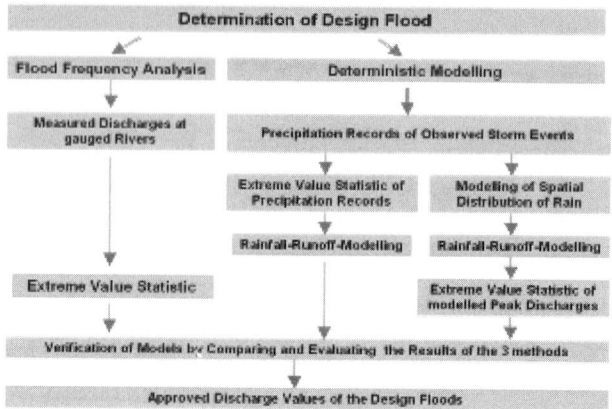

Figure 11.16. Work flow of determination of design flood.

for flood frequency analysis to the period 1996 to 2005. All updated reality data and model data have been delivered to the BSU-B. Also these data are stored on a CD which is attached in A-1 to this document.

11.5.2.2 *Determination of the design flood for the Kollau*

After a new calibration of the model data for the rainfall-runoff-model and the water surface profile model the discharge of the most relevant flood probabilities have been determined by applying the scheme in Fig. 11.16. First on the basis of daily rainfall data the long-term water balance has been modelled for the period 1996 to 2005. With the soil water content and ground water table resulting from this long-term simulation storm water and a temporal resolution of $\Delta t = 5$ min the flood hydrographs have been determined with KALYPSO-Storm for at least 4 storm events in each hydrological year. A flood frequency analysis provided finally the flood quantiles (peak discharge) for a given probability. Two distribution functions, the 3-paramater Log-normal (LN3) and the 2-parameter Log-Pearson (LP2) have been applied alternatively. But most often the LN3 distribution produced the best result that this function has been taken for all requested flood quantiles. Because of the short time-series of flood events the quantiles for large return periods (30 years and more) are of great uncertainty. This uncertainty is increased by the extreme flood event of 1997 which exceeds the other floods by far and thus might have a higher probability then determined by the Weibull formula. Therefore alternatively this flood has not been considered for the parameter estimation of the distribution curves and its return period deduced from this function (Fig. 11.17).

However the statistical extreme values are only affected by not more than 10% for the LN3-distribution. The results have been discussed in detail with the BSU-B and it has been finally agreed to take for the determination of the design flood the distribution curve in which all observed extreme events are included (Tab. 11.4,

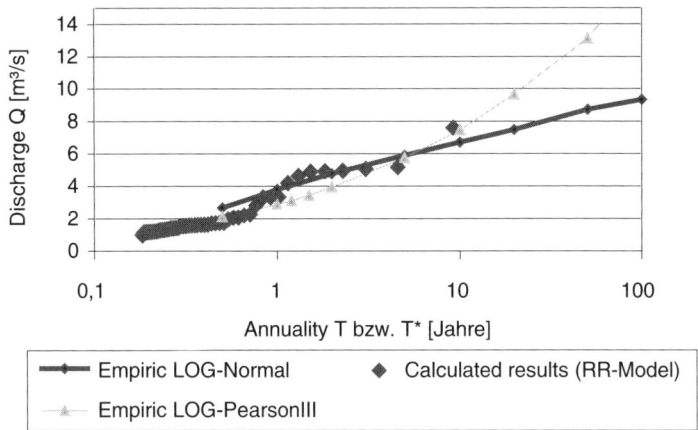

Figure 11.17. Statistical results.

Table 11.4. Statistical results.

T	Log-normal peak discharge [m³/s]	Log-Pearson III peak discharge [m³/s]
0.5	2.69	2.12
1	3.81	2.88
2	4.77	3.95
5	5.91	5.75
10	6.71	7.47
20	7.49	9.65
50	8.73	13.12
100	9.34	17.13

left column). All finally determined and officially approved design floods are documented for the Kollau in attachment A-2.

11.5.2.3   *Determination of the inundation maps for the Kollau*
The corresponding water stage and inundation area have been simulated with KALYPSO-WSPM and determined on the basis of the LIDAR DTM. In Fig. 11.18 the result of the inundation mapping is given for the lower section of the Kollau. The inundation areas for the whole watercourse are given in the attachment A-3.

Critical flooding begins at a 5-year flood and reaches extensive inundation at 20 year floods and more, confirming the observation during the extreme flood event of 2002. Especially along the downstream section of the watercourse developed areas are widely affected.

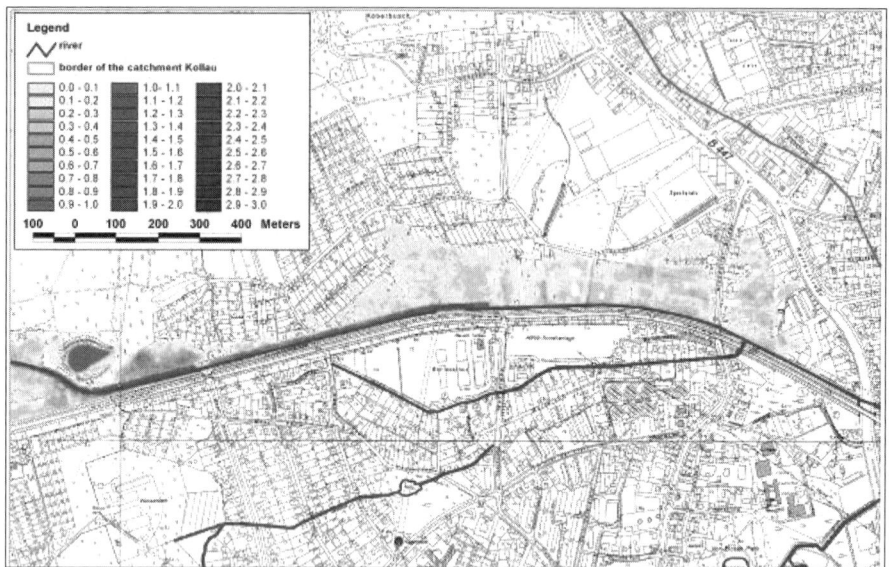

Figure 11.18. Extract of the inundation map of the Kollau.

11.5.2.4   *Determination of the risk maps for the Kollau*

With the methodology of section 4.3 finally the flood risk on the inundated areas along the Kollau has been determined. The required statistical data have been provided by the Statistical State Office Hamburg and Schleswig-Holstein and the German Federal Statistical Office in Bonn. The data reflect the situation in 2005. In the first step the specific asset of all land use categories have been determined:

Asset of residential buildings. The value of the buildings on private properties was specified by the German Federal Statistical Office with 3,300 billion Euro in 1995. With a total of 36.22 million households the average capital stock of residential buildings is 91,000 € per household in Germany. According to Pasche/Kräßig (2005) the specific asset of the inventory is in average 700 € per m^2 in Germany (derived from information of the insurance industry).

The 7 districts Bahrenfeld, Lurup, Lokstedt, Niendorf, Schnelsen, Eidelstedt and Stellingen of the City of Hamburg are sharing the area of the catchment Kollau. The regional statistical agency of Hamburg and Schleswig-Holstein give the number of households and the average living area of the households in these districts (Tab. 11.5). On this basis the total asset of the residential buildings and their inventory is determined to 5,053 Mio. Euro for the whole area of the catchment Kollau.

In the same way the capital stock for vehicles is determined. The regional statistical agency of Hamburg and Schleswig-Holstein give for each of the 7 districts the

Table 11.5. Statistical data and capital stock for residential areas within the catchment Kollau.

	No. of housholds[1] [HH]	Net asset value/ housholds[2] [1000 €/HH]	Total net asset value [Mio. €]	Living area/ apartment[1] [m²/Wo]	Value of inventory/ living area[1] [€/m²]	Value of inventory/ apartment [€/Wo]	No. of apartments[1] [Wo]	Total value of inventory [Mio. €]
Reference year	2004	1995		2004	2004		2004	
Bahrenfeld	14230	91.1	1296	66.20	700.00	46340	12728	590
Lurup	15139	91.1	1379	73.00	700.00	51100	14758	754
Lokstedt	13723	91.1	1250	70.00	700.00	49000	12656	620
Niendorf	13082	91.1	1192	79.00	700.00	55300	19879	1099
Schnelsen	15043	91.1	1371	80.60	700.00	56420	12118	684
Eidelstedt	20092	91.1	1831	70.60	700.00	49420	14899	736
Stellingen	12437	91.1	1133	66.10	700.00	46270	11898	551
Kollaugebiet	103746	91.1	9452	72.70	700.00	50890	98936	5035
Hamburg gesamt	910304	91.1	82938	71.50	700.00	50050	873645	43726
Column	(a)	(b)	(c)	(d)	(e)	(f)	(g)	(h)
Calculation key			(a)*(b)			(d)*(e)		(f)*(g)

Source: (1) district-database (www.statistik-nord.de and belonging pages).
(2) Pasche/kräßig (2005).

total number of motor vehicles. With an average value of 10,000 € per automobile and 3000 € per motor cycle a total capital stock of 811 Mio. Euro results for the catchment Kollau (Tab. 11.6):

The capital stock of settlements consists of the three components "building", "motor vehicle" and "inventory". They are summed up for each district in Tab. 11.7. Division by the total area of settlement gives the specific asset of settlement for each district. It varies in the investigated districts between 416 €/m^2 and 600 €/m^2, showing a high variability. In the end the assets of the districts are averaged over the whole catchment Kollau leading to an average specific asset of 500 €/m^2. In the damage assessment the specific asset has not been averaged over the whole catchment.

It varies from district to district according to Tab. 11.5.

Asset of industrial sector. The land use category industry is composed of the 3 industrial sectors "productive industry", "construction industry" and "Trade and commerce". For theses sectors the net asset for reproduction can be taken from the regional statistical agency of Hamburg and Schleswig-Holstein and is given for Hamburg and for the year 2005 in Tab. 11.8. The asset of the stock is only given as total value for Germany. Thus it can not be directly determined for Hamburg. Assuming that the ratio between the asset of the stock and the capital the asset it more or less the same throughout Germany the asset of the stock can be derived for Hamburg. In Tab. 11.8 the percentages for the stock derived from the national statistical data are given. With the net asset of industry for Hamburg the asset of stock results for Hamburg. Summing up the net asset and the stock value gives a total asset for industry of 57,556 Mio. € in Hamburg. With a total area of 43.4 km^2 for the land use category industry a specific asset of 1326 €/m^2 follows.

Asset of the traffic sector. According to the land use catalogue in Tab. 11.1 the category traffic is composed of the ALB-land use units recreation and traffic. Thus the asset for this sector is taken from the economic sector A6-6 "Public and private service"' which is reaching 25,450 Mio. € for the City of Hamburg. Adding 1% for the inventory stock ends up in 25,707 Mio. € for the total asset of traffic in Hamburg. With a total area of 161.3 km^2 for traffic the specific asset for this economic sector is 159,33 €/m^2.

Asset of the farming sector. Farming is composed of intensive farming with the products crops, vegetables and fruits, and extensive farming on meadows and pastures (cattle, milk and hay products).

The main products of intensive farming are vegetables and fruits in Hamburg. About 81% of this farm land is used for apple production, 3.4% for pear and appreciable parts for cherry and cabbage production (Statistisches Amt für Hamburg und Schleswig Holstein [2005]). The rest of the farm land is used for lots

Table 11.6. Statistical data about motor vehicles and their capital stock for the catchment Kollau.

	No. of registered automobiles[1] [car]	No. of registered motor cycles[1] [cycle]	Ave. value of automobile[2] [€/car]	Ave. value of motor cycle[2] [€/cycle]	Automobile-assets [€]	Motor cycle-assets [€]	Total motor vehicle-assets [Mio. €]
Reference year	2006	2006					
Bahrenfeld	8942	893	10000	3000	89420000	2679000	92
Lurup	11707	823	10000	3000	117070000	2469000	120
Lokstedt	8986	629	10000	3000	89860000	1887000	92
Niendorf	18046	1334	10000	3000	180460000	4002000	184
Schnelsen	11980	940	10000	3000	119800000	2820000	123
Eidelstedt	11466	866	10000	3000	114660000	2598000	117
Stellingen	8127	693	10000	3000	81270000	2079000	83
Kollaugebiet	*79254*	*6178*	*10000*	*3000*	*792540000*	*18534000*	*811*
Hamburg ges.	*633872*	*48242*	*10000*	*3000*	*6338720000*	*144726000*	*6483*
Column	(a)	(b)	(c)	(d)	(e)	(f)	(g)
Calculation key					(a)*(c)	(b)*(d)	(e)+(f)

Sources: (1) Statistical agency for HH and SH; registered automobiles and motor cycles referring to the districts; prepared tabular data.
(2) Paschen/Kräßig (2005).

Table 11.7. Specific asset of the land use settlement within the catchment Kollau.

	Total net asset value [Mio. €]	Total value of inventory [Mio. €]	Total motor vehicle-assets [Mio. €]	Total asset value [Mio. €]	Related area [m²]	Specific asset value [€/m²]
Reference year						
Bahrenfeld	1296	589.8	92	1978	4747171	416.76
Lurup	1379	754.1	120	2253	4348519	518.10
Lokstedt	1250	620.1	92	1962	2895500	677.67
Niendorf	1192	1099.3	184	2476	6477905	382.17
Schnelsen	1371	683.7	123	2177	4884345	445.69
Eidelstedt	1831	736.3	117	2684	4479902	599.15
Stellingen	1133	550.5	83	1767	2950626	598.86
Kollaugebiet	9452	5034.9	811	15298	30783968	496.95
Hamburg gesamt	82938	43725.9	6483	133147	241799549	550.65
Column	(a)	(b)	(c)	(d)	(e)	(f)
Calculation key				(a)+(b)+(c)		(d)/(e)

Table 11.8. Total capital stock of the industrial sectors in Hamburg.

Economic sector	Net asset[1] [Mio. €]	Inventory stock as percentage of net asset value[2] [%]	Inventory stock [Mio. €]	Total asset [Mio. €]	Area of the economic sector [m²]	Specific asset [€/m²]
Basic year	2003	1995				
Processing industry without building industry A6-2	12566	30.2	3791	16358	–	–
Building industry A6-3	575	1.0	6	580	–	–
Commerce, service and traffic A6-4	27524	47.6	13093	40618	–	–
Total	40665	41.5	16890	57556	43399080	1326

Sources: (1) Statistical State Office HH and SH – national accounting: prepared tabular data.
(2) Taken from Pasche/Kräßig (2005).

Table 11.9. Yield and profit from vegetable and fruit farming in Hamburg.

Vegetables and fruits	Area of cultivation[1] 2004 [ha]	Yield[1] 2004 [dt/ha]	Total production[1] 2004 [dt]	Wholesale prices[2] 2004 [Euro/dt]	Total Euro
Vegetables					
White cabbage	**10**	**490.0**	**4866**	**17.52**	**85252**
Red cabbage	**3**	**500.0**	**1460**	**26.63**	**38880**
Fruits					
	(a)	(b)	(b)	(c)	
Apples	997	324.2	323090	70.6	22810154
Cherries	11	98.1	1087	148.75	161691
Total	*1021*				*23095977*

(a) Survey of fruit free in 2002.
(b) Survey of the harvest of fruit trees in 2002.
(c) Apple price averaged over various species.
Sources: 1) Statistisches Amt für Hamburg und Schleswig Holstein (2005).
       2) Statistisches Amt für Hamburg and Schleswig Holstein (2006).

of different sorts of fruits. For the named fruits with exception of pears the prices for wholesale trading in Hamburg could in addition to the total production also be acquired from the Statistical State Office (Statistisches Amt für Hamburg und Schleswig Holstein [2006]) (Tab. 11.9). The Division of the monetary yield by the related agricultural area of 1021 ha leads to a specific production of 22,620 €/ha or 2.26 €/m^2. This value is taken as specific asset for intensive farming in Hamburg.

The specific asset of agricultural land with extensive farming has been taken from MERK, a research study of the German Ministry for Education and Research, published in 2002. They determined an average profit of 25 €/m^3 for each swath with an average yield of 15 m^3/ha. Since not the whole area is used for production of silage the average profit is reduced to 15 €/m^3. This leads to a specific asset of 225 €/ha or 0.023 €/m^2.

Forests are not commercially used in Hamburg. Therefore they are not considered in the asset evaluation.

Functionality between degree of damage and flood intensity. In case of flooding the asset is not totally destroyed. In dependence of the maximum water depth, the duration of flood and the current the degree of damage varies. As described in section 4.3 different functions have been derived in Germany during the last years which describe this functionality. However only the influence of the water depth has been considered. Despite of the deficiencies of these functions additional investigations on the damage sensitivity of the different land use categories could

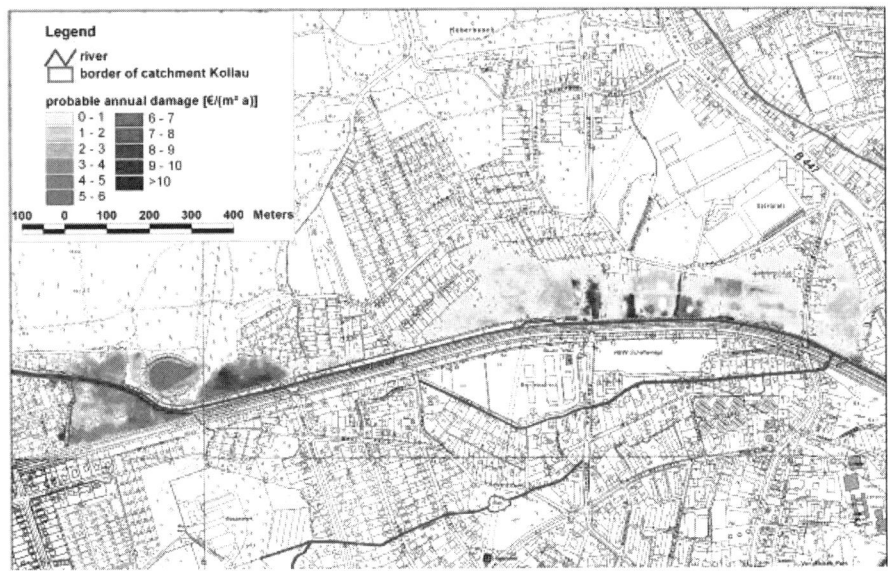

Figure 11.19. Extract of the damage map of the Kollau (probable annual damage).

not be accomplished in this project. The damage functions delivered from the International Commission for the Protection of the River Elbe (IKSE) have the closest geographical reference that they have been used for the damage assessment in this project (Fig. 11.11).

Spatial distribution of the statistically evaluated annual damage along the Kollau. By applying steps 4 and 5 of the damage assessment method described in section 4.3 the probable annual damage has been obtained for the inundation areas of the Kollau. The necessary parameters for the application of equation 11.26 have been described above. The result of this map overlay and intersection is given as an example in Fig. 11.19. The whole map is documented in attachment A4 and on CD.

Risk map of the Kollau. The final step of risk mapping has been the classification of the probable annual damage into different risk zones. The thresholds for the developed areas have been taken from section 4.3.

The flooded farmland need to be classified in a different way. These areas have not the same relevance as the developed areas and thus should differ from the developed areas. Therefore only two zones of risk have been introduced, moderate and medium risk. The threshold between these two zones was set in a way that meadows which are flooded a maximum of once in two years will be evaluated as

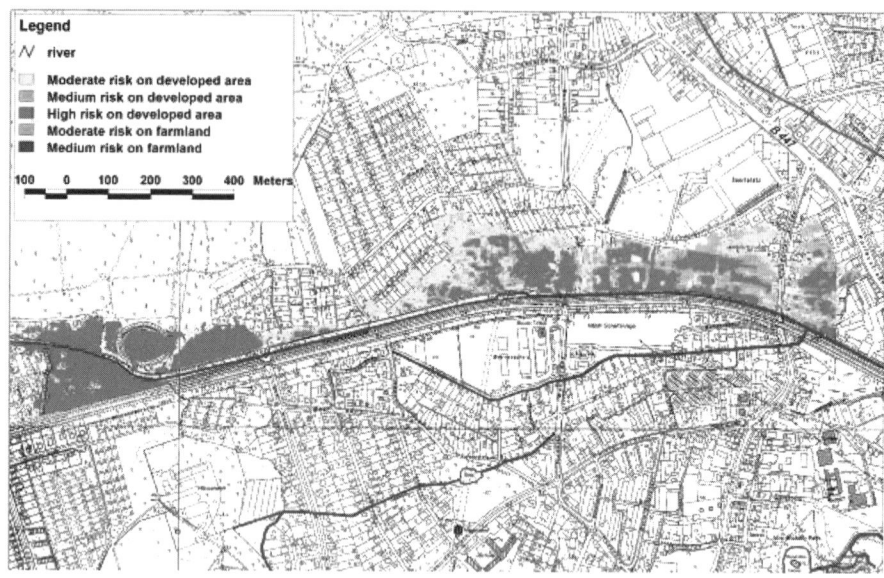

Figure 11.20. Extract of the risk map of the Kollau.

moderate risk. With this assumption the threshold between moderate and medium damage is:

Δp times specific asset of meadows $= 0.49 \times 0.023 = 0.012$ [€/m²a]

To present the flood risk in the map the following legend has been developed:

- Moderate risk on developed area
- Medium risk on developed area
- High risk on developed area
- Moderate risk on farmland
- Medium risk on farmland

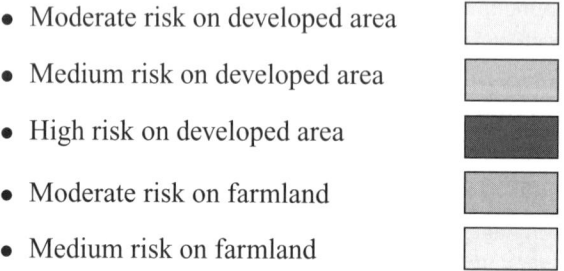

The colour of the legend is in the style of a traffic light. But in contrast to the known signs the risk map colour for a moderate risk is not displayed green but yellow, to avoid that on areas with moderate risk the risk is associated with acceptable risk. The farmland has been given a different colour palette to make transparent the difference to the developed areas.

The result of this risk mapping process is given as an extract in Fig. 11.20. The full map is documented in attachment A5 and in digital form on the attached CD.

Many developed areas which are flooded have been evaluated with a high risk indicating a high urgency for improved flood risk management.

REFERENCES

BWK (1999): Hydraulische Berechnung von naturnahen Fließgewässern. Merkblatt 1, Bund der Ingenieure für Wasserwirtschaft, Abfallwirtschaft und Kulturbau e.V., Düsseldorf.

DVWK (1985): Ökonomische Bewertung von Hochwasserschutzwirkungen. Editor: Deutscher Verband für Wasserwirtschaft und Kulturbau e. V., Mitteilung Heft 10. Bonn.

DVWK (1991): Hydraulische Berechnung von Fließgewässern, Editor: Deutscher Verband für Wasserwirtschaft und Kulturbau e.V., Merkblatt 220, Verlag Paul Parey, Hamburg.

IKSR (2001): Übersichtskarten der Überschwemmungsgefährdung und der möglichen Vermögensschäden am Rhein, Editor: Internationale Kommission zum Schutz des Rheins, Koblenz.

IKSE (2003): Aktionsplan Hochwasserschutz Elbe, Editor: Internationale Kommission zum Schutz der Elbe. Magdeburg.

Ostrowski, M. W. (1982): Ein Beitrag zur kontinuierlichen Simulation der Wasserbilanz. Mitteilungen 42. Institut für Wasserbau und Wasserwirtschaft, Rheinisch-Westfälische TH Aachen, Aachen.

Pasche, E. (1984): Turbulenzmechanismen in naturnahen Fliessgewässern und die Möglichkeiten ihrer mathematischen Erfassung, Mitteilungen Institut für Wasserbau und Wasserwirtschaft, RWTH Aachen.

Pasche, E., Schröder, G. (1994): Erhebung und Analyse hydrologischer Daten auf der Basis geographischer Informationssysteme. Seminarunterlagen zum 18. DVWK-Fortbildungslehrgang Niederschlag-Abfluss-Modell für kleine Einzugsgebiete und ihre Anwendung, Karlsruhe, DVWK.

Pasche, E., Brüning, C., Plöger, W., Teschke, U. (2004): Möglichkeiten der Wirkungsanalyse anthropogener Veränderungen in naturnahen Fließgewässern, published in Proceedings of the "Jubiläumskolloquium 5 Jahre Wasserbau an der TUHH, Amphibische Räume an Ästuaren und Flachlandgewässern", Hamburger Wasserbau-Schriften, Heft 4, Hrsg. Erik Pasche, Hamburg.

Pasche, E., Geissler, T. R. (2005): New strategies of damage reduction in urban areas proned to flood. In: Szöllösi-Nagy, A., Zevenbergen, C.: Urban Flood Management. A. A. Blakema Publishers. Leiden/London/New York.

Pasche, E., Kräßig, S. (2005): Regeneration der Stör. Abgrenzung der Überschwemmungsgebiete im Einzugsgebiet der Stör, Research Report, Institut für Wasserbau der Technischen Universität Hamburg-Harburg. Hamburg.

Pasche, E., Kräßig, S., Lippert, K., Nasermoaddeli, H., Plöger, W., Rath, S. (2006): Wie viel Physik braucht die Strömungsberechnung in der Ingenieurpraxis? Published in "Wasserbauliche Mitteilungen, Heft 32, Technische Universität Dresden.

Statistisches Amt für Hamburg und Schleswig Holstein (2005): Bodennutzung und Ernte in Hamburg und Schleswig Holstein 2004. Hamburg.

Statistisches Amt für Hamburg und Schleswig Holstein (2005): Erzeuger- und Großhandelspreise für die Landwirtschaft in Schleswig Holstein 2004. Hamburg.

12

Flood Repair Standards for Buildings

Stephen L. Garvin & David J. Kelly
Building Research Establishment Scotland, East Kilbride, UK

ABSTRACT: Flooding of buildings has become an increasingly important issue in Europe in recent years. This is driven in part by the impact of climate change, but also by the realisation that development in flood planes have increased and can be at risk if flood defences fail. In the urban environment issues of flash flooding from heavy intensity rainfall and sewer flooding add to the major risks from river and coastal flooding.

In recent years research has addressed the standards of repair required for buildings after a flood and the need for resilient repair. This builds on guidance produced in the early 1990s that was in response to specific high impact events. The latest guidance considers not just the material repairs, but takes the approach that a standard of repair is based upon proper assessment of the impact of a previous flood, the drying and decontamination process, the assessment of risk of future flooding and the specification and implementation of repairs. The standards or repair are related to the level of risk determined. The standards of repair consider not just the materials used, but the use of flood protection products and where necessary the relocation of services and rooms.

Further research is necessary for the development of better resilience. Research directed at new or improved materials, combined with better information on the performance of flood protection products is also required. In addition, the barriers and opportunities for flood resilient repair need to be addressed on an international basis.

12.1 INTRODUCTION

Flooding is a major problem for many people in Europe, posing a risk to health, safety and wellbeing, and resulting in widespread damage to property. The issue of flood damage to buildings has increased in importance in recent years (OST, 2004; FBE, 2001). Research projects have sought to understand the damage that occurs to properties and to seek effective means of protecting and repairing buildings (Scottish Office, 1996; BRE, 1997; CIRIA, 2000, 2001, 2001a, 2004). Planning and to a certain extent building control deal with the risk of flooding of new buildings (ODPM, 2001), however, there are many existing buildings that may be in flood prone areas and require careful consideration with regard to repair after a flood.

The focus of repair has in the past few years addressed increased resilience in the repairs carried out. This can involve a number of means of improving the performance of the building in a flood. Changing materials to less vulnerable types in the event of a flood is one means of achieving improved resilience. However, it may not always be possible to change the material or indeed the manner in which material is used. The performance of materials in flood has been addressed in previous publication (Garvin, 2003, 2004). Materials performance depends not just on the materials properties, but on the duration of flooding, the effectiveness of the recovery process (including drying and decontamination), the nature of the flood water and other factors that may result in impact damage to the building.

12.2   FLOOD DAMAGE AND REPAIR

12.2.1   Damage

A report from the Environment Agency (EA, 2001) provided statistics on the extent and severity of flooding to England in 2000. This year was the worst with regard to rainfall in recent years in the UK, this occurred particularly in the period of October and November 2000. Prolonged and repeated flooding was found to occur in several locations. In total over 10,000 properties were flooded in over 700 locations. It is however estimated that over 280,000 properties were spared flooding as a result of flood defences, and in addition 38,000 properties were protected by temporary sand bags. The report states that it is impossible to guarantee flood protection and there are many locations where an engineering solution is impractical and others where considerable damage would occur to the environment through the use of such defences.

The report of the Flood Foresight study (OST, 2004) indicated that the numbers of properties at risk of localised flooding could typically increase four fold under four future climate change and flooding scenarios. Estimates of more than 350,000 properties are given in the report. The report states that urban flood waters are invariably mixed with sewage, therefore future increases in urban flooding would be compounded by the additional risks to health, and higher costs of repair to properties.

As well as river and coastal flooding the urban environment can be affected by local intense storms which can overwhelm drains and sewers. The OST report states that it is important to manage the layout and functioning of cities to adapt for future climate change, particularly changes in rainfall pattern. Building away from flood risk areas is the key, however, if building in at risk areas to ensure that there is space for river and coastal processes.

The ODPM report of 2001 (ODPM, 2001) provides a guide on preparing for floods. This guide gives practical information on the measures that can be taken to improve the flood resistance of both existing and new properties at risk of flooding within the UK. The report discuses how flood water can enter buildings, which include the following:

- Ingress around closed doorways
- Ingress through airbricks and up through the ground floor
- Backflow through overloaded sewers discharging inside the property through ground floor toilets and sinks
- Seepage through external walls
- Seepage through the ground and up through the ground floor
- Ingress around cable services through external walls.

Measures that can be taken to improve the flood resilience of the building are described in terms of dry proofing and wet proofing. These are dealt with later in this paper.

A report produced by CIRIA for the Environment Agency (CIRIA, 2001) has covered the available flood protection products as a guide for homeowners. This includes the following:

- Temporary barriers such as those used to protect against water entry at doors and windows, up to a height of one metre.
- Temporary free standing barriers, these are located some distance from the building and are intended to protect the whole building. Pumps can be used to prevent flooding from water that seeps through or over the barrier.
- Household products such as those used to cover vents and air bricks.

Crichton (2005) has written on the insurance implications of flooding of properties. Over the years the UK insurers have provided voluntarily insurance to everyone no matter the level of risk involved. However, this situation is unlikely to be sustainable and insurers are likely to increase premiums or avoid insuring high risk properties in the future. The use of engineered defences has significant cost and there is competition for available government funding. The author argues for non structural measures to be used to deal with flood risk, particularly through planning controls and by using technology such as sustainable drainage. Experts around the world see non-structural measures as being more sustainable, an efficient way to tackle the problem rather than relieving the symptom.

Green and Penning Roswell (2002) also write on the issue of insurance and the need for a partnership between the insurance industry and government in order to make it viable for flooding. The authors argue that the objective of flood management is not to reduce flood losses, but rather to maximise the economic efficiency of the use of the catchment as a whole.

The report on *Learning to Live with Rivers* (ICE, 2001) was produced in response to the floods of 2000 and other flood events. The report argues that for existing towns and cities on flood plains, there will generally be a strong justification for continuing to provide a high standard of defence. In addition, there will still be some justifiable development on flood plains. There could be a low to medium risk, i.e. less than 1 in 100 chance of a flood in any one year where there is already an appropriate standard of defence and where the requirements of a flood risk assessment can be met.

Flood defences can give a false sense of security, as they can never give protection against all flood events. The key message of the report is that such defences should be worked as long as risk and vulnerability is understood. Development on inadequately protected areas should be discouraged.

Flooding occurred to the areas of Glasgow and Perth in Scotland in the 1990s with significant economic losses being experienced. In Glasgow there were economic losses of £100 million and over 700 homes affected by the flooding (Alexander Howden Group, 1995). The flood warning procedures operated by the then Clyde River Purification Board for the White Cart drainage system reduced the effect of flooding in the south of the city. Defences constructed to a 100 year standard saved properties, but some were affected by surface water ponding and inadequate sewerage facilities. In the paisley area a large housing estate was affected, although most of the impact was on newly built houses. Flooding of over one metre depth affected the properties to above ground floor window level. The damage involved loss of carpets due to damage, contamination by sewage and loss of electricity. Major concerns also included health risks due to hypothermia as heating could not be restored quickly and the risk of looting.

Worse damage was experienced in older properties that were constructed with timber lath boards and old plasterboard walls inside. All the walls and woodwork had warped in affected properties, with demolition of some properties being required.

Damage in the Perth floods of 1993 (Perth and Kinross Council, 1995) was more focussed into one area than the Glasgow floods. Out of an estimated 1500 homes flooded there were 1200 on one estate in the city. Of these there were 873 that were classified as extensively damaged. It took until October 1993 for some residents to be able to return to their homes. A flood depth of nearly two metres was reached in some buildings.

12.2.2   Repair

BRE guidance produced in 1991 covered the issues of dealing with flood damage (BRE, 1991). The guidance indicated that damage to the building is likely to range further than ruined carpets and furniture. The various steps involved in recovery and repair of the building include the following:

• Cleaning
• Drying

- Assessing damage to walls, timber, metals and decorations
- Repair methods.

It is stated that drying out may have to continue for months before the house is reinstated completely. The walls absorb large quantities of water during a flood, a slid one brick thick wall may take in as much as 55 litres per square metre. Damage to plasters and to timber as a result of flood is described.

The guidance recommends that suitably qualified surveyors and engineers are employed to assess damage and to specify repair requirements.

Guidance was produced by the then Scottish Office (1996) on the repair of buildings in response to flood events such as those that occurred to Glasgow and Perth. The guidance covers the issue of the impact of flood water on materials. The following issues are relevant:

- Masonry and concrete – unlikely to have serious effects, but impact on thermal performance when wet. Lightweight blockwork expands on wetting and shrinks as it dries with a risk of cracking. Masonry and concrete may take a long time to dry, which will impact on what can be done in repair.
- Timber – swells and may distort on wetting; even with reversible swelling the damage might be considerable. In timber frame structures the swelling of immersed members could cause damage in other parts of the structure.
- Wall finishes – gypsum plaster may soften, but may harden again to near its original state. Plasterboard will soften whilst wet and may delaminate, thus being unacceptable. External cement and lime based renderings are unlikely to be affected.
- Metals – could be affected by corrosive effects of sea water flooding. Stainless steel should be selected for components which may be subject to attack, including wall ties, nails and fixings.
- Insulation – floor and wall insulation can be affected depending on the type and the manner in which it fills the floor or wall.

A series of Good Repair Guides were developed by BRE in order to address repair of buildings following flooding (BRE, 1997). The guides included the following issues:

- Immediate action after a flood
- Repairs to ground floors and basements
- Repairs to services, secondary elements, finishes and fittings.

Flooding and Historic Buildings is the subject of a Technical Advice Note from English Heritage (2004). Anti flood protection products must be applied to a historic building with sensitivity, so that they do not damage the special interest or integrity of the historic structure. In particular the existing structural systems and materials must be retained and respected. The materials and techniques used in repair work should be traditional and compatible with the existing ones. The use of lime based mortars and renders for traditional stonework as opposed to cement based mortar is advised. Although the cement based materials harden faster they also cause damage frequently to the historic buildings.

Protection of the historic building can fall into three categories, as follows:

- Regular maintenance of the existing building and grounds
- Designing any additions or alterations to the property with flood proofing in mind
- Special measures designed specifically to combat the immediate effects of flooding.

The last of these points includes provision to protect buildings using flood door boards or permanent flood barriers formed from brick or stone perimeter walls.

Pasche and Geisler (2005) have written on the issues of new strategies of damage reduction in urban areas prone to flood. This comprises individual preventative and emergency measures at buildings and municipal infrastructure and a land use policy to adapt building activities to the risk. The flood resistance of buildings can be accomplished by three defence strategies, using waterproof materials, sealing and shielding the building. In the first strategy the floodwater is not held back, but the impact of water on fabric and fixtures is minimised. This strategy requires that all household products and furniture are removable. The second and third strategies try to keep the flood water out of the building; this they are dry proofing. The paper concludes that despite the great resilience potential of dry proofing and water proofing the readiness of the stakeholders has been low to apply these techniques. The reasons being the insufficient knowledge and information on the techniques and the economic implications. It is difficult for the stakeholders to assess these resilience measures. It was recommended that water policies should be created to encourage the use of resilience measures.

In recent years CIRIA (2005) and BRE (2006) have produced extensive guides to the flood repair of buildings. There is some overlap in approach between the documents, but together they represent the UK's state of the art with regard to flood resilient repair. The incorporation of flood resilient repair as a means to manage future risk will become increasingly important. Flood resilient repair has benefits for a range of stakeholders. For building owners it represents the opportunity to reduce the disruption and cost of a future flood. Insurers will face less claims costs in the future with regard to the next flood events. It is possible that the flood resilient repair will initially be greater than that of a repair to the existing standard. However, on a whole life cost basis the flood resilient repair will cost less.

Flood resilient repair involves not just the repair specifications and activities, but also the issues of drying, decontamination, post flood surveys and the assessment of the risk of a future flood. The repair methods include interventions in the design, methods of construction and materials used. However, flood resilient repair may also require the use of flood protection products, household products and temporary barriers; these methods of protecting buildings may be used where there is a high risk of flood of a building in subsequent years.

12.3   FLOOD WATER AND DAMAGE

Buildings will allow water to enter during a flood through the following routes:
- Through masonry and mortar joints where the natural permeability of both these materials, particularly the mortar, can be high.
- Through the brickwork/blockwork.
- Through cracks in external walls.
- Through vents, airbricks and flaws in the wall construction.
- Through or around windows and doors at vulnerable points such as gaps and cracks in the connection of the frames and walls.
- Through door thresholds especially where these have been lowered to the ground to allow level access.
- Through gaps around wall outlets and voids for services such as pipes for water and gas, ventilation for heating systems, cables for electricity and telephone lines.
- Through party walls of terraced or semi-detached buildings in situations where the property next door is flooded.
- Through the damp proof course, where the lap between the wall damp proof course and floor membrane is inadequate.
- Through underground seepage which directly rises through floors and basements.
- Through sanitary appliances (particularly WCs, baths and showers) caused by backflow from flooded drainage systems.

The amount of damage caused to buildings will depend, amongst other factors, upon the water depth of exposure, as follows:
- Below ground floor: basement damage, plus damage to any below ground electrical sockets or other services, carpets, fittings and possessions. Minimal damage to the main building. Deterioration of floors may result if the flood is of long duration, and/or where drying out is not effective.
- Above ground floor: In addition to the above, damage to internal finishes, saturated floors and walls, damp problems, chipboard flooring destroyed, plaster and plasterboard. Services, carpets, kitchen appliances, furniture, electrical goods and belongings are all likely to be damaged to the point of destruction. Services such as water tanks and above ground electrical and gas services may be damaged.

The duration of the flood can make a significant difference to the extent of cleaning and repair that is required. Generally, short duration flash flooding will be quickly remedied, and will be less costly to repair than a flood of longer duration. A flood of longer than 24 hours duration can cause serious damage to the building elements. The nature of the flood water is also significant, for example, saltwater from coastal floods results in corrosion to metal components, and water containing sewage requires extensive cleaning and decontamination.

12.4   POST FLOOD ACTIONS

Making safe, decontamination and drying should be undertaken as soon as possible after the flood water recedes and prior to the post flood survey. Floodwaters can bring both the structure and contents of a property into contact with a wide range of silt, debris and other contamination. It is likely that the building fabric will absorb moisture, the degree of saturation is dependant upon the duration of the flood and materials properties. Although it is desirable to reduce the moisture content in all locations as quickly as possible, care is required so that the process is controlled and the situation is not made worse.

A post flood survey should be undertaken to identify damage as soon as the flood waters have receded and decontamination and initial drying has been completed. It is essential to identify and check parts of the building fabric that could potentially retain significant volumes of water and thus potentially promote corrosion or the growth of moulds. It is recommended that the type of building is first identified, this is particularly important if the building is of non-traditional construction. In such cases specialist advice may be required.

The first step in assessing the impact that flooding has had on a structure is to identify the exact form of construction of the building. The following list describes these activities in more detail:
- Assess for structural damage
- Assess for settlement damage
- Assess the surface material condition of the wall (internal and external)
- Assess for material damage
- Assess the internal condition of the wall
- Assess for staining
- Measure moisture content
- Assess for corrosion
- Assess for blocked ventilators.

12.5   RISK ASSESSMENT

The first step toward defining risk is to identify whether or not there are factors that indicate the likelihood of a future flood. Desk studies and site surveys are used to assess the risk of future flooding and are briefly described here.

12.5.1   Desk study

Table 1 lists some factors that may help to determine the likelihood of a future flood occurring. The factors can mostly be assessed as a desk exercise and should help to identify the potential flood hazards. These hazards will be relevant later when determining the level of risk and the corresponding standard of repair required. There may be specific circumstances that are not covered by this generic list and that will need to be considered by the assessor.

Information on the flood history of a building may also be available through discussion with building owners, neighbours, the fire service, wastewater undertaker, Environment Agency, and/or local authority, who may all be able to provide valid information. It is also recommended to check newspaper articles, local history societies or library records for information on previous flood incidents.

12.5.2   Site survey

Any information that is required but not obtainable from the desk study should be gained through a site survey, such as topography of land and likely routes of surface water run off. Site surveys can serve a dual purpose to also determine what happened during previous flood events e.g. the consequences. Relevant information should be collected during the post flood survey (see Section 3.1) and a record kept with all the other relevant information.

12.5.3   Consequences

Table 2 lists factors that need to be considered in order to assess the consequences of a future flood occurring (CIRIA, 2005). An assessment of the consequences of future flooding is essential to understand and prepare the requisite standards of repair. It enables the building owner (or other interested party such as insurance companies) to make decisions on how best to manage their exposure to flood risk using appropriate standards of repair.

12.5.4   Levels of risk

For each flood risk there is a corresponding standard of repair.

$$\text{Level of risk} \Rightarrow \text{Standard of repair}$$

These levels of risk and standards of repair are used throughout the remainder of the document.

There is no single level of risk for a building. More frequent flooding may cause small amounts of damage, whereas less frequent flooding to higher depths may result in much larger, longer term damage and disruption.

To assess the level of risk, first consider the factors in Tables 1 and 2. Determine as much information as possible from the flood event that has occurred and assess

Table 12.1. Factors to be considered to help determine the likelihood of a future flood.

Risk likelihood factors

Historic factors
Previous flood history of the building. Information will be held by insurers, householders, building owners, the local authority, fire brigade or local flood groups
   Note: these organisations may have previous survey reports; alternatively information may be available in newspaper reports and/or from local societies
Flood warnings have been issued for the area
Previous floods have caused closure of buildings, movement of people from their homes or disruption to business
Previous floods have been of greater than 24 hours duration
Sewer flooding has occurred previously
   Note: Could be a result of overloaded or blocked sewers. The responsible authority may have subsequently solved the problem

Exposure
The building is located within a flood risk zone, as indicated on flood maps
   Note: Maps are available from the Environment Agencies, Ordnance Survey or insurers.
The building is located within a flood risk zone, as indicated on flood maps, and is unprotected by a flood or coastal defence, or is unprotected by a flood control structure (e.g. flap valve, sluice gate, tidal barrier, etc.)
The building is protected by existing flood defences; in this case it is important to understand the extent and standard of the existing protection and its anticipated effectiveness over time
The building is protected by current flood defences, or there are plans to construct or repair flood defences
The building is protected with temporary defences
   Note: Emergency deployment plans are essential
The development is unprotected by a flood control structure (e.g. flap valve, sluice gate, tidal barrier, etc.)
The building is covered by the Environment Agency AFW system
The site of the building has characteristics that indicate it may be prone to flooding
   For example, a building situated in a natural or artificial hollow; at the base of a valley or the bottom of a hillside
The building is underlain by a chalk aquifer
The building is close to an intermittent stream
There are prolonged periods of wet weather in winter and a propensity for intense rainfall events
   Note: Climate change is likely to increase winter rainfall across the UK and annual the rainfall in northern part.

Adjacent installations and facilities
The building could impede natural or artificial land drainage flow paths
The building is located upstream of a culvert
   Note: culvert can be prone to blockage
Water levels in watercourses near to the building are controlled by a pumping station
The building is downstream/down-slope of a reservoir or other significant water body

(Continued)

Table 12.1. Continued.

Risk likelihood factors

Local ground conditions
There is a relatively high proportion of hard standing to soft ground
The soil is impermeable or often near saturation point
There are signs of soil erosion upslope of the site
There are artificial drainage systems on, adjacent to or upslope of the site

Table 12.2. Factors to be considered to help determine the consequences of a future flood (CIRIA, 2005).

Risk consequence factors

The value of the property
The type of property e.g. detached, terraced, basement, prefabricated
Building design e.g. timber, modern brick, suspended floor etc.
The costs of repair from previous floods
The availability and cost of insurance
The depth of the flood relative to the building and any differences between the external
  depth and internal depth of the flood
The duration of the flood
The type of flood that occurred (or is likely to occur)
The elements of the building affected and the amount of damage caused with
  differentiation made between material and structural damage (gather as part of the
  post flood survey, Section 3.1)
The sources of flooding e.g. tidal, rising groundwater, sewer backflow, riverine, etc.

the damage caused to the building using the information in the post flood survey. The building owner's attitude toward the levels of risk that are acceptable to them in the future will play an important role in decision making. For example, an owner may prefer to implement a higher standard of repair than recommended to reduce the trauma associated with a future flood event. Using all the information available (on the likelihood of a future flood and the potential consequences of a flood event) it should be possible to make an assessment of the level of risk for the building. Three levels of risk have been recommended and these are associated with the standard or repair.

12.5.5   Little or no risk

Even where a flood has occurred, there may be an insignificant risk of this returning and this should be recognised. An example of this may be where the source of flooding has been reduced or removed e.g. sewer blockage removed or flood defence installed. If this is the case, it is unlikely that any of the risk factors in Tables 1 and 2 would be relevant. If little or no risk is determined, the risk assessment may be limited to a particular time frame, e.g. the lifetime of a flood protection measure.

12.5.6   Low to medium risk

Most previously flooded buildings will have some level of risk of future flooding. If there are some identified risk factors from the lists in Tables 12.1 and 12.2 then low to medium risk is likely. This level of risk should also take into account the depth and duration of exposure, and the damage caused to the building. Although not prescriptive, the following may apply:

- The depth of previous floods was above floor level.
- The duration was up to 12 hours.

12.5.7   High risk

If a significant number of the risk factors in Tables 12.1 and 12.2 apply, then the building should probably be assessed as high risk; likewise if the previous floods caused significant damage. The following points may be useful in determining high risk:

- The depth of previous floods was above floor level.
- The duration was in excess of 12 hours.
- Significant work was required to dry and decontaminate the building.

12.6   STANDARDS OF REPAIR

12.6.1   Definitions

Achieving the appropriate standard of repair requires more than simply improving resilience of the building and the design of the existing building. The standards should include the following:

- A proper regime for decontaminating and drying the building after the flood water has been removed.
- A thorough post flood survey of the building.
- An assessment of the risk of future flooding.

The appropriate standard of repair can then be determined from the likelihood of a future flood occurring. This approach has recently been developed in the UK in a research project funded by public and private bodies. The research was undertaken by the Building Research Establishment (BRE) and published by the Construction Industry Research and Information Association (CIRIA).

The standard of repair is defined as follows:

- *The standard of repair is the extent to which repair work is carried out and the extent of measures undertaken so that damage from future flooding is minimised. The standards of repair are determined through risk assessment. The flood resilience or flood resistance measures in the standard of repair can include dry proofing and wet proofing of the property.*

Three standards of repair are defined in this report and relate to the level of risk determined. They can be summarised as follows:

- Standard of Repair – Level A: The risk assessment shows that there is little to no risk of a future flood. It is recommended to repair the building to the original specification, although some minor upgrades may be incorporated to improve the flood resilience.
- Standard of Repair – Level B: The risk assessment shows that the likelihood of a future flood is low to medium i.e. it is considered sufficiently high to recommend repairs to increase the resilience and/or resistance of the property above the original specification.
- Standard of Repair – Level C: The risk assessment shows that the risk of a future flood is high. It is recommended to instigate repairs that will increase the resilience and resistance of the property significantly. Such repairs involve dry proofing and/or wet proofing of the building.

It is recommended that the standards of repair are determined as follows:

- Little or no risk = use standard of repair Level A
- Low to medium risk = use standard of repair Level B
- High risk = use standard of repair Level C.

The type and use of a building may affect the standard of repair required. For example, a building that has been found to have little or no risk of the flooding returning, but is supplying an essential service (e.g. hospital, electrical sub station) should have its level of risk increased so that potential future disruption is reduced.

In the domestic building sector it is essential that the type of housing present be taken into account. For example, in some cases where the housing supplies sheltered accommodation to elderly or disabled people, it may be necessary to change the use of the buildings to avoid future risk to life.

In some instances, including terraced or semi-detached housing, joint action is required to prevent water entering a property from the neighbouring building.

All information associated with the risk assessment should be kept for future reference, including details of the activities undertaken and the decisions made. Information may be maintained within a formal risk assessment report, or by the homeowner through personal notes. The following should be included:

- List of factors from Tables 12.1 and 12.2 that are relevant to the building, plus any additional information that has emerged during the assessment.
- All potential sources and causes of flooding to the building.
- The depths and durations of the recent and any previous historical flood.
- The assessed level of risk of a future flood.
- Any additional relevant factors.

Where there is a risk of future flooding, then the standards of repair will suggest improving the resilience or resistance of the building.

Table 12.3. Particular flood damage vulnerability of typical construction elements.

Construction	Vulnerability to flood damage
Masonry walls	Damage to plasterboard Damage to gypsum plasters Damage to applied plaster
Timber frame walls	Water damage to frame structure (including rot and warping) Damage to plasterboard Damage to lath and plaster
Steel or concrete frame walls	Corrosion to Steel Frame, or reinforcement and fixings Damage to plasterboard
Timber suspended floors	Damage to floor cover and structure (including rot and warping) Damage to timber boarding and decking
Ground supported or suspended concrete floors	Damage to sand/cement screeds Damage to insulation, particularly in floating floors
Fenestration	Glass breakage Sealed insulating glass unit failure Timber frame warping Leakage through window frames and door leafs Corrosion of metal frames or steel reinforcement in PVC-U frames

Improved resilience: means improving the ability of the materials to recover from flooding.

Improved resistance: involves the use of building materials, components and elements that are undamaged or unaffected by flood water.

For refurbishment of a previously undamaged property where a risk assessment shows there is a flood risk, use repair standard level B or C in any remedial work.

The particular vulnerabilities to flood damage associated with common forms of construction are illustrated in Table 12.3. By adopting the preferred standards of repair for these forms of construction, subsequent flooding should be easier and less costly to repair.

Improving the flood resilience and resistance of buildings is likely to include either the dry proofing or wet proofing of the existing buildings, which are defined as follows:

• Dry proofing involves the use of flood protection barriers to prevent water entering the building through walls, floors, doors, windows, air-bricks, ventilation holes and other openings thus preventing wetting and damage to internal building materials. Dry proofing measures can be used for floods of up to one metre depth but should not be used above this depth of water. Where flood waters rise above one metre depth of the wall, dry proofing measures become undesirable as the pressure of holding the floodwater back can cause structural damage to the building. Therefore, water should be allowed to enter the building. Dry proofing can also involve repairs such as repointing damaged brickwork walls that prevent water from entering the building through the brickwork, and raising floor levels to above a predicted flood level.

• Wet proofing is based on the acceptance that some water will enter the building, so the intention is to design and/or use materials in the construction/repair of the building that will help to prevent damage to the building when this occurs. It should make it easier and quicker to re-instate the building once the flood has passed. There are a range of potential wet proofing measures possible including the use of flood resilient or resistant building materials and the raising of electrical wiring above flood levels.

Often a combination of dry proofing and wet proofing measures will offer the most effective solution, especially where there is a high risk.

12.6.2   Brickwork repair

Whilst this paper does not intend to list all aspects of the repair standards that have been developed in recent years, it does take the example of brickwork.

Brickwork and other forms of external masonry are designed to allow a degree of moisture penetration/movement so it is unlikely that they will be waterproof if they are exposed to flood waters for long periods. Water always finds the route of least resistance so appropriate attention to detail can slow the rate of moisture ingress and offer protection against shorter duration flood events. Table 12.4 details weaknesses in external masonry, including brickwork, against moisture penetration and outlines possible waterproofing options relating to both new build and repairs.

If the risk of a future flood is defined as little or no then a level A repair is required. This may involve simple repairs to mortar joints, finishes or ancillary components, to in some cases repointing where the mortar has failed. In effect the resilience of the brickwork will be substantially as it was prior to the flood, but the condition should be such that it is improved and will not be as damaged in the event of a future flood.

For a low to medium risk the resilience of the wall should be improved as per a level B repair. This level will require at least repointing and possible rebuilding of areas of brickwork, especially if structural damage has occurred. One means of improving resilience and the resistance of passage of water into the building is to apply a render finish. Although render is not intended to hold back flood water if it is in good condition and there is a certain duration of flood then it can effectively act as a barrier to water entry.

For a high risk a level C repair is required. It is likely that same type of repair standard as level B will be applied. In addition, there will be a move towards

Table 12.4. Moisture penetration and waterproofing of external masonry.

Issue	Moisture pathway	Waterproofing options
Weep holes in the brickwork	Through the weep hole to the cavity	Use proprietary covers where available, when flood warning is given
Air bricks and ventilators	Through the ventilation holes	Use proprietary covers when flood warning is given
Service penetrations and openings, including pipes, flues and tumble dryer vents	Through the junction between pipe-work and brickwork, and through the pipes/vents themselves	Clean and renew sealant materials and maintain an effective seal at these junctions Consider moving to higher level in the wall For services below ground level consider using additional protection measures
Cracks in brickwork	Through the crack.	Repair excessive cracks: – Water repellent can cover and protect fine cracks for short periods, but little water pressure resistance – Application of render or external insulation system may be appropriate for dealing with extensive cracking
Cracks in the brick-mortar interface and cracks in the mortar	Between the brick and mortar joint interface	Inspect the mortar joints for cracking, remove and re-point the mortar where evident
Through the mortar	Mortar is generally much more permeable than brick. Water can quickly saturate a mortar joint and, through the pressure of a flood, form a moisture pathway	This can only be stopped by application of external waterproofing, such as a render coating, or a rendered external insulation system Consider the mortar mix and the benefits of increased lime content
Through the brick	Bricks will absorb water and under the pressure of flood water may saturate and allow water to pass through them. This will normally take longer than through mortar.	Low permeability bricks, such as engineering bricks, will reduce the speed of water penetration. However, to reduce the risk of the problem in existing brickwork, barrier techniques will be necessary (e.g. including render and external insulation systems)
Through movement joints in external masonry	Through the joint, especially where external sealant is in poor condition	Apply good quality sealant (two part gun grade) to the joint Joints should be backed with suitable filler

improving the flood resilience by using household products and/or individual temporary flood defences to protect against the passage of water into the building (BSI, 2003a, b, c). The use of these types of products allows greater flexibility in the repair approach, especially when building type and design issues mean that there is limited scope to amend the building structure. The protection products can be used to protect not only brickwork walls, but also floors, basements and internal parts of the building.

Where there is a high risk and level C is determined, then the use of further protection in addition to materials or design changes is required. In this case the specifiers and those involved in risk assessment should carefully consider the performance limits of the protection products and barriers that are used. It is possible that in floods where there is rapid flow of water both into and around buildings will destabilise some of the protection methods and lead to a risk of failure. This may be seen as acceptable however if the flood water causes limited damage to the inside of the property. The specification of a level C repair cannot therefore rely solely on either external flood protection measures, or the use of resistant and resilient building materials in the repair. The use of both approaches is essential for success of the repair.

There are further additional approaches that can be introduced in level C and possibly level B repair. For example, the relocation of services or of a kitchen, bedroom, living room or bathroom to a higher level in the building. This approach should be targeted at particularly vulnerable rooms as determined from the amount of damage caused by previous floods. The cost involved in relocating rooms or services such as electricity to a higher level may result in a initial high cost in comparison with repair of the service or room in the same location. However, locating rooms and services above the potential flood level will reduce substantially the opportunity for damage to be caused at a later flood.

12.7   DISCUSSION AND CONCLUSIONS

The issue of flood resilient repair has grown in interest in the United Kingdom in recent years. The interest has grown from the amount of properties that were flooded, primarily in major events since 2000. In the urban environment the flooding of property can occur rapidly and with little warning through flash floods and sewer floods at almost any time, i.e. in summer or winter, as a result of heavy rainfall and overloaded infrastructure.

12.7.1   Current guidance

Guidance has been prepared by various organisations since the early 1990s that helps those involved in the repair of buildings to specify and implement changes.

The early guidance on this topic was not so much concerned with flood resilient repairs and not to the extent that is set out in the most recent guidance. Current thinking is that flood resilient repairs involve not just the use of materials or a change of design, but the whole process of investigative the extent of flood damage to a building, the drying and decontamination process and the use of risk assessment to set the standard of repair required. The standard of repair will vary with the risk; a higher risk will be likely to involve greater amounts of changes to materials, designs, the use of flood protection products and the possible relocation of services and rooms within a building.

The guidance is limited, as is the case with all guidance, in that the specific cases of individual buildings require an individual consideration. The generic guidance given by BRE (2005) and CIRIA (2005) covers certain types of buildings, materials and services in detail. A number of new types of construction are currently being advanced in the United Kingdom and other parts of Europe, so called Modern Methods of Construction (MMC). This can include a variety of modularised or framed construction. It is not possible to produce guidance that reflects all potential designs and materials. However, encouragement needs to be given to the suppliers of MMC systems to provide information of flood resilience and repair. This is especially important where known flood risk areas are planned for MMC buildings.

12.7.2   Climate change impacts

The increasing interest and indeed concern over flood damage and repair to property is at least in part driven by the impact of climate change. For much of Europe there is an increased risk of higher winter rainfall and for some areas an overall increase on an annual basis. This increased rainfall will in turn increase the risk of flash floods and sewer floods. Even in periods, such as the summer, where rainfall overall is predicted to decrease there is still an increased potential for flash floods from severe intensity rainfall events. Building Regulations and Standards are based on historical data of the probability of a particular significant climatic event happening. A margin of safety is added and the recommended strength or size of a building component is defined by its being able to withstand that 1-in-X years event. X may vary from standard to standard.

Climate change will have significant influences on return rates of 'severe' climatic events. For example, hot summers will increase in frequency from 1 in 50 years now to 1 in 2.5 years by 2080. It is likely that for many standards the margin for safety is no longer adequate to reduce risks to acceptable levels. The current building regulations in the United Kingdom do not provide significant assessment and management of risk, which is generally considered to be an issue for planners. However, the construction of new buildings in areas of risk that are flood resilient or indeed where refurbishment or extension work is taking place may in the future require coverage by the building regulations and associated guidance.

A report produced for the Foundation for the Built Environment (now BRE Trust) provided a methodology for assessing climate change impacts can be summarised as follows:

- What is the current position?
- How is the current position likely to change with climate change?
- Will the changes be significant?
- What can economically be done to adapt?

This type of methodology can be expressed as a climate change adaptation route map. In the case of new construction or adaptation or repair of existing buildings an improvement on the resilience of the construction is required. For a new-build project, adaptation strategies to account for predicted climate change may be made at any stage in this process. In general, the further down the line the adaptations are made, the more risk there is of health and safety or economic damage occurring before the adaptation is made. There is also a good change that the whole life costs will be greater as the building has not been designed for the average conditions during its lifetime.

Adaptation strategies taken early in the cycle, i.e. at the planning and design stage, will improve the building and be more effective in the longer term. The initial capital costs of taking these measures will be slightly higher, but will be much lower than having to do retrofit strengthening at a later stage.

For existing stock the earliest point at which this process may be effected is at the maintenance stage. However, even at this stage there is a choice between future-proofing or not. All maintenance may be done considering one of two criteria:

- Lowest price
- Best value for money

By choosing the latter, a good-quality repair or replacement at the best price, climate change adaptation can be made at minimal extra cost.

12.7.3   Further research and innovation

There are number of areas where further research and innovation is required in the area of flood resilient repair. These areas include technical aspects of creating flood resilient buildings and considering the barriers and opportunities to improve resilience.

The technical issues include the following:

- The properties of materials related to flood conditions. This includes the impact of the duration of the flood and the type of water involved on the integrity of the materials and their ability to perform their intended function. The recovery of materials from saturation by flood also needs to be considered and better guidance produced on the potential for recovery of repair of materials after a flood.

- The design impact of buildings on flood resilience. In recent years better accessibility standards have driven the ground floor levels of buildings lower in order to achieve level thresholds. This provides a greater risk of water penetration at from doors and through items such as air bricks. It is therefore necessary to look at the whole design of flood resilient buildings that do not compromise their other essential functions and properties.
- The performance of flood protection products and temporary barriers requires further consideration in order to produce guidance on design and performance limits. Such research is necessary in order to be able to define level C types repairs as set out in previous sections.

Further research is required into the implementation of flood resilient repair. This should involve the consideration of stakeholder views and investigation of organisational and regulatory resistance to flood resilient repair. The research should also seek to address the issue of the costs and whole life costs of flood resilient repair against repair to original specification. Examples are required to set out in detail the potential benefits for building owners, insurers and governments. This is particularly important in the United Kingdom where insurance is entirely private for all buildings. In other countries the situation varies considerably and the roll of government in compensating individuals and businesses that are flooded may provide a greater incentive for flood resilient repair.

12.7.4   Recommendations

This paper has set out the issues surrounding the repair standards for buildings as currently apply in the UK, but which could be extended to the whole of Europe. Indeed, much of the issues regarding building technology and flood protection products are already well known in different countries. There does however exist considerable scope for further research and the production of guidance. The research should address the following factors:
- The actual performance of materials in floods of different durations and under different climatic conditions.
- The production of materials with improved flood resilience and at the same time high sustainability credentials.
- Providing better knowledge on the benefits of improved flood resilience and the incentives required for insurers and building owners to invest.
- Improve the industry standard by encouraging greater specialisation in flood repair amongst the contracting industry.
- Consider other building types that are not covered by current guidance, and the specific issues involved for some modern methods of construction.
- Investigate the risks involved in refurbishing or extending buildings, especially minor additions that homeowners make such as conservatories or garages.

- Investigate the role of planning and building regulation on flood resilience of buildings.

REFERENCES

Alexander Howden Group Ltd (1995) *Flooding in Glasgow and Ayrshire, A report on the December 1994 floods in Glasgow and an assessment of the residential and commercial property damage*, AH, Glasgow.

Building Research Establishment (1991) *Dealing with flood damage*, BRE Press Ltd, Watford.

Building Research Establishment (1997) BRE Good Repair Guide 11, *Repairing Flood Damage* Part 1–4, BRE Press Ltd, Watford.

Building Research Establishment (2004) BRE Report BR466, *Understanding dampness*, BRE Press Ltd, Watford.

Building Research Establishment (2006) *Repairing flooded buildings: an insurance industry guide to investigation and repair*, BRE Report EP69, BRE Press Ltd, Watford (copyright – Flood Repairs Forum).

Building Standards Institution (2003a) PAS 1188-1, *Flood protection products. Specification. Building apertures*, BSI, London.

Building Standards Institution (2003b) PAS 1188-2, *Flood protection products. Specification. Temporary and demountable products*, BSI, London.

Building Standards Institution (2003c) PAS 1188-3, *Flood protection products. Specification. Building skirt system*, BSI, London.

Construction Industry Research and Information Association (2001) Environment Agency, *Flood products: using flood protection products – a guide for homeowners*, CIRIA, London.

Construction Industry Research and Information Association (2001a) Environment Agency, *After a flood – how to restore your home*, CIRIA, London.

Construction Industry Research and Information Association (2000) Environment Agency, *A review of testing for moisture in building elements*, CIRIA Publication C538, London.

Construction Industry Research and Information Association (2004) C624, *Development and flood risk – guidance for the construction industry*, CIRIA publication C624, London.

Construction Industry Research and Information Association (2005) C623, Standards for the repair of buildings following flooding, CIRIA publication C623, London (authors Garvin S L et al).

Crichton D (2005) The role of private insurance companies in managing flood risks in the UK and Europe, in Urban Flood Management (eds C Zevenbergen, A Zllosky), Taylor and Francis Group plc, London.

English Heritage (2004) Flooding and historic buildings, Technical Advice Note, English Heritage, London.

Environment Agency (2001) *Lessons learned: Autumn 2000 floods*, EA, London.

Foundation for the Built Environment (2001) *Climate Change and buildings* (eds M Phillipson, H Graves), FBE, Watford, UK.

Green C and Penning-Rowsell E (2002) *Flood risk and insurance: Strategic options for the insurance industry and government*, Flood Hazard Research Centre, University of Middlesex, London.

Garvin S L, *Flood damage to buildings*, World Water Forum III, Japan, 16–23 March 2003.

Garvin S L (2005) *Flood damage to buildings*, in Urban Flood Management (eds C Zevenbergen, A Zllosky), Taylor and Francis Group plc, London.

Institution of Civil Engineers (2001) *Learning to live with rivers*, ICE, London.

Office of the Deputy Prime Minister (2001) *Planning Policy Guidance Note 25 (PPG25), Development and flood risk*, ODPM, London.

Office of the Deputy Prime Minister (2003) *Preparing for floods: interim guidance on improving the flood resistance of domestic and small business properties*, ODPM, London.

Office of Science and Technology (2004) *"Future Flooding"*, Report of the OST, London, 2004.

Pasche E and Geisler T R (2005) *New strategies of damage reduction in urban areas proned to flood*, in Urban Flood Management (eds C Zevenbergen, A Zllosky), Taylor and Francis Group plc, London.

Perth and Kinross Council (1995) *Flood damage in Perth 1994*, Private communication to BRE.

The Scottish Office (1996) *Design guidance on flood damage to dwellings*, The Scottish Office, Edinburgh.

13

Economic Feasibility Study of Flood Proofing Domestic Dwellings

Chris Zevenbergen[1,2], Berry Gersonius[2], Najib Puyan[1] &
Sebastiaan van Herk[3]

[1] *Dura Vermeer Business Development, Hoofddorp, The Netherlands*
[2] *UNESCO-IHE Institute for Water Education, Delft, The Netherlands*
[3] *Willems & van den Wildenberg España sl, Spain*

ABSTRACT: The present paper describes the main results of an economic feasibility study on flood proofing domestic buildings using a priori analysis. Functional relations between hypothetical flood damage on the construction and their content and inundation depth (FDC) were established for different types of flood proofing dwellings.

Loss probability curves were constructed for a new built-up area on the basis of slow-rise flood depths and losses were provided in monetary terms. Two typical Dutch flood prone areas have been selected for this study. These are a low lying polder (case: Hollands Noorderkwartier) and a flood plain (case: Dordrecht) in The Netherlands. Economic assessments of different types and combinations of flood proofing options where then conducted on the basis of these characteristics using a cost-benefit analysis. This study is limited to the tangible flood damage to buildings.

13.1 INTRODUCTION

Deltaic and other low lying regions are subjected to flooding. Due to a combination of urbanisation, land subsidence and poor drainage conditions both flood frequency as well as flood impact are increasing. Despite of large investments in traditional, collective protection measures, the recurrence of floods and subsequent rise in economic flood damage have shown that this strategy fails to a large extent to be effective in these vulnerable areas world-wide. It is increasingly recognised that new, more risk-oriented flood management strategies are warranted. These strategies call for a holistic approach, considering both measures to reduce flood probability as well as flood impact. This certainly holds true for urban areas which are particularly susceptible to the impact of floods. Accepting and preparing for some degree of flooding may be more effective from financial perspective, but also from social and environmental perspectives Flood proofing of new and existing buildings is one of the measures to reduce the potential flood damage to building structures in flood prone areas.

The focus on preventing floods in urban areas may in some cases be counter productive on the longer term as shown by Hoes et al. (2005). They concluded on the basis of a cost-benefit approach that damage prevention will be likely more cost-effective than flood prevention in anticipating on climate change in regional water systems of The Netherlands. Along similar lines Lamothe et al. (2005) concluded that in many European countries decisions on major investments in flood protection works are still taken without a detailed insight in costs and benefits of alternative strategies. A comprehensive risk analysis should consider all relevant flooding scenario's, their associated probabilities and potential flood damages. The latter is usually estimated on the basis of flood damage curves (FDC). FDC can be developed from (i) data synthesis over real damage obtained from sample surveys of historic events (a posteriori analysis) or (ii) by estimation of hypothetical damage (a priori analysis). Most flood damage assessments focus on direct economic losses: damage that occurs due to physical contact of flood water with humans and property. A cost-benefit analysis (a comparison between a reference or base-line option and alternative options) is used for the economic evaluation, ranking and selection of appropriate flood management options. Spatial scales can also influence the choice between different flood management options, for example measures to reduce flooding may ultimately lead to increased flood risks elsewhere.

Flood proofing has demonstrated for many centuries to provide an effective strategy for communities to cope with floods. These indigenous techniques still exist and have been a major source of inspiration for the development and improvement of modern flood proofing constructions. Despite of the numerous literature, however, the widespread implementation of these techniques is still hampered. This is partly due to a lack of information on their technical performance and economics of implementation and maintenance.

13.2   EFFECTS OF FLOODING TO HOUSING

The aim of the research described in this paper is to provide insight in the feasibility of flood proofing domestic buildings in new built-up areas in the Netherlands. The impact of flooding to housing is generally extensive and typically represents a large share of the total flood damage after flooding of urban areas (Lamonte and Görlach, 2005). However, the authors of this publication recognise that for the final consideration of flood proofing of domestic buildings damages to other sectors will have to be taken into account as well.

Damages to all sectors (e.g. housing, public services and networks) can be classified on two levels: as direct and indirect, and as tangible and intangible. Direct, tangible damage to housing is physical damage to its structure and content.

Structural damage includes damage to walls and floors, whereas content includes installations, windows, doors, and other fixed supplies (utilities). Expenditures for rehousing and cleaning are in general considered as indirect tangible damages. Examples of intangible damage are diseases and casualties. This research focuses on direct, tangible damage, which is most relevant for assessing of potential damage reduction as a result of implementing flood proofing technologies.

Flood damage to buildings is influenced by many factors, which, in broad terms, can be classified as follows:

- Flood actions as acts which a flood could directly do to a building (Kelman, 2004), depending on its characteristics (water depth, flow velocity, flood duration, contamination, etc.)
- The resistance of the building to the flood actions, which according to Kreibich et al. (2005) is related to its permanent resistance (type of building, building material and precautionary measures) and temporal resistance (flood warning and preparedness)
- Exposed capital among which are the real-estate values, value and location of personal properties.

Flood actions determine to a large extent the applicability of flood proofing measures. The following flood actions are generally considered relevant in this respect (Kelman, 2004; Roos, 2003).

13.2.1   Hydrostatic actions

Two forms of hydrostatic actions exist:

- lateral pressure from flood depth differential between the inside and outside of a building
- capillary rise: elevation of water within building components such as walls, causing damage beyond the flood level.

The difference in lateral pressure inside and outside the building could cause failure of walls and even structural failure of buildings.

13.2.2   Hydrodynamic actions

Hydrodynamic actions result from the water's motion:

- velocity: moving water flowing around a building imparting a hydrodynamic pressure
- velocity's localised effects, such as at corners or through gaps (building could succumb due to local variations)
- velocity: turbulence: irregular fluctuations in velocity, either magnitude and direction
- waves changing hydrostatic pressure (non-breaking)

- waves breaking in, over, through or near a building lead to peak dynamic pressures.

13.2.3   Erosion

Flood water may cause erosion by souring away soil from the sides or bed along which the water flows, also around buildings. Erosion is especially hazardous for buildings with slab – on grade (damage to building foundation) and for buildings on slopes (landslips destabilise buildings or damage them from direct impact). The occurrence of erosion depends on the top soil layer: a grass covered soil is less vulnerable to erosion than a asphalt surface.

13.2.4   Buoyancy action

Buoyancy action is caused by the buoyancy force, an uplift force that could damage the building or even cause it, or parts of it, to float. Therefore it is important to anchor buildings adequately and to apply the appropriate foundation.

13.2.5   Debris actions

Debris refers to solids in the flood that may cause damage by crashing into the buildings. Think of wooden balks, large rocks, but also ice or even cows. The forces depend on the weight of the debris and the velocity of the water flow. The damage can range from material damage to a collapse of the building. Roos (2003) also refers to debris inside houses, e.g. heavy furniture that starts to float and crash into the walls because of wave action. Kelman (2004) distinguishes 3 debris actions: static (sediment accumulated to building), dynamic (impact of debris crashing into building as described above) and erosion.

13.2.6   Water quality (salinity, contaminated water)

The water quality contacting the building could impact the material properties and cause damage. Salt water can make brickwork chip off because of salt crystal-lizations (NaCl and Na_2SO_4), whereas reinforced concrete could be affected by chloride-induced corrosion leading to cracks and brown colouring of the concrete. Flood water may be contaminated with sewage, petrol, oil, paint, household clean-ers or industrial chemicals. Any corrosiveness or flammability in the contaminants could result in chemical damage to residences, apart from the possible conse-quences on public health. Kelman (2004) refers to non-physical actions: chemical, nuclear and biological.

13.2.7   Frost

Water could be absorbed in the construction material and freeze in winter. The increased pressure (expansion of water) could cause cracks and eventually cause bricks to break.

13.3   FLOOD PROOFING TECHNOLOGIES

Important options to reduce the impact of flood actions on housing are flood proofing technologies. These types of measures generally fall within five categories:
- Elevated configuration, e.g. building with an elevated entrance, building on columns, or building on pillings
- Dry flood proofing, which are measures to keep water out of the building
- Wet flood proofing, improving the ability of the property to withstand the effects of flooding once water has entered the building
- Construction of permanent or mobile flood walls
- Floating or amphibious homes

The ICPR have evaluated the effectiveness of various flood proofing measures depending on their capability to reduce the existing damage potential or the increase in damage potential (ICPR, 2002). The ICPR publication indicates that flood proofing measures are mainly effective in areas with frequent flood events and low flood depths. However, Kreibich et al. (2005) mention that it remains unclear on which data basis these estimates rely. Therefore they have interviewed about 1,200 private households, which were affected by the 2002 flood at the river Elbe, about their flood damage as well as about their flood proofing measures. This survey has shown that even during this extreme flood event many of the measures led to significant mean damage reductions of up to 53% for buildings and contents. The most effective ones were wet proofing measures, namely flood adapted use and flood adapted interior fittings. Dry proofing measures and private flood walls had no or only little effect, because many of the structures were overtopped.

On the basis of the information available from IPCR, Kreibich and FEMA (1998) the importance of flood proofing measures can be qualitatively indicated as proposed in Table 13.1, taking into account flood frequency, depth of flooding and flow velocity:
- Frequency influences cost-effectiveness and acceptance
- The cost of the measures will vary depending on the flood depth
- Velocity because it may cause severe damage (e.g. as a consequence of failure of walls or scour of the foundation)

It should be noted that the grey squares illustrate the main fields of application of the type of measures.

To enhance the knowledge about application of these technologies the objective of this publication is to conduct an economic assessment on the use of flood proofing measures based on their costs and benefits. The three flood proofing techniques that have been investigated in this study are:
1) Dry flood proofing until 0.9 meter;
2) Wet flood proofing until 1.2 meter;
3) Elevated structure until 0.6 meter.

Table 13.1. Importance of flood proofing measures.

	Flood proofing technologies						
	Elevated entrance	Building on columns	Building on pilings	Dry proofing	Wet proofing	Private flood walls	Floating structure
Frequent events	■	■	■	■	■	■	
Very rare events					■	■	
Low flood depth	■			■	■	■	
High flood depth		■			■		■
High velocity			■			■	
Low velocity	■	■	■	■	■	■	■

A typical Dutch single-family dwelling (SFD) has been used as a starting point and reference to investigate the effectiveness of the different flood proofing techniques. Damage reduction for each alternative has been computed by estimating the difference in expected annual damage between the reference (traditional SFD) and a traditional SFD upgraded to a flood proofing building by adopting the above mentioned techniques (flood proofing SFDs). In Table 13.2 main design features and specifications of the traditional and flood proofing SFD's are summarized. A general technical description and a cost estimation are given in Tables 13.2–13.3.

13.3.1   Reference: Single-family dwelling (traditional SFD)

A typical Dutch single-family dwelling (PCS[1] – type standard, see Table 13.2) comprises a two story structure (with a height of 2.7 m per floor) with an attic and without a basement (see Figure 13.2). The bearing walls and floors are constructed of concrete. The exterior walls are covered with an impermeable brick masonry panel with in between an insulation layer (10 cm thick). The structure is founded on concrete piles (10 m in length). This standard dwelling includes a kitchen, sanitary facilities, floors and wall tiling, skirting boards and painting.

---

[1] PCS stands for Pre Choice System, a typical Dutch single family dwelling composed of standardized, partly pre-fabricated components.

Table 13.2. Specifications reference SFD flood proofing SFD variants.

Flood proofing alternatives	Features and specifications (general description)
Reference 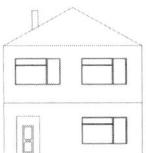	A typical Dutch single-family dwelling (PCS-type standard) comprises a two story structure (with a height of 2.7 m per floor) with an attic and without a basement (see Figure 13.2). The bearing walls and floors are constructed of concrete. The exterior walls are covered with an impermeable brick masonry panel with in between an insulation layer (10 cm thick). The structure is founded on concrete piles (10 m in length). This standard dwelling includes a kitchen, sanitary facilities, grad and wall tiling, skirting boards and painting Building (construction) costs: € 66,000 (PCS, 2005)
On posts or columns 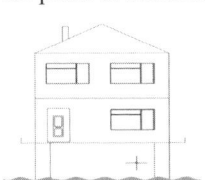	Posts or columns could be made of wood, steel, masonry or precast reinforced concrete The posts are not founded deep into the ground (as opposed to piles) Suited for low to moderate water depth and velocity, otherwise the posts could be damaged (even though are used). Suitability depends on local flood conditions, such as salinity (could cause corrosion of posts), resistance to debris (e.g. strong and stiff reinforced concrete to avoid damage to posts) Extension of access and utilities needed, e.g. utilities are installed at higher level, requiring longer cables and adequate coatings/insulation to avoid electrical shortcuts References: USACE (1998); FEMA (1998)
On piles 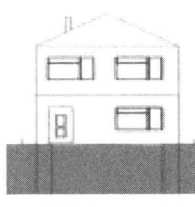	As "on posts or columns" but suited for high-velocity flooding as piles are better founded (to greater depths) compared to posts. Typically made from precast reinforced concrete, but also steel and wood (no masonry). Extension of access and utilities is needed References: BfVBW (2002); USACE (1998); FEMA (1998)
Dry flood proofing 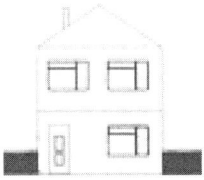	Covering and protecting openings watertight up to a height of 0.9 m to withstand flood forces and avoid entrance of water in the building. Note that dry proofing above 0.9 m would be very expensive and make the house considerably less user friendly. Dry proofing is recommended in frequently flooded areas with low flood water depth

(Continued)

Table 13.2. Continued.

Flood proofing alternatives	Features and specifications (general description)
	Dry proofing normally includes backflow prevention by pressure sewerage systems (valves) to prevent backflow of floodwater, closure system for ventilation grids, etc. Additionally adequate drainage systems can reduce flood duration and thus damage
	Dry proofing elements should be inspected periodically and repair works might be necessary after floods. Guide walls in front of doors are not recommended to avoid most damage by debris (small elements, small force below 0.9 m) as they are to expensive and would make dry flood proofing not cost-effective compared to other options
	References: BfVBW (2002); CIRIA (2001); USACE (1998); FEMA (1998); LSU AgCenter (2002); A-F-S (2005)
Wet flood proofing 	Wet proofing minimises damage to building and its contents although flood water is allowed to enter the building. Hence, the hydrostatic force is equal outside and inside the building. Utilities are installed at a higher level and people relocate vulnerable items permanently or temporarily (if sufficient flood warning time). Therefore wet flood proofing is not recommended in areas threatened by flash floods (little warning time)
	Also wet proofing should be accompanied by pressure sewerage (backflow prevention) and adequate drainage systems
	Wet proofing entails health risks, risk of contamination, impact of debris (guide walls are not cost effective for wet proofing either) and the risk of electrical shortcuts. Additionally cleanup and repair works are envisaged after flooding
	References: BfVBW (2002); USACE (1998); FEMA (1998); LSU AgCenter, USA (2002)
Amphibious structure	Floating homes in the Netherlands can accommodate a difference in water level of 5.5 meter, but is the most costly flood proofing technology. Pressure sewerage (for dirty water) and water pumps (drink water) and flexible cables are used to ensure proper functioning during floods. Conduction works are applied to reduce/avoid damage by debris or ice
	Inspection for erosion and settlement are recommended
	Reference: DVG (2005)

Table 13.3. Flood proofing costs estimation.

Category	Wet proof (in €)	Dry proof (in €)	Elevated (in €)
Total cost	1,300	8,000	4,400

13.3.2   Dry flood proofing SFD

The dry proofing SFD is designed to prevent flood water from entering the building up to a design level of 0.9 meter. When a flood exceeds this design level damage occurs as it normally would without any flood proofing modification. In this study a flood design level of 0.9 meter has been chosen because brick masonry structures generally withstand a hydrostatic pressure up to this flood level (Kelman, 2004; & FEMA, 1998). The SFD has been made dry flood proof using barriers to prevent water entering doors, windows, air-bricks, ventilation holes and other openings up to the flood design level. Removable flood shields have been selected for doors and windows, impermeable foil and waterproof mortar for small openings, valves for backflow prevention and special waterproof floors to watertight the floor construction (Garvin et al., 2005).

13.3.3   Wet flood proofing SFD

The wet proofing SFD is designed to accept flood water entering the building up to a certain design level in order to equalize hydrostatic forces. Allowing the interior and exterior hydrostatic pressures to equalize greatly reduces the likelihood of wall failures and structural damages. To prevent flood damage and to reduce cleanup costs all constructions of the building below this level consist of flood resistant materials (e.g. water-resistant stained concrete floors and walls). In addition, the electrical wiring has been raised above the flood design level and a watertight cupboard has been used for the electricity/gas meter. In this study a design level of 1.2 meter has been used. It should be noted that a higher design level is feasible because the loads on walls and floors will be less than in a dry flood proofed home.

13.3.4   Elevated SFD

The elevated SFD is designed to prevent flood water from entering the building by elevating the entire structure up to a certain flood design level. Elevating is generally considered as an appropriate and cost-effective measure for buildings facing frequent floods with a long duration (at least 72 hours), but with relatively low flood depths (less than 0.6 m). In order to comply with a flood design level of 0.6 meter in this study the first floor of the standard SFD has been placed on a 0.3 meter thick wall along each side. A concrete staircase has been constructed for the elevated entrance.

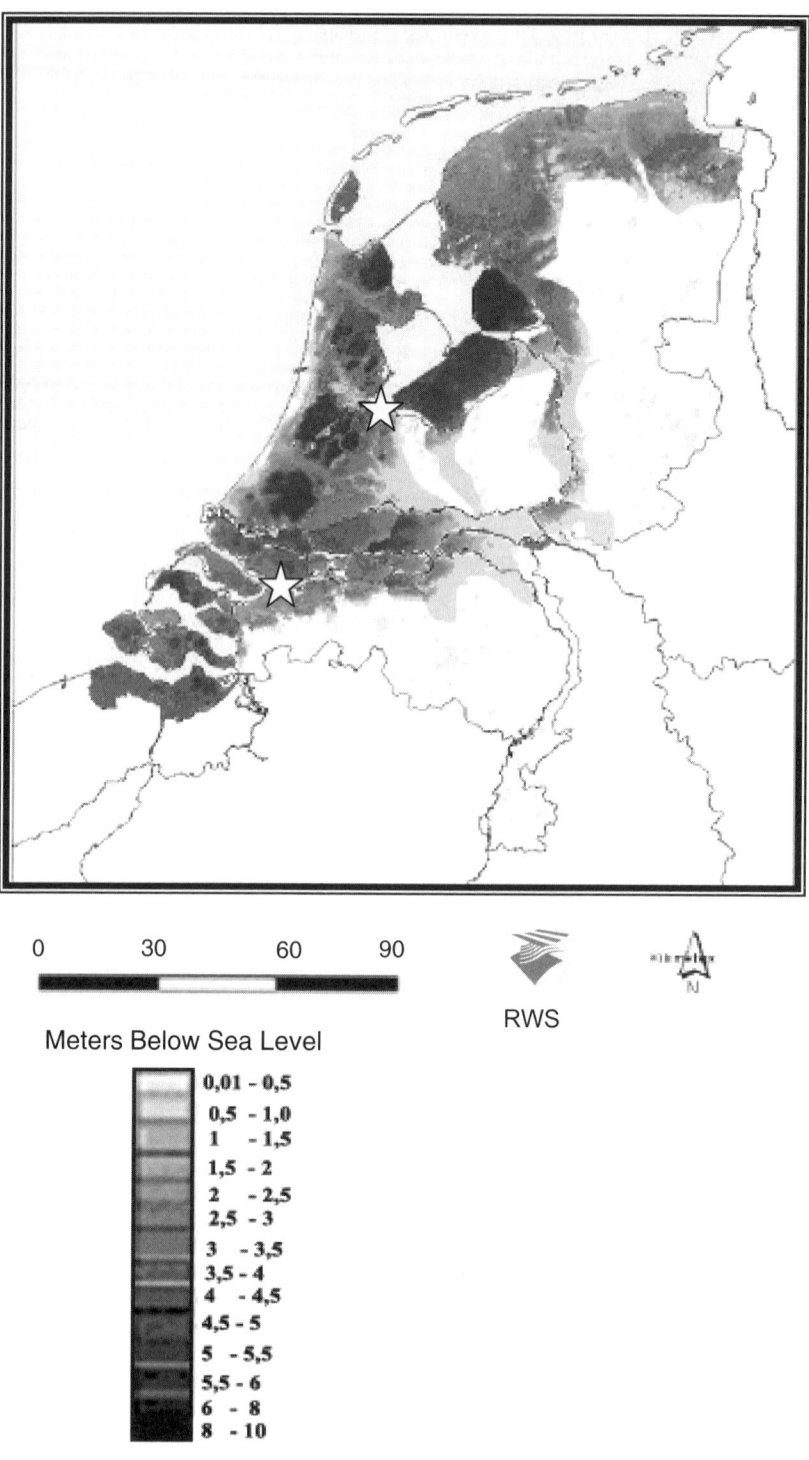

0   30   60   90

Meters Below Sea Level

RWS

0,01 - 0,5
0,5 - 1,0
1   - 1,5
1,5 - 2
2   - 2,5
2,5 - 3
3   - 3,5
3,5 - 4
4   - 4,5
4,5 - 5
5   - 5,5
5,5 - 6
6   - 8
8   - 10

Figure 13.1. Areas which have been selected for this study.

Figure 13.2. Typical examples of a Dutch single-family dwelling (PCS-type).

13.4   FLOOD PROOFING COSTS ESTIMATION

The costs of flood proofing have been estimated for each flood proofing technology by comparing the difference in total building costs of the flood proofing SFD variants with the traditional SFD. The total building costs estimate includes materials and labour.

13.4.1   Reference: Single-family dwelling (traditional SFD)

The average building costs of a traditional SFD (net floor surface of 106 m^2) built under competitive conditions in the Netherlands are estimated at €66,000 per PCS dwelling (DVG, 2005).

13.4.2   Dry flood proofing SFD

The additional building costs of dry flood proofing SFD are estimated at €8,000 per PCS dwelling. Materials include: flood barriers, a pump, 2 backflow valves, and a drainage system.

13.4.3   Wet flood proofing SFD

The additional building costs of wet flood proofing SFD are estimated at €1,300 per PCS dwelling. Materials include: waterproof floor and, electricity meter cupboard, waterproof insulation (rigid, closed-cell foam insulation).

13.4.4   Elevated SFD

The additional building costs of an elevated SFD are estimated at €4,400. Materials include: concrete walls, sand, concrete staircase, extra cables, 2 backflow valves, and a drainage system (see Table 13.3).

13.5   FLOOD DAMAGE COSTS ESTIMATION

Building damage and failure arise from a range of flood actions such as: hydrostatic, hydrodynamic, erosion, buoyancy, debris, water quality and frost (Kelman, 2004 & Roos, 2003). In order to assess flood damage it is important to identify the failure mechanisms and the combination of factors that result in damage. In addition, a relation between those factors and the consequences (damage) must be established. Notably flood parameters like depth, duration and velocity, building features like the type of structure, and the load and pollutants carried by the flood water are important factors.

However, in many studies the damages are determined as a function of the depth of flooding only and the magnitude of losses is estimated using damage-probability curves. More comprehensive approaches assume that losses are dependant on the likelihood of building failure when exposed to flooding. Clausen (1989) identified three types of damages based on a research of the Dale Dyke dam failure in Sheffield: inundation damage, partial damage and total destruction. Where the flood actions (for example, flood velocities) are great enough to cause structural failure, the replacement value of the building and its contents are adopted. If partial or complete failure is not likely, the application of damage-probability curves is appropriate.

In a recent study the damage to buildings caused by a wide spectrum of failure mechanisms such as scour of foundation and failure of walls has been considered using a deterministic model (Roos, 2003). According to calculations using this model, (partial) collapse due to scour will occur if the affected building is built on a shallow foundation and the water velocity initiates the top layer of the soil to wash away. In order to calculate the failure of walls in this study, the loads on the building have been compared to the capacity of the structures in terms of bending moments and shear forces. It has been concluded on the basis of these results that the collapse of walls due to loads applied by pounding debris will cause most damage.

Such magnitudes of flood forces (associated with partial or total collapse) are usually experienced only in extreme events, for example, where flooding will result from failure of protective structures. However, natural floodwaters usually involve low velocities and short durations so that the primary determinant of the extent of losses is the depth of flooding. For such conditions where structural failure is not likely, in this study flood damage curves (FDC's) have been constructed for the traditional SFD and the flood proofing SFD variants based on a priori estimates. The following steps have been taken for the construction of flood-damage curves:

1. Detailed survey of the individual costing for replacement or repair of structural properties and contents. Other relevant data include the costs of cleaning, disinfection and re-housing.
2. Identification of the damage costs from flooding of the privately owned property and real-estate inflicted by a range of flood magnitudes (water depths).

3. Construction of linear damages functions on the basis of the identified damage categories.

   It should be noted that in this study direct, tangible damage being physical damage to the building's structure and content have been considered only. Indirect or intangible damage, such as casualties or disturbance of daily life have not taken into account.

13.5.1   Step 1

The costing of replacement or repair of structural properties and damage to household inventory items are estimated using reported empirical data on direct tangible damage made after the floodings of the Meuse river in The Netherlands in 1993 and 1995, in conjunction with data obtained from the insurance industry (CBA, 2005) and expert judgement. The costs are categorised by walls, floors, interiors and complete installations. The value of total contents (including contents of outbuildings and sheds) (Economic benefits et al., 2002) is estimated at €43,300. Other relevant factors that are included are the cost of cleaning, repair of windows, painting work and external items and expenditures on temporary accommodation. The costs for these factors are estimated at €16,000.

13.5.2   Step 2

The damage to both building and contents is estimated against a range of water depths. The damages experienced to the contents are maximal when the depth of flooding reaches the second floor. Many of the costs for replacement or repair works don't vary noticeably in relation to the flood depth. In this respect one may consider, for example, the necessary repair works to walls, tiling and painting due to a 1 m and 1.5 m flood event; clearly the floodwater, humidity and capillary rise affect the entire wall of that floor.

13.5.3   Step 3

The flood-damage curve (Figure 13.3) is established based on the calculation of total direct and indirect damage cost inflicted by a range of water depths. The depth of flooding above the first floor is used as reference.

   The application of stage damage curves allows a direct estimate of the losses by a known flood height from a single flood event. When the purpose is to calculate the damage which can be expected in one year, then the damage corresponding to each flood size is weighted by the percent change of exceedance for that flood event. For rarer flood events, like the probable maximum flood, the damage caused by the event is weighted less. The sum of the weighted flood damages represents the annual average damage cost. The monetary value is equal to the area under

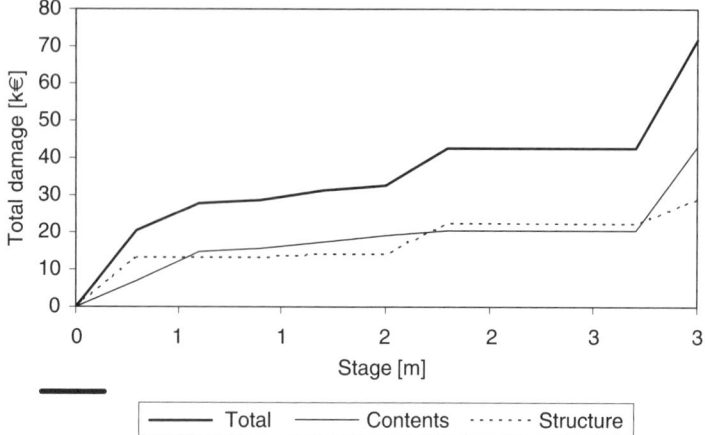

Figure 13.3. Flood damage curves for a traditional SFD showing damage to structure, contents as well as total damage as a function of inundation depths, respectively.

the damage-probability curve, i.e. the line in the graph of flood damage estimates versus exceedance probability.

13.6   IDENTIFICATION OF FLOOD DAMAGE REDUCTION

For the three flood proofing SFD variants the potential reduction in damage against a range of flood depths was simulated. For the elevated entrance and dry proofing techniques it was assumed that: (i) below the design level o no damage occurs to the structure and contents of the adapted SFD and (ii) that above these flood levels damage is occurs as it would for the reference SFD. After the observations of Kreibich, it was assumed that wet proofing technologies reduce the existent damage potential by about 50% for contents and structural damage. In Figure 13.4 the flood-damage curves for the standard SFD and the three flood proofing SFD variants are given. The potential damages include both structure and contents. The amount of damage reduction achieved varied depending on the flood proofing technique and the level of protection.

13.7   COST-BENEFIT ANALYSIS

Current flood standards in The Netherlands are designed to reduce the probability of flooding, and do not take the impacts of flooding into account. For new built-up areas which are currently envisaged in low lying polders and flood plains, these flood standards may, as a consequence lead to cost-inefficient measures (Hoes et al., 2005). In addition, they may increase flood risk down stream.

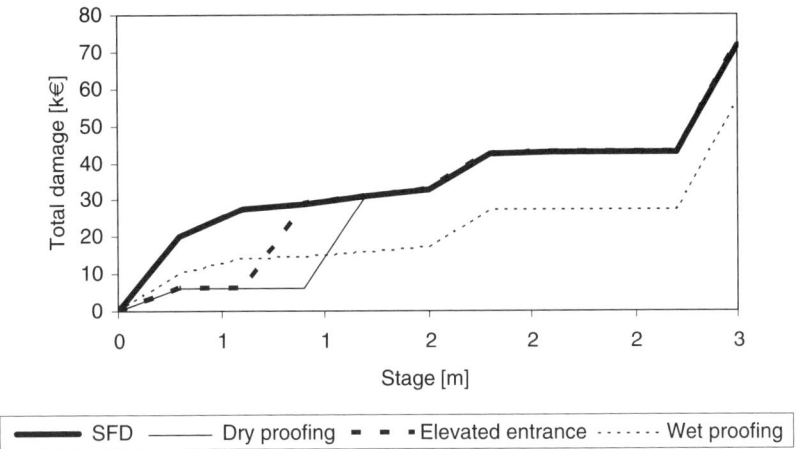

Figure 13.4. Flood damage curves for a standard SFD and elevated, dry proof, and wet proof SFD.

In this study a cost benefit analysis has been conducted to evaluate the economic feasibility of the three flood proofing SFD variants in a new built-up area in a regional water system (polder) in the province of North Holland (case: Hollands Noorderkwartier) and in a flood plain along the river Beneden Merwede (case: Dordrecht), respectively. The costs are determined by the investments made to flood proof SFD (see Table 13.3) and the benefits are equal to the reduction of the annual flood damage. Annual flood damage is computed as the integral of the damage-probability function:

$$E(S) = \int S(x)dP(x)$$

Where E(S) is the expected annual damage, S(x) is the flood damage caused by flood depth x, and P(x) is the probability of flood level x. Probability distribution functions of flood levels are calculated using simulation models. For the case Hollands Noorderkwartier model calculations from Hoes et al. were used. They applied a combined rainfall and hydraulic model (SOBEK) to simulate flood levels. For the case Dordrecht model calculations from Delft Hydraulics were used. The exceedance probability of extreme water levels was simulated by Delft Hydraulics using the probabilistic model Hydra-B. The model takes into account the discharge of the river Rhine, the sea water level, wind velocities and the control situation of Maeslant- and Hartelbarrier. For a range of flood-exceedance probabilities, the damage corresponding to that water level is obtained from the flood-damage curve. Integration is applied to deal with the full range of increased water levels. All future losses are discounted over a period of 50 years to obtain the present value of the expected flood damage. In this study the effects of future climate change are not taken into account.

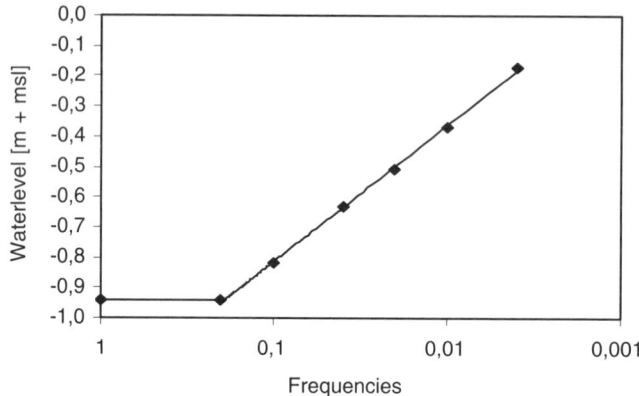

Figure 13.5. Water level-frequency relationship Hollands Noorderkwartier (*Source*: Hoes et al., 2005).

13.7.1   Case: Hollands Noorderkwartier (low lying polder, regional system)

Polders are the result of a step-by-step reclamation of land, where there once was sea. In the western part of The Netherlands behind the coastal protection zone the reclaimed land consists of a patchwork of polders. These polders separated by ring dykes warrant continuous drainage through an intricate system of ditches because the land surface within the polders is below sea level. The land surface level varies from less than 1 m up to 7 m below sea level (see Figure 13.1). In this study the exceedance probability of extreme water levels for the polder systems of the Hollands Noorderkwartier Waterboard were used. For each polder the flood stage and associated exceedance probability can be derived from the probability distribution function of water levels and the surface elevation for different locations in the polder (Figure 13.5). If a water level exceeds the land surface level then the area is flooded.

To evaluate the cost and benefit of the three flood proofing techniques a cost-benefit analysis has been conducted for the flood proofing SFD variants. The benefit has been calculated as the present value of the losses avoided by implementing the measure (c.q. applying the standard SFD). Benefit-cost ratios were computed as a function of the land surface level at different locations in the polder system. The results are given in Figure 13.6.

Not surprisingly the results indicate that the BC-ratio's of the three flood proofing SFD variants are increasing with decreasing land surface level. Based on the results, the following comments can be made in relation to the efficiency of flood proofing new-build dwellings in the investigated very low lying polder system:
• Wet proofing seemed to be the most efficient technology in reducing flood damage. Moreover, the calculations indicated that the measure should be considered a worthwhile mitigation option until a return period of approximately 1/100

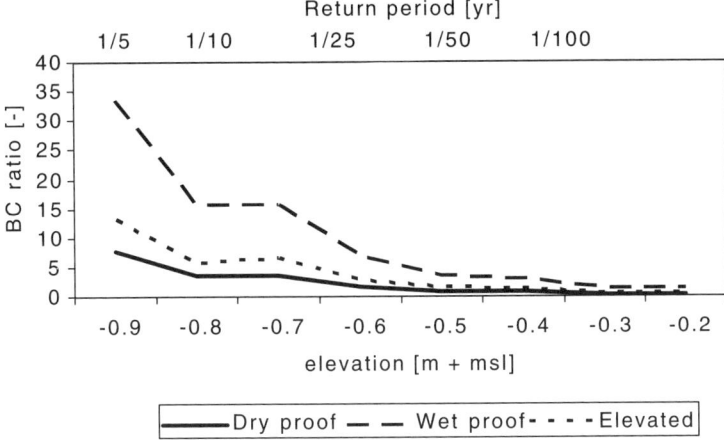

Figure 13.6. Cost-benefit (BC) ratio's as a function of land surface level in m + msl.

years. We recognise however that the largest uncertainties in total building costs and damage reduction potential relate to this technology. Therefore it is recommended to minimise the uncertainties for this option, in particular, since it proved to be the most important measure.

• Each measure has the potential to reduce the flood risk: the effectiveness of the measures is determined by the flood frequency. This is accordance with the observations from the ICPR (2002). Dry proofing is cost efficient (defined by a benefit-cost ratio greater than 1.5) until a return period of 1/30 years, elevated entrance until 1/50 years and wet proofing until 1/100 years.

13.7.2   Case Dordrecht (flood plain)

Despite being part of the densely populated Dutch delta-metropolis with its serious housing shortages, Dordrecht has large areas of abandoned land and buildings. One of these areas is Stadswerven (Figure 13.7). Stadswerven is an island in the heart of the city outside the primary defences which will be redeveloped in this decade. Stadswerven has a built-up area of 30 hectares and is located in a flood plain along the river Beneden Merwede. Stadswerven is in the design phase, a first masterplan comprising 1,600 new homes (for 2,500–3,000 people) has been developed, but could be subject to changes. For Dordrecht the biggest threat comes from a high discharge by the rivers Rhine and Meuse in combination with storm at sea. Probability of design flood for the island of Dordrecht is 1:2,000 year. Flood risk outside the primary defence (e.g. Stadswerven) is much higher, in some areas up to 1:5 year. In the present study an assessment of the cost-effectiveness of the three selected individual flood proofing techniques has been conducted for a new built-up area in Stadswerven. In this new built-up area a residential development

Figure 13.7. The city of Dordrecht along the river Beneden Merwede with his historical centre and floodplains (areas outside the primary defence). Arrow indicates location new built-up area used in this study.

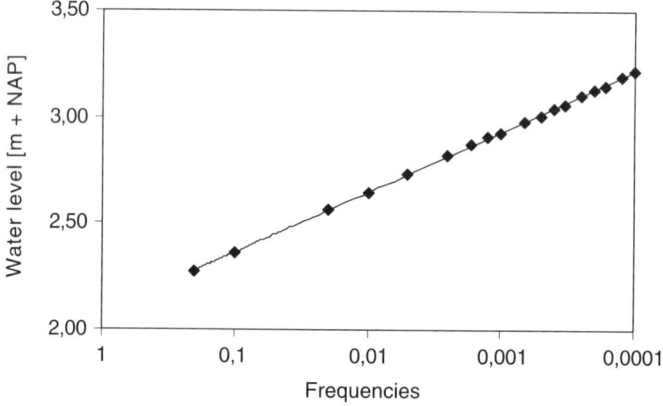

Figure 13.8. Water-level-frequency relationship for Stadswerven, Dordrecht (*Source*: Delft Hydraulics, 2006).

of approximately 30 hectare is planned. The actual land surface level in the area varies between 2.3 and 3.8 meter +NAP.

For this location the exceedance probability of extreme water levels are given in Figure 13.8. In Figure 13.9 the cost and future flood damage (discounted over

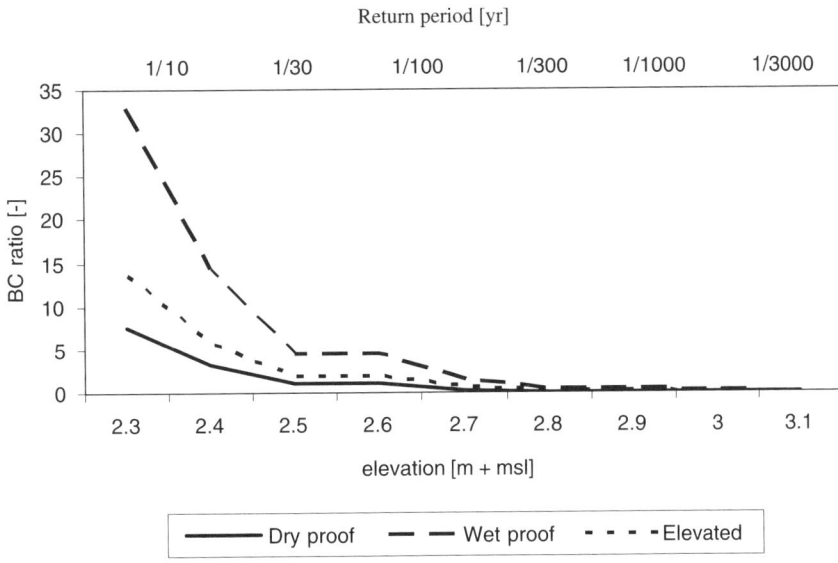

Figure 13.9. Cost-benefit (BC) ratio's as a function of land surface level in m + msl. Stadswerven, Dordrecht.

a period of 50 years) are given for the standard SFD and for the different flood proofing SFD variants of the land surface level at different locations in the new-built area.

The results indicate that for the new-built-up area in Stadswerven the most cost effective measure is wet proofing, which is appropriate until a return period of 1:200 year. Dry proofing and building with an elevated entrance are economically worthwhile until a return period of 1:30 and 1:100 year respectively.

13.8   CONCLUSIONS

Based on the results of this study two fundamental points are apparent. First that flood proofing of new standard dwellings (e.g. PCS-type) using relatively simple and known techniques may provide a cost-effective measure to mitigate flood impacts to domestic buildings in low lying areas. The option of flood proofing new built-up areas should, therefore, be taken into account in any economic evaluation underlying a master plan in these areas. This specifically holds true for the Dutch context in which new urban developments are projected in low lying polders and flood plains in the coming decade. Moreover, flood proofing of buildings could contribute to the implementation of the new policy of "Making space for Water", which is more and more embraced in Europe.

The second fundamental point is that an *a priori* analysis using flood depth damage functions provides useful information for the evaluation and optimization of the different flood proofing options for relatively standard, well defined objects (such as PCS-type of dwelling) in new built-up areas. However, apart from the flood impact to the structure (building fabric) and the contents (household inventory items) of the individual dwellings, the flood impact to the exteriors (garden) and (public) infrastructure has till date received little attention. The latter thus warrants further detailed study e.g. in the form of specific damage estimates through a *posteriori* flood damage assessments.

REFERENCES

A-F-S (Affordable Flood solutions), 2005

BfVBW (2002), Bundesminister fur Verkehr, Bau-und Wohnungswesen, Germany

CIRIA (2001), Damage Limitation, Guide how to make your home flood resistant

Clausen, L.K. (1989). Potential Dam Failure: Estimation of Consequences, and Implications for Planning. Unpublished M.Phil. thesis at the School of Geography and Planning at Middlesex Polytechnic collaborating with Binnie and Partners, Redhill

CBA (2005), Insurance industrie, Central Beheer Achmea en Interpolis, the Netherlands

DVG (2005), Dura Vermeer Group, Pre Choicesystem (PCS) "Single family Dwelling", the Netherlands

Economic benefits of land use planning in flood management, Victorian Department of Natural Resources and Environment by URS Australia Pty. Ltd, Department of Natural Resources and Environment and Emergency Management Australia, EMA Projects Program, 2002, pp 14–1. Appendix 4

FEMA (1998), Homeowner's guide to retrofitting – six ways to protect your house from flooding

Garvein, S.L., Reid, J.F., (2005), Scott M, Construction Industry Research and Information Association C623, Standards for the repair of buildings following flooding, CIRIA publication C623, London

Hoes, O.A.C., Schuurmans, W., Strijker, J. (2005), Water systems and risk analysis. Water Science & Technology, 5 (51), 105–112

ICPR (2002), Non Structural Flood Plain Management – Measures and their Effectiveness

Kelman, I., Spence, R. (2004), An overview of flood actions on buildings. Engineering Technology, 279–309

Kreibich, H., Thieken, A.H., Petrow, Th., Müller, M., Merz, B. (2005), Flood loss reduction of private households due to building precautionary measures – lessons learned from the Elbe flood in August 2002. Natural Hazards and Earth System Sciences, 5 (2005), 117–126.

Lamonte, D.N., Neveu, G., Görlach, B. (2005), Evaluation of the impact of floods and associated protection policies. European Commission DG Environment

Lamothe, D.-N., Neveu, G., Görlach, B., Interwies, E. (2005), Evaluation Of the Impact Of Floods and Associated Protection Policies; WRc plc (2005), Impacts of coastal flooding, flood mapping and planning, Office International de l'Eau and Ecologic

LSU AgCenter, Lousiana House, Hosted by the, USA, 2002

Roos, W. (2003), Damage to buildings. Delft Cluster-publication: DC 1-233-9

The State of Queensland – Department of Natural Resources and Mines (2002) Guidance on the Assessment of Tangible Flood Damages

U.S. Army Corps of Engineers (1998), Flood Proofing Performance, Success & Failures

USACE (1978), Physical and economic feasibility of nonstructural flood plain management measures

Internet sites

– http://www.louisianahouse.org/flood
– http://www.contractorguides.com/natbuilcosma.html
– http://www.degoudenham.nl
– http://www.goudenkust.nl.nu
– http://www.waterbouw.tudelft.nl/; b.stalenberg@citg.tudelft.nl
– www.drijvendestad.nl
– www.ciria.org.uk/flooding
– www.bouwkostenonline.nl
– www.wippseystem.com
– www.noort-novations.nl/SCWindex.ht
– www.floodcontrolam.com/flood_proofing
– www.psdoors.com/flooddoors.htm
– www.fema.gov
– www.a-f-s.biz

14

Local Flood Defence Systems in Europe

Mitja Brilly

University of Ljubljana, Faculty of Civil and Geodetic Engineering, Ljubljana, Slovenia

ABSTRACT: The paper provides an overview of local flood defence measures in the urban environment. The practice and implementation of measures have a long tradition in Europe and they are subject to climate and social conditions. The results of the data collected through a questionnaire among respondents from different countries are also presented.

14.1   INTRODUCTION

The intensive urbanisation has changed water regimes resulting in the increase of maximum flow and thus the increase of flood hazard and flood risk on urban areas (Grigg, 1976). Due to economic development and high vulnerability of urban flood areas the flood-induced damage is high. Urban areas are highly sensitive and vulnerable to thorough structural measures that would help to maintain the high security standards, especially in historic city cores, during the rare 100-year return period events. Often, only mitigation measures can be undertaken after these events, using different non-structural measures.

An important concern is urban areas flooded with their own waters, since storm sewers are designed for two-year return periods or less. During events of return period of ten years or more the water is collected inside urban areas (especially in underpasses, Brilly, 1991) or flows along roads (as if in streams). Floods of urban areas have particular characteristics due to the extraordinary effects of urbanisation to surface runoff. These effects can be mitigated, but not eliminated. The effects of urbanisation are complex and should be treated comprehensively (Gardiner, 1991; LAWA, 1995).

Due to the fast urbanisation of agricultural land, new settlements emerge under hydrological changed conditions that we do not know much about. In a short period of time, a new settlement is not subjected to the rare phenomena that would testify in what way we had actually changed the water regime and increased the flood hazard. Also, we do not know whether the protective measures would work. Time and again we are surprised when floods occur and when human settlements suffer direct and indirect damage.

As a rule, the local protection of urban areas is undertaken after the floods occur, when due to the large costs incurred by the damage the politics steps in, and the results are often poorly thought out or involve costly and inefficient measures. Besides, there are not enough data available on performance and efficiency of local measures. Large systems are well documented and presented in relevant monographs, however, the local systems lack a strong technical support and monitoring.

Large systems of dykes and levees protect extensive areas against flood hazards. They are technically well designed, monitored and under the supervision of the central government or well organised agencies. The central agency, however, does not care if urban areas, which are well protected against floods from the main stream or the sea, are flooded by the water collected in the urban area itself or from nearby watersheds. Local flood defence systems are as an important part of an integrated flood mitigation system as is the completion of a large system protecting the area against the flood. These take place inside a very vulnerable urban area.

Local systems for flash flood protection are developed inside the urban area on small watersheds. They are often excluded from the national monitoring system and poorly technically supported. Small-urbanised watersheds have strong anthropogenic impacts on hydrological processes and hydraulic characteristics of water flow (Figure 14.1). Impervious surface coverage increases the runoff amount

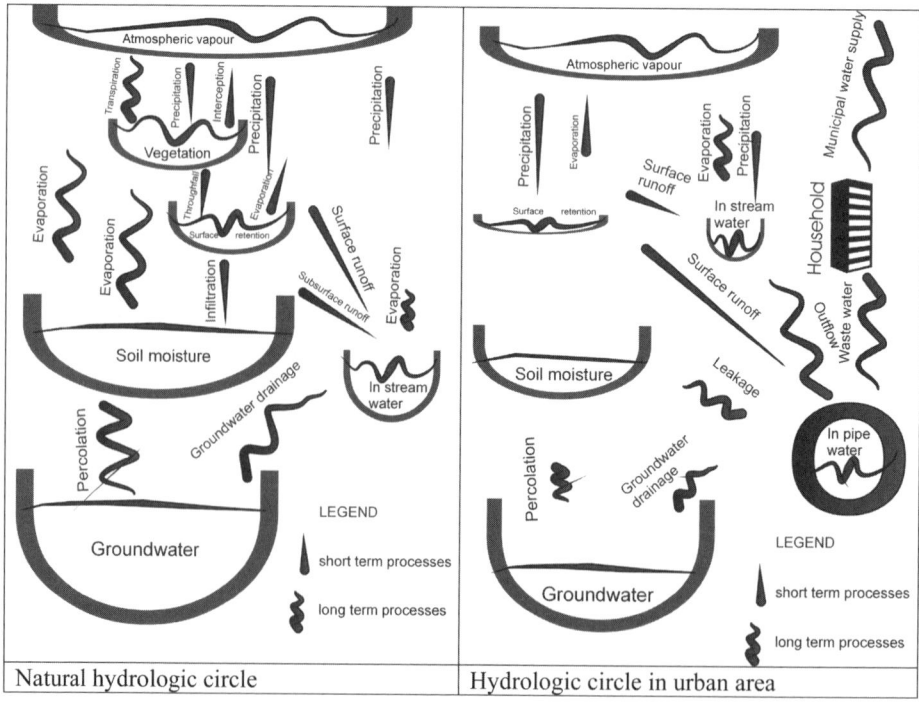

Figure 14.1. Hydrologic circle.

tremendously, roads increase the runoff and water velocity and culvert natural streams diminish the capability of streams to carry water out of the watershed. All these processes are the cause of flood events in areas where they are not expected and did not occur before. Local flash floods are thus events produced by urbanisation and are becoming more and more frequent. The integration of local flood defence systems into urban planning is essential for successful management (Brilly, 2001).

14.2   FLOOD HAZARD IN URBAN AREAS

The implications of urbanisation on runoff processes depend on the scales of the watershed and urban development. Small, densely urbanised river basins are more strongly affected by the urban runoff flows than large rivers where local urban runoff peaks contribute only a very small proportion of the flow to the river (Maksimovic, 2001). As stream channels are largely a product of the upland watershed area, urbanisation affects not only the local runoff but also produces effects downstream, where flood peaks may increase. During the process, the discharge of small and frequent yearly floods increases and changes the curve of discharge duration, so that the flood discharge is increased and flood duration is reduced (Booth *et al.*, 1997). The influence on large floods with a longer return period is much smaller and can even be deemed as insignificant.

Storm duration, intensity and frequency, including spatial and time variability of rainfall, are climate characteristics of specific areas. These characteristics are very important as the hydrological input for an integrated urban storm water management approach and for the urban flood management. There is frequently a lack of data and the solution could be the development of design storms as input data for hydrological modelling of environmental impact of proposed measures. The main impact of climate on runoff is the form and intensity of the precipitation. In a cold climate water accumulates, in the form of snow cover, on the vegetation and soil surface and then is subsequently released in the process of snowmelt. The process of snowmelt varies depending upon the energy budget. The snowfall could have been accumulated during colder periods but may be melted in just a few days. Such snow melt events can produce more runoff than very intense rainfall. On the other hand, short, but intense thunderstorms are characteristic of the Mediterranean climate. Flood hazard characteristics differ significantly due to climate conditions, which is why the same standards and practices cannot be used in the same way in the different European countries.

In urban areas the concept of watershed becomes more complex and difficult to define because the natural topography has been disturbed; the water may be drained through storm drains and in some cases may be diverted by drains into other basins (Riley, 1998). The area contributing to runoff may be completely different in an urbanised watershed compared to the previous natural condition. The impact of urbanisation on the hydrological cycle is complex and affects almost all-hydrological processes. The overall belief is that urbanisation alters the watershed response to rainfall, increasing the volume, peak flow and flood risk downstream, decreasing low flows, increasing pollution and reducing stream corridor habitat. Although this is generally the case, it may not always be true due to the nature of complex runoff processes involved and therefore each case should be carefully analysed (Maksimovic *et al.*, 2001).

The urbanisation has some impact on precipitation. Air pollution in urban areas slightly increases precipitation. Small particles of dust in the air produce a similar effect to cloud seeding. Impervious urban areas without transpiration and soil moisture can result in increased air temperatures near the ground during the day that can produce connective precipitation with showers in summer time. Energy emission in urban areas, as waste of heating energy, slightly increases the air temperature in cold climates or during the cold seasons of the year. The results can be increased rainfall and less snowfall in urban areas than in the surrounding rural areas. Watering of plants during summer in deserts has a converse effect (Pickett *et al.*, 2001).

The impact of urbanisation on the hydrologic cycle is most significant on interception losses (Figure 14.1). The precipitation that falls on the ground is dispersed in several ways depending upon the form of precipitation, the intensity of storms, properties of land cover, geology and pedology. Interception is part of precipitation stored in the vegetation cover and is evaporated during and in few hours immediately after the storm thus returning to the atmosphere and not participating in the runoff processes. Trees have very high interception capacity, while that of grass is lower. Land surfaces without vegetation do not provide in general any interception. Rainfall on surfaces without vegetation, and throughfall below vegetation is collected in surface retention. The water from the surface retention then evaporates or infiltrates in the soil in few days after the storm. The infiltrated water, collected in the soil as soil moisture, disperses in subsurface runoff, percolates to the groundwater, and evaporates and transpirates during the weeks following the rainfall event. Runoff occurs only if the rainfall exceeds a certain value. Interception is first/primary precipitation loss and depends on the form of vegetation and rainfall amount. The low rainfall events can be completely intercepted in natural conditions (Figure 14.1). The increase in the amount of impervious land surface without vegetation, as a result of urbanisation, can have a large impact on the hydrologic abstractions. On the impermeable surface, there is no interception and no infiltration, just some retention on the surface that takes a few millimetres of precipitation and no more. The hydrological processes in green areas: parks, gardens and similar recreation areas, are normally similar to those in the nature. And even individual trees vegetated over the impervious surface have a strong impact on runoff.

Evapotranspiration is a continuous process. It depends on the energy needs for transforming water from the liquid to vapour state, the humidity of air that receives the vapour and the amount of water for evapotranspiration. Potential evapotranspiration is the amount of water that evaporates if there is no deficiency of water in soil. Water evaporates not only on the soil surface but also inside the soil pores, where the air is saturated by water vapour, and then moves out of soil by air movement. The vegetation roots also collect water from the soil pores and transpirate it. The potential reservoirs of water for evapotranspiration are both the surface water and the soil moisture in depth up to the capillary rise. Evapotranspiration is a complex and non-linear process, which is highly variable in both space and time and with strong cyclic seasonal variation. Urbanisation has a strong impact on the process of evapotranspiration. Reservoirs of soil moisture have decreased or even disappeared. Planting of any kind of vegetation is highly recommended to increase evaporation. In highly urbanised areas, evaporation exists only a short period of time immediately after the rainfall. The absence of evaporation increases the temperature in urban areas significantly.

Precipitation and through fall below the vegetation cover can be trapped in depression storage and then they subsequently evaporate or infiltrate into the soil surface for days after rainfall. The depression storage depends on the micro-topography and properties of soil surface. Micro-topography with features of up to few meters in length and few centimetres in depth forms small depressions. The water collects in the depression flushing small soil particles from the soil surface that then settle in the depression, clogging the bottom and decreasing the infiltration rate. Depression storage is almost absent in urban areas.

Infiltration is normally the most important process for continental water balance and the environment. Infiltrated water constitutes soil moisture, provides subsurface runoff and percolation into the groundwater. The infiltration rate depends mainly on the permeability and porosity of soil, and the soil moisture content. Impervious coverage of roofs and roads are the main characteristics of the urban landscape, so pervious pavements and green roofs are highly recommended.

The hydrology of urban areas deals mainly with storm runoff from small areas with a few-minute time lag of the hydrograph. Almost the total amount of rainfall quickly collects on the impervious surface of roofs, roads, pavements, parking, etc. The runoff from green urban areas is often neglected and is frequently not taken into consideration in the design of urban drainage systems. The peak discharge and time concentration are the important parameters for the design of structures for urban storm drainage and there is often less interest in other parameters associated with the runoff processes, such as base flow or subsurface flows.

Urban areas present complex features, resulting from modified land uses, different development densities and ages of development, and variable proportions of catchment occupation. As the land is urbanised, the natural land surfaces are replaced by artificial ones, such as paved roads, parking lots, and roofs, which usually implies vegetation clearing and soil compaction. The drainage system is also affected as gutters, drains and storm sewers are laid in the urban areas to convey

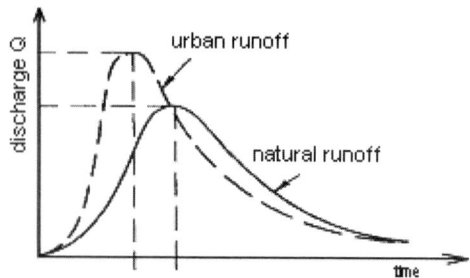

Figure 14.2. Urban flood hydrograph.

runoff rapidly to stream channels. These stream channels may be extremely modified and relegated to underground culverts. Also, urban development may expand
to river flood plains, reducing storage and conveyance area for floodwater.

The structures of urban drainage and road drainage systems have a limited capacity. They are often designed on the basis of storm events with return periods of
order of 10 years. In France, the recurrence interval, which is to retain rainfall or
combined sewer pipes, varies from 5 to 100 years depending on the surrounding
vulnerability to inundation. The drainage system may take a small percentage
of total peak surface runoff for an event with a hundred-year returns period or
greater. During storm events that exceed the design event, flooding and damage
may occur. Urbanisation can have a strong impact on discharge of a watercourse
also by reshaping of the watershed area and rearrangements of the stream network.

The effects of urbanisation on the runoff hydrograph are experienced in the short
time of collection of water at the surface and in larger discharge (Figure 14.2).
Because of limited discharge capacities of the sewage system, during events of
longer return periods the city streets take on the role of stream channels and they
have to be designed accordingly. The streets are deepened and the houses along the
road are raised so that the water current does not endanger the houses along the
roads (Marco, 1994). The road must be designed as a dry channel that is usually
filled with water. The road design must include an adequate longitudinal slope,
which diverts water into a stream of flood plains outside urban areas. We must be
careful to control the torrential water of the road, thereby reducing the threat to the
vital buildings (Figure 14.2).

The main impact on the urban environment is runoff produced by man-made
watercourses constructed for water supply, sewage water collection and storm water
collection and detention. They consist of piping networks and heavily regulated
watercourses. The system is mainly used for household waste collection and disposal. The water in the sewage system is highly polluted and loaded with sediments.
The flooding of the sewage system or its improper function during floods causes
additional pollution hazards in urban areas. The problem is the limited capacity of
man-made watercourses that produce flash floods by highly polluted water. Water

quality in urban storm water systems is a very important issue challenging the urban water drainage of today. In the conclusions of the NOVATECH conference it was recognised that urban storm waters are the most frequent source of pollutants (Chocat *et al.*, 2004).

14.3   CHARACTERISTICS OF DESIGN OF LOCAL FLOOD DEFENCE SYSTEMS

It is highly difficult to predict flash floods and operation of local defence systems. Urban areas undergo fast changes, the built-up areas spread out into the surrounding non-urbanised areas, and the urban areas themselves are often reconstructed, including the drainage systems, which has effects on runoff and the hazard level. Unfortunately, we do not hold enough historical data on floods of longer return periods in urban areas. Even when they exist, they cannot be used under the changed circumstances. Only the results of modelling simulations are available, which lack proper calibration. In the planning of measures we must take into account the high level of uncertainty.

Time lags between the precipitation and runoff peaks are quite short in urban flash flood events. In this respect, all measures should be carefully planned and prepared in advance before the event. After the alert has been activated, there is usually very little time, allowing only for evacuation or retreat. The increase of flood defence measures, such as enlargement of levees or increased flood proofing during flood alert, is not possible in the short time. The alert mainly helps saving lives and some personal property. The measures must be robust and designed so that their operation is possible without temporary measures and human control. If special mechanical equipment is designed, such as locks, valves, sluice gates, we must allow for the possibility that the equipment will not function flawlessly.

The local flood defence systems are closely connected to the urban drainage system and pollution. There is a great difference between mixed and separate drainage systems, especially with respect to the pollution of the urban runoff. Each system has its own strengths and weaknesses. The traditionally recognised advantage of the separate system was that only used and polluted water from the industry and households is collected into waste water treatment plants, where the water can be properly treated and released back into streams. The system does not collect surface runoff during precipitation and is therefore not additionally burdened with increased discharge during floods. There is also no special danger of pollution caused by slipovers. During heavy precipitation events, the seemingly clean water freely runs into streams, without being the cause of pollution. However, the surface runoff carries polluted water from urban surfaces, due to different types of emissions, especially traffic. In this way the problem of polluted water occurs also in separate sewage systems and, if untreated, cannot be released into surface streams. Today, we cannot no real advantage of one system over the other can be identified. The problems and solutions to the problems, however, have changed.

In design of measures, special attention should be given to public amenities: traffic, energy supply, media, water supply, health care, food supply, safety, etc. Traffic infrastructure is especially significant, since it is closely linked to other aspects of public supply. Recently, Europe has seen an increase of victims that died when driving on flooded roads. Quite often, warnings and safety-illuminated signs do not help. Drivers are simply unaware that a sunken vehicle will start to float and that the driver will lose control over the vehicle or even lose his life if he will fail to abandon the vehicle in time. Another case is railway, which during floods turns, with road underpasses, into an impassable barrier, and parts of the city may be temporarily completely cut off. Or the importance of finding the proper location of fire brigade premises or hospitals in the flood area. The same holds true for storage of perishable foodstuffs and medicines. Flooded medicines are of no further use and must be considered as hazardous waste in further steps (Brilly *et al.*, 1999).

From the economical aspect, the local defence systems are part of the municipal infrastructure. Similar to other services (police, education, water supply) they can be provided only within public's willingness to pay (Grigg *et al.*, 1976). In design of measures, the costs of implementation of measures are easy to assess; however, it is much more difficult to identify the benefits. Economical calculations of benefits are defined with vague starting points and are impossible to test. The implementation of measures itself is in the hands of the local community, which can expect a certain degree of support from the central administration. There are no known cases where individuals would participate in financing of measures for protection of their own property against floods. Also, public funding of measures against floods does not entail payment of flood protection fees, which was, however, customary in some parts of the Austro-Hungarian Monarchy. Which is why local systems or more often particular measures are driven by particular flood events. They are constructed with local funding, but without a strong enough institutional support, since countries set up large systems under the supervision of central and technically well equipped water agencies. In some cases, directly after floods, the local politics proposes measures designed for a systematic and long-term solution of flood-related issues. The systems are constructed based on experience in flash floods, however decisions are made under strong political pressure and have to be supported by the central water authority. The system for defence against floods must be a comprehensive one, taking into account the operation of centrally run systems and protection of interests of stakeholders downstream.

Local defence systems differentiate according to specific climate and social conditions (Maksimović, 2001). The climate conditions are governed by the intensity of rainfall, annual rainfall, temperature, sediment yield etc. Floods in urban areas are governed by the natural conditions. In colder climates, floods occur after spring snow melt and melting of ice-covered and clogged canals, in hot sea-side resorts

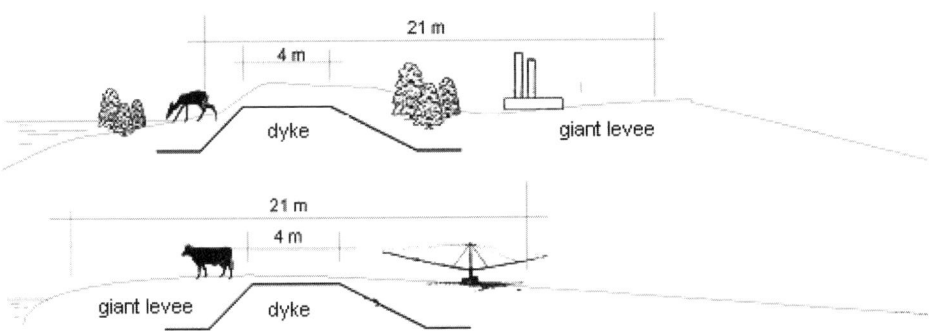

Figure 14.3. Giant levee.

floods are caused by intensive tropical and subtropical cyclone related rainfall. The floods are also the result of high tides. Each of these events has its characteristic intensity and duration and best-fit measures.

The level of consideration of technical, economical and environmental standards governs social conditions. This is basically the height and level of development of social capital of the local community prepared to contribute financially or support the implementation of measures in spirit of solidarity. Local floods are locally based and they do not affect everyone in the same manner.

Urban landscape planning is crucial for the development of the system. Some crucial public services should be more protected than other should. Also, there should be enough space for the water regime for the benefit of nature and society.

Real estate owners are highly important stakeholders and land prices are a strong driving force in the municipal policy. The solutions ask for additional space necessary for prevention measures. Without financial participation of stakeholders, the implementation of defence measures is quite difficult.

Today, the construction of a giant levee can be a very good solution. Giant levees are several times larger than ordinary dykes (Brilly et al., 1999). Dykes are constructed of minimum size and should be carefully cared for and maintained there is no activity other allowed. A giant levee is an extraordinary surface elevation that gives protection against almost maximum possible floods and, on the other hand, provides free space for land users. Protection by giant levees is not in risk by some landowner activities (Figure 14.3).

14.4   STRUCTURE OF THE LOCAL DEFENCE SYSTEM

The local defence system could incorporate the same measures as the regional system, but at a smaller scale. The measures range from structural to non-structural measures, from flood preparation to flood recovery works. The optimal solution varies in relation to the standards, vulnerability and economical development.

The local defence system may be a very complex one and consists of structures that include the sewage system, roof drainage system, road drainage system, and stream network and road network. System works depend on design criteria. The man-made drainage system is not capable of carrying the surface runoff with a return period longer than 10–20 years. Detention of water in depressions and stream flow along streets is unavoidable in catastrophic storm events.

Urbanas and Roesner (1999) derived elements of Local Urban Storm Drainage Criteria in US that consist of:

- Governing legislation and statements of policy and procedure,
- Initial and major drainage provisions,
- Data required for design,
- Detention requirements,
- Water-quality criteria,
- Special consideration.

The Criteria give a very good overview of the complex systems. Policies driven by legislation and procedures very often create separate systems for drainage of water from household properties and roads, which then divert runoff with increasing or decreasing watersheds and increase the runoff from impervious surfaces. Those changes are present during low and medium flows rather than during flood events. The low flows could be diverted completely, but the drainage system cold not carries high water with the same efficiency as low flow. In a flood event, when the flood drainage system is overflowed, the runoff water is retained in low detention areas and flows along streets. Improper design and blockage of intake structure by debris flow makes the situation even more complicated (Roso, 2004; Gomez *et al.*, 2004). For example, an important obligation is that water flow from neighbours' property cannot be restricted or diverted.

Detention of excess water is very important for floods in urban areas. The surface of detention ponds takes some part of the total area, for example in the lle-de-France region (Paris and its Department) detention ponds with surface between 0.5 and 20 hectares take 4% of the total urbanised surface (Tassin *et al.*, 2004). The ponds are connected into green corridors with high ecological quality. Ponds and green corridors have also pedestrian and bicycle ways and recreation facilities. It was recognised that ponds are a very important part of the local surface management system with a high ecological value. Increasing of surface elevation was very popular fifty years ago when the new parts of the city of Ljubljana were elevated by one metre and the new suburbs of Belgrade were built on an elevated area of more than $10\,km^2$ in surface.

Recently, best management practice has promoted control of runoff in its source, meaning that the construction of pervious cover of pavements, roads and court-yards has been introduced. Rainfall infiltrates in the ground in such surfaces and decreases the runoff. A similar solution is the construction of vegetated roof covers that detain rainfall water collected on roofs. The detained water partly evaporates from the vegetated roof and decreases the total runoff. Those measures have a stronger impact on the low and medium runoff situation. In catastrophic events of long return periods the impact of such solutions is less important.

There are some risks inherent in the development of scenarios today. We do not know the potential long-term effects of the "green solutions" and also climate change. Uncertainties of hazard estimation, future vulnerability development and impact assessment of the solutions are very high. Also, the local authorities could understand this as the way to free them from costly maintenance of flood defence infrastructure. Finally, we do not know if green scenarios are really sustainable (Chocat, 2004).

14.5   LOCAL FLOOD DEFENCE SYSTEMS

In France, Belgium, the Netherlands and Germany, local systems inside the municipality are under the responsibility of the mayor and municipal council that approve the municipal disaster plan (Roenthal and Hart, 1998). A similar solution has been adopted in other countries in Europe, like Switzerland (Gillet and Zanolini, 1999). When trying to establish and maintain a system, the main issue is then the technical and financial capability of a small municipality.

From 1982 to 2003 more than 75% of municipalities in France had at least one flood experience caused by runoff (Gaber and Balades, 2004). This raised the importance of local defence systems for protection from inner water, that is, local runoff. Local flood defence systems in urban areas are used for protection against water collected inside the urban area. These systems have specific characteristics, contrary to the large regional and countrywide systems. Large systems that include levees and large dams are well known and documented in journal papers and country reports. The local governance performs maintenance of local systems and often these systems are not integrated into the national system. Regional systems and management of these systems is well documented (Correia, 1998).

14.5.1   Dimensioning of sewerage and drainage systems in Slovenia

Cost benefit relations related to the sewage system are typically random relations that involve high uncertainty. The costs of construction works can be well estimated, but damage cannot be estimated. Damage caused by floods is a combination of direct damage – damage to structures and other facilities; and secondary or indirect damage – damage to business production or loss due to disturbances in everyday life. The indirect costs are quite often higher than the costs of direct damage; however, they are very difficult to estimate. For example, a flooded highway could have no direct damage, but the indirect costs of closed traffic may be high.

The practical solutions are then related to the "willingness to pay" for the measures preventing the disturbances caused by flood events. We find that the optimal return period for protection by construction standards has increased in recent years. Further on, practice has shown that the pipe diameter is not large enough. The increase of urbanisation and growth of impervious land slowly, but surely, result in higher runoff, which increases the "willingness to pay" for higher security.

In Slovenia, there are no official standards or guidelines for sewerage or drainage systems. The practice is related to examples of "good practice" and soft recommendations developed by professional associations, or guidelines accepted by municipal companies responsible for development and maintenance of the drainage systems.

Up-to-date guidelines of the Ljubljana municipal water supply have adopted the SIST EN 752-2 standard for return periods for dimensioning rains and floods. The additional safety rules are the following:

• Water from roofs should infiltrate into underground if possible;
• Minimal depth of the drainage system for storm runoff is 80 cm and sewage system 120 cm, respectively;
• Minimal pipe diameter is 250 mm.

The formal guidelines and the existing system are designed for two to ten years of return period of rainfall.

In Slovenia, the local measures are introduced under the responsibility of the central government (on streams) and by the municipality (on the sewage systems). The measures implemented on the sewage systems are driven by the EU standards. The measures developed on streams are poorly documented and monitored and are derived case-by-case without special guidelines or manuals.

14.5.2   Responds to the short questionnaire on development of the local defence system

A questionnaire on the development of the local defence system was sent to the representatives of the COST C22 countries. The country representatives were asked to provide a summary (half- to full-page long) describing the policy in local flood protection systems in their country. They were asked to address the following questions:

• Who is responsible for development and maintenance of local systems: government, regional or local authority?
• Are local measures systematically developed by guidelines based on standards and legal acts or are they developed as post-flood event measures?
• Do you have any systematically collected databases for local systems, and are these systems monitored?
• Do you have a well-monitored local system that could be used as an example of good pan-European practice? Could you provide a short description?

The responds were received from Hungary and Norway. We also received the corresponding paper from Cyprus (Toumazis A.D. and Toumazis D.). The answers were highly interesting in terms of characterisation of flood hazard and administrative development in the particular countries.

14.5.2.1   *Respond from Hungary*

The responds were sent by Mr. Lajos Szlávik from the Lower-Danube Valley Directorate for Environment and Water, situated in Baja, Hungary (http://adukovizig.baja.hu/.

In Hungary, the flood hazard is caused by large rivers, that is, the Danube River and the Tisa River, and those tributaries that represent risk over huge flood plain areas protected by about 4000 km of levees. This system is under the responsibility of the central government and supervised by local directorates. The local measurements are in the responsibility of the local government but also supervised by the local directorates. The existing monitoring system is insufficient.

The Directorate as the administrative public institution is responsible for flood defense. It is an independent legal entity under the ministerial control and centrally budgeted body. Its activity in water damage prevention, environmental and water quality damage prevention is controlled by the Ministry of Environment and Water. The under-secretary of state approves the annual work plan, organizational and operational regulations of the Directorate.

The Directorate has a country-wide task to be performed within its basic scope of activities on flood control: continuing the use of the classical principles of river regulation, Hungary has to continue the maintenance and development of flood protection structures and protection of the areas endangered by floods. The development of the water run-off regulation on the lowlands is also being continued.

The Directorate
a) Co-ordinates and co-operates in the preparation of concepts and plans affecting the water management of its territory;
b) Establishes the co-ordination of developing and operating public, state, municipality and proprietary waterworks;
c) Co-operates
   a) in the performance of tasks related to international and especially border-related water treaties (the most important field of international co-operation is the transboundary co-operation with the neighbouring countries in the subject of frontier waters),
   b) in the preparation and execution of national and regional programs relating to the cleaning and harmless discharge of wastewater of settlements (drinking-water supply, as for sewerage and sewage water treatment),
   The directorate has to cooperate

    c) in the tasks related to the Environment protection Funds Targets and water management goals, the utilisation and supervision of the targets as specified in separate measures by the minister, and also

    d) in environment protection and water management research, education, teaching and informative activities;

d) keeps the records as specified in separate measures;

e) performs – based on the task division determined by the minister – the regional tasks required for the operation of the Informative system within its scope of power;

f) gathers and forwards to the Information System all of the competent data that are necessary for its operation, and co-operates with other observation and information systems;

g) cooperates with the regional municipalities in the completion of water management tasks.

The responsibility for development and maintenance of local systems is shared between the central government and the local authority. Essentially, in Hungary two types of local systems exist:

1. managed by the local government,
2. managed by the local directorate for environment and water.

All summer dikes managed by the local government, and the main flood control system is managed by the local directorate for environment and water.

All the measures are based on standards and legal acts. Of course after or during floods these measures are reconsidered. If needed, the local measures can be amended.

Local systems are monitored systematically. The maintenance is based one flood control plans. These plans can be divided in two parts:

1. Flood control plans at the level of the local government.
2. Flood control plans at the level of the local directorate for environment and water.

Every year (mainly in autumn) the directorate check the flood control systems (condition, quality, etc.).

Unfortunately the monitoring system is still poorly developed. The water level and ice forecasting service is not dense enough. In the region, there are only four remote monitors. In other cases the dam keeper system accomplishes monitoring.

In connection to the implementation of the EU Water Framework Directive the efforts are focused on taking into consideration the special circumstances of flood control. During the implementation we endeavour to take into consideration the aspects of flood control. The directorate has applied for different EU funds for the development and maintenance.

14.5.2.2   *Respond from Norway*

In Norway the local measures are implemented according to the standards for development of sewage systems and national guidelines for those measures. They are widely used but there is no supervision and no database of local measures. The local municipality is responsible for the development and maintenance of local systems in Norway.

The local measures are developed based on the European standard EN-752, and national guidelines issued by the association of Norwegian water and sewerage works. Legal acts have caused the municipalities to take more action to avoid being sued by insurance companies. There is no central databank concerning flooding problems on this.

14.5.2.3   *Respond from Cyprus*

In relation to the high flush flood hazard, Cyprus has well developed systems with a high number of dams and ponds that are used for water supply. Development, oversight and maintenance od ponds are in the hand of the Water Development Department (WDD) of the Ministry of Agriculture, Natural Resources and the Environment. The system of dams is used also as flood prevention structures and is very efficient. There was no experience with floods downstream of dams but there is no special ALERT system or monitoring of dams during the flood event. The responsibility for storm drainage systems is in the hand of several local and governmental administrative organizations. Co-ordination and operation of the constructed systems are in the hand of Sewerage Boards. The borders of regions, under the responsibility of the Sewerage Board, do not correspond directly to the borders of municipalities. The main problem is institutional straitening and lack of proper co-operation between different administrative organisations and municipalities with limited responsibility and improper legal support. Useful legal procedures for integrated decision making processes are missing. The problems in relations between neighbouring municipalities seem to be the same as those between countries on international watercourses (Toumazis A.D. and Toumazis D.).

14.5.3   Local flood defence in the US

There is a low number of well developed, maintained and documented systems, such as in Boulder and Denver Colorado, US. In the US, a special interest association for the use of alarm systems ALERT has been set up. ALERT is an acronym for Automated Local Evaluation in Real Time, which is a method of using remote sensors in the field to transmit environmental data to a central computer in real time. The numerous systems for local flood forecast were developed in the past thirty years. The users of such systems developed common standards for equipment and manuals for implementation and maintenance. Yearly, several national or regional meetings are organized for exchange of experience and promotion of good practice.

There are also international members and from Europe only Spain has joined the organization (http://www.alertsystems.org).

The local flood defence systems for Denver and Boulder are very well documented. The user manuals are derived for public purposes and provide the necessary information and data sources. The manuals cover integrated implementation of non-structural and structural measures.

14.6   CONCLUSION

Local flood defence measures are widely used in Europe and, in some countries, these measures are well developed. Unfortunately, no proper overview has been made, containing information about the experience, performance and effectiveness, as well as providing guidelines or examples of good practice. The measures differ in relation to climate and hydrological conditions that cause floods, and social conditions (legal frame and economy).

The flood protection measures on the sewage system are driven by EU standard, which are directed into protection against pollution rather than flood protection. There are no proper procedures that would enhance the co-operation between neighbouring municipalities related to management of joint facilities.

Local measures are mainly the responsibility of the local municipality; however, the supervision and information are insufficient. They are also poorly monitored. There is a lack in engineering knowledge and support that would be supervised from the relevant central agencies.

Case studies of good and bad practices are also missing. There is a lack of knowledge, especially on the long-term impacts. Also, the measures that are successful in one place may fail in another.

ACKNOWLEDGEMENT

The author highly appreciates the correspondence with, and responds to the questionnaire by, Lajos Szlávik, Oddvar Lindholm and Antonis Toumazis.

REFERENCES

Booth, D.B. and Jackson, D.R., 1997, Urbanization of aquatic systems: degradation thresholds, storm water detection, and the limits of mitigation. J. American Water Resources Association, 33 5, pp. 1077–1090.

Brilly, M., 1991, Varnostni ukrepi pri obrambi pred poplavami. 1Ujma 5, Ljubljana, str. 150–151.

Brilly, M., Mikoš, M. and Šraj, M., 1999, Vodne ujme (Water hazard), University of Ljubljana, Faculty of Civil and Geodetic Engineering, Ljubljana.

Brilly, M., 2001, The integrated approach to flash flood management, GRUNTFEST, Eve and HANDMER, John W. (eds.), Coping with flash floods (NATO science series. 2, Environmental security, v. 77). Boston: Kluwer Academic Publishers, pp. 103–113.

Chocat, B., Ashley, R., Marsalek, J., Matos, M.R., Rauch, W. Schilling, W. and Urbonas, B., 2004, Urban drainage – Out-of-sight-out-of-mind? NOVATECH – Sustainable Techniques and Strategies in Urban water Management. Proceedings 5th International conference, GRAIE, Lyon, France.

Correia, F.N., 1998, Selected Issues in Water Resources Management in Europe, A.A.Balkema.

Gaber, J. and Balades, J.D., 2004, The risk prevention of flood by runoff in France, NOVATECH – Sustainable Techniques and Strategies in Urban water Management. Proceedings 5th International conference, GRAIE, Lyon, France.

Gardiner, J.L., 1991, River Projects and Conservation – A Manual for Holistic Appraisal. John Wiley & Sons, Chicester.

Gillet, F. and Zanolini, F., 1999, Mountainous natural hazards, Cemagref, Grenoble.

Grigg, N.S., Botham, L.H., Rice, L., Shoemaker, W.J. and Tucker, L.S., 1976, Urban drainage and flood control projects – Economic, legal and financial aspects, Hydrology papers, Colorado state University, Fort Collins, Colorado.

Gomez, M., Guirado, v., Nania, L. and Dolz, J., 2004, Join analysis of street flow and inlet efficiency on the risk associated to runoff, NOVATECH – Sustainable Techniques and Strategies in Urban water Management. Proceedings 5th International conference, GRAIE, Lyon, France.

LAWA, 1995, Guidelines for Forward-Looking Flood Protection, Floods – Causes and Consequences. Stuttgart, Länderarbeitsgemeinschaft Wasser.

Maksimović, Č., 2001, Urban drainage in specific climates, UNESCO IHP, Technical Documents in Hydrology, No 40, Vol I, II and III.

Marco, J.B. in Cayuela, A., 1994, Urban flooding: the flood-planned city concept, Coping with Floods, NATO ASI, Kluwer Academic Publishers.

Riley, A.L., 1998, Restoring streams in cities. A guide for planners, policymakers, and citizens. Island Press, Washington, D.C.

Roesner, L.A. and Brashear, R.W., 1999, Are BMP criteria really environmentally friendly?, Proceedings of the 8th international conference on urban storm drainage, Sydney Hilton Hotel, Sydney, Australia, 30 August–3 September 1999, Joliffe, I.B. and Ball, J.E, (eds.) pp. 1366–1373.

Rosenthal, U. and Hart, P., 1998, Flood Response and Crisis Management in Western Europe (A comparative analysis), Springer.

Roso, S., Boyd, M., Rigby, T. and Van Drie, R., 2004, Predestine of increasing flooding in urban catchments due to debris blockage and flow diversions. NOVATECH – Sustainable Techniques and Strategies in Urban water Management. Proceedings 5th International conference, GRAIE, Lyon, France Tai Kon Chin, 1975, Analysis and synthesis of flood control measures, Hydrology papers, Colorado state University, Fort Collins, Colorado.

Tassin, B., Mouchel, J.M. and Aires, N., 2004, A posterior analysis of the design and the maintenance of retention ponds in the lle-de-France region. NOVATECH – Sustainable Techniques and Strategies in Urban water Management. Proceedings 5th International conference, GRAIE, Lyon, France, Urbanos, B.R. and Roesner, L.A., Hydrologic design for urban flood control (Maidment, Handbook of hydrology), McGraw-Hill.

Toumazis, A.D. and Toumazis, D., Urban flood management in Cyprus – case study in Nicosia urban area.

15

European Flood Strategies in Support of Resilient Buildings

David J. Kelly & Stephen L. Garvin
Building Research Establishment, East Kilbride, UK

ABSTRACT: Approaches to limiting disruption and damage from flooding have changed significantly in recent years. Worldwide, there has been a significant move from a strategy of flood defence to one of flood risk management (Gouldby et al., 2005). This includes the use of flood defences, where appropriate, but recognises that managed flooding is essential to meet the requirements of a sustainable flood strategy. The success of this approach is dependent on integrating enhanced defences and warning systems with improved understanding of the river systems and better governance, emergency planning and disaster management actions.

The basis of flood risk management and the development of regional policies throughout Europe are described in this document. This includes a detailed review of current legislation and policy within the UK, as well as the cross-border approach that is often employed in Europe to account for the diversity and magnitude of the river network.

15.1   INTRODUCTION

In 1817, the Governor General of Australia, Lachlan Macquarie, after serious flooding of the Nepean and Hawkesbury rivers stated (Emergency Management Australia, 1998):

"... when the too fatal experience of years has shown the sufferers the inevitable consequence of their wilful and wayward habit of placing their residences and stock yards within the reach of the floods ... the compassion excited by their misfortunes is mingled with sentiments of astonishment and surprise that any people could be found so totally insensible to their true Interests, as the settlers have in this instance proved themselves."

The risk of a disaster can be considered in the form of a triangle with the three sides being hazard, vulnerability and exposure (Crichton, 1999). The growth in the cost of global disasters so far is mainly due to an increase in exposure, with an unprecedented increase in population leading to more people living in areas exposed to natural hazards. It is also due to a growth in vulnerability, especially in developing countries which account for 97% of casualties. It is the poor, the elderly and the young who suffer most.

So the exposure and vulnerability sides of the triangle are increasing. As climate change increases the hazard side of the "risk triangle" as well, the consequences could be very serious. From a reinsurance point of view there is the prospect of a "catastrophe clash" where different continents have different disasters at the same time, potentially resulting in global economic meltdown. As Professor David King, the UK Chief Scientific Advisor, recently stated (King, 2004):

"In my view, climate change is the most severe problem we are facing today – more serious even than that of terrorism."

Within Europe, more than 10 million people live in the areas at risk of severe floods along the Rhine, and the potential damage from floods amounts to € 165 billion. Coastal areas are also at risk of flooding. The total value of economic assets located within 500 metres of the European coastline, including beaches, agricultural land and industrial facilities, is currently estimated at € 500 to 1000 billion (EUrosion).

Approaches to limiting disruption and damage from flooding have changed significantly in recent years. Worldwide, there has been a significant move from a strategy of flood defence to one of flood risk management (Gouldby et al., 2005). This includes the use of flood defences, where appropriate, but recognises that more managed flooding is essential to meet the requirements of a sustainable flood strategy. The success of this approach is dependent on integrating enhanced defences and warning systems with improved understanding of the river systems and better governance, emergency planning and disaster management actions.

The basis of flood risk management and the development of regional policies throughout Europe are described in this paper. This includes a detailed review of current legislation and policy within the UK, as well as the cross-border approach that is often employed in continental Europe to account for the diversity and magnitude of the river network.

15.2   FLOOD RISK MANAGEMENT – AN OVERVIEW

Over the years, flooding and flood risk management are becoming increasingly important issues. As the impact of severe events is felt by increasing numbers of the population, the cost from associated damage increases dramatically. The cost of flood cannot only be measured in monetary terms as the human costs are often impossible to calculate. Between 2000 and 2004, floods killed 185 people within the EU and affected half a million. The geographic and demographic spread of these events are described in Table 15.1 (Lamothe et al., 2005).

Table 15.1. Consequences of floods in EU between 2000 and 2004 (extract from Lamothe, David-Nicolas, Neveu, Gilles, Gorlach, Benjamin and Interwies, Eduard (2005). 'Evaluation of the impact of floods and associated protection policies', European Commission, Brussels).

Country	Number of floods	Human casualties			Estimated financial damages ($\times$1000 US$)
		Dead	Injured	Affected	
Austria	1	7	0	60,000	2,046,407
Belgium	5	2	0	2,400	0
France	16	42	13	49,661	2,832,350
Germany	2	34	128	331,000	10,460,447
Greece	11	12	0	8,490	656,218
Ireland	1	0	0	300	0
Italy	12	59	22	57,550	2,506,960
Portugal	3	7	0	296	0
Spain	3	21	0	1,250	14,285
UK	9	1	8	2,460	6,918,150
Total	63	185	171	513,407	25,434,817

The flood event itself is not the only source of concern for those who have experienced an event or live within a risk area. Feelings of fear often remain long after the actual event and storms at sea, heavy rain and high river levels can lead to a state of alarm for many of the population. It is necessary, therefore, to manage the risks associated with flooding in a consistent and effective manner.

At the centre of any flood risk strategy is the fact that flooding, as a natural event, cannot always be predicted or prevented. The effects of climate change may make the occurrence of extreme weather events and sea level rise more likely, however it is difficult to give any guarantee that the impact of flooding will be managed and minimised effectively. This has led to the development of flood risk management strategies that have been developed in both the UK and beyond.

In the UK, DEFRA (Department for Environment Food and Rural Affairs) has developed a flood risk strategy in consultation with stakeholders (DEFRA, 2006). This follows a holistic approach to the management of risk from all forms of flooding (river, coastal, groundwater, surface run-off and sewer) and coastal erosion and is seeking to make sure that the programme allows the sustainable development of the flood risk areas. Within this strategy a number of key objectives have been identified as follows:

- Identify the most effective ways of tackling the causes of urban flooding.
- Ways to help people adapt to the changing risk of flooding and coastal erosion.
- Introduce a pilot grant scheme to make individual properties more flood resistant and resilient where public-funded improvement may be impractical.

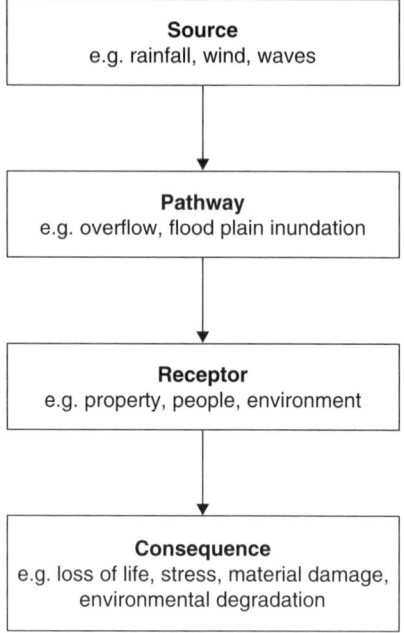

Figure 15.1.

- Work with natural resources and developing approaches to flooding and erosion that achieve many objectives simultaneously.

In developing a flood risk management strategy, the hazards and risks have to be identified. To understand the relationship between "hazard" and "risk" the *source-pathway-receptor-consequence* model can be adopted (Lamothe et al., 2005). This model, shown in Figure 15.1, is a simple conceptual model for representing systems that lead to a particular consequence. For a risk to arise there must be a hazard that consists of a *source* or initiator event (e.g. high rainfall); a *receptor* (e.g. flood plain properties); a *pathway* between the source and the receptor (e.g. flood routes including defences, overland flow or landslide).

A hazard does not automatically lead to a harmful outcome, but identification of a hazard does mean there is a possibility of harm occurring, with the actual harm depending on the exposure to the hazard and the characteristics of the receptor. Therefore, to evaluate the risk, a number of components have to be considered including the following:

- The nature and probability of the hazard;
- The degree of exposure of the receptors to the hazard;
- The susceptibility of the receptors to the hazard;
- The value of the receptors.

In terms of flooding, a description of the hazard in an area is often measured by the annual probability of flooding or the return period of the flood which would overwhelm the area concerned. It is also important to note that flood "risks" are primarily associated with people or the society in which an event may occur. There may be some associated natural damage to the environment but this can be viewed as an entirely natural occurrence. The mitigation of flood risk can be accomplished through managing any of the hazard, exposure and vulnerability. Flood hazard may be reduced by engineering or structural measures which will alter the frequency (or probability) of flood levels in an area. The exposure and vulnerability of a community to flood loss can be mitigated by non-structural measures, for example through the change and regulation of land use, through flood warning and effective emergency response, and through flood resistant construction methods.

Flooding can have many consequences, some of which can be expressed in economic terms. However, consequences can also include fatalities, injuries, damage to property or the environment. Consequences of a defence scheme can include environmental harm or benefit, improved access and many others including reduced risk. Placing a value on the consequences of flooding is subject to much research and investigation. This is often hindered by the lack of agreement on the terms used to identify the consequence. An important part of a flood risk assessment method is to decide on how the impacts are to be evaluated. Some descriptions of consequence could be as follows:

- Economic damage (national, community and individual);
- Number of people/properties affected;
- Harm to individuals (fatalities, injuries, stress, etc.);
- Environmental and ecological damage (can be expressed in monetary terms).

A clear understanding of the hazards and consequences in relation to flooding are key to its effective management. Various flood risk strategies have been developed throughout Europe. These often involve cross-border agreements and policies to manage the many risk areas which have been affected in recent times.

15.3   UK PLANNING GUIDANCE ON FLOOD MANAGEMENT

Approximately 10% to 15% of the land mass within the UK lies in areas where flood risk is apparent (Garvin et al., 2005). There are often social and economic reasons for continual development within these areas and this requires a carefully considered approach to planning and flood risk management. The percentage of land mass affected by this risk may not seem excessive, however when considering the amount of property and assets this could affect, it presents a very different perspective.

Regional and local planning authorities are responsible for assessing flood risk in specific areas. Development plans should be prepared which will ensure developers assess flood risk for all construction projects within the area. ODPM have the overall responsibility for the planning system which is designed to prevent inappropriate new developments in flood risk areas, and to direct developments away from areas at high risk. The current planning policy guidance is PPG 25: Development and flood risk (ODPM, 2001). This was published in 2001 and advises local planning authorities on how to manage flood risk as part of the planning process. This will be replaced in 2006 by Planning Policy Statement (PPS) 25 and accompanying guidance.

Local planning authorities must currently adhere to PPG 25 when developing new communities and regenerating existing ones. This policy guidance ensures that flood risk is considered at all stages of the planning and development process, from regional plans, to local plans and individual site development. In flood risk areas, local planning authorities should consult with the Environment Agency about development proposals. These proposals should always be accompanied by a flood risk assessment.

The introduction of PPG 25 in 2001 has raised awareness and the profile of flood risk in the planning process. ODPM stated that the number of applications permitted by local planning authorities against sustained advice from the Environment Agency has halved since PPG 25 was introduced. Approximately 8% of decisions made by local planning authorities are not in line with Environment Agency advice. This reduction has been welcomed, however the number is still considered too high. This prompted ODPM to strengthen and clarify the policy in the new PPS 25 (ODPM, 2006).

In December 2005, the UK Government published a consultation on Planning and Policy Statement (PPS) 25: Development and flood risk. The intention was that PPS 25 and accompanying "Practice Guide" would replace the Planning and Policy Guide (PPG) Note 25. PPS 25 was under consultation until February 2006 and a summary of comments will be published by ODPM in late 2006.

PPS 25 will replace PPG 25 as the Government's policy document on dealing with flood risk. It will be used by local authorities throughout England and Wales and will be supported and complemented by other national planning policies. PPS 25 describes some key planning objectives that regional planning bodies and local planning authorities should consider. These regional and local bodies are tasked with implementing planning strategies to deliver sustainable communities. This should include the following objectives:

- Identifying land at risk and the degree of risk of flooding from river, sea and other sources.
- Preparing regional or strategic flood risk assessments as required.
- Framing policies for the location of developments which avoid flood risk to people and property where possible and manage any residual risk, taking into account the effects of climate change.
- Reducing flood risk to and from developments through location, layout and design, including the application of a sustainable approach to drainage.

- Using opportunities offered by new developments to reduce flood risk to communities.
- Only permitting developments in areas of flood risk where there is no suitable alternative sites in areas of lower flood risk and the benefits of the development outweigh the risks from flooding.
- Working effectively with the Environment Agency and other stakeholders to ensure that best use is made of the expertise and information so that decisions on planning applications can be made expeditiously.
- Ensure spatial planning supports flood risk management and emergency planning.

The new PPS 25 has been developed to provide a stronger, clearer policy statement that aims to ensure flood risk is taken into account at all stages in the planning process. It will also do the following:

- Focus on core policies that are clearer and easy to understand.
- Provide a more strategic approach, emphasising the need to consider flood risk as early as possible in the planning process.
- Clarify the sequential test that matches types of development to degrees of flood risk.
- Strengthen guidance on the need to include flood risk assessments at all levels of the planning process.

The policy statement also includes proposals as follows:

- To make the Environment Agency a statutory consultee for planning applications in flood risk areas, and
- For a "Flooding Direction", providing greater security for major developments proposed in flood risk areas.

The introduction of PPS 25 will ensure all planning applications for flood risk areas are thoroughly scrutinised. This will help to promote appropriate and sustainable developments that are in keeping with the Government's policy on flood management. Under the new policy, all local planning authorities will have to take account of Government policy and Environment Agency advice on minimising flood risk. This will help to create developments that will have taken into account future increases in flood risk due to the expected impact of climate change. Within PPS 25 there will also be opportunities to reduce flood risk in existing communities. This will be carried out using techniques such as re-creating and safeguarding functional flood plain and wash lands, designing "green" spaces and introducing sustainable drainage systems.

There will, however, be limitations in the issues controlled under PPS 25. It will not have the power to:

- ban all development in flood risk areas.
- require all planning applications in flood risk areas are called-in for decision by a Minister.

A general ban on developments in flood risk areas would be unsustainable. It is also expected that this would result in economic stagnation and deprive existing communities with much-needed homes and services.

Scotland has its own guidance which is concerned with construction in flood risk areas. Scottish Planning Policy 7 "Planning and flooding" was introduced by the Scottish Executive in 2004 (Scottish Executive, 2004). The policy is based on the following principles:

- Developers and planning authorities must give consideration to the possibility of flooding from all sources.
- New development should be free from significant flood risk from any source.
- In areas characterised as "medium to high" flood risk for watercourse and costal flooding, new development should be focussed on built up areas and all development should be safeguarded from the risk of flooding.
- New development should not:
  - materially increase the probability of flooding elsewhere;
  - add to the area of land which requires protection from flood prevention measures;
  - affect the ability of functional flood plains to attenuate the effects of flooding by storing water;
  - interfere detrimentally with the flow of water in the flood plain;
  - compromise major options for future shoreline or river management.
- Flooding from sources other than watercourses and on the coast must be addressed where new development is proposed, if necessary through a drainage assessment. Any drainage measures proposed should have a neutral or better effect on the risk of flooding both on and off the site.
- Alterations and small-scale extensions to buildings are generally outwith the scope of SPP7 provided they would not have a significant effect on the storage capacity of the functional flood plain or affect the local flooding problems.

15.3.1 Risk-based approach to flood management

A risk-based approach to flood management has often been used as a viable means of assessing the diversity of locations and developments that could be affected by flooding. DEFRA have been developing a new strategy entitled "Making space for water". This is a risk-based strategy for managing flood and costal erosion in England and Wales. The vision for the strategy is to (DEFRA, 2004):

"... allow space for water so that we can manage the adverse consequences for people and the economy that can result from flooding and costal erosion while achieving environmental and social benefits in line with wider Government objectives."

There are a number of themes that are described in this strategy which include:

- Risk management – The strategy calls for a more integrated and holistic approach to risk. This should include environmental and social factors as well as economic damage. Multi-criteria approaches will also be developed so that the appraisal of potential schemes are given due consideration even though they may not be expressed in monetary terms. Included in this are management schemes to deal with intra-urban flooding. The development of trunk road design and maintenance guidance is considered, as is the risk management role of the rail networks. Consideration is also given to the role that individual parts of the road network might play in the management of extreme events.
- Sustainable approach – Flood management and costal erosion solutions that work with natural processes to provide more space for water should be identified and used where possible. Targets for wetland habitat creation should be imposed to fulfil biodiversity commitments. Also under consideration is the role of rural land management and the need for improved water level management.
- Planning and building – It is Government policy to ensure land use reduces, and certainly does not add to, the overall level of flood risk. Of concern in this area are the number of developments that proceed against the advice of the Environment Agency and the possibility of producing flood risk assessments. For individual buildings, many issues regarding flood resistance and resilience are considered for new and existing buildings. The strategy recognises the role the Building Regulations can play for new buildings, it asks how owners of existing buildings can be encouraged to incorporate resistance and resilience measures.
- Awareness – Raising the awareness of flood and costal erosion risk is an important element of this strategy, which has a long-term goal of promoting more informed decision making and understanding of risk. Issues such as flood warning systems and the impact of the Civil Contingencies Bill on emergency response is also addressed.
- Coastal issues – Some flood issues are common to both coastal and inland areas. There are, however, some issues that are exclusive to the coast. These should be dealt with using long-term strategic planning and decision-making processes such as Shoreline Management Plans.

These strategy and policies inform the Government and local authorities on how to deal with flood risk issues in a sustainable and efficient manner. In dealing with existing developments or new housing projects, it is often the Building Regulations which inform decision making at lower levels.

15.3.2   UK building regulations

Building regulations in the UK do not specifically deal with the risk from flooding to domestic or non-domestic buildings. The situation is unique in that different building regulations exist for the separate countries that make up the United Kingdom. Flooding risk is dealt with (indirectly) as follows:

1. England and Wales: Approved document C "Site preparation and resistance to contaminates and moisture" (2004) and Approved document H (2002) "Drainage and waste disposal" (ODPM, 2004).
2. Northern Ireland (ODPM, 2004).
3. Scotland: Scottish Building Standards (2005) Section 3.3 "Flooding and groundwater" (SBSA, 2005).

The regulations for England, Wales and Northern Ireland are concerned only with the preparation of construction sites and the quality and workmanship of the completed building. Little or no consideration is given within these regulations to the flood risk management of the site or surrounding areas. Government guidance in this area is covered by the PPG 25 (and the new PPS 25).

The Scottish Building Standards are more advanced in that they make direct reference to the issues of flooding and ground water. Section 3.3 of this Standard states that:

"Every building must be designed and constructed in such a way that there will not be a threat to the building or the health of the occupants as a result of flooding and the accumulation of groundwater."

Where development is to take place on land where flood risk is apparent, advice should be sought from the local planning authority, the Scottish Environmental Protection Agency (SEPA) and those responsible for costal defences. There is also a clause in which developers are encouraged to account for the potential effects of climate change and how that could impact on the rainfall and temperature variations.

For individual properties, the ground below the structure and immediately adjacent that is liable to accumulate floodwater or groundwater, requires treatment against possible harmful effects. The ground immediately adjoining a dwelling relates to the area where the groundwater would affect the structural stability of the dwelling. The drainage of groundwater may be necessary for the following reasons:

• To increase the stability of the ground.
• To avoid surface flooding.
• To alleviate subsoil water pressures likely to cause dampness to below-ground accommodation.
• To assist in preventing damage to foundations.
• To prevent frost heave of subsoil that could cause fractures to structures such as concrete slabs.

The selection of appropriate drainage layout will depend on the nature of the subsoil and the topography of the ground. With the removal of topsoil from a development site, there are possible dangers from surface run-off from the building site to other properties. Scottish Building Standards suggest the use of field drains, rubble drains, channels or small dams to divert run-off where the conditions are potentially serious.

15.4   MAINLAND EUROPEAN APPROACH TO FLOOD MANAGEMENT

At the national level, there are variations in the regulations and responsibilities for flood protection and flood risk management for most European countries. There are some issues relating to how flood protection and risk management are dealt with as they are often regulated by different laws (planning, water, housing, environmental, civil, nature conservation and agriculture). This can lead to a conflict of interests when dealing with the different aspects of flooding. Some countries are focussed mainly on precautionary measures, while others put more effort into flood management.

There has been a common strategy on flooding development throughout Europe (EU, 2003). This follows a river basin approach and describes the following aspects that should be taken into account when developing effective flood risk management programmes:

- Prevention: preventing damage caused by floods by avoiding construction of houses and industries in present and future flood-prone areas; by adapting future developments to the risk of flooding; and by promoting appropriate land use, agriculture and forestry practices.
- Protection: taking measures, both structural and non-structural, to reduce the likelihood of floods and/or the impact of floods in a specific location.
- Preparedness: informing the population about flood risks and what to do in the event of a flood.
- Emergency response: developing emergency response plans in case of a flood.
- Recovery and lessons learned: returning to normal conditions as soon as possible and mitigating both the social and economic impacts on the affected population.
- Research: more research is required to better understand the climatic, hydrological, ecological and landscape context of floods.

An EU action programme for flood protection has also been proposed. This includes a list of features that should be included to address flooding:

a) improving cooperation and coordination through the development and implementation of flood risk management plans for each river basin and coastal zone where human health, the environment, economic activities or the quality of life can be affected by floods.
b) Developing and implementing flood risk maps as a tool for planning and communication.
c) Improving information exchange, sharing of experiences and the coordinated development and promotion of best practice.

d) Developing stronger linkages between the research community and the authorities responsible for water management and flood protection.
e) Improving coordination between relevant community policies.
f) Increasing awareness of flood risks through wider stakeholder participation and
   more effective communication.

Flooding and flood damage are a major problem in North Western Europe. It is
estimated that the effects of climate change will worsen the situation in the coming
years. Improved flood prevention and protection are increasingly important and this
trend will continue into the foreseeable future. This imposes a greater emphasis
on the development and implementation of integrated flood management policies
in many European countries. These policies and strategies are often specific to
individual countries, however, along many of the main river systems (e.g. the
Rhine) a cross-border integrated approach has been established.

   The EU is progressing on an action programme on flood risk management, and
a proposal for an EC directive on flood mapping and flood risks management plan
(Floods Directive) is an important feature of these developments. The objective of
the Floods Directive is to reduce and manage the risks of flood to people, property
and the environment by concerted, coordinated action at river basin levels and in
coastal zones.

15.4.1   The Water Framework Directive

Flood management plans are being developed within the Water Framework Directive (WFD) (Wood et al., 2005). This Directive was introduced in December 2000
and has been implemented by many European countries. It intended to achieve
good status for bodies of water (rivers, lakes, coastal waters and groundwater)
throughout Europe and to prevent any further deterioration.

   The WFD does not directly address precautionary flood protection. This is
addressed through the requirement that there should be no further deterioration
of the river systems and good ecological and chemical status shall be maintained.
A further goal of the Directive is that the impact of floods should be reduced though
precautionary flood protection measures are not specifically described.

   The development of policies and strategies for sustainable flood prevention and
protection is controlled through UN/ECE guidelines as follows:
a) All appropriate action should be taken to create legal, administrative and economic frameworks that are stable and enabling. These frameworks should also
   allow the public, private and voluntary sectors to contribute to flood prevention,
   dam safety and the reduction of the effects of floods to human health, safety,
   property and the aquatic and terrestrial environment.
b) Priority should be given to integrated water management measures for the whole
   catchment area rather than the management of floods in specific locations.

c) The impact of all major human activities concerning flood protection in the catchment area on society should be considered. All activities with the potential to adversely affect human health, ecological, environmental or architectural features should be subjected to an environmental impact assessment. This assessment should not be limited to the immediate area, but should also account for any potential effect on surrounding zones (including cross-border effects).

d) Physical planning, construction, urban and rural development should take into account the requirements of flood prevention and reduction, including the provision of retention areas. The development should be surveyed by monitoring urban settlement in areas that may be affected by floods.

e) These frameworks should account for local problems, needs and knowledge. Local decision making mechanisms and policy should also be considered.

f) An information policy should be developed that covers risk and communication and also facilitates public participation in the decision making process.

The EU is encouraging cooperation and sharing of experiences among international river basins. This is an important measure to ensure flood defence measures on a catchment area basis are extended across regional boundaries and country borders. This should be achieved through cooperation with relevant organisations working in the areas of regional planning, urban development, transport, river control, hydrology and meteorology. Existing flood protection plans should be examined to assess their effectiveness and, where necessary, further development and action plans should be completed within a reasonable timescale. The preparation of risk analysis and flood forecasts at trans-national level is one of the components of this anticipatory approach. This will call for solidarity between upstream users and downstream users across national borders.

15.4.2   The Netherlands

The Netherlands has a long history of flood management. In addition to the four major rivers that flow through the country, two thirds of the land surface is below sea level. The management of water is, therefore, central to the economic and social well-being of the Netherlands.

In 1993 and 1995 there were heavy floods in the Rhine and Meuse rivers due to heavy rainfall in Europe. This prompted a change in policy by the Dutch Government which included a plan to secure the land and to prohibit buildings and private dwellings outside of the protective dyke system. A new policy of "space for rivers" was launched which is founded on the following principles (Frijters et al., 2001):

- Anticipating potential problems in a pro-active approach.
- Tackling water management problems by following a three-step strategy; retaining, storing and draining.

• Allocating more space to water in addition to implementing technological measures.

Around the main rivers (Rhine, Waal and Meuse) more space for the watercourse was to be created to allow the increasing drainage volumes to be handled more effectively. This approach was not restricted by historic nature of flooding in the Netherlands and included scope for accommodating the unpredictability of climate change effects.

Traditionally, a policy of flood resistance i.e. keeping floodwater away from certain areas by way of barriers and defences, had been utilised but its became increasingly apparent that this approach was not sustainable. It was felt that this policy may have even been affecting the natural dynamics of the water system as well as damaging the environment. Alternative solutions were explored and the concept of resilience was introduced in flood risk management. Resilience has been defined as the ability of a system to persist if exposed to an event by recovering after the emergency measures have been initiated. Therefore, resilience is the opposite of resistance, the ability of a system to persist if disturbed, without showing any reaction.

Strategies for flood risk management in which resilience is used focus on reducing the impacts of floods. This traditional strategy of "fighting floods" is in contrast to the innovative approach of "living with floods" (Vis et al., 2003). Therefore, the "resilient" approach to flood risk management is aimed at giving the flood areas the capacity to accumulate and disperse water but with minimal impact. This implies that the consequences of flood also have to be taken into account and that safety mechanisms and legislation must be specified in relation to land use and spatial planning.

In the Netherlands, flood management based on resilience has been viewed as a better policy to that of resistance which has, traditionally, involved increasing defences. Resilience strategies are more flexible and offer more opportunities for nature and landscape development.

15.4.3   Belgium

The main regional Governments of Belgium, Flanders, Wallonia and the Brussels Region have responsibility for all environmental issues, including flooding. This means that three different strategies for flood management have been developed (Schneidergruber et al., 2004).

The Flemish region is one of the most densely populated in Europe and also happens to be located within the Scheldt river basin. This is the most industrialised in Europe and this, in conjunction with its high population density, makes it a high risk area in relation to the impact of flooding. The length of the main river is 355 km, of which 165 km are tidal. Storm tides and heavy rainfall can cause flooding due to erosion and runoff from drained agricultural areas and sealed urban surfaces.

The approach followed by water managers has evolved over recent years and is known as the Sigma-Plan. This plan involves making "room for the river" via the creation of new floodplains which has resulted in some twelve controlled flood areas being developed. These areas cover 550 hectares around the Scheldt and its contributaries. The largest of these has been developed near Antwerp. The concept was to make this polder into a natural area with only a slight tide. By letting a limited volume of water into the polder when the water level is normal, the lowest parts will change into freshwater tidal area. The higher parts of the polder which flood only rarely could be used for grazing of livestock. This method of developing freshwater tidal areas has become rare in western European estuaries, but in this case a large area is being created through the revised Sigma-Plan.

In the Wallonia region, the main river is the Meuse which is fed from the low mountain ranges in Western Europe. This often results in high water peaks in winter but low levels in summer. The Governments of Belgium and the Netherlands have searched for solutions to the problems in areas surrounding the Meuse. Plans to widen the river over a 40 km stretch will increase the safety of some 1400 residents along the Meuse, however this may not be sufficient to address all of the concerns. This has prompted a consortium involving representatives from the Netherlands, Belgium and the Wallum Nature Organisation to initiate a new project to increase the retention capacity of upland areas. The improved functioning of the "natural sponge" upstream, the higher the amount of rainwater that can be retained and the higher the chance that a given rainfall event will end before the sponge becomes saturated.

The Brussels area has had problems with excessive rainwater run-off and these have not been addressed. Some conventional methods of dealing with these problems have been implemented, however a long-term and sustainable approach has not yet been identified. Under the Flemish Environmental Plan, every municipality or province has to work towards sustainable local water management and focus on the sub-basin approach. In this way widespread cooperation between the municipalities is required in the immediate future. If this is not undertaken, current and future water problems may not be resolved effectively and in a sustainable manner.

15.4.4   Germany

At present the prime objective of German water management is the practical implementation of the EC WFD in the ten river basins which are partially or wholly on German territory. The largest river system in this area is the Rhine. This system is one of the largest in Europe and is more than 1000 km long. The Rhine is susceptible to many hydrological events such as flash flooding, fluvial floods and sea influence in the deltas. Many regions will be prone to a combination of these events which makes flood policies on the Rhine important issues (German Ministry of the Environment, 2006).

The International Commission for the Protection of the Rhine (ICPR) has been created in order to ensure international cooperation for the management of the Rhine (Lamothe et al., 2005). As far as policy is concerned, objectives are determined by ICPR and the individual member states must act independently to achieve these.

Following several flood events during the 1990s, ICPR decided to establish an action plan on floods. This document was adopted by the member states in 1998 and it identified four key objectives:

1. reduce damage risks – no increase by 2000, reduce by 10% by the year 2005, and reduce by 25% by the year 2020.
2. reduce flood levels downstream of the regulated Upper Rhine by up to 5 cm by 2000, by up to 30 cm by 2005 and by up to 70 cm by 2020.
3. increase flood awareness – drafting risk maps for 50% of the floodplains and flood prone areas by the year 2000 and 100% by 2005.
4. improve the flood warning system – increase warning lead time by 50% until 2000 and by 100% until 2005.

This action plan set objectives by 2020 (at the latest) in terms of flood policy. It also stated that interim reviews should be made to assess the degree of implementation of these measures. In 2000, the first assessment consisted of information provided by the member states which described measures taken to fulfil the objectives of the plan.

In 2005, a formal evaluation also took place. This was based on the specific approaches, the levels of reduction of the risks and long-term objectives. In comparison to the assessment carried out in 2000, this evaluation looked at the following:

• the scale – instead of national reports, member states were requested to elaborate reports to a smaller scale. To do this, a dedicated working group identified fifteen zones within the Rhine basin. Information on the implementation of the plan was tailored to this scale, which allowed a more precise report on the specific activities to be carried out.

• The methodology – the working group described a common methodology which would be implemented by member states.

This information will be used to implement and manage the flood management in the Rhine basin in the coming years. The plan implemented by ICPR is typical of the cross-border coordination which is essential if Europe's large river systems are to be managed effectively.

15.5   DISCUSSION

It is becoming increasing popular for flood management strategies to be focussed on a risk-based approach. For many years policies have been developed on the basis of flood protection and creating barriers or diversions as a method of flood defence. The sustainability of such an approach may not be suitable as the effects of climate change become more pronounced and Governments become more environmentally aware of land quality and usage. This, coupled with high density population and housing shortage in some areas, has led to a shift in the approach to flood management.

Building higher barriers or erecting greater defences has been overshadowed recently by a new approach which is focussed on managing floods in a more sustainable manner. This includes "making space for water" and creating areas which can be used to accommodate flood water during and after an event. This change in strategy has had a knock-on effect in other areas such as urban planning, construction methods and design and social awareness. Innovative housing building techniques and urban planning have now been the focus of integrated flood management systems that are being developed and implemented across Europe.

The implementation of these strategies across Europe is continuing at various levels and using different methodologies, dependent upon the requirements and issued faced by the different regions. In the UK, DEFRA have developed a holistic approach to flood risk management which includes all forms of flooding. This can be implemented across the UK and is relevant in a variety of situations. The UK operates a system of devolved Government which allows each parliament to determine a number of policies which will be focussed on the needs of the population, the geography of the land and the specific issues in relation to flood management.

The UK Government has been actively improving their policy documents in relation to flood risk. In England and Wales the revised PPS25 document "Development and flood risk" will be introduced in 2006. This document supports the use of flood risk assessment and provides the platform for a more strategic approach to flooding to be developed. The Scottish Parliament has developed a similar strategy which is described in SPP7 "Planning and flooding". These strategies also inform the revisions of the Building Regulations and Standards within the UK. These regulations are enforced at the practice level and are particular to the devolved regional areas. This can often lead to some confusion for specifiers and designers who are planning developments in various areas of the UK. They have to be aware, not only of the differences in Building Regulations and Standards, but also of the subtle differences in the planning policies in relation to flooding.

The approach implemented in the UK could be viewed as segregated due to the devolved Government system. However, the lack of any large river systems on the scale seen in mainland Europe does not often present difficulties in implementing the various policies.

The situation in Europe is, however, markedly different. Large river system and estuaries which can pass through two or more countries are more common and this requires close collaboration in the management of flood risk in these areas. The Rhine, for example, has its beginnings in the Swiss Alps and flows north and east through Austria, Germany, France and the Netherlands. The management of flood risk along the route of this river involves careful planning and consideration of potential knock-on effects downstream.

The Water Framework Directive has been introduced and is viewed as the most substantial piece of EC water legislation to date. It requires all inland and coastal waters to reach "good status" by 2015. It will do this by establishing a river basin district structure within which demanding environmental objectives will be set, including ecological targets for surface waters. After its introduction in 2000, the Directive sets out a timetable for both initial transposition into laws of Member States and thereafter for the implementation of requirements.

The WFD states that the best model for a single system of water management is management by river basin, instead of according to administrative or political boundaries. Initiatives taken forward by the Countries concerned for the Maas, Schelde or Rhine river basins have served as positive examples of this approach, with their cooperation and joint objective-setting across Member State borders, or in the case of the Rhine even beyond the EU territory. While several Member States already take a river basin approach, this is at present not the case everywhere. For each river basin district, some of which will traverse national frontiers – a "river basin management plan" will need to be established and updated every six years, and this will provide the context for the co-ordination requirements described in the WFD.

This integrated approach is key to the long-term, sustainable management of flood risk. In areas where large river systems are present, the objectives of the WFD are essential and should be embraced by the members of the EU. These objectives should, however, not be used in isolation but in combination with regional strategies and methods of flood management at a local level. This is particularly relevant for countries like the Netherlands who have specific land use issues and methods of construction that are suited to their region.

15.6 CONCLUSIONS

This paper has identified the strategies and policies being adopted in the UK and Europe with respect to the management of flood risk. There exists some subtle differences in the approach employed in the UK, where the devolved system of

Government has produced specific policy documents based on regional variations, methods of construction and land use.

This is in contrast to the integrated approach being supported in Europe through the WFD. The large river systems in mainland Europe require specific plans for managing their associated flood risk. This involves cross-border collaboration in many cases, together with regional methods of flood preparedness and management.

REFERENCES

Approved document C (2004). 'Site preparation and resistance to moisture', ODPM, UK.

Crichton, D. (1999). The Risk Triangle, in *Natural Disaster Management*, Ingleton, J. (ed), Tudor Rose, London, England.

Department for Environment and Rural Affairs (DEFRA) (2006). 'Flood and coastal erosion risk management', UK.

Department for Environment Food and the Rural Affairs (DEFRA) (2004) 'Making space for water – Developing a new Government strategy for flood and coastal erosion risk management in England', DEFRA, UK.

Emergency Management Australia (1998). *Floodplain Management in Australia. Best practice Principles and Guidelines.* Commonwealth of Australia, Canberra.

EUrosion: http://www.eurosion.org.

EU Best practice document (2003). 'Best practices on flood prevention, protection and mitigation', EU, Brussels.

Frijters, D. and Leentvaar, J. (2001). 'Participatory planning for flood management in the Netherlands', Ministry of Transport, Public Works and Water Management, The Netherlands.

German Ministry of the Environment, Nature Conservation and Nuclear Safety (2006). 'Water management', internet factsheet.

Gouldby, Ben and Samuels, Paul (2005). 'Language of risk – Project definitions', Floodsite – Integrated flood risk analysis and management methodologies, Floodsite, UK.

King, David, A. – Climate Change Science: Adapt, mitigate or ignore; Science, Vol 303, January 2004.

Lamothe, David-Nicolas, Neveu, Gilles, Gorlach, Benjamin and Interwies, Eduard (2005). 'Evaluation of the impact of floods and associated protection policies', European Commission, Brussels.

Planning and Policy Guidance note 25 (2001). 'Development and flood risk', Office of the Deputy Prime Minister (ODPM).

Planning and Policy Statement 25 (2006). 'Development and flood risk', Office of the Deputy Prime Minister (ODPM).

Garvin, S., Reid, J. and Scott, M. (2005). 'Standard for the repair of buildings following flood', CIRIA, London, UK.

Scneidergruber, M., Cierna, M. and Jones, T. (2004). 'Living with floods: Achieving ecologically sustainable flood management in Europe', WWF European Policy Office, Brussels.

Scottish Executive (2004). 'Scottish Planning Policy 7: Planning and flooding', Scottish Executive, UK.

The Scottish Building Standards (2005). 'Section 3.3: Flooding and ground water', SBSA, UK.

Vis, M., Klijn, F., de Bruijn, K.M. and van Buuren, M. (2003). 'Resilience strategies for flood risk management in the Netherlands', International Journal of River Basin Management, Vol. 1, No. 1, pp. 33–40.

Wood, John and Gendebien, Anne (2005). 'The impacts of coastal flooding, flood mapping and planning', European Action Programme on flood risk management, EU, Brussels.

16

FloReTo – Web Based Advisory Tool for Flood Mitigation Strategies for Existing Buildings

Nataša Manojlović & Erik Pasche
Technical University Hamburg-Harburg, Hamburg, Germany

ABSTRACT: Recent widespread floods due the presumed changes in the climate caused paradigm shift form "flood fighting" to "living with floods" in flood management. The governmental institutions worldwide are redefining their flood policies. In Europe, the EU Flood Directive has been released and brought forward through the national laws. More importance is given to non-structural or resilience measures, where the capacity building of stakeholders is gaining more relevance. It can be achieved by defining the appropriate mitigation measures and by raising awareness among the stakeholders. The communication between experts and the stakeholders is gaining more importance. The mitigation measures to be applied can be divided into two main groups: dry proofing and wet proofing. Combined solutions, such as controlled flooding of the basement and dry proofing of the ground floor, are usually applied.

This paper deals with the problem of measure selection and definition of, so called, technically justifiable solutions (TJS). The selection criteria are given and discussed. Selection of TJS is followed by the damage assessment and the cost benefit analysis (CBA) or multi criteria analysis (MCA) in order to find the optimal solution.

FloReTo, a three-tier web based advisory tool for flood mitigation strategies for existing buildings that manages flood protection on a micro scale level, has been introduced. The optimal mapping function of the business logic on the test data set is the Nested Generalised Exemplas data mining method. The results are discussed and evaluated together with the ideas for further development.

16.1 INTRODUCTION

The growing probability and severity of the recent flood events in Europe and worldwide showed that the existing flood defence structures, primarily dikes and walls, do not guarantee sufficient protection level of people and properties. Due to this increased risk of flooding Europe-wide, the Directive of the European Parliament and of the council on the assessment and management of floods (EC, 2006), has been adopted with amendments by the European Parliament and of the Council on 13 June 2006. This Directive follows the objectives of the Water Framework Directive (WDF), promoting coherent and interdisciplinary cross border, catchment based approach in flood management.

Driven by the EU initiative for integrated flood management, many countries adapted or released new water policies to cope with floods. For example, in Germany those regulations are brought forward within the Flood Control Act (FCA). Instead of fighting floods, the preference is given to strategies of living with floods which cover the measures of integrated flood management that minimize the consequences of flooding. Those strategies are called flood resilience strategies or non-structural flood mitigation (Pasche et al., 2006). This paradigm shift in flood management triggers a demand for more interactivity among the experts of different fields such as spatial planning, social sciences, structural engineering or geology, but also for the active involvement of the public in integrated flood management. The public is not a passive receptor any more, it becomes an important stakeholder of flood management and has to be actively involved in decision making as decisions made influence the way of living in flood prone areas. But also, the citizens in flood prone areas can themselves reduce the flood risk of their properties by protecting their properties in an adequate way.

Capacity building of stakeholders is a resilience strategy devoted to promote the importance of the concept of "living with floods" instead of "flood fighting" on the microscale level, focusing on the individual flood mitigation measures and increasing public awareness of flood hazard so that the people are able to adjust their homes by applying appropriate flood mitigation measures. Further, for successful implementation of adequate mitigation measures, it is necessary that the people in flood prone areas have adequate knowledge on it. Therefore, the communication between the experts and the stakeholders is of high importance.

16.2   OVERVIEW OF THE FLOOD MITIGATION MEASURES AND STRATEGIES FOR EXISTING BUILDINGS

The mitigation measures applied to the existing buildings with the aim to reduce the flood damage potential can be summarised under the term retrofitting. Retrofitting can be defined as "a combination of adjustments or addition to features of existing structures that are intended to eliminate or reduce the possibility of flood damage" (FEMA, 2005). By applying flood-retrofitting measures, the resilience of the buildings can be improved either by preventing floodwater entering the building or by applying waterproof materials and elevating the inventory above the expected flood level. The main resilience strategies for the existing buildings are dry proofing and wet proofing, often combined with using of the inventory in lower parts of the building that is easily removable as shown in Table 16.1.

Table 16.1. Overview of the main flood mitigation (retrofitting) strategies.

Strategy	Measure		Main characteristics	Techniques available
Dryproofing	Sealing		The building is sealed i.e. the external walls are used to hold back the flood water	Water resistant concrete "Weisse Wanne" Polymer bituminous seal "Schwarze Wanne" Protection (sealing) of the openings
	Shielding		The floodwater does not reach the building itself. Barriers are installed at some distance from the building or a group of properties	Flood barriers, flood skirts Installing non-return valves (anti-flooding device) within the private sewage system as a protection from the sewerage backflow
Wetproofing	Controlled flooding of building using water resistant materials		Flood damage potential of a building is reduced by applying water resistant materials. Due to economic or technical reasons, the lower parts of the building (basement) is partly flooded and not used for living. At the same time, the ground floor can be dry-proofed	E.g. water resistant paints and coating, lime based plaster, plasters of synthetic resin, mineral fiber, insulation tiles, oil based paints
	Adapting the occupancy of the building			Raising the inventory, heating tanks and electrical appliances above the expected flood level
	Elevating the inventory		Using easily removable pieces of inventory in lover parts of the building	E.g. removable furniture (e.g. with wheels), rugs rather than fitted carpets

16.2.1   Dry proofing

In the case of dry proofing floodwater is kept out of the building. Those techniques are generally more cost-intensive than the wet proofing ones and always carry a certain failure risk, as the stability of the building can be jeopardised through the increased water pressure. According to Table 16.1, the main dry proofing techniques are sealing where the building itself i.e. its external walls are used to hold back the flood water and shielding where the flood water is kept out of the building by installing flood barriers or flood skirts at some distance from the building or a group of estates (Pasche and Geissler, 2004).

16.2.2   Sealing

By applying the techniques of sealing floodwater does not reach the interior of the building as the external walls and openings are sealed and used to hold back the floodwater. The main techniques are application of water tight concrete, application of polymer bituminous seal, protection of openings and installation of non-return valve to protect the building from the sewerage back flow.

Application of watertight concrete is a technique usually applied to protect new buildings from flooding. In Germany it is considered as the state of the art in protecting the buildings (usually basements) from the floodwater coming through the ground. In the case of existing buildings, this application is more complicated as the concrete slab has to be replaced or covered with a reinforced concrete cap (Pasche and Geissler, 2004). Due to increased water pressure, it might be necessary to enforce the external walls to fulfil the static requirements. In general, this is necessary if the water level exceeds 1m above the lowest part of the building. Although considered as a reliable technique for flood protection, the problems still occur especially in the realisation phase. In order to regulate design and building of waterproof constructions, guidelines for watertight constructions in Germany, (WU-Richtlinie) are realised. Before starting with the design of the construction, it is to be decided which usages class the building should have. Two main classes are distinguished (Zement-Merkblatt, 2006):

Class A: water in liquid form and moisture spots on the surface are not allowed. In order to protect the construction from condensed water, additional measures are to be undertaken such as ventilation, heating or thermal insulation. Class A is a standard for dwelling rooms and storerooms containing valuable goods.

Class B: water in liquid form, moisture spots on the surface are allowed. They can appear in the area of the joints. Class B is a standard for storerooms with less valuable goods and garages.

Additionally, special attention should be paid to the joints, as they are the main cause for the leakages through the construction. Besides, this construction is cost-intensive and thus not always feasible. In many cases, it is necessary to look for other solutions.

Apart from the watertight concrete, a building can be sealed by polymer bituminous seal (in Germany known as "Schwarze Wanne"). For new constructions the sealants are applied form the outside face. For existing buildings, however, the sealants are placed inside, what it is technically demanding and cost intensive. In this case, additional enforcement of the construction is necessary.

In order to manage the implementation of the sealing measures in a better way, many countries include regulations for sealing measures in their building standards (e.g. Germany DIN 18195).

Doors and windows can be closed by temporary barriers. One technique often used is to close doors and windows with the aluminium beams piled up and pressed together to form a stable construction. The lowest beam is sealed to the ground disabling water to percolate through the ingresses between them. For the windows, the similar systems can be applied but often using plastic or metal boards. If the wooden beams are used, oil based coating are to be applied in order to protect the wood from damaging. For windows of the lower parts of the building, special metal coverings are used as depicted in Figure 16.1. Other openings, such as light wells, or ventilation ducts should be closed with watertight caps.

In order to prevent sewerage backwater entering the building, non-return valves are to be installed, that should be placed within the private sewer of a property upstream of the public sewerage system. The device disconnect the property form the public sewerage and the appliances within the property can not be used until the flooding device is removed, otherwise, the property can be flooded by its own waste water. Those devices require careful installation and must be regularly maintained (CIRIA, 2003).

16.2.3 Shielding

By applying shielding measures, flood barriers are installed at some distance from the building or a group of properties. Freestanding barriers or anchored barriers

Figure 16.1. Protection for openings.

Figure 16.2. Flood barriers during flooding and storage of the beams after flooding.

(pillars temporarily fixed at a concrete plate) are usually used. For removable barrier solution, the beams are ending at pillars, which are temporarily fixed at a concrete bearing plate, which surrounds the building. Freestanding barriers are cost effective solutions, which need no bearing plate (Pasche and Geissler, 2004). They resist the floodwater pressure by creating friction between the barrier itself and the ground. On the other hand, this principle can cause hydraulic erosion due to the seepage, and consequently the stability of the barrier can be endangered. Flood barriers should be used up to 1 m height. There is variety of the systems used such as rubber belts, fibre-plastic coats, inclined barriers or flood protection walls (demountable walls) connected to a fixed support construction. An important aspect of shielding are logistic requirements i.e. mounting/dismounting, storage. In some communities, the fire brigades are informed about households that are to be protected by barriers and take the responsibility of mounting the barriers in case the homeowners are absent or not able to do it on their own. They also make sure that the barriers are properly installed. After being used, the barriers should be cleaned and stored. One solution is depicted in Figure 16.2.

In general, dry proofing measures should guarantee "dry homes" during flooding if installed or applied properly. On the other hand, they can be very cost intensive for existing buildings and often need additional stability check or enforcements. In general, the factors considered as unfavourable for the construction stability are high water level, low weight of the construction, construction is close to a water course, shallow foundation. If the application of dry proofing measures is not feasible or cost demanding, other strategies are to be considered.

16.2.4   Wet proofing

By applying wet proofing measures floodwater is not prevented from entering the building, but the building is adapted to flooding in a way that potential damage is reduced by applying water resistant building material or adapting the occupancy of the building (e.g. basement is not used for living). Additionally, the inventory can be raised above the expected flood water level e.g. by placing it on a sufficiently high platform. Which materials are to be applied depends on the building type and the construction parts to be protected. At existing solid buildings walls and ceiling are often made of porous material and thus need to be coated by water resistant materials. At the outside face of the external walls cracks should be repaired and the brickwork coated by a water resistant paint or render. At the inside face of an external wall an appropriate means would be a render out of sand/cement between wall and plaster or polymer bituminous seal (damp proof membrane). Gypsum based plaster or plasterboards, which are filled with gypsum, have to be avoided. All coatings at the inner and outside face of the external walls should be applied to 500 mm above the maximum expected level of flooding in order to prevent soaking up of the walls. As wall coverings, tiles, hydraulic lime coatings, low permeability oil based paints or emulsion should be used. Regarding the insulation, self-draining mineral fibre batts or boards rigid plastic insulation are to be used rather than mineral wool. For the cavity walls the use non-absorbing closed cell insulation is recommended (CIRIA, 2003). In general, any measures to improve water resistance at walls must allow adequate water vapour transmission to avoid trapping moisture within the wall (Pasche and Geissler, 2004).

In case of solid concrete floors, rigid boards with low water absorption should be applied as insulation (below or above the concrete slab). Cracks need to be repaired on the concrete slab. Suspended concrete floors should be protected by applying the polymer bituminous seal such as damp proof membrane between the concrete slab and the screed (CIRIA, 2003). At the joint between the masonry wall and the floor the seals must overlap to acquire effective connections. Additionally a damp proof course has to be installed in the walls and to be connected with the outside and inside seal of the external walls to prevent water seeping up from the ground (Pasche and Geissler, 2004). Solid concrete floors should be covered by ceramic tiles or stone, while for suspended concrete or timber floors cost effective materials (e.g. lino) are to be applied as it has to be removed after flooding to gain access to sub floor void. Chipboards, parquets and fitted carpets should be avoided. An underfloor heating system may be damaged by flooding and should not be applied. Building openings and staircases are normally not affected if they are made of solid timber, PVC, aluminium or stainless steel. In case the light timber boards are used it is necessary to seal all faces of the board using oil-based paint or waterproof stain paint. Services and fitting should be removable or water resistant. For example, electrical installations can be protected by moving the ground floor ring main cables and elevating the socket above the expected flood level to first floor level or by usage of plastic cables. Also, telephone lines and internal control boxes should be elevated above the expected flood level (CIRIA, 2003). For heating systems, the attention should be paid to boilers and tanks. Gas and oil fired boilers and the equipment including pumps should be installed above the max. expected flood level. In case of oil heating, the tanks should be fixed to the masonry, which has to be able to take the buoyancy forces in case the tank gets empty. If it is not possible the tank should be elevated to a higher level or the oil heating system should be replaced by gas driven system. The application of watertight materials is often combined with the adaptation of the occupancy of the building. In the case that floodwater is allowed to enter the basement, it might be used as a storage for non-valuable goods that are also easy to remove in case of emergency. The inventory remained in the room should be elevated above the expected water depth as shown in Figure 16.4. The wall and floor coverings should be made of water resistant materials such as tiles, which after flooding should be cleaned up and dried. If the inventory has to be removed out of the reach of the expected floodwater during the case of emergency, easily removable peaces of inventory are to be chosen (e.g. rugs should be used rather than fitted carpets).

Examples of flood resilient meterials for floor covering and masonry are given in Figure 16.3.

Figure 16.3. Flood resilient materials for floor covering and masonry after flooding (left: Stans, Switzerland, Aug 2005; right: Hamburg, Germany, 2006).

Figure 16.4. Examples of elevation of the inventory above the expected flood level, Stans, Switzerland.

Wet proofing measures are generally used when dry proofing measures are not technically feasible or they are cost intensive. In case of wet proofing, the major costs are for rearrangement of utility items, installation of flood resistant materials, manpower and equipment to relocate items and organization of cleanup after flooding.

16.3   RETROFITTING MEASURES – MODULES OF APPLICATION

The strategies presented above have some shortcomings when applied solely. Most of the deficiencies are related to the limitations of the current construction technologies, design and materials, that are, to lesser or greater extent, subjected to the impact from flood. To design a so-called "resilient house", which would recover fast and with minimal damage after being disposed to flood is one of the research challenges. An overview of the main characteristics of flood mitigation strategies is depicted in Table 16.2.

In order to overcome those deficiencies, combined solutions are applied. One of the commonly used strategies is the combination of controlled flooding and dry proofing techniques. It is suitable for houses with basements. For the lower part of the buildings (basements) the controlled flooding strategy is applied, while the upper potentially affected parts of the building are dry proofed. The basement is filled with floodwater through a pipe. The valve, a pump in the sump and pressure sensors regulate and control the water level within the building that the water pressure is kept below the critical load of the external walls and ground floor.

Table 16.2. Characteristics of the main retrofitting strategies.

	Dryproofing	Wetproofing
Advantages	(1) The homes are kept dry (2) The occupancy of the building is not affected	(1) Cost effective solution (2) As people face floodwater in their homes, risk awareness is kept alive
Deficiencies	(1) Cost intensive solution (2) Stability of the building can be jeopardized (3) Logistic requirements	(1) Materials used are not fully resilient (2) Limited occupancy of the building parts reached by the floodwater
	Failure due to insufficient communication between the homeowners and the experts (engineers and architects) and Limitation of the existing technology	

The targeted water level in basement should not reach the level of the electrical appliances or power sockets, as it can cause power black out and further damage. Waterproof materials are applied in the basement, in order to minimise the damage to building fabric.

To assess the targeted water level, it is necessary to analyse the masonry of the building, the elevation of the basement level compared to the design flood level, pathways of flooding (groundwater or flood water is also coming through openings).

Controlled flooding of the basement is recommendable if the difference between the basement level and the floodwater is more than 1.5 m. For flood water level greater than 1.5 m in most cases the stability of a building cannot be guaranteed due to strong buoyancy forces.

This concept is considered as a good alternative to total dry proofing or total flooding solution, with the main advantages being:

1. It is a cost effective solution comparing to total dry proofing strategies:
Using this combined solution, the basements are not dry proofed applying cost intensive dry proofing strategies such as waterproof concrete or polymer bituminous seal that considerably reduces overall costs.

2. Stability of the building:
The water is allowed to enter the building (lower part of the building), which reduces buoyancy and lateral pressure to floor and walls. In this way, the stability of the building is improved. The water can enter through gaps and cracks in walls and floors or gaps between openings or fittings and the masonry.

3. Flood awareness is kept "alive" as people experience "wet homes":
This concept has also some shortcomings:
1. Logistical aspects:
    – Additional effort for operating the equipment is necessary (critical in case of vulnerable groups such as old people, disabled people, etc.).
    – This strategy strongly depends on the electricity used for pumps and it is necessary to provide constant power supply during the flood event.
2. Basement can not be used as a storage of valuable goods or inventory or as a living area.
    Basements of the buildings that are protected by controlled flooding strategy are to be used as a temporary storage and in case of flooding the inventory has to be brought to upper floors. Therefore, total damage strongly depends on the efficiency of the flood warning systems (warning time has to be sufficient for removal of the inventory from the lower parts of the building).
3. If the flood level is higher than the terrain, it is difficult to reach the building.
4. Damage can still happen during as well as after the flood event.

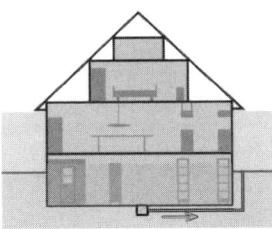

Advantages:
– const effective solution as
  there are no additional
  expenses for measures
– flood awareness is kept
  active as people experience
  "wet homes"
– stability of the house is
  improved in comparison with
  the dry proofing solution

Disadvantages:
– Additional effort for operating
  the equipment is necessary
  (critical in case of vulnerable
  groups such as old people,
  disabled, etc.)
– Basement cannot be used
  as a store room or living area
– If the flood level is higher than
  the terrain, it is difficult to reach
  the building
– Damage can still happen

Figure 16.5. Combination of controlled flooding and dryproofing strategy with pros & cons.

Although the basement should be wet proofed, damage to fabric and inventory can still occur. The moisture can remain trapped in the walls, which can cause damage to building fabric and can through capillary.

The main features of this combined solution are depicted in Figure 16.5.

In case that this concept of controlled flooding of basements is applied for a city quarter or a number of houses, it can reduce the amount of floodwater as in that case the basements are considered as controlled polders and are flooded to reduce the pressure of floodwater to the fabric and infrastructure. The condition of foundations is to be assessed. Old buildings such as timber-framed constructions are especially vulnerable. In some cases it is necessary to apply geotechnical measures.

16.4   DEFINING APPROPRIATE MITIGATION STRATEGY – CRITERIA ANALYSIS

Experience has shown that combined solutions, such as the combination of controlled flooding and dry proofing techniques, performed well if were applied and maintained properly. Nevertheless, defining appropriate mitigation strategy is a complex task and can hardly be generalised. Before any decision is made, a pre-selection of technically justifiable solutions is to be performed. Technically justifiable solutions are applicable measures or combination of measures form the technical point of view for the given conditions. The main criteria for selection of technically justifiable solutions can be summarised as following:

1. Location parameters refer to the following main parameter groups:
    - geo data and terrain configuration
    - source of flooding (e.g. water course, lake)
    - proximity to the source of flooding
    - geologic profile of the examined area (including the stability of the soil after flooding).

2. Pathways of flooding describe the way the flood water can enter the building. In the case of a flood, the water can enter the building either through the rising of groundwater, or by flowing into the building from the surface.
   - Above ground
     - openings in the building, such as windows and doors
     - floors and walls (permeable brickwork)
     - gaps and cracks in the brickwork and around openings.
   - Below ground
     - seepage through solid ground floors where the seal between the floor and the walls is not sufficient
     - seepage through the below ground structure where the seal between the masonry and the fittings is not sufficient
     - seepage underground into the unprotected basements
     - seepage at gaps in the below ground building structure
     - backflow through sewer system (Environment Agency).
3. Parameters describing the property (estate), composing of the building(s) description and the user description.
   - Building description
     - Type of building (single house, duplex, multi storey building)
     - Purpose of the building (e.g. dwelling house, adjoining building)
     - Type of Foundation
     - Period (Year) of construction
     - Walls (Wall type, Outside face, Inside face, Covering)
     - Floors (Floor type, Floor covering)
     - Ceiling (Ceiling type, Ceiling covering)
     - Staircases (Type, Covering)
     - Openings (Windows, Doors)
     - Services and Fittings (Electro, Heating system, Sewerage system)
     - Inventory (Movable assets, Fixed Assets).
   - User Description refer to parameters describing the user profile as well as his flood awareness the main ones being:
     - User type (private, trade and commerce, public)
     - User profile (affiliation, age, vulnerability level)
     - User status (Owner, Lodger)
     - User's experience with flooding
     - Existing (already applied) mitigation measures
     - Insurance
     - Commonly used flood information sources.
   - Hydrological conditions (Physical, Chemical, Biological)
     - Physical parameters describe hydrostatic and hydrodynamic condition of the flood water that are water depth, velocity, duration of flooding, sediment transport, ice transport, temperature

- Chemical parameters refer to organic and inorganic contamination of the flood water such as salt, bases or acids concentration, organic (oil) contamination
- Biological parameters describe the presence of the micro-organisms or animals in flood water that can cause further damage to properties.

5. Logistical parameters and the availability of resources describe available manpower, storage capacity or possible escape routes in case of flooding. This aspect is especially important in case of dry proofing measures such as shielding of a group of properties using mobile elements that have to be installed before and stored after the flood event.

Before any detailed analysis for single properties is performed, it can be decided that an integrative concept for the project area is to be applied. In this case, for the single properties the measures should be applied that are consistent with the proposed concept. The possible technically justifiable solutions for single properties in this case are selected using the criteria listed above with the boundary conditions defined by the integrative concept.

This analysis can also be done for the group of properties or city quarters, trying to identify in order to define optimal technical solution for the examined area.

16.4.1   From technically justifiable solutions to adopted solution

Once the technically justifiable solutions have been selected, the next step is to assess the benefit of applied solutions, by assessing the damage potential.

Damage potential expresses the value of possible loss of the property in the event of a flood.

Damage matrix comprises the laboratory results and experience from the past flood events on building elements behaviour when being exposed to flooding. The main flood parameter considered for the analysis of the building materials are described in chapter Defining appropriate mitigation strategy – Criteria Analysis, section Hydrological conditions. They can be summarised under the three main groups of factors: physical, chemical and biological. Further, it should be distinguished between primary and secondary effects of hydrological conditions on different building elements.

Primary damage is the one that is caused directly by floodwater such as creation of cracks and gaps due to dynamic pressure to walls.

As secondary damage, the consequences caused by too quick pumping out of water from a basement can be considered.

For description of those dependences both, quantitative as well as qualitative analysis are to be applied. Those single dependencies are defined in impairment functions.

Impairment function represents a relation between the potential damage to an item of an estate (simple or complex) in dependence on the relevant flood parameter. As simple item a piece of inventory can be considered while a complex item is, for example, a floor construction.

At the moment, the impairment functions describing dependences between flood depth and to a certain extent the flood duration are being applied. Definition of all dependences including the functions of the combination of different flood parameters is still a research challenge.

As the main objective of the mitigation measures is to mitigate the flood damage, one of the aspects of their efficiency is the improved damage potential after the measure has been applied.

Improved damage potential is the value of the potential damage resulting from the total estate value and impairment functions after a technically justifiable solution has been applied.

Damage assessment can be performed for the design flood level (e.g. HQ100) or expressed in terms of annual values and referred to as annual damage potential.

Annual damage potential can be calculated as following (DVWK, 2000):

$$\overline{D} = \int_{P_0}^{P_{max}} D(P)dP$$

whereby:
$\overline{D}$ – annual damage potential
$P(x)$ – probability of flooding
P_0 – probability of the lowest flood event that causes damage
P_{max} – probability of the highest considered flood event.

Annual damage potential is the input for the cost benefit analysis.

16.4.2   Cost Benefit Analysis (CBA)

Which (technically justifiable) solution is to be selected depends on both, monetary and non-monetary criteria. The selected technically justifiable measures, considered as different scenarios, are processed and evaluated to yield the cost benefit analysis (CBA).

The monetary criteria are comprised within the Cost Benefit Analysis (CBA). It relates the difference between the improved and initial damage potential (benefit) to the costs of selected measure(s).

Costs include all expenses for applying the selected measure (material, labour and maintenance costs) discounted to the present values.

Benefit is defined improvement of the annual damage potential after applying the selected measure (combination of measures).

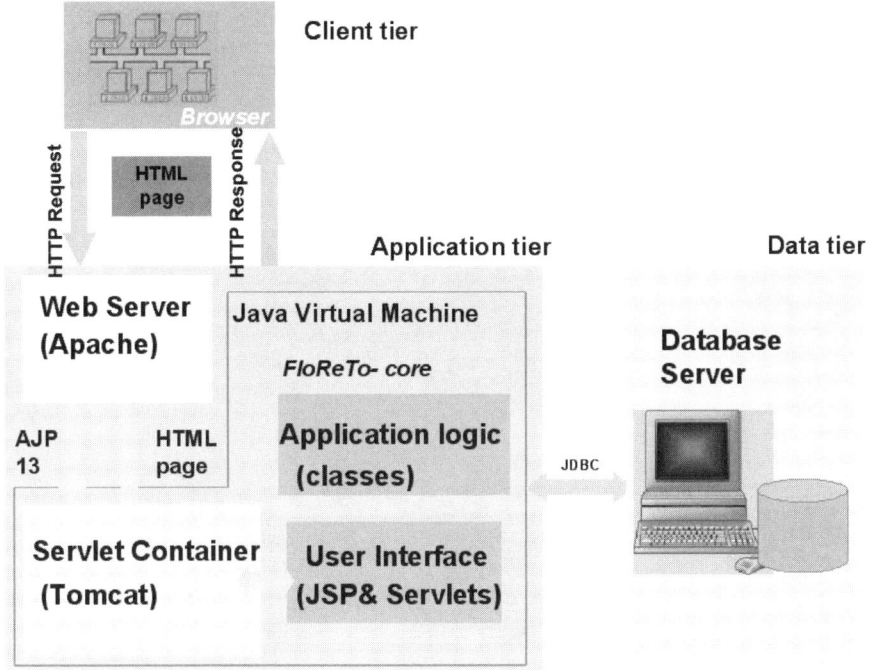

Figure 16.6.   Controlled flooding with pros & cons.

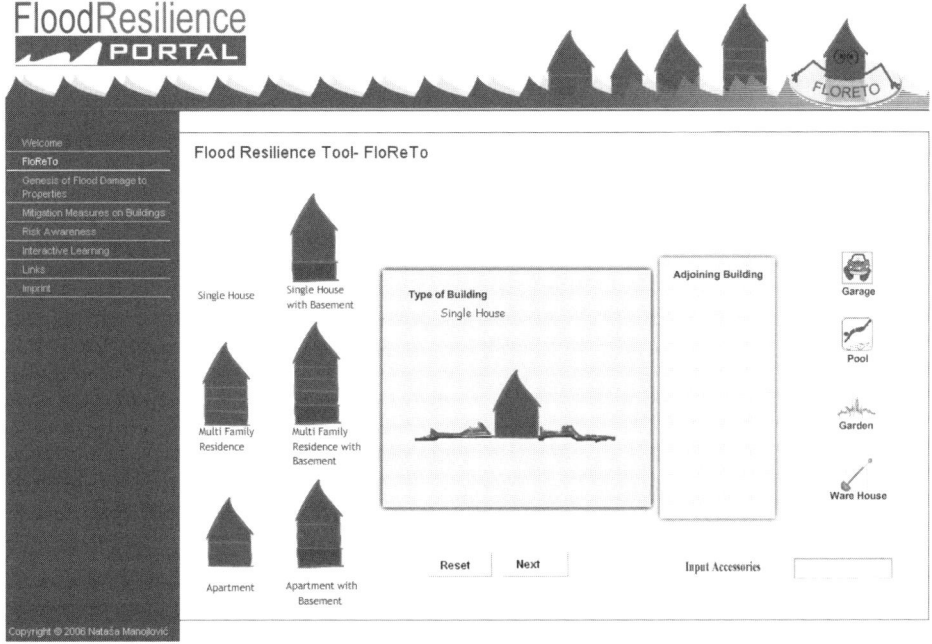

Figure 16.7.   An example of the FloReTo input module.

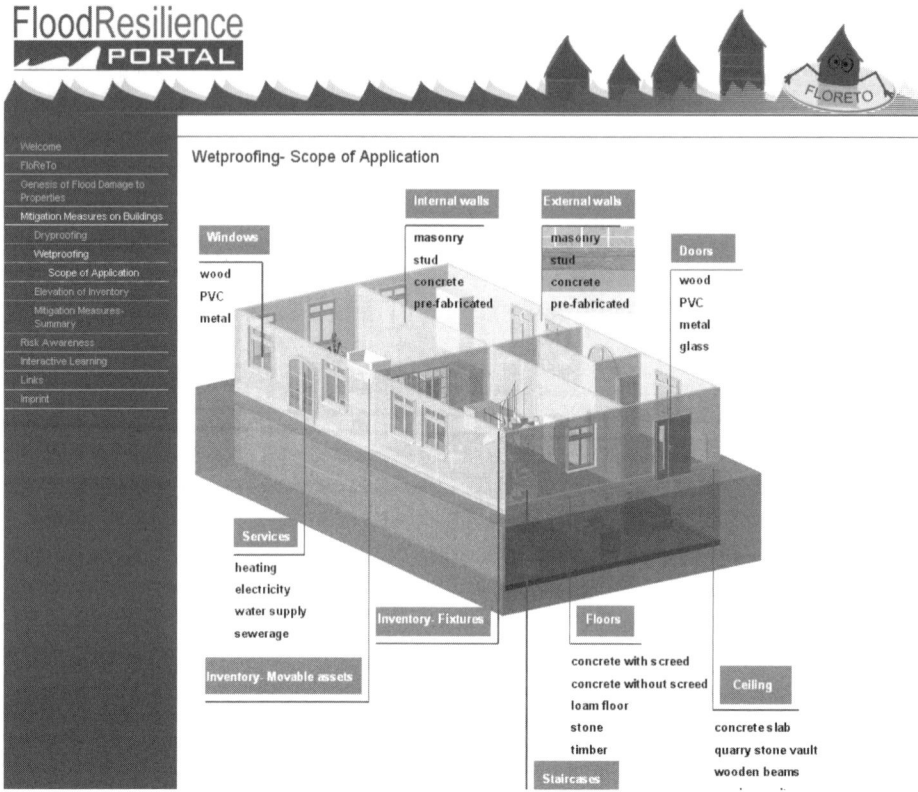

Figure 16.8. An example of the FloReTo educational module.

16.4.3   Multi Criteria Analysis (MCA)

In case that the decision on mitigation measures depends on both, monetary and non-monetary criteria, the Multi Criteria Analysis (MCA) is applied. MCA is a decision-making tool developed for complex problems including qualitative and/or quantitative aspects of the problem in the decision-making process. It "describes any structured approach used to determine overall preferences among alternative options, where the options accomplish several objectives." (http://unfccc.int/). In MCA, the objectives are specified and corresponding attributes or indicators are identified. As key output a single most preferred option, ranked options, short list of options for further appraisal, or characterization of acceptable or unacceptable possibilities is obtained (http://unfccc.int/).

On the microscale level i.e. in the case of individual protection, the main objectives can be summarised as technical, economic and social as depicted in Figure 16.9. By making a decision on which measure is the most convenient for his own case, the user confronts different criteria and creates scenarios that have to be

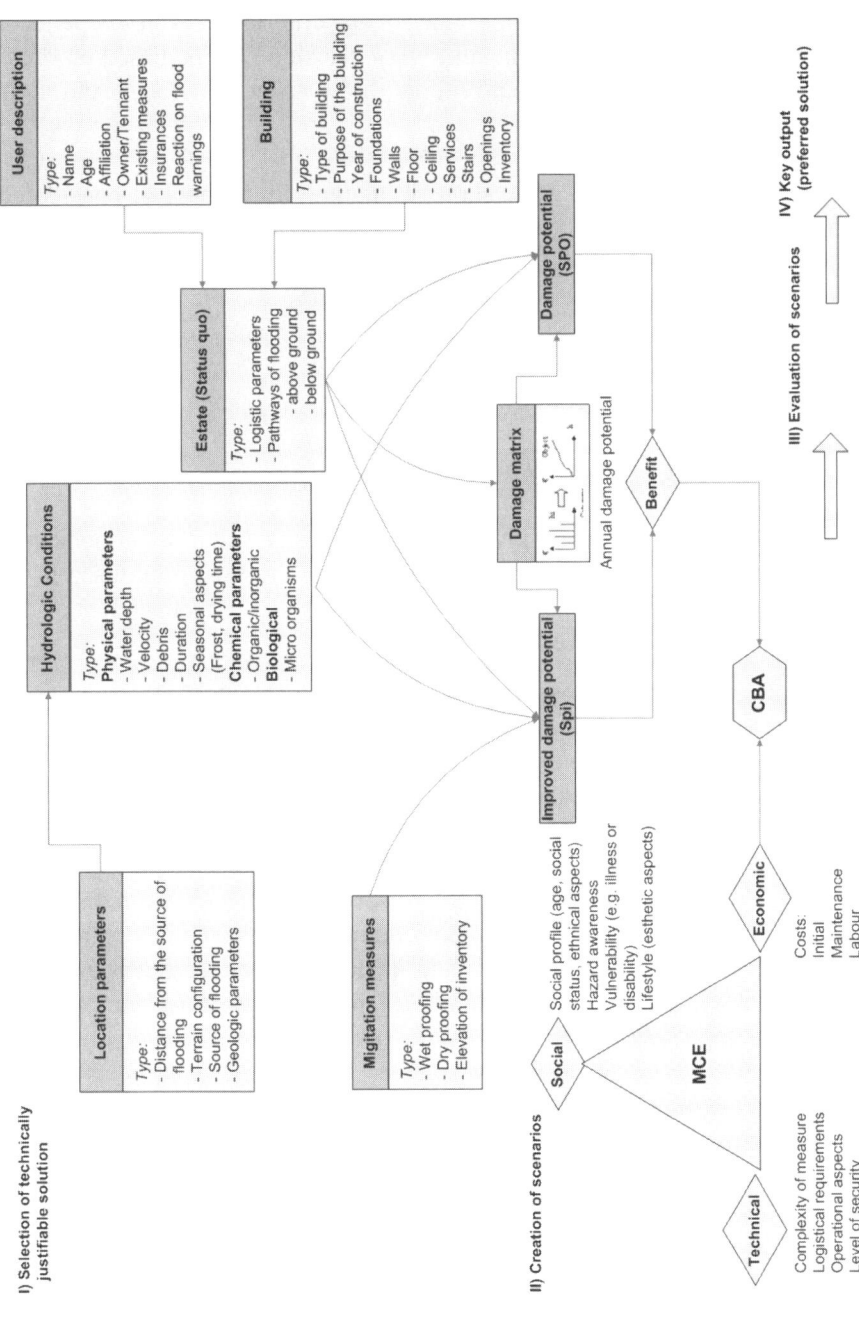

Figure 16.9. Decision making on microscale level as a basis for the FloReTo structure.

processed applying MCA tools. Each criterion can be assigned a specific weight that shows its importance relative to other criteria under consideration. The weight value is not only dependent on the importance of any criterion; it is also dependent on the possible range of the criterion values. For decision making based on multi criteria value assignment for different criteria, different methodologies and tools are used, the main ones being:

- Ranking and Rating
  Each criterion is assigned a value showing its rank in the pool of considered criteria (e.g. 1–100).
- Analytic Hierarchy Process (AHP) and pair wise comparison
  Decision making process is performed by arranging the important components of a problem into a hierarchical structure. This method reduces complex decisions into a series of simple comparisons, called Pair wise Comparisons, between elements of the decision hierarchy. and the score at each level of the hierarchy can be calculated as a weighted sum of the lower level scores (http://www.cifor.cgiar.org/acm/methods/toolbox9.html#top).
- Goal programming
  Each of conflicting criteria is given a goal or target value to be achieved. (Unwanted deviations from this set of target values are then minimised in an achievement function. This can be a vector or a weighted sum dependent on the goal-programming variant used.)

More detailed description of those tools is given in attachment.

As microscale approach i.e. the individual solutions are a part of city or regional planning, it is the question how to integrate those solution in the medium/macro scale approach. Further, the scalability of the solutions i.e. to which extent it is reasonable to perform the protection on the microscale level rather than protecting groups of properties or even a whole city quarter, is to be discussed. In order to make sound solutions, both bottom-up and up-bottom approach are required. It would start with the improvement of the communication between the experts and the public, especially with the people living in flood prone areas with the main objective to build capacity of the communities (spatial, planners, experts and the stakeholders) to cope with flooding.

16.4.4   Existing resources on integrative flood mitigation strategies on microscale level

Having recognised this importance of integrated approach in flood management, many environmental agencies or governmental institutions offer consultancy or publish material to inform citizens about possible mitigation measures for their homes. Taking advantage of the Internet, a considerable part of provided information is available via Internet (a list of examined web sites is given in http://tuhh.de/wb/cost2006). In Germany, for example, a great number of pages is offered by the communities. Those web sites usually provide the citizens with the information whom to contact or who carries the responsibility for certain flood related issues. In some cases they offer forums, such as a forum about oil heating where the citizens can get information about approved manufacturer or actual laws regarding oil tanks and corresponding equipment.

Those sites are usually region (area) specific and partially cover the aspects of integrative flood management on microscale level.

An integrative capacity building system that would involve expertise on appropriate mitigation measures for each single case on the microscale level, offered that at the same time maximise the interaction between the public and the experts is not available.

16.5 FloReTo – WEB-BASED FLOOD RETROFITTING TOOL

FloReTo is web-based advisory system on mitigation measures on microscale level that can be used for capacity building of stakeholders that enables tailored approach for the user's own property data. The system at the same time:

- enables determination of suitable retrofitting measure(s) which improve flood resilience of single households and consequently of the whole area. As a final output of the system, the user gets different flood retrofitting scenarios for his own property, based on the cost benefit analysis (CBA).
- It used for capacity building of stakeholders, it can improve flood hazard awareness of the stakeholders by providing them with the information related to flood management and helps them understanding the main terms and concepts relevant for flood protection and mitigation. Contained information covers the range from basic definitions, e.g. HQ100, up to detailed description of measures and materials used.
- facilitates decision making on the microscale level and enables continuous exchange of information between the experts and the stakeholders by giving the stakeholders enough information to make their decisions and at the same time retrieving the feed back of the preferable solutions and use it as an input for the spatial planning and further flood management decision making process.

FloReTo targets the public (especially the population potentially affected by flood), the experts, decision-makers e.g. local authorities and other parties involved in flood protection on microscale levels (e.g. construction companies).

Development of FloReTo was initiated within the EU funded FLOWS project, that focuses on finding innovative flood management solutions for an adequate response of the communities to the climate change, considering both, technical and social aspects. The system has a client/server architecture consisting of three well-defined and separate processes (three-tier design), each running on a different platform, as depicted in Figure 16.3 (link: http://floreto.wb.tu-harburg.de available from February 2007, contact address: natasa.manojlovic@tuhh.de).

The three-tier design has many advantages, the chief one being the modularity of such a system, enabling modification and replacement of one tier without affecting the other ones. The tiers discuss in more detail are as follows:

1. User interface (UI), which runs on the user's computer with the browser (the client).

    As the FloReTo platform targets different user groups, the main ones being the experts and the public, two main views regarding the user interface are offered: expert view and public user or "stakeholder" view.

    Expert view is meant for the people with substantial experience in urban flood management.

    The main tasks this user group is to perform are:

    (I) Input data gathered during the "on site" data collection process

    (II) Evaluate data entered by the public user, that can further be used for development of flood management concepts on the medium scale level.

    The main idea by the design of the user interface of the stakeholder view was to provide the users with an easy to handle intuitive tool, using graphical elements and functionalities such as drag & drop or resizing the objects for guiding them through the process of retrofitting measures selection that are:

    (I) description of the property (preprocessing)

    (II) selection of technically justifiable solutions (processing)

    (III) performing the cost benefit analysis (processing)

    (IV) evaluation of scenarios (postprocessing).

    The technologies used that support user-friendly functionalities are Flash and AJAX (Java Script). Data input by the users are stored in the database and are required input for the analysis performed on the application server.

    Examples of the GUI are depicted in Figures 16.7 and 16.8.

2. The functional module (application server) is the part of the system that actually processes data. For the purpose of FloReTo two main business logic modules (strategies for measure selection) are defined:

    (I) The user can create his own solution for home protection. In this case, a validation module is activated dismissing technically unjustifiable solutions or illogical combination of measures.

    Such system is designed as rule-based system that represents an expert system in a classical sense.

    Expert system is "an intelligent computer program that uses knowledge and inference procedures to solve problems that are difficult enough to require significant human expertise for their solution" (Giarratano and Joseph, 2005). Rule based system represent knowledge in terms of multiple rules that specify what should or should not be concluded in different situations (Giarratano Joseph, 2005).

Although this approach offers an appropriate and applicable mapping of the matching function to the basic components of an expert system some major shortcomings limit the full realization of FloReTo's matching function as an expert system.

These shortcomings fall into two broad groups that are The Knowledge Acquisition Bottleneck and The Robust learning limitation Bottleneck (Owotoki et al., 2006).

In order to overcome those deficiencies, the data mining approach has been introduced.

Due to the modularity of the system it is possible to switch between those two approaches.

(II) The system recommends the measures to the user. In this case, the "intelligent measures estimator" is activated based on the data mining algorithms.

The disadvantages of the expert system in a classical sense can be mitigated by using computational intelligence (CI) models in a data mining process to define the optimal mapping function for matching the input parameters describing the design criteria as given in chapter Defining appropriate mitigation strategy – Criteria Analysis, to flood mitigation measures that can be applied to protect the estate from flooding (technically justifiable solutions).

The input parameters consist of vectors of categorical and/or numerical attributes, which constitute the design criteria described in the chapter Defining Appropriate Mitigation Strategy – Criteria Analysis. For the test version, the parameters describing the property and the flood water parameters (water depth) of the specific location are considered. (The input parameters are further represented as X with the complete set of all possible design criteria represented as X^{∞}, and the cardinality $|X|$ being the number of attributes of the input parameters.) (Owotoki et al., 2006).

The methods that have been used for testing are: a multilayer perceptron neural network with back propagation learning (MLP), the nested generalized exemplar method (NGE), Bayesian or belief networks and an implementation of the C4.5 decision tree.

For the model testing WEKA, an open source platform for machine learning has been used (Ian H. Witten and Eibe Frank, 2005). Two approaches for training and validation were used with each of the methods on both data sets – the split and cross validation methods of validation.

The initial dataset, originating form the City of Kellinghusen data collection, has been used for the testing of the four data mining methods. This dataset contained 210 examples with 12 non-classes and 1 class attribute with 23 distinct values. The 12 non-classes attributes consist of the selected

Table 16.3. Results of the model testing (Owotoki et al., 2006).

Model	Predictive accuracy			
	Split			10-fold CV
	60%	70%	Average	
NGE	77.38	71.42	74.4	82.86
MLP	72.62	65.08	68.85	79.05
BayesNet	69.05	58.73	63.89	74.28
J48	76.19	77.78	76.98	82.86

criteria for technically justifiable solutions regarding the building description and flood parameters. The results are given in Table 16.3. The class attribute describes the mitigation measures to be applied for different combinations of non class parameters. For the given data set, the best results are obtained by applying the NGE with the 10 fold cross validation (82.86%). In the NGE, the induced concepts take the form of hyperrectangles in an n-dimensional Euclidean space. The axes of the space are defined by the attributes used for describing the examples. New data points are classified by computing their distance to the nearest "generalized exemplar" (i.e. either a point or an axis-parallel rectangle) (Salzberg, 1991).

The results obtained although satisfactory for an initial deployment of the matching function of FloReTo, still leave room for improvement. As improvement potential, the collection of more and high quality data, refinement of the criteria for measure selection (attributes) as well as refinement of the measures to be more specific is considered (Owotoki et al., 2006).

3. A database management system (DBMS) that stores the data required by the middle tier. This tier runs on a second server called the database server. The database used for FloReTo design is an open source database system – MySQL. The database is optimised for different input modules, for the public user and for the experts (data as results from polls, interviews, etc.).

16.6   OUTLOOK AND DISCUSSION

The system as such presents a novel approach to the decision making on the microscale level as it enables the citizens to get the solutions tailored to their own situation at the same time gaining knowledge about the basic concepts of integrated flood management. On the other side, they populate the database with the real data, which helps the spatial planners and water managers for better design and implementation of the policies within integrated flood management. In order to get the preliminary feed back, FloReTo was presented to a group of students of the Hamburg University of Technology, which resulted in a thorough, and helpful feed back. Related to the content of the educational part it was assessed as a very useful concept that offers understandable and condensed theoretical background for each of introduced topic. Also, very positive feedback was given to the logical structure. Still, the feed back on real estates is missing and further research is needed.

As the system integrates extensive knowledge on several disciplines such as damage assessment and material science, flood mitigation and construction and data mining and machine learning science, it relies on the results and findings of all included disciplines. Regarding further development of the system the following main aspects are considered:

1. Novel research results in the field of constructive mitigation and material science. As presented in this paper, the measures used have some deficiencies when applied singly. One of the objectives of this system is to find the most appropriate solution for given input data. For reliable outcomes, the measures used for defining algorithms are to be described in detail identifying prerequisites for their best performance.

   An exhaustive database that would contain detailed explanations of measures reinforced with practical experience and hints for successful application in a form of best practices would bring a benefit to the existing knowledge.

   Those novel findings and improvements regarding the existing measures can be integrated into the existing algorithms.

2. Improvement of data mining algorithms: In order to have optimal working performance of data mining algorithms, a certain "learning time" of the machine is required. The availability of flood damage data is rather low which makes this phase longer and more difficult. Further development of this work can go in two directions. The general one, should also involve improvement of the existing business logic to "general cases" i.e. to data acquainted form any public user. It involves refinement of the existing attributes where the dominant influence is taken by the building description criteria.

   The second, more specific approach should focus on adapting the business logic of FloReTo to cover other locations by similarly collecting data and mining it and considering area specific criteria, together with the general ones. Further, the transparency in the data mining algorithms is to be created, that would improve the trust between the user and the machine.

3. Final implementation of damage potential assessment and cost benefit analysis (CBA).

4. Integration of MCA modules: Finally, the existing business logic is to be enhanced with the multi criteria analysis algorithms.

5. Involvement of more relevant parties

   An interested party for such a system is also seen in private construction companies. They would participate by sharing their experience and knowledge on retrofitting measures and in return they can post their contact and expertise on the FloReTo platform, so that interested stakeholders can contact them and ask for help. In this case, FloReTo might be a kind of collaborative platform among experts, decision makers, practitioners and stakeholders.

Due to its modularity, user oriented nature and accessibility via internet, FloReTo offers as a good basis for developing and maintaining systems for capacity building and interaction between the stakeholders and the experts, that can be a valuable contribution to the integrative flood management decision making.

REFERENCES

Pasche E., Kraus D., and Manojlovic, Kalypso Inform – A Web-Based Strategy for Integrated Flood Management, Proceedings of the Hydroinformatic Conference 2006, Nice, Vol. 4, pp

Pasche E., Geissler T. R., New Strategies of Damage Reduction in Urban Areas Prone to Flood, Proceedings in: Urban Flood Management; Editor: A. Szöllösi-Nagy, C. Zevenbergen, 2004

FEMA, Engineering Principles and Practices of Retrofitting Flood Prone residential Structures, 2005

CIRIA, Improving the Flood Resistance of Your Home, CIRIA, 2003

Environment Agency, Flood products – A guide for homeowners

Ministerium für Umwelt, Raumordnung und Landwirtschaft des Landes Nortrhein-Westfalen (MURL), Hochwasserfiebel, MURL, 1999

Bundesamt für Bauwesen und Raumordnung, Bauliche Schutz- und Vorsorgemaßnahmen in hochwassergefährdeten Gebieten, 2004

www.flows.nu

http://tuhh.de/wb/cost2006

Giarratano, Joseph C., "Expert Systems" – Principles & Programming, Thompson Course Technology, 2005

Owotoki P., Mayer-Lindenberg F., "Degree of Confidence Measure for Integrating Non Monolithic Intelligence Models", in Proc. of the 3rd International Conference on Computational Intelligence, Robotics and Autonomous Systems (CIRAS), Singapore, Dec. 2005

Salzberg S. Machine Learning, Springer Netherlands, Volume 6, Number 3, May, 1991

http://unfccc.int/

http://www.defra.gov.uk/environment/airquality/mcda

http://www.cifor.cgiar.org/acm/methods/toolbox9.html#top

http://tuhh.de/wb/mca

http://en.wikipedia.org/

Zement-Merkblatt Hochbau H10, Wassertundurchlässige Baubetonwerke, 2006

Ian H. Witten and Eibe Frank, "Data Mining: Practical machine learning tools and techniques", 2nd Edition, Morgan Kaufmann, San Francisco, 2005

17

New Approaches to Flood Risk Management – Implications for Capacity-Building

Joanne Tippett & Emma J. Griffiths
Centre for Urban and Regional Ecology, School of Environment and Development, University of Manchester, Manchester, UK

ABSTRACT: New approaches to flood risk management will require new behaviours from a wide set of stakeholders and practitioners. Wide scale participation in planning and decision making can help achieve these new behaviours. There is a lack of practitioners with the skills to facilitate such participatory and integrated planning. An understanding of capacity building in relation to urban flood management is developed here. Little empirical research has been conducted on the effectiveness of capacity building in the skills required to deliver the meaningful participation and integrated planning called for in recent shifts in policy, as in the European Union Water Framework Directive.

This chapter develops a conceptual framework for new methods in capacity building using systems thinking principles. These were tested in action research in England. The research highlighted several key themes for effective capacity building:
- Create an action learning approach
- Actively encourage reflective practice
- Develop peer networks and mentoring opportunities
- Create supportive conditions for learning
- Develop learning from participants' own understandings
- Incorporate a diversity of approaches and perspectives
- Encourage a holistic view and working across scales
- Incorporate learning about systems thinking and sustainability

There is a clear need to develop new approaches to capacity building if integrated solutions for flood risk management are to be realised and implemented. Further challenges for capacity building are elaborated, including: developing effective mechanisms for sharing experience and pooling knowledge amongst practitioners and finding ways to encourage an open, yet critically reflective, approach to learning.

17.1   INTRODUCTION

"The skills, aptitudes and attitudes necessary to industrialise the earth, however, are not necessarily the same as those that will be needed to heal the earth or to build durable economies and good communities. Resolution of the great ecological challenges of the next century will require us to reconsider the substance, process, and purpose of education at all levels" (Orr 1994).

An international survey of experience in implementing Local Agenda 21 following the Rio Earth Summit found that "Water resource management is the common priority issue for municipalities in all world regions and regardless of economic situation" (International Council for Local Environmental Initiatives 2002). Serious problems arise from every aspect of society's current interaction with water, problems which span environmental, social and economic spheres. Water flow provides a framework for thinking about environmental impacts throughout the landscape and atmosphere, as well as across political boundaries (Flournoy 1995).

   Major flooding incidents in urban areas in the last decade have focused attention on the need for new approaches to urban water management (Howe and White 2004). Increased awareness of possible climatic variability, with increased storminess and urban run off has highlighted the need for adaptive approaches to climate change in the urban environment (Evans et al. 2004).

   New approaches to flood risk management will require new behaviours from a wide set of stakeholders and practitioners. New ways of working, considering geo-ecological river management have made great strides. We are developing a better awareness of ways of managing the water cycle, but there is an implementation gap, even for relatively small scale solutions, such as Sustainable Drainage Systems (Best Management Practices, BMPs) on new developments (White and Howe 2005). In this context, "changing people's thinking is becoming at least as important as gaining new scientific knowledge" (Douglas 2000).

   Participation in decision making and developing a shared understanding of problems and options may increase the likelihood of changing behaviour with regards to flood mitigation and developing flood resilient solutions (Allen, Kilvington and Horn 2002). Two different types of stakeholders can gain value from participation: communities of place, or people who live and work in an area; and communities of interest, people who have a connection through their shared interests and concerns. Through participation these two types of community are able to learn about changes happening in their areas and in the policies that will affect them, and are able to learn from other stakeholders. If the engagement is meaningful, they gain a sense that their ideas and input make a difference, which in turn helps build the confidence to act.

   Reflecting a global shift in awareness of the value of community and stakeholder participation in planning, several significant changes in policy have recently been enacted. The *Convention on Access to Information, Public Participation in Decision-Making and Access to Justice in Environmental Matters* entered into force in October 2001, and now has 40 signatories (UN ECE 1998). Recent environmental legislation, including the *European Union Water Framework Directive (WFD)*, calls for more effective stakeholder and public participation in water management.

The European Sustainable Cities & Towns Campaign (2003) recognises that the "lack of integration at many levels, [including a] lack of common sustainability vision and interest" is a key challenge for sustainable urban management. In order to achieve sustainable water management, water systems need to be treated as holistic entities, incorporating supply (or input), use, and disposal (or output). Sustainable management of water requires integrated planning, recognising the interconnections between systems operating at different levels of scale, and the dynamic nature of interactions in a complex environment.

Thus, new approaches to flood risk management will require both enhanced participation of all stakeholders in developing options and managing resources, and integrated approaches. The limited implementation of the many advances in flood risk management over the last twenty years points to the need for institutional changes, "underpinned with capacity-building interventions targeted at enabling a learning culture that values integration and participatory decision making" (Brown 2005). There is a lack of practitioners skilled in new approaches to participation and integrated planning (e.g. Commission for Architecture and the Built Environment 2003; Office of the Deputy Prime Minister 2004). Addressing this lack will require a new approach to capacity building.

Capacity building is defined as "strengthening people's capacity to determine their own values and priorities, and to organize themselves to act on these" (Gensamo 2002). It consists of developing awareness, knowledge, skills and operational capability, so that individuals and groups can achieve their purpose. Capacity building in the area of facilitating participatory processes and integrated planning could help to achieve more integrated solutions to flood risk management and mitigation in urban areas. It will also be necessary to help achieve the ambitious goals of the WFD.

17.2 BACKGROUND TO THE NEED FOR CAPACITY BUILDING

This review takes an historical approach to develop an understanding of the need for capacity building in new approaches to flood risk management. Internationally there is an increasingly recognised need for thinking about water management in terms of integrated systems. This awareness has arisen from an awareness of the limitations of trying to solve problems in water management in isolation. This has a long pedigree in ecological thinking. In the early 1970s Odum (1971) stated "it is the whole drainage basin, not just the body of water, that must be considered as the minimum ecosystem unit when it comes to man's [sic] interests". What is new is the widespread political acceptance of this concept. The World Bank, the US Environmental Protection Agency and the United Nations Economic Commission for Europe have recognised catchments as important units for integrating land use planning and water management (UN ECE 1993; USEPA 1992, 1996; World Bank 1993). In recent years, this understanding has been extended to urban areas, where the functioning of water flows through the landscape may be less obviously related to ecosystem processes, but are still intimately connected to land use and cover (e.g. Baschak and Brown 1995; Hellström, Jeppsson and Kärrman 2000; e.g. Rijsberman and van de Ven 2000).

Shifts in thinking about water management have largely arisen as responses to crises and catastrophes (e.g. Douglas, Hodgson and Lawson 2002). The period from the Industrial Revolution to the burgeoning environmental movement in the 1970s was characterised by taking resources for granted in development. This is echoed in the title of a report about Britain's waterways *Taken for Granted* (HMSO 1970). Early approaches to environmental problems were characterised by single objective solutions, and an attempt to fix the symptoms of the problem, not necessarily its causes. As Tait et al. (2000) suggest, "in the past, the response has often been to manage one problem (unknowingly) at the expense of others". Planning for single sectors and single solutions has been exacerbated by funding for projects coming from government departments responsible for only one aspect of water management, such as water supply, with little integration between departments (e.g. Cate 1999).

Integrated Catchment Management (ICM) is the "integration of land and water management" (Gardiner 1996). The term was first coined by Gardiner, and for-malised in the late 1980s. It is seen as a way to apply the concept of sustainability to the water sector and aquatic environment (Gardiner 1984). The concept of ICM has developed in tandem with a shift to "soft engineering" in flood man-agement and replacement of the idea of eliminating flood risk with the concept of "elegant failure". This shift has been given impetus by the acceptance of a climate of inherent risk and uncertainty in flood management stemming from a realisation that flooding *will* occur, and engineered structures are both limited in their ability to mitigate damage and are capable of failing under pressure (Riley 1998).

European countries have a long history of developing strong environmental policies, partly in response to population pressures, and to the impacts of envi-ronmental problems becoming apparent after centuries of industrial development (e.g. Robert 1991). The WFD offers an unparalleled opportunity for improving the way that river basins are managed. Shifts in European policy have taken place within a context of changing ideas of government. Instead of governments acting directly to impose the provisions of plans and policies, their role is seen as shifting to one of facilitators. Government thus shifts towards governance, a process of facilitating partnerships and enabling actors to implement policies.

Bureaucratic departments have traditionally been organised around particular functions, such as agriculture or transport, with a high degree of vertical integration, but few functional links between departments. Such an organisation is effective in terms of delivering single objectives in policy, for example, increased agricultural production. The realisation of the interlinked nature of problems arising from the sustainability debate, and the related need for more integrated approaches to managing flood risk, however, has resulted in a call for increased integration between departments. This is seen as important both in order to resolve cross-cutting problems, and to develop plans and policies which deliver multiple objectives, with minimal negative unintended consequences (Selman 2000).

There are however, many problems with such an integrated approach. Administrative structures and social groups may have different boundaries than hydrological units. Kolavalli and Kerr (2002) suggest that catchments may be a "cumbersome unit for a socio-economic programme" and note the need to work within existing social organisations in a catchment context, paying particular attention to NGOs involved with local communities.

River catchment authorities often play no more than an advisory role, lacking the ability to enforce decisions. Inadequate resources often limit their scope and effectiveness. They may not have a remit over the entire basin, which means that activities in one part of the basin can threaten other areas, especially downstream. The sheer complexity of the sectors, actors and issues mean that coordination can be onerous and frequently insufficient. In particular, conflicts and lack of understanding amongst groups can lead to a lack of cooperation and sharing of vital information (Barrow 2000).

Insufficient support from local communities, engendered by a lack of involvement and understanding of key issues, can undermine efforts to improve the aquatic environment. Overly centralised bodies can become rigid and unresponsive to changing circumstances, unable, in particular, to communicate with NGOs and communities (Barrow 2000). Whilst stakeholder representation on river catchment authorities is seen as essential for a full understanding of the issues at stake, special interest groups can exert undue influence on the direction of the planning.

Active involvement of stakeholders in developing options and plans in flood risk management is still uncommon. Existing participatory processes often fail to help experts and citizens learn from each other, which can lead to disillusionment and participants becoming disengaged from the process. In addition, in many participatory methods, sustainability is seen as a "bolt-on" at the end of the process, which inhibits developing integrated plans. There is an opportunity to develop new approaches to participation that deliver environmental benefits through participatory planning. This will, however, require new skills and approaches from practitioners.

Despite these difficulties, there are many benefits to be gained from implementing innovative and participatory approaches to water management in general, and flood risk management in urban areas in particular. These include:

- increased knowledge base amongst a range of actors involved in flood risk management;
- enhanced ability to look at the water as an integrated system, including the way in which water is supplied, used and disposed of;
- increased likelihood of developing integrated solutions to flood risk management;
- increased support for the implementation of new measures;
- increased likelihood of changing behaviour in a multitude of stakeholders and actors in order to achieve long-term, sustainable management of water and develop flood resilience;
- strengthening partnerships in order to achieve the broad changes necessary for sustainable water management and flood resilient solutions.

In order to develop integrated approaches to flood risk management, changes will be necessary not only in the water sector, but also in urban planning, industrial design, architecture, infrastructure planning and landscape management (Braga 1999). This will require an innovative approach to planning and to pulling together the insights of these different disciplines, and developing new ways of thinking about water management. Engineers and environmental managers will be required to develop innovative approaches and to integrate complex technical information with the meanings and concepts that emerge from participatory planning into achievable plans and strategies. To do this, they will need to develop their skills in communication and integrated planning. In addition, practitioners in the fields of spatial planning, design and community development, who have not traditionally been as involved in flood risk management, can contribute to this task. To make an effective contribution they will need to learn new ways of working with technical experts.

Learning is an essential component of management. Whilst it is individuals who learn, they do so in social groups. Social learning can be seen as a "combination of adaptive management and political change" (Lee 1993). It is "a dynamic process which enables individuals to engage in new ways of thinking together to address problems such as the unsustainable use of water" (SLIM Project 2003). Social learning can be seen as a desirable outcome from participation in planning, and a necessary component of political change (e.g. Rees, Searle and Tippett 2005; Sabatier and Jenkins-Smith 1999; Tippett et al. 2005).

Soderqvist et al. (2000) suggest that sustainable water management will require different "social norms, codes and other informal constraints for human behaviour". These will be required in addition to formal means of changing behaviour, such as laws and regulations. Capacity building can thus be seen to be necessary for both individuals and organisations. It is necessary for professionals involved in facilitating participatory approaches and stakeholder involvement, as well as for practitioners involved in all aspects of changing the environment. It is also essential that capacity building takes places at the community level, to enable community members to effectively contribute to flood risk management and developing flood resilient solutions.

17.3   CAPACITY BUILDING IN INTEGRATED, PARTICIPATORY PLANNING

The previous section developed an understanding of capacity building in relationship to urban flood management and integrated water planning. As this is a relatively new area, little empirical research has been conducted on the *effectiveness* of capacity building programmes for practitioners in the skills required to deliver meaningful participation and integrated planning.

The remainder of this review develops an understanding of the challenges of capacity building for flood risk management. This is developed from two sources, a theoretical application of systems thinking principles to develop new approaches to capacity building, and the results of action research in the North West of England sponsored by the Environment Agency, which tested these approaches in a pilot project. This project aimed to develop understanding of capacity building to support holistic participatory facilitation. The objectives were:

1. **Capacity Building** – develop and evaluate new approaches for building capacity and skills in an ecologically informed participatory planning process;
2. **Pooling Knowledge** – develop new approaches to support ongoing learning amongst participants, exploring ways to pool knowledge and experience so it can be easily shared.

An assumption was made that moves towards sustainable development and new approaches in flood risk management will require a holistic approach, and thus that systems thinking would be useful in this endeavour. A set of skills and capabilities that could be used as a yardstick to evaluate the effectiveness of capacity building programmes were developed from the review, and is outlined below. Systems thinking concepts were then used to develop methods for training and capacity building, which might help to develop these skills and capabilities. The first author worked with trainee facilitators in a capacity building programme, which enabled them to engage stakeholder participation in a large scale workshop. This tested the new methods for training and capacity building developed from systems thinking insights. Sources of data included:

- In-depth, semi-structured "before and after" interviews with trainee facilitators;
- Participant observation during the action learning phase of the research; and

- Ideas developed by participants in workshops, which were entered into a database.

An ability to facilitate participatory, integrated planning, leading towards new solutions for flood risk mitigation and resilience, will require the following set of skills and capabilities, which can also be seen as a set of desirable outcomes against which capacity building programmes could be evaluated (initially developed in Tippett and Griffiths 2006).

17.3.1   Awareness (learning how to learn from what is happening around us)

- Observation and analysis of water as an integrated system and ecological processes;
- Spatial awareness and knowledge of factors that create a sense of place;
- Awareness of group interactions and social processes;
- Ability to ask critical questions;
- Knowledge of statutory, regulatory issues, plan making process – levers for, and barriers to, change.

17.3.2   Critical thinking

- Uncover and explore assumptions – both your own and those of others;
- Explore and be aware of differing cultural norms;
- Ability to think of projects in terms of the bigger picture – global issues and possible implications for economic vitality, social equity and ecological integrity;
- Awareness of the many ways that power is enacted – formally and informally;
- Ability to think through different scenarios and their potential ecological, social and economic impacts;
- Developing awareness of, and ability to manage, uncertainty and risk.

17.3.3   Creativity

- Asking questions at a deeper level to generate new solutions;
- Ability to use imagination;
- Cultivating different ways of looking at issues and using many types of knowledge;
- Adopting a flexible approach to process design in response to situations;
- Ability to link vision and general principles to context – the particulars of a place/situation;
- Aesthetic sensibility.

17.3.4   Facilitating and enabling

- Encouraging others to develop their skills and knowledge base
- Creating the conditions for dialogue and questioning of underlying frameworks and assumptions
- Mapping stakeholders, with an awareness of diversity in communities and stakeholders
- Communication and facilitation skills
- Planning and managing stakeholder events and workshops
- Managing relationships between stakeholder groups and clients
- Entrepreneurial skills, to develop new opportunities

17.3.5   Integration

- Develop shared understandings between disciplines
- Ability to encourage social learning
- Ecological design skills
- Multi-criteria decision making skills
- Adaptive management approach – assessment and readjustment of management in light of change and uncertainty
- Leadership and governance skills (e.g. institutional analysis)
- Ability to link planning and thinking across different levels of scale
- Ability to develop synergies and positive benefits between different areas of work and programmes

17.3.6   Ethical development

- Ability to reflect on own values
- Ability to reflect on role as a practitioner and citizen
- Creating the conditions for learning cycles – being open to feedback and creating opportunities for reflection
- Humbleness and openness to admitting error
- Openness to learn from people with different backgrounds and cultures
- Using sustainability criteria to inform decision making, asking about long term implications of actions

The following section develops a framework for new approaches to capacity building, aimed to develop the skills listed above. This looked in particular at developing skills for practitioners who are called upon to engage in integrated planning and facilitate participatory planning.

17.4   APPLYING SYSTEMS PRINCIPLES TO DEVELOP IDEAS FOR CAPACITY BUILDING

The report, The Law of Sustainable Development produced by the European Commission, explores the "legal theory of sustainable development" and states: "today, no serious study and application of the principles of sustainable development is possible without the help of systems science" (Decleris 2000).

Concepts of systems thinking have also had a profound influence on the field of ecosystem management, in helping to understand "the complexity of ecological and organizational systems" (Wondolleck and Yaffee 2000). "Whole ecosystem approaches" have yielded valuable insights in applications ranging from forestry to fishery management, to integrating indigenous concepts of nature with "modern" modelling and management of wildlife populations and in the awareness of environmental hazards, such as flooding (e.g. Kendrick 2003; Knott and Haywood 2001; Mitchell 2005; Seixas and Berkes 2003; e.g. Trosper 2003).

A conceptual model based on systems thinking principles and their application to developing new methods for capacity building is developed below. This approach was chosen as it was considered that the new demands on practitioners would require new approaches, and the field of systems thinking could provide a useful starting point for developing these new concepts. This builds on work to explore the teaching of ecological principles in design education (Johnson and Hill 2002; Karr 2002; Pulliam 2002), Capra's (2002) development of living systems thinking in relationship to ecological design and eco-literacy, and recent applications of complexity theory principles to learning and innovation in organisations (Van der Walt 2006; Webb, Lettice and Lemon 2006) Eight major concepts of systems thinking were explored, and are discussed below.

17.4.1   Holism, nested systems of wholes

A central tenet of systems thinking is that "the whole is greater than the sum of its parts". This has emerged from a realisation of the limitations of reductionist thinking, which has produced much valuable knowledge. Recent discoveries of the interconnected and complex dynamic nature of the world have suggested, however, that there is a fundamental limit to knowledge derived in this way. An attempt to find principles applicable at multiple levels of scale has been central to the development of systems thinking. This endeavour aims to reduce the duplication of effort and to promote improved communication between people working in different fields (e.g. Checkland 1991; Maiteny and Ison 2000).

Over the last three decades there has been an increased awareness that local actions have regional and global effects, and in turn local environmental issues can be affected by regional and global environmental change. Many environmental problems have only become apparent over time, due to delays between cause and discernible effect. This lag is further complicated by the fact that global climate change, pollutants and ecological problems cross boundaries of scale, such that effects from a source of pollution or a human activity may be manifested at a different level of scale than its cause (Gibson, Ostrom and Ahn 2000). Such awareness points to the need to develop capacity in practitioners to link analysis and action across levels of scale.

Managing for flood resilience should not only be the domain of the engineer, planner, developer and policy maker, working at the strategic level of scale, but also community members and stakeholders working at the local level of scale, where individual actions have a profound effect on the flow of water through the landscape. Broadening the responsibility for flood risk management is essential to increase resilience and decrease vulnerability to flooding. Facilitators of planning processes need to learn to consider the projects they are working on as parts of larger systems, and to think of ways to encourage participants to do the same.

Capacity building processes that actively seek opportunities to bring together actors working at different levels of scale would promote the skills of linking planning across levels. This would be enhanced by developing communication and training materials that show similarities of principles and ideas across levels of scale, and which enable participants to think of differences and similarities between the levels.

17.4.2   Networks of relationships

The study of ecology has led to an understanding of ecosystems as networks of relationships, including the relationships of food webs, bio-geological nutrient flows and species' interactions with habitat. It involves understanding hierarchies of networks, in which an organism (a network of relationships of cells and organs) cannot be seen as a separate entity from the ecosystem in which it is embedded, which is a part of the larger network of relationships in the biosphere. Integrated approaches to flood risk management require an awareness of the interconnected relationships of the various elements of urban and natural systems.

Castells (1996) writes of the "rise of the network society". This rise is influenced by the impact of the internet on communication, and the increased role of partnership working in many aspects of public life (e.g. Carley and Christie 2000). Developing an awareness of connections amongst practitioners is encouraged by bringing together people from different backgrounds and creating conditions where they are able to learn from each other. This would be encouraged through developing incentives for such learning and support mechanism to encourage it, for instance through mentoring relationships, and the development of networks of learners who are able to support each other in developing new skills and areas of knowledge. Principles for developing "communities of practice" (Wegner, McDermott and Snyder 2002) may help in this endeavour. To help meet the challenge of integration, facilitators need to enable participants to explore connections between different areas of knowledge. This requires new approaches that make such connections visible and accessible to people from different backgrounds.

17.4.3   Value of diversity

Participatory planning will require work in a variety of settings and contexts. A wide range of techniques to encourage people to interact have been pioneered, including new types of events and support frameworks. Appropriate methods for capacity building for both practitioners and participants will vary according to the nature of the work, size and type of the project and the stage in the process.

Capacity building needs to take into account the fact that people have a range of ways of learning, and that effective training will tap into and use a variety of means of promoting this learning. The originator of the theory of "multiple intelligences", Gardner (2000), suggests, "humans possess a range of capacities and potentials". Gardner (2001) writes that using multiple intelligences in learning means that lessons are "much more likely to remain with us, embedded in our neural networks, and to be usable in flexible and innovative ways". This insight has profound implications for capacity building, emphasising the need to develop new skills in ways that tap into these different intelligences. In addition, capacity building processes can encourage participants to become more aware of their own strengths and weaknesses in learning styles, and to think how this affects their approach to facilitating the involvement of others.

17.4.4   Self-organisation and emergence

A major concept of systems thinking is that of self-organisation, in which the interaction of different elements creates emergent properties, which cannot be predicted from the individual properties of the parts of the system. As the concept of self-organisation has been explored with more sophisticated mathematical modelling, new understandings of "the creation of novel structures and modes of behaviour in the process of development, learning, and evolution" have been developed (Capra 1996).

The trend of applying systems thinking to organisational management was seen in early work of Beer (1980, 1995) and de Geus (2002) and popularised in books such as The Fifth Discipline (Senge 1990) and the work of Wegner (2002). The concept of self-organisation plays an important role in this work, and management is seen as an arrangement so that people can manage their own learning process, and create learning organisations through their dynamic interactions.

17.4.5   Pattern in systems open to flows

Physical forms are like ripples created from dynamic change, chimeras which may be relatively stable, but which are derived from the maintenance of self-organisation in a state far from equilibrium. Patterns are configurations of relationships, which are expressed as repetition and similarities in space (form) and time (development).

The concept of process and pattern implies the form of the training and communication tools used to encourage capacity building can have an effect on its effectiveness. For example, graphic, interactive toolkits which encourage participants to explore and make visible connections between ideas, in a process of learning that allows patterns to emerge and be made visible, can help deepen learning of principles.

We are not always suffering from a lack of data in attempting to "plan for Sustainability" in the water environment; rather we suffer from a relatively "data-rich but information-poor syndrome" (de Pauw 1996). This is a general phenomenon of recent development. Wilson (1998) says:

"We are drowning in information, while starving for wisdom. The world henceforth will be run by synthesizers, people who are able to put together the right information at the right time, think critically about it, and make important choices wisely".

In this flow of information, ways of allowing meaningful patterns to emerge are essential. Capacity building programmes would benefit from mechanisms which encourage synthesis and critical reflection on ideas. A recent increase in interest in open source forms of knowledge creation may have profound impacts on the way we organise and share knowledge (e.g. Behlendorf 1999; Peizer 2003; Weber 2004). Open source intellectual property allows people to use ideas without locking them up as proprietary intellectual property. This makes ideas easily accessible and can encourage many people's creativity to be harnessed in ongoing development. Open source models are extended to domains outside software development, encouraging new models for developing standards in information and enamouring peer review (Keats 2003; Schweik, Evans and Grove 2005). These innovations raise questions as to how best to develop structures that maintain consistency and encourage sharing of improvements and innovation, which could make important contributions to the sharing of knowledge and development of capacity in flood risk management.

17.4.6 Uncertainty and triggers of change

Ecological systems cannot be fully described and understood from a description of the interaction of simple particles in a Newtonian field of forces (e.g. Capra 1996). Systems which are open to a flow of energy, information and materials self organise in unpredictable ways, dependent on the interaction of the parts, and on the context in which this interaction takes place (Kay et al. 1999).

Systems thinking has arisen in part in response to three problems in science: "complexity in general, the extension of science to cover social phenomena, and the application of science in real world situations" (Checkland 1991). Ackoff (2003) stresses the need to be able to make mistakes, or to admit that unexpected consequences have arisen from actions, and to learn from them. Changes in approaches to flood risk management will require practitioners to be willing to admit to mistakes and to discuss potential learning in new areas, which is not always a comfortable process. Change needs to be supported from directors and key people in the higher ranks of organisations, helping to create a culture in which admission of mistakes is seen as not only acceptable, but a necessary part of learning. An attitude of "zero tolerance of failure" needs to be replaced with zero tolerance of rigid attitudes.

This understanding leads to a realisation that capacity building to enable facilitators to deal with complex situations cannot in and of itself be rigidly regimented. Learning through action, in which unusual situations arise, allows participants to learn to respond to an evolving complex system. This points to the need for action learning interspersed with periods of reflection.

17.4.7 History and context

Any process of capacity building will involve working with people who have their own history, having developed knowledge and skills in different contexts. These diverse starting points need to be taken into account in capacity building, respecting the experiences of the learners, whilst encouraging reflection on how people's experiences and prior learning affect their ability to develop new skills and capacities. It is important to allow time towards the beginning of the process for participants to reflect on what they already know and ways in which their prior experience might influence their learning. Allowing opportunities for participants to discuss their own experiences is important in providing a learning environment in which people are able to connect new learning to their own context. Through sharing stories of their own learning process and development, trainers can help ground abstract principles in real examples. Trainers are thus demonstrating in their own behaviour the processes of change and reflection that are essential for developing new skills.

17.4.8 Adaptive capacity and learning

Learning from living systems can be related to the process of rethinking the nature of development. Living systems are organisationally closed. Changes in the environment can trigger changes in the organism, but cannot determine the changes. The organism cannot be separated from the environment with which it interacts, it is in a real sense embedded in a "circular pattern of interaction through which it is defined" (Morgan 1997). It is not possible to impose an integrated plan for flood resilience on an area through applying a formula or imposing a development plan. It is only possible to create the conditions in which such resilience can emerge.

Equally, it is not possible to teach practitioners the skills necessary to encourage such resilience, it is only possible to create the conditions in which they may experience and learn them for themselves. In this way they are more likely to gain in their own adaptive capacity, and ability to respond to new situations. In this endeavour, Kolb's (1973, 1984) insight into the value of learning cycles is important. In any capacity building programme, it is important to allow for hands on experience, reflection, conceptualisation and further active experimentation in a cycle of learning.

Effective deployment of participatory planning requires careful consideration of how to tailor the planning process to the particular context. This implies both attention to the needs and interest of the participants, and an understanding of the reasoning behind the process, so that it can be adapted without losing its essence. This may need to be done "on the fly" during workshops, requiring a degree of flexibility on behalf of the facilitator. Training should encourage an awareness of the underlying principles behind participatory tools, preferably in a context of practice. Facilitators should be offered support as they learn to apply new approaches in different contexts, as they learn when and how it is appropriate to adapt the approaches appropriately. Developing skills in reflection on practice will help practitioners' in this process.

17.5 METHODS FOR CAPACITY BUILDING DEVELOPED IN ACTION RESEARCH

This section describes the capacity building methods developed from this systems thinking framework, and which were tested in the action research. The effectiveness of these methods for capacity building was explored through analysis of the interview data, participant observation and ideas developed by the participants in workshops. The discussion section summarises key themes that emerged from this analysis.

17.5.1 Training in stakeholder engagement and facilitation

A one-day training session was held for the cohort of trainees. There were four main components to this training. The first session provided an opportunity for participants to reflect on their own understandings and experiences. At the beginning of the process, participants were engaged in an exercise to explore their understanding of engaging stakeholders in planning. This involved using a hands-on interactive toolkit, shown in Figure 17.1, in a process of engagement, so that trainees actually experiencing a hands-on workshop session as participants use the kit to brainstorm ideas. This exercise was carried out early in the process to encourage linking of participants' current understanding to new ideas.

Figure 17.1. Hands-on toolkit, ThinkIT, used in training (photograph courtesy of Joanne Tippett 2006).

This stage was followed by a more "traditional" session in which the trainer gave a presentation, discussing the key principles of an integrated, participatory planning process, and examples of its application.

This was followed by an interactive session, in which trainee facilitators used the hands-on toolkit to ask the question, "What do we need to do to plan a workshop?" In this session, it was seen as important that the trainee facilitators learn from each other, and the trainer's role was to guide the knowledge building process, with occasional input. In this case, there were also people with a mixed range of skills attending the workshop, who were able to provide a diverse range of inputs. The discussion amongst the trainee facilitators and the process of struggling with the question itself was seen as important to the learning process. This was supplemented by a discussion of key points and examples of planning workshops from past experience, both of the trainer and the trainee facilitators.

The trainee facilitators were then led through a session in which they planned the stages and process of the upcoming large scale workshop (for Manchester City Council). The cohort were to facilitate sessions at this event, which was seen as

Figure 17.2. Trainees facilitating at Green City Network – Manchester conference (photograph courtesy of Joanne Tippett 2006).

an opportunity to put their new skills into action and to begin applying them to a real-life problem. The day was closed with a discussion of the key points that had been learned.

17.5.2   Action learning, opportunities to apply knowledge

There were two opportunities for the trainee facilitators to apply their learning in practice. The first was during a half-day workshop on sustainability frameworks. The second was running workshops for the launch of the Green City Network – Manchester for the City Council, shown in Figure 17.2. The conference was an all-day event, with 150 delegates.

The outcomes of the workshops included a detailed database, showing all of the ideas and linkages developed in the workshop, and a report summarising the ideas which participants identified as most important (Tippett et al. 2006). Both of these events meant that the trainee facilitators were able to test their learning in a real-world context, an important component of the learning process.

17.5.3   Reflection and feedback session

Following these opportunities to apply new skills, the cohort attended a feedback session to review learning and to further develop their skills. This session was seen as important to help build skills of critical thinking and reflective practice, allowing participants the space and time to reflect on their learning and ways they could improve their skills. It was also seen as an important way to develop the network of peers that could offer ongoing support and advice in the learning process.

17.5.4   Results and analysis

In-depth interviews were conducted with each member of the cohort, both before and after the capacity building process. The interview questions explored participants' understanding of stakeholder engagement and sustainability, and shifts in this understanding as a result of the training. They were used to gather information about the cohort's experience of the training and how they learned new skills. They explored ways of improving the capacity building programme, issues which were further developed in two workshops, one with the cohort and one with delegates at the 2006 UK Systems Society Conference.

The total training time for this process was 1.5 days, with an additional day of applied work and an additional half-day workshop planning for the large scale facilitation event and developing ideas to improve the process. It has to be noted that this research started with a group of trainee facilitators who had some experience of running workshops. Any capacity building programme would have to take into account the starting point of the trainees, and the desired level of skill to be attained. Dreyfus and Dreyfus (2000) have developed a model of the five stages of skills acquisition, namely:
1. Novice
2. Advanced beginner
3. Competence
4. Proficiency
5. Expertise

The analysis in this research suggests that a brief training programme, such as that developed in this project, could enable trainees with some prior experience to reach an advanced beginner stage as a facilitation. With an opportunity to practice and to reflect upon the learning of that practice, a level of competence could be achieved quite quickly. This research only looked at the process of engaging stakeholders in workshops. Further training in integrated approaches would require more time. A lack of opportunities to practice was cited as an important barrier to improving skills in the interviews. This finding needs to be validated with a wider group of trainees.

The aim of a full capacity building and accreditation programme would be to achieve the level of expertise, where the practitioners are able to 'critically reflect on their intuitions' and achieve "holistic similarity recognition" in different contexts (Dreyfus and Dreyfus 2000). This level of expertise allows a flexible response to situations, whilst still maintaining an understanding of the key principles and ideas informing the approach, thus encouraging quality and integrity in the application of the approach.

Compared to other training the participants had experienced, this approach was seen as enjoyable, practical, engaging, empowering and open-ended, in the sense that it helped participants to open up new areas of inquiry and ways of applying what they had learned. The action learning approach, with opportunities to reflect on learning, was seen as key to learning new skills.

17.6 DISCUSSION

The key themes to emerge from this analysis can be summarised:
- Create an action learning approach
- Actively encourage reflective practice
- Develop peer networks and mentoring opportunities
- Create supportive conditions for learning
- Develop learning from participants' own understandings
- Incorporate a diversity of approaches and perspectives
- Encourage a holistic view and working across scales
- Incorporate learning about systems thinking and sustainability

These are discussed below.

17.6.1 Create an action learning approach

One of the key findings to arise from the research was the importance of an action learning approach to capacity building. In the training itself, it was seen as important that the processes that were being taught were actually used in the workshop, so that trainees were able to experience them as participants. The trainees were asked to help in planning the upcoming workshop (the action learning part of the programme) and to think of the potential difficulties and solutions to these, during the training session. This active approach to exploring the issues concerned in facilitation was seen as very important in the learning process. The fact that the training used interactive techniques and encouraged dialogue was seen as key to effective learning. Several of the participants mentioned that they learned the most from interacting with the other participants. Within the training it was seen as important to keep a balance between action orientation and some more informational sessions (more focused on "transfer" of knowledge), which provided the participants with key information and principles to help them understand the process.

The fact that the cohort had an opportunity to be actively involved in a "real-world" project to apply their skills and develop their confidence was seen as a valuable part of the learning. If it was not possible to arrange the training around a "real-world" project, it was seen that a role-play that allowed participants to actually facilitate a workshop that they had planned earlier in the day, by taking turns to be facilitators or participants, would be helpful. The workshops could also be used as an opportunity to plan something that the trainees (or some of the trainees) were about to do in their work, even if the event itself was not part of the capacity building programme for all of the trainees.

The need for more opportunities to practice, in a supportive environment, was emphasised by all of the trainees. Ways this could be achieved include: being invited to help at events, having opportunities for some paid work as assistant facilitators while being trained, or having opportunities to practice their skills in the workplace. Several of the cohort mentioned that they would appreciate opportunities to observe events being facilitated.

17.6.2   Actively encourage reflective practice

The importance of building in time for reflection on understanding and learning was emphasised by the participants, as was the role of an experienced trainer in being able to draw such reflection out of the group process.

The need for facilitators and planners to maintain a critical stance, and to hone the skills of asking difficult questions, was discussed in the interviews. This was related to the need for periods of reflection, as being one way of developing this critical faculty. Periods of reflection can also inculcate a sense of humbleness and openness to learning more, which is invaluable in developing skills of facilitation and integration of ideas across disciplines and projects.

It was suggested that it would be helpful in an accreditation process to make a reflective journal a requirement for facilitator accreditation, as this would act as an incentive for trainees to take the time to reflect on their learning, a process which is all too often pushed aside by more immediately pressing matters.

17.6.3   Develop peer networks and mentoring opportunities

Several of the participants mentioned that they learned the most from discussing their ideas with the other participants. Being able to explore their learning with people who had different experiences of applying the process was seen as very valuable, especially in terms of thinking through how they would be able to apply the approach in their own contexts. Encouraging the development of networks of peers, with discussion of learning and issues that arise in the application of the process was seen as essential for the trainees' ongoing learning. If this was done in a way that encouraged critical reflection and peer review, it was felt that this would help maintain the quality of the process whilst also encouraging innovation and growth of ideas.

A discussion group was seen as a key way to promote further learning. Whilst opportunities to meet were emphasised as important, access to an online discussion group was also seen as a valuable support. It was seen that a peer support group should consciously include opportunities to discuss trainees' experiences of applying the skills they have learned, and to receive feedback on their experiences and possible ways to improve them.

It was seen as particularly helpful to have a few more experienced people available for support during the capacity building process, in particular during the action learning phase, where trainees are applying their skills in a "real-world" situation. This led to the idea of cascading learning, where once there is a group of people with some training in an organisation, they can act as mentors to new trainees. This would enable the mentors to develop their own skills, whilst providing support to people who are learning new approaches. Several members of the cohort discussed the need for mentoring as they started to apply the ideas in their work, they felt that having access to a more experienced person's advice and support would help them build confidence and help them develop their capacity to use the ideas learnt in different ways.

17.6.4   Create supportive conditions for learning

The need to develop a comfortable environment for learning was emphasised. Whilst the need to create challenging circumstances to stretch the trainees was acknowledged, and seen as helpful, it was also seen as important that people were allowed to progress through the learning at their own pace, with "scaffolding" in place to support them in testing and acquiring new skills. The need to allow trainees sufficient time to reflect on their learning and ask questions was emphasised. Proceeding in small steps, so that trainees were able to start with thinking of working with small groups as a facilitator, and then proceed in stages to the more complex area of overall programme design, was seen as helpful in creating the conditions for optimal learning.

Developing a more participatory and integrated approach to planning will require new skills. This will require a greater tolerance of experimenting and making mistakes than has often been allowed in a managerial system driven by targets. The trainers' attitude and ability to admit to her/his own mistakes and difficulties with applying the process helps create a climate in which trainees feel comfortable to explore new ways of working. The fact that the trainer was able to discuss her own difficulties with applying the principles during the training was seen as helpful in creating an atmosphere open to critical reflection.

The interactive processes of the training were seen as important in creating the conditions for trainees to learn from each other within the workshop, which in turn will encourage further learning within a peer network. The training approach itself used tools and processes that encouraged the use of "multiple intelligences", allowing participants to think of the same ideas in different ways, e.g. using verbal, visual, spatial ways of exploring the ideas (Gardner 2003; Gardner and Hatch 1989). This was seen as key to encouraging learning, as it allowed people with different ways of learning to come at the information in varied ways.

Having a well-structured day, with clear aims and a sense of direction in the training, was seen as important to facilitate learning. At the same time participants valued the fact that there was some flexibility in the schedule and training approach to respond to different circumstances and learning opportunities as they arose. Providing good meals and refreshments was considered an important factor in creating a good atmosphere for training, along with sufficient time for breaks between the sessions.

17.6.5   Develop learning from participants' own understandings

The fact that the training started with an exercise to draw out what participants already know was seen as helpful in building confidence as well as providing hooks on which to hang the new learning. At the same time, this provided an opportunity to explore the different understandings of stakeholder participation within the group, which led to better understanding and communication. The value of developing a common language was mentioned in the interviews, and this can only be achieved by allowing for time to explore different meanings and participants' ways of viewing the world, so that they can be related to this emerging common language.

Trainers and mentors are also participants in this process. The fact that the trainer shared personal experiences with learning and using the process was seen as helpful in terms of helping the trainees to engage with the material. It also made them feel more comfortable exploring their own understandings of the material and discussing difficulties and concerns. Telling the story of the development of the process used in the action learning was seen as helpful for participants to understand its context, and to gain a better sense of its possible evolution.

17.6.6   Incorporate a diversity of approaches and perspectives

Having trainees with a diversity of backgrounds was seen as a positive way to encourage learning, as participants could learn from each other and the different perspectives raised. Even if the training was being carried out with participants drawn from just one organisation, it was seen as helpful to endeavour to include people from different departments, and to try to include a few people from outside the organisation (it was suggested that these could be other trainees who have some experience, using this as an opportunity to practice facilitation).

Participatory planning will require work in a variety of settings, with many different types of organisations. It will require ability to adapt to the context and the emergent dynamics of a complex situation. The trainees felt it was important to demonstrate a diversity of approaches and to show how the principles of the process could be adapted in different contexts, to help facilitators learn how to manage such complexity.

17.6.7   Encourage a holistic view and working across scales

The approach used for training in this research strives for integration, so that ideas and plans developed by participants achieve multiple benefits. This can be achieved partly through encouraging trainees to think of whole systems in every aspect of the training – e.g. asking "who are the stakeholders?", "what larger systems are they part of?", "what are the underlying goals and dynamics of these systems?". Learning is strengthened if the trainer makes this process of thinking of wholes explicit in the discussion. Part of the role of the trainer is to set up exercises that lead learners through a process of thinking of connections, then drawing this out to bring that knowledge to the conscious foreground in the discussion after the exercises. This helps trainees to apply a more holistic approach to planning workshops, so that they learn how to help others think more holistically in their own facilitation work.

Linking planning and action across different levels of scale is essential for integration and effective delivery of change. If possible, it is helpful to include people who are working at different levels of scale in the training process. This encourages discussion about the general applicability of the principles and processes, and what similarities and differences there are in working at different levels of scale. In this training programme, the cohort was drawn from people with experience working at the local level of scale, in community parks, to the strategic level of river catchments.

To support this learning, case studies and examples should be developed for projects from the local to the regional scale. The trainees discussed the value of seeing how the process was applied at different levels of scale, aiding their learning of how to apply the process in their own work.

17.6.8   Incorporate learning about systems thinking and sustainability

It was seen as important that information about systems thinking and sustainability were discussed and explored in the context of the applied work. Several participants said they thought that the session in which these principles were introduced was essential to their learning, and many felt that this should be expanded. At the least it was considered important to make it clear in training that there are more levels and aspects to be considered in a full integrated process, if a short training programme looking only at the process of running participatory workshops was being held.

17.7   FUTURE CHALLENGES

Suggestions to improve the capacity building approach were derived from this analysis. These included developing a process leading to formal accreditation, with enhanced professional status. Participants saw this as a valuable step forward, especially if it could be linked to external professional bodies and continuing personal professional development.

A skills assessment for the trainees was seen as a potentially useful addition, especially if the training was to lead to an ongoing programme of capacity development. Asking for a reflective journal as part of the accreditation process was seen as a useful way of encouraging the reflective learning style of the training to be carried into the ongoing development of skills.

Developing a structure to provide experienced mentors to support trainees was highlighted as an important element of future capacity building. This would mean that trainees would have some access to support in planning workshops and events, and someone to provide feedback on their ideas. In the early days of applying the skills, it was seen as important to have experienced help on hand at events (perhaps in the background to offer support), to help build confidence in skills whilst having a fall back position to ask for help.

There is a clear challenge to develop new approaches to capacity building if we are to achieve integrated solutions for flood risk management. This capacity building will be necessary for practitioners in the more traditional engineering fields of water management, to enable them to better develop integrated solutions and work with different stakeholders to develop new approaches. It will also be necessary for planning, design and community development practitioners, in order to enable them to better engage with community members and to better integrate their concerns with environmental and technical issues.

In addition, capacity building will be necessary for community members and stakeholders with an interest in the urban and water environment. Community members are often primarily concerned with their immediate environment, without necessarily being aware of more strategic concerns that may impact on their environment (Burnigham and Thrush 2001/2002). Through participation in planning for their area, they can be convinced, however, to change their behaviours and develop and implement holistic approaches to wider issues such as flood risk management and mitigation. Disaster management at neighbourhood levels and mitigation of hazards such as flooding will only be effective with the full participation of the local community. This will require a very different approach to design and planning than that currently experienced in the majority of flood mitigation projects.

A programme of capacity building for integrated flood management needs to ask the following questions:

• Who needs capacity building, at what levels?

- How can people outside of the traditional remit of flood risk management be engaged?
- What skills and knowledge areas need to be developed?
- What levels of capacity building are needed – e.g. enhancing awareness through to fully skilled practitioners?
- What are the existing resources that can help to achieve capacity building?
- What types of training and learning mechanisms have been tried?
- Have they been evaluated?
- What new approaches and mechanism need to be developed?
- How can they be integrated with existing activities to develop beneficial synergies?

Further research should evaluate the wider applicability of the capacity building approach developed in this chapter, in particular testing its application in work-based learning for practitioners engaged in participatory activities. Evaluating how to apply the learning from this project in work-based training in public sector agencies would help them to better meet their increased requirement to engage public and stakeholder participation in planning and environmental management. A related strand of the research should look at ways of building skills and capabilities in community members, to better equip them to manage their local environments effectively, developing resilience to flooding at the local level. The research should explore effective implementation of the capacity building approach, and evaluate its impacts both in the short and long term.

A particular challenge will be developing effective mechanisms for sharing experience and pooling knowledge amongst practitioners from different fields. A further challenge will be finding ways to encourage an open, yet critically reflective, approach to learning.

This research would thus provide practical benefits, with practitioners and community members better equipped to meet the challenges of developing and implementing new plans and approaches, as well as in the form of a growing body of knowledge about integrated flood management and developing flood resilient solutions.

17.8   CONCLUSION

We need to carefully consider the application of scientific findings in the field of water management. Flood resilience should not only be the domain of the engineer, planner, developer and policy maker, but also of the individual home owner and wider stakeholders, such as environmental NGOs, and insurance and water companies. Such a process of broadening the responsibility for flood risk management is a prime requisite in increasing resilience and decreasing vulnerability to flooding.

There is a key need to ensure that technical research influences policy and practice. Putting plans into action requires an understanding of the "implementation gap", which can be reduced by closely involving a forum of stakeholders able to implement the findings over differing spatial areas, from the region or river catchment scale to the local planning level and street level.

Developing and implementing integrated approaches to flood risk management will require a new set of skills and capacities in practitioners. Participatory processes to develop solutions to urban flooding should consider the changing urban ecosystem as a whole, and be geared to the implementation of the Water Framework Directive. Developing capacity in a range of practitioners to facilitate such integrated processes would therefore enhance vertical integration between national, regional and local scales, and strengthen the potential of resilience to influence differing spatial levels of scale. It would also improve horizontal linkages between actions in one part of an urbanised catchment and impacts elsewhere.

ACKNOWLEDGEMENT

The Environment Agency NorthWest (England) provided financial support for the research, with matched funding from Manchester City Council. The Economic and Social Research Council and the Mersey Basin Campaign (UK) provided financial support for the earlier research that formed the basis of this approach. Participants in the action research were generous with their time and ideas.

REFERENCES

Ackoff, R. L. and Strumpfer, P., 2003, Terrorism: A Systemic View, Systems Research and Behavioral Science, Vol. 20, pp 287–294.

Allen, W., Kilvington, M. and Horn, C., 2002, "Using participatory and learning-based approaches for environmental management to help achieve constructive behaviour change", Contract Report LC0102/057, Lincoln, New Zealand, Landcare Research, Ministry for the Environment. http://www.landcareresearch.co.nz/research/social/par_rep.asp

Barrow, C. J., 2000. "Global Experiences of Integrated River Basin Development in Planning and Management", Swansea, School of Social Sciences and International Development, University of Wales Swansea.

Baschak, L. A. and Brown, R. D., 1995, An ecological framework for the planning, design and management of urban river greenways, Landscape and Urban Planning, Vol. 33, No. 1–3, pp 211–225.

Beer, S., 1980, Preface to the article Autopoiesis: The Organization of the Living, accessed: July 28, 2003. http://www.cogsci.ed.ac.uk/~jwjhix/Beer.html

—, 1995, Beyond Dispute: the Invention of Team Syntegrity, Chichester, John Wiley & Sons.

Behlendorf, B., 1999, Open Source as a Business Strategy, in. Open Sources: Voices from the Open Source Revolution (editors, O'Reilly Online Catalogue. http://www.oreilly.com/catalog/opensources/book/brian.html).

Braga, B. P. F., 1999, Non-structural flood control measures – introductory notes for a special issue of urban waters, Urban Water, Vol. 1, No. 2, pp 112. http://www.sciencedirect.com/science/article/B6VR2-40B851V-2/1/e1fa182b0 e038e3d5037e53ba4339830

Brown, R. R., 2005, Impediments to Integrated Urban Stormwater Management: The Need for Institutional Reform, Environmental Management, Vol. 36, No. 3, pp 455–468.

Burnigham, K. and Thrush, D., 2001/2002, Rainforests in perspective – exploring the environmental concerns of disadvantaged groups, Ecos, Vol. 22, No. 3/4, pp 20–26.

Capra, F., 1996, The Web of Life, New York, Anchor Books.

—, 2002, The Hidden Connections – Integrating the Biological, Cognitive and Social Dimensions of Life into a Science of Sustainability, New York, Doubleday.

Carley, M. and Christie, I., 2000, Managing Sustainable Development, Second edition, London, Earthscan Publications.

Castells, M., 1996, The Rise of the Network Society, Massachusetts, USA, Blackwell.

Cate, F. M., 1999, River Basin Management in Lower and Upper Austria: Beginnings and Future Prospects, Water Science and Technology, Vol. 40, No. 10, pp 185–194.

Checkland, P., 1991, Systems Thinking, Systems Practice, Chichester, John Wiley and Sons.

Commission for Architecture and the Built Environment 2003. "Building Sustainable Communities: Developing the Skills We Need", London, Commission for Architecture & the Built Environment. www.cabe.org.uk

de Geus, A., 2002, The Living Company, Boston, Harvard Business School Press.

de Pauw, N., 1996, Conclusions and recommendations of the International Symposium "Environmental Impact Assessment in Water Management", Bruges, 15–17 May 1995, European Water Pollution Control, Vol. 6, No. 1, pp 68–69.

Decleris, M., 2000. "The Law of Sustainable Development – General principles, A report produced for the European Commission", Luxembourg, Belgium, European Commission, Environment Directorate-General: 147. http://europa.eu.int/comm/environment/pubs/home.htm

Douglas, I., 2000, Fluvial geomorphology and river management, Australian Geographical Studies, Vol. 38, No. 3, pp 253–262.

Douglas, I., Hodgson, R. and Lawson, N., 2002, Industry, environment and health through 200 years in Manchester, Ecological Economics, Vol. 41, No. 2, pp 235–255.

Dreyfus, H. L. and Dreyfus, S. E., 2000, Mind Over Machine, The Power of Human Intuition and Expertise in the Era of the Computer, New York, Free Press.

European Sustainable Cities and Towns Campaign, 2003. "Thematic Strategy On The Urban Environment, Consultation Report of Local Authorities, Coordinators: Council of European Municipalities and Regions (CEMR) and EUROCITIES", Brussels, ACRR, Climate Alliance, CEMR, Energie Cities, Eurocities, Healthy Cities, ICLEI, Medcities, UBC, UTO: 12. http://www.eurocities.org/epurbanpolicy/EP_frameset.htm

Evans, E., Ashley, R. M., Hall, J., Penning-Rowsell, E., Saul, A., Sayers, P., Thorne, C. and Watkinson, A., 2004. "Foresight Future Flooding: Scientific summary: Volume 1 – Future Risks and their Drivers", London, Office of Science and Technology.

Flournoy, L. S., 1995, Cumulative Impacts of Development in the Big Bear Lake Catchment: A Comparison of EIA, SEA and ICM (Integrated Catchment Management) from a Builder's Point of View– Extended Abstract. International Symposium on Environmental Impact Assessments in Water Management, Bruges, Belgium, May 15–17, 1995.

Gardiner, J. M., 1984, Sustainable development for river catchments, Journal of the Chartered Institution of Water and Environmental Management, Vol. June 8, pp 308–319.

—, 1996, The use of EIA in delivering sustainable development through integrated water management, European Water Pollution Control, Vol. 6, No. 1, pp 50–67.

Gardner, H., 2000, Intelligence Reframed: Multiple Intelligences for the 21st Century, New York, Basic Books.

—, 2001, An Education for the Future: The Foundation of Science and Values. Paper presented to The Royal Symposium Convened by Her Majesty, Queen Beatrix, March 13, Amsterdam. http://www.pz.harvard.edu/PIs/HG_Amsterdam.htm

—, 2003, Multiple Intelligences After Twenty Years. Paper presented at the American Educational Research Association, April 21, Chicago, Illinois. http://www.pz.harvard.edu/PIs/HG.htm

Gardner, H. and Hatch, T., 1989, Multiple intelligences go to school: Educational implications of the theory of multiple intelligences, Educational Researcher, Vol. 18, No. 8, pp 4–9. not read, pla reading, added nov 2006.

Gensamo, M., 2002, Capacity Building: Experiences and Challenges of the Furra Institute. "Beyond the Development Workshop" – New approaches to government and community capacity building using community-based learning in Ethiopia, Addis Ababa, Ethiopia, 28th & 29th January, DFID.

Gibson, C. C., Ostrom, E. and Ahn, T. K., 2000, The concept of scale and the human dimensions of global change: a survey, Ecological Economics, Vol. 32, No. 2, pp 217–239.

Hellström, D., Jeppsson, U. and Kärrman, E., 2000, A framework for systems analysis of sustainable urban water management, Environmental Impact Assessment Review, Vol. 20, No. 3, pp 311–321.

HMSO, 1970. "Taken for granted", London, Working Party on Sewage Disposal; Ministry of Housing and Local Government: 65.

Howe, J. M. and White, I., 2004, Like a Fish Out of Water: The Relationship between Planning and Flood Risk Management in the UK, Planning Practice and Research, Vol. 19, No. 4, pp 415–425.

International Council for Local Environmental Initiatives, 2002. "Second Local Agenda 21 Survey – Background Paper no. 15", DESA/DSD/PC2/BP15, Department of Economic and Social Affairs, Commission of Sustainable Development: 29. http://www.iclei.org/rioplusten/final_document.pdf

Johnson, B. R. and Hill, K., 2002, Ecology and Design, Frameworks for Learning, Washington D.C., Island Press.

Karr, J. R., 2002, What from Ecology is Relevant to Design and Planning?, in. Ecology and Design, Frameworks for Learning (editors B. R. Johnson and K. Hill), Washington D.C., Island Press, pp 133–164.

Kay, J., Regier, H. A., Boyle, M. and Francis, G., 1999, An ecosystem approach for sustainability: addressing the challenge of complexity, Futures, Vol. 31, No. 7, pp 721–742.

Keats, D., 2003, Collaborative development of open content: A process model to unlock the potential for African universities, First Monday, Vol. 8, No. 3, pp accessed Nov. 4, 2004. http://firstmonday.org/issues/issue8_2/keats/index.html

Kendrick, A., 2003, Caribou co-management in northern Canada: fostering multiple ways of knowing, in. Navigating Social-Ecological Systems, Building Resilience for Complexity and Change (editors F. Berkes, J. Colding and C. Folke), Cambridge, Cambridge University Press, pp 241–267.

Knott, A. and Haywood, J., 2001, A systems strategy for flood forecasting and warning – an integral part of the Environment Agency's approach to integrated river-basin management. River Basin Management, Cardiff, Wales, WIT Press, Southampton. 325–337.

Kolavalli, S. and Kerr, J., 2002, Scaling up Participatory Watershed Development in India, Development and Change, Vol. 33, No. 2, pp 213–235.

Kolb, D. A. 1973. "On management and the learning process", Cambridge, MA, MA: Massachusetts Institute of Technology.

—, 1984, Experiential learning, Englewood Cliffs, NJ, Prentice Hall.

Lee, K. N., 1993, Compass and Gyroscope: Integrating Science and Politics for the Environment, Washington, Island Press.

Maiteny, P. T. and Ison, R. L., 2000, Appreciating systems: critical reflections on the changing nature of systems as a discipline in a systems learning society, Systems Practice & Action Research, Vol. 14, No. 4, pp 559–586.

Mitchell, B., 2005, Integrated water resource management, institutional arrangements, and land-use planning, Environment and Planning A, Vol. 37, No. 8, pp 1335–1352.

Morgan, G., 1997, Images of Organization, Thousand Oaks, Sage Publications.

Odum, E. P., 1971, Fundamentals of Ecology, 3rd Edition, Philadelphia, W. B. Saunders.

Office of the Deputy Prime Minister 2004. "The Egan Review – Skills for Sustainable Communities", 04UPU1892, London: 106. www.odpm.gov.uk/eganreview

Orr, D., 1994, Earth in Mind, on Education, Environment and the Human Prospect, Washington DC, Island Press.

Peizer, J., 2003, Realizing The Promise of Open Source in the Non-Profit Sector, accessed: Nov. 6, 2004. http://www.soros.org/initiatives/information/articles_publications/articles/realizing_20030903

Pulliam, H. R., 2002, Ecology's New Paradigm: What Does it Offer Designers and Planners?, in. Ecology and Design, Frameworks for Learning (editors B. R. Johnson and K. Hill), Washington D.C., Island Press, pp 51–84.

Rees, Y., Searle, B. and Tippett, J. 2005. "Good European Practices for Stakeholder Involvement – Lessons from Real Planning Processes, Case-studies and Experiments: WorkPackage 5 report of the HarmoniCOP project – Harmonising COllaborative Planning", Swindon, UK, WRc. www.harmonicop.info

Rijsberman, M. A. and van de Ven, F. H. M., 2000, Different approaches to assessment of design and management of sustainable urban water systems, Environmental Impact Assessment Review, Vol. 20, No. 3, pp 333–345.

Riley, A., 1998, Restoring streams in cities: a guide for planners, policymakers, and citizens, Washington DC, Island Press.

Robert, K.-H., 1991, Educating A Nation: The Natural Step, In Context, No. 28, pp 10–14. http://www.context.org/ICLIB/IC28/Robert.htm

Sabatier, P. A. and Jenkins-Smith, H. C., 1999, The Advocacy Coalition Framework: An Assessment, in. Theories of the Policy Process (editors P. A. Sabatier), Boulder, Colorado, Westview Press, pp 117–166.

Schweik, C., Evans, T. and Grove, J. M., 2005, Open Source and Open Content: a Framework for Global Collaboration in Social-Ecological Research, Ecology and Society, Vol. 10, No. 1, pp. http://www.ecologyandsociety.org/vol10/iss1/

Seixas, C. and Berkes, F., 2003, Dynamics of social-ecological changes in a lagoon fishery in southern Brazil, in. Navigating Social-Ecological Systems, Building Resilience for Complexity and Change (editors F. Berkes, J. Colding and C. Folke), Cambridge, Cambridge University Press, pp 271–298.

Selman, P., 2000, Environmental Planning: The Conservation and Development of Biophysical Resources – Second Edition, London, Thousand Oaks.

Senge, P., 1990, The Fifth Discipline, Great Britain, Century Business.

SLIM Project, 2003, Home Page of the SLIM Project – Social Learning for the Integrated Management and sustainable use of water at catchment scale, accessed: July 29, 2003. http://slim.open.ac.uk/page.cfm?pageid=publicsocial

Soderqvist, T., Mitsch, W. J. and Turner, R. K., 2000, Valuation of wetlands in a landscape and institutional perspective, Ecological Economics, No. 35, pp 1–6.

Tait, J. T. P., Cresswell, I. D., Lawson, R. and Creighton, C., 2000, Auditing the Health of Australia's Ecosystems, Ecosystem Health, Vol. 6, No. 2, pp 149–163. www.nwlra. gov.au

Tippett, J. and Griffiths, E. J., 2006, Applying Systems to Capacity-Building in Participatory, Ecologically Informed Planning. Complexity, Democracy and Sustainability: The 50th Annual Meeting of the International Society for the Systems Sciences, Sonoma State University, Rohnert Park, California, USA, July 9th–14th, International Society for the Systems Sciences.

Tippett, J., Le Roux, E., Griffiths, E. J. and How, F. 2006. "Green City Network Launch Conference – Defining the Future of the Green City Network", Manchester, Centre for Urban and Regional Ecology (CURE), University of Manchester: 42. http://www.manchestergreencity.co.uk/site/index.php?option=content&task=view& id=201&Itemid=192

Tippett, J., Rees, Y., Searle, B. and Pahl-Wostl, C., 2005, Social Learning in Public Participation in River Basin Management – Early findings from HarmoniCOP European Case Studies, Environmental Science and Policy, Vol. 8, No. 3 – Special Edition – Research and technology integration in support of the European Union Water Framework Directive, pp 287–299.

Trosper, R., 2003, Policy transformations in the US forest sector, 1970–2000:implications for sustainable use and resilience, in. Navigating Social-Ecological Systems, Building Resilience for Complexity and Change (editors F. Berkes, J. Colding and C. Folke), Cambridge, Cambridge University Press, pp 328–349.

UN ECE, 1993. "Protecting Water Resources and Catchmen Ecosystems", Water Series 1, ECE/ENVWA/31, Geneva, United Nations Economic Commission for Europe.

—, 1998. "Convention on Access to Information, Public Participation in Decision-making and Access to Justice in Environmental Matters", ECE/CEP/43, entry into force 30 October 2001, Aarhus, United Nations Economic Commission for Europe. http://www.unece.org/env/pp/

USEPA, 1992, The Watershed Protection Approach. EPA/503/9-92/002, Washington D.C., US Environmental Protection Agency.

—, 1996. "Watershed Approach Framework", EPA810-S-96-001, Washington D.C., US Environmental Protection Agency, Office of Water.

Van der Walt, M., 2006, A framework for knowledge innovation, Emergence: Complexity & Organization, Vol. 8, No. 1, pp 21–29.

Webb, C., Lettice, F. and Lemon, M., 2006, Facilitating learning and innovation in organizations using complexity science principles, Emergence: Complexity & Organization, Vol. 8, No. 1, pp 30–41.

Weber, S., 2004, The Success of Open Source, Cambridge, MA, Harvard University Press.

Wegner, E., McDermott, R. and Snyder, W. M., 2002, Cultivating Communities of Practice, Boston, Harvard Business School Press.

White, I. and Howe, J., 2005, Unpacking Barriers to Sustainable Urban Drainage Use, Journal of Environmental Policy and Planning, Vol. 7, No. 1, pp 27–43.

Wilson, E. O., 1998, Consillience, The Unity of Knowledge, New York, Alfred A. Knopf.

Wondolleck, J. and Yaffee, S. L., 2000, Making Collaboration Work – Lessons from Innovation in Natural Resource Management, Washington D.C., Island Press.

World Bank 1993. "Water resources management, A World Bank Policy Paper", Washington D.C., World Bank.

18

Towards Integrated Approaches to Reduce Flood Risk in Urban Areas

Richard Ashley[1], John Blanksby[1], Jonathan Chapman[2] &
JingJing Zhou[1]
[1] *Pennine Water Group (PWG), University of Sheffield, Department of Civil and Structural Engineering, Sheffield, UK*
[2] *Environment Agency, Bristol, UK*

ABSTRACT: The approach to flood risk management in Europe has changed in recent times. There has been a move away from "defence" to risk-based cost benefit approaches. In addition, there has also been a widespread recognition that it is not possible to provide universal flood protection for human habitation and activities, due both to the rate of change in environmental systems, such as climate change, but also in view of the high costs involved. It has therefore become increasingly necessary to engage with stakeholders and communities in general so that they are better prepared for flooding, and so that they can also take an active role in decision making for where, when and how investments and measures are taken. Across Europe and in many other parts of the world, the advent of detailed computer based models for predicting flooding and the effectiveness of response measures has led to a more integrated view being taken. Integrated flood management, particularly in urban areas, is essential if the complex flood related processes are to be properly understood and dealt with. Integration does, however, need to be addressed in all aspects of flood management are dealt with including: policy, regulation, decision making and engagement, as well as in the technical approaches. The process of adopting an integrated approach view has only just begun and much more needs to be done, with integration ultimately spanning the entire water cycle, not only flood risk management.

18.1   INTRODUCTION

Flooding and consequent impacts are global problems that have been increasing in recent years due to: increasing population; lifestyle and expectations; indiscriminate urbanisation and environmental change. For example, the National Land Agency in Japan in 1994 found that worldwide flood-related damage increased some 43 times from around €500 billion in the mid 1960s to some €23,000 billion by 1994. During the same period, the number of people stricken by floods increased almost 100 times to 356.2 million (Todini, 1999).

Europe should implement the Water Framework Directive (WFD) by 2015 (Commission of the EC, 2000), providing for the protection of inland surface waters, transitional waters, coastal waters and groundwater. The fundamental aim of the WFD is to promote the sustainable use of water, while progressively reducing or eliminating pollutants for the long-term protection and enhancement of the aquatic environment. A new proposal for a Directive on the Assessment and Management of Floods (Floods Directive) (Commission of the EC, 2006) sets out to reduce and manage the risks which floods pose to human health, the environment, infrastructure and property. The Floods Directive and measures taken to implement it are to be closely linked to the implementation of the WFD. The Commission proposes to fully align the organisational and institutional aspects and timing between the Directives, based on river basin districts defined in the WFD. In the draft Floods Directive it is reported that between 1998 and 2004, Europe suffered over 100 major floods, causing some 700 fatalities, the displacement of about half a million people and insured economic losses totalling at least €25 billion. In the UK, nearly 2 million properties in floodplains, along rivers, estuaries and coasts are potentially at risk of flooding. Some 80,000 properties are at risk in towns and cities from flooding caused by heavy downpours that overwhelm urban drains – so-called "intra-urban" flooding. In England and Wales alone, over 4 million people and properties valued at over €300 billion are at risk. In 2004, flooding and managing it, was estimated to cost the UK around some €4 billion each year. Some €1200 million was then spent on flood and coastal defences per annum and even with the present flood defences, on average damage exceeding some €2000 million occurs every year (Evans et al., 2004a).

Climate change in particular is expected to increase the risk from flooding across the EC (e.g. Eisenreich, 2005; Commission of the EC, 2006). Estimates of the likely costs of this for the UK alone indicate that by the 2080s the costs could rise at least by an order of magnitude (ibid). Responses to this cannot simply be technological (Evans et al., 2004b) and will require changes in lifestyle and expectations. However, recent studies in the UK suggest that investing in responses as soon as possible is likely to be more cost-effective than waiting (Stern, 2006).

Studies into the origins, effects and responses to flooding are generally segregated into the main areas of flooding: rural (mainly fluvial); coastal (due to marine phenomena) and urban (pluvial, but may also be as a result of coincident fluvial and coastal impacts). They also tend to be phenomological, technological, economic, environmental or sociological. Rarely are the coincident aspects considered in conjunction (but see: Hankin, 2006). For example, the EU's flagship project on flood risk management, FLOODsite, whilst adopting an integrated approach, has so far concentrated on the mainly non-urban phenomenological, technological and economic aspects of flood risk assessment and management (Samuels, 2006). Even though the majority of urban areas are not affected by coastal or fluvial flooding, they are affected by overland flow from surrounding rural areas and urban greenspace, and they are also affected by groundwater and local watercourses. Therefore, integration is essential if the phenomena are to be better understood and the most effective responses made. Segregation into component areas with differing main institutional or stakeholder responsibilities and resources is not likely to produce sustainable outcomes and ways of managing flood risk into the future.

An integrated approach to flood risk management in urban areas may be taken using a framework as outlined in Figure 18.1. Within this, the scenarios refer to visions of the future (e.g. Evans et al., 2004a). The stakeholder groups A, B and C are simply illustrative of groups who may have different responsibilities, capabilities or resources. It is clear from Figure 18.2 that the key aspects of an integrated approach begin with policy and strategy; include institutional arrangements and funding; scientific understanding and knowledge; technology capability and innovation; and social systems and acceptabilities, especially resilience and adaptability. Also the need to ensure good water quality means that any integrated approach should also include quality issues under the WFD in the EU and increasingly the reuse and utilisation of excess or wastewater for potable and other purposes.

The historical development of Integrated Water Resources Management (IWRM) is described in a recent report from the Newater project (Pahl-Wostl, 2006) and is a concept that has been talked about for a number of decades. The Global Water Partnership (GWP) defines IWRM as "a process which promotes the coordinated development and management of water, land and related resources in order to maximise the resultant economic and social welfare in an equitable manner without compromising the sustainability of vital ecosystems" (GWP-TAC, 2000). Therefore, as an ambition, IWRM seeks to simultaneously address two highly complicated and complex problem sets; sustainable development and cross sectoral planning. Water is one of two enabling factors for sustainable development, with the dual challenges of water scarcity and water quality. If water is to be managed in a sustainable way then future developments in water management and technology should make it possible to meet the needs of current and future generations without compromising the sustainability of vital ecosystems. Such developments are of vital importance, in particular for developing countries, but also offer immense opportunities for an innovative water sector. IWRM in a European context means the integrated delivery of all Directives and Regulations related to water, in particular the WFD and the Floods Directive. Nonetheless (Phol-Wostl, 2006) argues that: "IWRM as currently practiced has not yet overcome the predict and control paradigm which may (be) a barrier for its successful implementation". They go on to state that "contemporary concern over this lack of success in application is such that the United Nations Environment Programme was recently prompted to classify the conversion of the concepts of integrated water resources management into practice as 'Unfinished Business' (IWA/UNEP, 2002)". The EU's Newater project sets out to address this and to finish the job through the characterization of

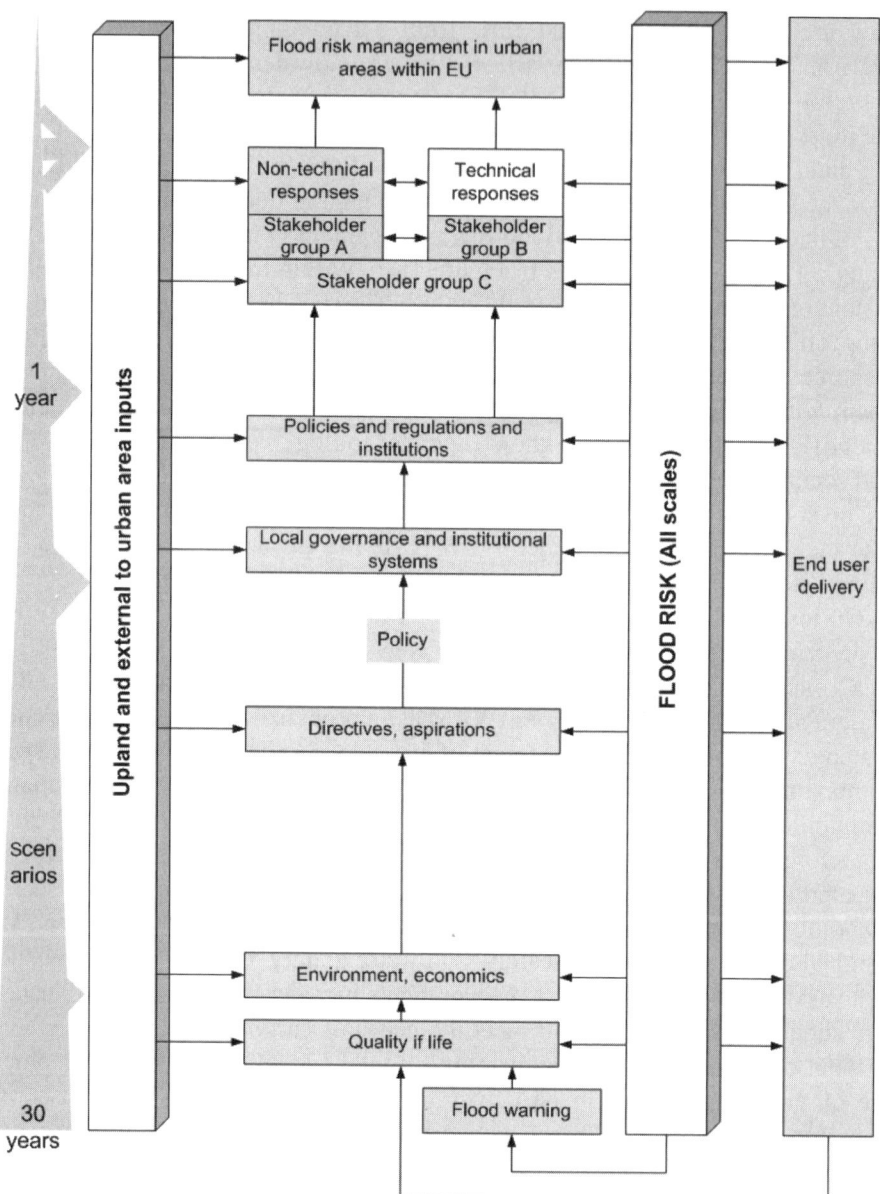

Figure 18.1. Outline of an integrated approach to flood risk management for urban areas in the EC.

a framework for effecting the transition from the current approaches to one which is predicated on adaptive management.

Worldwide there are a number of examples of attempts at integrated approaches being taken to water and flood risk management, however, these tend to be

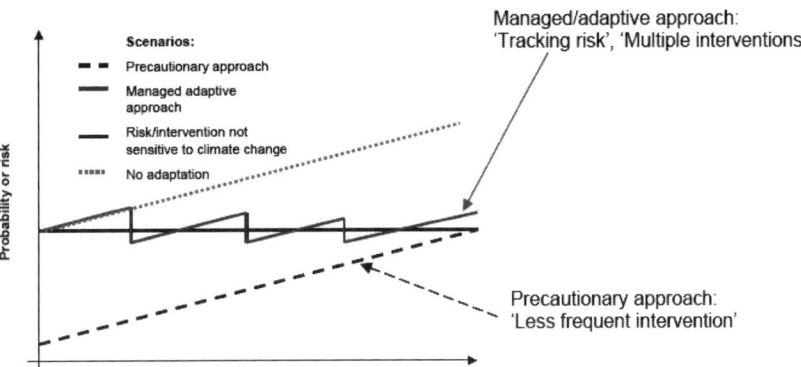

Figure 18.2. Precautionary approach compared with an adaptive approach (Defra, 2006a).

case-by-case and as yet there is no evidence of any formalised structural approaches that may be widely applicable. Although the recently started SWITCH project sets out an ambitious programme to develop Integrated Urban Water Management (IUWM) for future cities (SWITCH, 2006). Here, a review is made of the need for, the possibilities for, and examples of, taking an integrated approach to flood risk management focusing on urban areas.

18.2   THE NEED FOR EFFECTIVE POLICY AND STRATEGIES

Flooding is a legacy problem, generally caused by errors of omission due to lack of understanding or collective community memory. Because of current awareness of flooding and its causes, new flooding problems can only be considered to be caused for reasons of commission. In order to protect citizens and communities from such errors, policy and regulation are at the heart of delivering effective flood management.

Despite the growing significance of flood risk as an issue across Europe, there are few publications describing and comparing Urban Flood Management (UFM) policy in more than one country. The ultimate goal of UFM is Sustainable Development, and policy is one of the tools necessary to formalise the process. Policy is a means to an end, not an end in itself. Benefits from UFM policy are only gained when end-users (be they in Government, public or private organisations) implement or apply these policies.

Policy may be defined as "a plan of action adopted or pursued by an individual, government, party or business etc.". Policies are focussed on what is to be achieved (the outcomes), and not how these outcomes will be achieved. It is common for policy to be legally adopted by a Government or Organisation. Five principles of good policy are: transparency, accountability, proportionality, consistency and targeting.

Regulation is "a rule, principle, or condition that governs procedure or behaviour". This may be a governmental or ministerial order with the force of law. Regulations normally follow policy as in the way in which EU member states implement European Directives. Policy is formulated at various levels in government and at European Community level may lead to both the issuing of Directives, whereas their subsequent delivery and implementation at member state level.

It should be noted at this point that at European Community level, there is a need to address cross border issues and this is of great importance to flood management of rivers. However, urban flooding is generally a local issue and occasionally a regional issue within countries. Therefore it is appropriate that approaches to urban flooding are developed by member nations rather than the community as a whole, although there are strong arguments for the sharing of experience and knowledge.

Policies can only be formulated once the system to be regulated has been defined. The flood risk system may be considered as comprising the following constituent elements:
• Physical scale
• Scope
• Timescale
• Drivers
• Policy
• Strategic planning
• Asset management
• Event management (pre and during)
• Post event management

Policy options fit into the way in which the system above is defined and UFM policy cannot be considered in isolation from other aspects of organisational role, but the extent to which this is integrated can vary. It may be necessary to define the most appropriate Governance and Institutional arrangements to address the challenge of increasing flood risk so as to effectively deliver UFM Policy.

An understanding of the physical, institutional and scale contexts are required as a pre-requisite to developing policy for UFM. The achievement of an integrated approach requires, in very general terms:
• Definition of objectives
• Identification of the tasks involved
• Identification of the organisations and other stakeholders involved
• Agreement on who does what
• Matching resources to the tasks undertaken by each organisation and stakeholder

The three main aspects of policy are:
1. Policy Development or setting
2. Policy Implementation
3. Policy evaluation

The need for integration is more urgent in Policy Development. Implementation can be integrated in different ways. For example, a number of separate actions may be planned together, and delivered separately. However, overall, this could still deliver an integrated approach. The European Commission (EC) has taken a greater role in Flood Management Policy across Europe. This has built on the work of the International River Commissions (e.g. Rhine), and the approach in Member States. A "best practice document" on Flood prevention, Protection and Mitigation was produced in 2003. This document was strategic rather than technical, and considered:

- Basic principles and approach
- How to translate and implement principles and approaches

The EC originally published the draft Floods Directive in January 2006. Individual Member States will have 2 years to transpose the Directive into National Legislation, and then some 10 years to deliver.

There appears to be agreement on the high level principles of flood management across Europe, and more broadly. As an example, the recommendations from a recent International Symposium (Nijland and Menke, 2005) embody the approach outlined by the EC. The ensuing declaration emphasised:

- the need for an appealing common vision;
- that for urban areas, innovative solutions can be made with redevelopment schemes;
- the need to enhance public and stakeholder participation, especially on the local level to improve awareness, decision making as well as acceptance, maintenance and operation of these measures.

UFM policy is generally subject to incremental development. Historically, the loss of life and flood damage has acted as a spur to develop policy. Similarly, major floods provide opportunities for new policies to be adopted or implemented. Major changes in policy are relatively rare, but a number of paradigm shifts in the way in which UFM is approached have occurred in many European countries. Examples include the shift from Land Drainage to Flood Defence/Control/Protection, to flood risk management, to integrated or sustainable flood risk management. This is seen in the approaches from the Dutch "Room for the Rivers", and the English "Making Space for Water". In general, there is a move to more integrated solutions, and more complex decision making processes. Supporting this, the role of evidence and science in policy making is becoming more crucial.

Compared with entire river basins, urban areas are more complex in terms of the physical, institutional and scale dimensions. Successful delivery of UFM policy can only be achieved when all actors agree that there is a problem and also how they will work together to overcome constraints (real or perceived) provided by the Institutional settings e.g. public/private.

Some flood problems in urban areas may not feature particularly highly in the priorities of, say, the manager of major system (such as rivers) risk, minor system risk (drains and sewers) and the Spatial Planning Authority. However, the flood victim is not interested in Institutional complexities and there is an onus on all of the key stakeholders to collectively manage floods better in the future. This includes all steps from risk assessment, awareness raising, through to provision of information and advice before and after a flood. Effective coordination of the repair and recovery effort, generally between multiple organisations, is an area where shortfalls have been highlighted in the past (CBMDC, 2005).

Improving Inter-institutional/stakeholder working will be key to improving UFM overall.

Policy should always reflect the way in which it will be implemented, reviewed and its success evaluated. Failure to do this can lead to unrealistic expectations, complexity and cost. Within flood management there is a move away from single, to multi-objective policies, to multi-objective management (MOM). An example of such a policy may be to reduce flood risk and concurrently provide environmental enhancement in an integrated way.

Floods are only managed when policy is implemented, and this is usually by means of an intervention. In a risk-based approach, interventions may be to either reduce the probability of a flood (usually structural) and/or to reduce the consequence of a flood (mostly non-structural).

Although interventions can both reduce or (in some cases) increase, risk, they cannot eliminate risk altogether. This aspect is not always appreciated, and those protected sometimes expect flood management measures to provide absolute protection. It is also necessary to make plans to manage the consequences of flooding.

Planning for extreme events is becoming more commonplace amongst authorities. In many countries this fits well with more general agendas on incident (or event) Management (contingency planning). Whatever the risk management strategy is (probability, consequence or both), a range of floods should be considered, together with the interaction between different flood types.

18.3   EXAMPLES OF ATTEMPTS TO ADOPT INTEGRATED APPROACHES

Differences worldwide and even across the EC, in Institutional arrangements and the way in which policy (which can include European Directive implementation) is formulated and delivered, mean that it is unlikely that a "universal" integrated approach to flood risk management can ever be developed, however, it may be possible to develop a best practice template.

18.3.1   UNESCO IFI/P

A major initiative in addition to FLOODsite, outlined above, is the UNESCO International Flood Initiative/Programme (IFI/P) that sets out to establish a more holistic approach within and as part of the recommendations arising from the various world summits on sustainable development and the World Conference on Disaster Reduction in Kobe in 2005 (Price, 2006). The objective is to contribute to flood damage mitigation by developing and integrating scientific, operational, educational aspects of flood management, including the social response and communication dimensions of flooding and related disaster preparedness (UNESCO/IHP, 2005).

The key elements here are:

- Living with floods – proactive multi-hazard risk based approaches to develop culturally sensitive and sustainable ways of living with and managing floods
- Equity – this is equitable distribution of the burdens and benefits of flood risk management – appropriate policy and management processes across generations
- Empowered participation – of all stakeholders through appropriate institutional frameworks and governance mechanisms, with cleverly designed communication technologies as part of social development
- Interdisciplinarity – integrate and exploit better across disciplines to promote emergence (see Tippett & Griffiths, 2006)
- Trans-sectorality – to include all levels of stakeholder as well as national and international bodies
- International and regional cooperation – clearly important across physical boundaries but also across socio-political boundaries as well

The key component responses suggested (Price, 2006) comprise four areas: monitoring and observation of floods and the responses and recovery thereto; archiving and analysis of data in globally accessible databases for early warning especially for the vulnerable; modelling of the natural systems, of integrated systems for management plans, and of long term system and driver changes, such as due to climate. Finally, the development of decision support systems to empower and engage all stakeholders is envisaged; these to include institutional systems as well as the technical, economic and environmental systems. Essentially, it is suggested that development in these areas should take place within a hydroinformatics framework.

Ultimately, the approach is expected to lead to:

- Improved understanding of complex flood defence systems, their failure modes and their interaction with natural processes
- Improved understanding of vulnerability of the public and assets to flood damage
- Improved disaster preparedness, evacuation and emergency management procedures and social resilience
- Incorporation of the integrated holistic framework into management practice
- Consequent reduction in human flood casualties and economic loss

Other international initiatives, such as SWITCH will comform to this approach. Elsewhere, suggestions for integrated approaches in developing countries have already been set out. For example, Parkinson & Mark (2005) provide a significant blue-print for integrated stormwater management within Integrated Water Resources Management (IWRM).

18.3.2   UK Foresight

A futures project looked in detail at Flooding and Coastal Defence up to the 2080s and covered each of the sectors (river, urban and coastal) in which flood risk may be assessed (Evans et al., 2004a, 2004b). The study suggested that in the UK, potential increases of up to 40% in rainfall intensity in urban areas could lead to a 100% increase in flood volumes, a 130% increase in the number of properties affected and a more than 200% increase in flood damage by the 2080s. Consequent economic damage could rise by at least an order of magnitude and the costs associated with managing flood risk in the UK, even to current day levels will rise prohibitively by the 2080s, probably being unaffordable even in the 2025s when using traditional protection measures. Despite these findings, rising land values and housing needs in the UK have led developers to look increasingly towards flood prone areas as suitable areas for development. Similar pressures on land and the need for housing and other developments are apparent in the Netherlands, where innovative ways of building and planning for UFM are evolving (Kabat, 2006).

18.3.3   Making space for water in England

In the UK, as a consequence of the Foresight review of future flooding (Evans et al., 2004a), Defra (2004, 2005, 2006) have reviewed the traditional approach to flood risk management in England and proposed a new perspective for managing flood risk in urban areas: "integrated urban drainage management" (IUDM) within a programme perspective known as "Making Space for Water". A major driver for this has been the complex and dispersed institutional responsibilities for managing flood risk in England and Wales, comprising different responsibilities residing in the various stakeholder communities. As a consequence, the management of flood risk arising from more than one source is problematic and responses to floods or the threat of flooding may not necessarily be delivered in the most cost-effective or sustainable way (Blanksby et al., 2005). Nonetheless in England and Wales there are no plans to change the complex institutional and governance structure; rather to promote better partnering between the various major stakeholders.

An integrated approach is emerging in the UK, and although embryonic has successfully been applied in a limited number of cases such as in Scotland in devising the Glasgow Drainage Strategy (Page & Fleming, 2005) and for urban water management in Renfrewshire in an EU funded INTERREG project known as Urban

Water (Urban Water, 2006), both studies being in Scotland within the Greater Glasgow area. Elsewhere, the City of Bradford Metropolitan District Council (CBMDC) have recently undertaken an inquiry into water management that was driven by local flooding and the need to regenerate the City and the major river valleys within which Bradford sits (BMDC, 2005). In parallel, a national project, known as AUDACIOUS (adapting local drainage systems to cope with climate change) has, as part of a programme "Building knowledge for a Changing Climate" (Built Environment, 2007) devised adaptation guidance for coping with changing flood risk from local rainfall events (Ashley et al., 2007). From these, an integrating Water Management Liaison and Advisory Group representing the key stakeholders has been established in order to: Improve communication between organisations; help the understanding of the roles, responsibilities and limitations of organisations; improve joint strategic and operational working between organisations; and share knowledge and problem solving between organisations. Elsewhere similar approaches have been taken although not necessarily all aimed at better flood risk management (e.g. the SMURF project, Birmingham City Council, 2006; various in: Nijland & Menke, 2005).

The vision set out in the Making Space for Water approach comprises 11 sequential steps, once a problem has been identified (Digman et al., 2006):

1. Identify the issues and categorise spatially
2. identify and engage stakeholders
3. identify sources of flooding
4. identify data/information requirements
5. identify and undertake appropriate diagnostic study
6. confirm/amend understanding: assess risk
7. identify stakeholder responsibilities
8. agree target objectives and develop integrated solution
9. identify individual stakeholder contribution to solution
10. confirm agreements for solution delivery
11. monitor performance and provide feedback.

The identification of a lead stakeholder to drive the project team and solution delivery is seen to be crucial. Such "champions" are also recognised as essential in the delivery of sustainable solutions to water related services and issues (Brown et al., 2006).

Digman et al. (2006) highlight barriers and change needs for delivering integrated urban drainage in England and Wales, some lessons from which may also be applicable elsewhere. These have been adapted from the UK context and are outlined below.

- There may be legal definitions which restrict the ability of designated stakeholders to manage, maintain, own or apply certain types of technology. The sewerage undertakers in England and Wales are not able to utilise Best

Management Practices (BMPs) as they legally can only manage "sewers". Traditional approaches to regulation presume that defined terms or standards are likely to be applicable for a considerable time into the future. With the pace of change of external influences (drivers) such as climate change, regulations need to be formulated in such a way that they can allow options to adapt to changing knowledge and circumstances. It should be noted that the unfortunate designation of "Sustainable Drainage Systems" (SUDS) in the UK as applying to non-sewered techniques (essentially BMPs) for the management of stormwater confuses this issue. As the sewerage undertakers *are* in fact encouraged to use sustainable techniques (which may include pipe systems), i.e. "SUDS" – in the proper sense of the acronym, but cannot own and maintain BMPs as if they were providing a stormwater sewerage function. This point is lost in the discussion by Digman et al. (2006).

- Local/national arrangements in place for the effective drainage of new developments may unwittingly encourage options that may not be the best in an integrated approach. For example, there is a right to connect new drainage systems into existing sewers in England and Wales. This may necessitate (allow) the connection of stormwater into a sanitary (or wastewater) sewer if no suitable local separate stormwater sewer exists. Even where a BMP system has been installed, a property owner may subsequently seek to reconnect into a sewer instead. This makes planning alternatives to sewers very difficult and risky for the sewerage undertakers. Changing this right requires amendments to primary legislation (Water Industry Act, 1991).
- The planning processes may not be sufficiently open to water related stakeholders to allow them to influence strategic decisions, or for the management of the details for new developments. Water systems have traditionally been seen in planning processes as subordinate to other aspects such as road layouts, the fitting of a maximum number of properties on to the plot etc. This has led to difficult situations for drainage layouts and arrangements, often resulting in their being no opportunity for any other options than traditional piped drainage.
- Creeping urbanisation often occurs in established urban areas where "soft" surfaces are replaced by informal paved surfaces. Typically for driveways and patios in individual properties. This has been a major problem in London where many front gardens have been paved over for car parking (London Assembly, 2005). Not only does this lead to increased runoff, but also the loss of green areas which can help alleviate climate change impacts (Gill et al., 2007). Such informal and potentially major changes to urban areas require better control than is traditionally taken.
- Many flood risks are attenuated or the effects minimised in existing urban areas by informal flood pathways, where excessive flows might pond, pass through or around areas where they might cause problems. These "exceedence flood

pathways" (Balmforth et al., 2005) may need to be protected to ensure they are continue to function. They may even need enhancement to cope with increasing use and growth in flood magnitude in the future. This is a new area of planning for which rules still need to be formulated. These new rules may, for example, include the need to ensure that highways will continue to act as surface drains during times of heavy rainfall. Thus the stakeholders responsible for highways will need to work with those dealing with flooding to ensure this happens. There are other examples from exceedence flow design and operation that require each of the various stakeholder groups to work together in an integrated way. Ironically, it is those working in the constituent parts of the same stakeholder organisation who most often fail to achieve this. In the UK, usually in a single municipality one group is responsible for highways and another for drainage. These groups often fail to communicate effectively.

- In most developed countries and particularly in Europe, sewerage systems are combined. However, there are moves to "separate" or disconnect the stormwater systems and deal with the stormwater separately from the wastewater. In many planned systems built in the last 30 years or so, these systems are already separate, although they frequently recombine at a point further downstream. Sewer separation is clearly a quality as well as a quantity issue and a major consideration in the WFD (Ashley et al., 2006). As yet there are no simple rules that can help provide guidance as to when, where and under what circumstances, sewer separation is the more sustainable option for existing and also new systems. This is an area that requires research, although it is the subject of the on-going EU INTERREG IIIb project No Rainwater in Sewers (NORIS, 2006). For example, sewer separation has not been seen to be an effective option for managing the current problems of CSO discharges into the river Thames in London (Thames Water, 2005), although this has been the approach used in several cities in the USA such as Boston.

- Disproportionate resourcing to the various stakeholder groups involved in flood risk management may unintentionally result in less efficient solutions being proposed and implemented. By providing greater or lesser resources to a particular key stakeholder, this stakeholder will preferentially adopt certain approaches to managing flood risk that fall under their aegis. Where flooding is due to more than one cause this may lead to partial solutions that, given a more effective allocation of resources, may have achieved a complete solution at no or little, additional cost.

- The way in which flood risk is defined and perceived varies between stakeholders. It also varies as to how it is addressed, with certain mores being adopted by particular stakeholder groups (Gouldby & Samuels, 2005). Traditionally for example, the definition and management of risk due to main river flooding has differed from the equivalent definitions and approaches used to manage sewer related flooding (Blanksby et al., 2005). Approaches need to be harmonised if integrated and the most effective solutions are to be delivered.

- Data, information and knowledge need to be shared between the key stakeholders in a readily accessible format. This is problematic as certain key stakeholders, such as the insurance industry often do not wish to share information for business purposes. Individual property owners may wish to keep confidential any incidents of flooding in order to preserve property values. As far as practicable, data and information should be maintained centrally by a lead stakeholder.

From this it is apparent that there is a momentum, evident in a number of global, national and EU projects, to develop an integrated approach to flood risk management, which may also integrate with the wider aspects of water management in urban areas.

Recently, a large number of research topics or projects have been linked to integrated urban drainage management (IUDM); however, the definition of IUDM is still unclear. A definition of IUDM from the English Defra programme provided the outline below (Defra, 2006).

The effective drainage of urban areas is needed both to reduce the risk of flooding to land and properties and to prevent the inappropriate discharge of waste waters causing pollution. As well the timely removal of rainwater, the criteria used to judge success include the elimination of foul sewage flooding and a reduced frequency of operation of Combined Sewer Overflows (CSOs); surface water quality will inevitably become a priority issue as a result of the introduction of the Water Framework Directive. Whilst the provision of adequate sewerage infrastructure will always be necessary, careful whole catchment planning can minimise the extent to which it is required, and it is increasingly apparent that the design, construction and operation of combined sewerage systems should take into account all sources of water, both foul and surface. For example, it is now recognised that wetlands and greenspace might be utilised for flood alleviation, including short term storage and conveyance of excess flows, and that Sustainable Drainage Systems (SUDS) can reduce the impact of intense rainfall events and can also be part of conveyance systems (e.g. Tourbier & White, 2007). This combination of well-rounded engineered solutions and landscape utilisation is termed Integrated Urban Drainage (IUD).

18.4   COPING WITH CLIMATE CHANGE

Climate change is now recognised as a substantial future risk across the EU (Eisenreich et al., 2005). Guidance on how to factor this into UFM varies across the member States. In the England, Defra have recently recommended using both a precautionary and an adaptive approach (Defra, 2006a). This is illustrated in Figure 18.2. The precautionary approach is recommended for changes that may reasonably be considered foreseeable, and is recommended for responding to rising sea levels, whereas the adaptive approach is recommended for changes that are less certain, such as rainfall intensity. The latter is considered to lead to incremental responses, following the approach recommended in Glasgow (Akornor & Page, 2004). It is clear that these approaches need to be set within a scenarios perspective (Ashley et al., 2006). As the ability to respond is a function of a wide range of factors, including socio-economics (Evans et al., 2004b).

The ideal IUDM approach can be adapted from Defra (2006a) should thus comprise:

- A long term perspective (e.g. Camphius, 2007)
- More holistic approach
- Sustainable development
- Stakeholder engagement (e.g. Teppett & Griffiths, 2007)
- Partnership working (e.g. Defra, 2006)
- Clear definition and assumption of ownership and responsibility
- Measures for climate change adaptability (e.g. Defra, 2006a)

It is arguable that the above are attainable through adopting a resilient approach (Gallopin, 2006; De Bruijn, 2005). However, the definition of resilience, like sustainability, is open to debate and seems to remain elusive other than as an aspirational concept or a principle.

18.5   CONCLUSIONS

Integrated urban drainage management (IUDM), Urban Flood Management (UFM) and Integrated Urban Water Management (IUWM) have become frequenftly used terms expressing concepts for the better management of water systems. Although a number of comprehensive methodologies for IUDM have been proposed, there is a need to draw on the growing number of finished and on-going projects that are more or less linked to IUDM to develop generic rules by which urban flood risk can be managed and better urban drainage planning achieved. Possibly due to the constraints and barriers in terms of policy, strategy, legislation, regulation, funding, technology, data, public perception and skills, IUDM still has not yet been successfully applied other than in a limited number of cases at local scale. Whilst there may be different barriers to effective delivery in the various countries across the EU, there is an urgent need to resolve the problems and deliver in particular, more effective urban drainage management. In the UK, the publication of Planning Policy Statement 25 (DCLG, 2006), which addresses issues of urban flood risk management, and the commencement of the Making space for water integrated urban drainage pilots in December 2006 represent two more steps in the development of an integrated approach which recognises the need to cope with the complexity of the problems faced.

ACKNOWLEDGEMENTS

This paper has drawn on a number of activities, discussions and endeavours of the COST C22 participants. Nonetheless the views expressed are entirely those of the authors and do not necessarily reflect those of the University of Sheffield or the Environment Agency for England and Wales.

REFERENCES

Akornor A., Page D W. (2004). Glasgow Strategic Drainage Plan Stage 1, Overview and Case Study. WaPUG Scottish Meeting, Dunblane, http://www.wapug.org.uk/past_papers/index_pages/SplusNI2004.htm

Ashley R M., Tait S J., Styan E., Cashman A., Luck B., Blanksby J R., Saul A J., Hurley L., Sandlands L. (2006). 21st Century Sewerage Design. WaPUG Spring conference. Coventry.

Ashley R M., Blanksby J R., Cashman A., Jack L., Wright G., Packman J., Fewtrell L., Poole A. (2007). Adaptable Urban Drainage – Addressing Change In Intensity, Occurrence And Uncertainty Of Stormwater (AUDACIOUS). J. Built Environment. In press.

Birmingham City Council (2006). Sustainable Management of Urban Rivers and Floodplains (SMURF) Supplementary Planning Document (SPD) – Sustainability Appraisal and SEA Scoping Report. Draft. October.

Balmforth D., Digman C., Kellagher R B B., Butler D. (2006). Designing for exceedance is urban drainage – good practice. CIRIA report C635. ISBN 0-86017-635-5.

Blanksby J R., Ashley R M., Saul A J., Eadon A., Poole A., Wilson D. (2005). Development of a framework for integrated storm and surface water management. WaPUG November. Blackpool.

Brown R R., Sharp E., Ashley R M. (2006). Implementation impediments to institutionalising the practice of sustainable urban water management. Wat. Sci. Tech. Vol. 54, No 6–7. pp 415–422. ISSN 0273-1223.

Built Environment (2007). Special edition on the UK programme Building knowledge for a Changing Climate (March edition).

Camphius N-G (2007). Flood risk management on the Loire river: a case study, COST C22.

CBMDC (2005). Review to consider the future of water management and the associated problems of flooding in the Bradford District. City of Bradford Metropolitan District Council. Ashley R M., Melling D. [www.bradford.gov.uk].

Commission Of The European Communities (2000). Directive 2000/60/EC of the European Parliament and of the Council of 23 October 2000 establishing a framework for Community action in the field of water policy, OJL 327 of 22.12.2000.

Commission Of The European Communities (2006). Proposal for a Directive Of The European Parliament And Of The Council on the assessment and management of floods {SEC(2006) 66} (presented by the Commission). Brussels, 18.01.2006 COM(2006) 15 final 2006/0005(COD).

De Bruijn K M. (2005). Resilience And Flood Risk Management – A Systems Approach Applied To Lowland Rivers. PhD thesis. Technische Universiteit Delft,

DCLG (2006). Planning Policy Statement 25, Development and Flood Risk, Department for Communities and Local Government web site, December 2006, http://www.communities.gov.uk/index.asp?id=1504639

Defra (2004). Making Space for Water. Developing a new government strategy for flood and costal erosion risk management in England. Defra web site.

Defra (2005). Improving data management in practice. Flood and Coastal Erosion Risk Management – Research News, Issue 7, Feb.

Defra (2006). Defra Integrated Urban Drainage Pilots Project title: Scoping Study. March. Author(s): Balmforth D., Digman C J., Butler D., Shaffer, P. http://www.defra.gov.uk/environ/fcd/policy/strategy/scoperev.pdf

Defra (2006a). Flood and Coastal Defence Appraisal Guidance FCDPAG3 Economic Appraisal. Supplementary Note to Operating Authorities – Climate Change Impacts. October.

Digman C J., Balmforth D J., Shaffer P., Butler D. (2006). The challenge of delivering integrated urban drainage. WaPUG conference Blackpool. November. Eisenreich S J. (Ed.) (2005). Climate change and the European water dimension. EC – Joint Research Centre. EUR 21553 EN. ISBN 92-894-9005-5.

Evans E., Ashley R M., Hall J., Penning-Rowsell E., Saul A., Sayers P., Thorne C., Watkinson A. (2004a). Foresight Future Flooding: Scientific summary: Volume 1 Future Risks and their Drivers. Office of Science and Technology, London.

Evans E., Ashley R., Hall J., Penning-Rowsell E., Sayers P., Thorne C., Watkinson, A. (2004b). Foresight. Future Flooding. Scientific Summary: Volume II – Managing future risks. Office of Science and Technology, London.

FLOOD*site*. http://www.floodsite.net/ (accessed 22/11/06).

Gallopin G C. (2006). Linkages between vulnerability, resilience, and adaptive capacity. Global environmental change. 16, 293–303.

Gill S., Handley J., Ennos R. (2007). Greenspace to adapt cities to climate change. COST C22.

Gouldby B., Samuels P. (2005). Language of Risk – PROJECT DEFINITIONS. March. Report: T32-04-01. FloodSite. GOCE-CT-2004-505420.

GWP-TAC (Global Water Partnership – Technical Advisory Committee), 2000, Integrated Water Resources Management. TAC Background Papers No. 4 (GWP, Stockholm, Sweden).

Hankin B. (2006). Flooding from other sources. Scoping Review. Environment Agency HA4a. Final report. (draft) August. Jba consulting.

Kabat P. (2005). Climate Proofing the Netherlands. NATURE Vol. 438, 17 November.

IWA (2002) Industry as a partner for sustainable development: Water Management. IWA/UNEP. Beacon Press. London.

London Assembly (2005). Crazy Paving. The environmental importance of London's front gardens. September. ISBN 1 85261 772 1.

Nijland H., Menke U. (Eds.) (2005). Flood risk management and multifunctional land use in river catchments. ISBN 90-369-5730-3.

Parkinson J., Mark O. (2005). Urban Stormwater management in developing countries. IWA pub. ISBN 1843390574.

Pahl-Wostl (Ed.) (2006). Framework For Adaptive Water Management Regimes And For The Transition Between Regimes. NeWater Report Series No. 12. May. [www.newater.info].

Price R K. (2006). Research, Education and information systems in the context of a framework for flood management. In: Transboundary Floods: Reducing risks through flood management. Ed. Marsalek J., Stancalie G., Balint G. NATO Science Series. IV Earth & Environmental Sciences – Vol. 72. Springer. ISBN 10 1-4020-4901-3.

Samuels P. (2006). Flood Risk Analysis and Management – Achieving Benefits from Research. Proc. European Conference on Floods. Vienna, 17–18 May 2006.

SMURF (2003). SMURF Project Methodology and Techniques. http://www.smurf-project.info/index.html

SWITCH (2006). http://www.switchurbanwater.eu (accessed 6th December 2006).

Stern N. (2006). The economics of climate change. Cambridge University Press. ISBN: 0-521-70080-9 (also downloadable from Defra website: http://www.hm-treasury.gov.uk/independent_reviews/stern_review_economics_climate_change/stern_review_report.cfm).

Tippett J., Griffiths E. J. (2007). New approaches to flood risk management – implications for capacity-building. COST C22.

Thames Water (2005). Thames Tideway Strategic Study. Steering Group Report. February.

Todini E. (1999). An operational decision support system for food risk mapping forecasting and management. Urban Water 1 (1999) 131–143.

Tourbier J T., White I. (2007). Sustainable Measures for flood attenuation: Sustainable Drainage and Conveyance systems (SUDACS). COST C22.

UNESCO/IHP   (2005).   http://unesdoc.unesco.org/images/0013/001397/139769e.pdf (accessed 11th December 2006).

Urban Water (2006). Sustainable Water Management in Urban Space. INTERREG III B NWE Project, Renfrewshire: Spatial Management of Water Infrastructure Regeneration (http://www.urban-water.org/).

19

Hydrological Modelling of Floods

Mira Kobold

University of Ljubljana, Faculty of Civil and Geodetic Engineering, Ljubljana, Slovenia

ABSTRACT: Precipitation is a common cause of floods and types of precipitation determines the scale of modelling. Although the processes which generate runoff are well understood, it is normally possible to incorporate them into flood forecasting procedures only in a generalised and largely empirical manner because of their spatial and temporal complexity. The experiences of flood forecasting in Slovenia are presented in the contribution. The applications of the HEC-1 model are described as well the experiences with the HBV model which was tested for runoff simulation on the Savinja basin. The uncertainty of simulated river discharge is mainly the result of precipitation uncertainty associated with the average basin precipitation. To find out the measure of runoff uncertainty regarding to precipitation error, the analysis of sensitivity of conceptual models to rainfall error has been performed showing that the accurate representation of precipitation in time and space scale is essential for rainfall-runoff modelling and flood forecasting. Flood forecasting with predicted precipitation of the ALADIN/SI model demonstrates the uncertainty of precipitation forecasts.

19.1   INTRODUCTION

Models are increasingly used in hydrology to simulate runoff and changes in catchment management, to extend data sets and to evaluate the impacts of external influences such as climate change. Many models have been developed for river catchments throughout the world with various degree of complexity, from simple empirical formulae or correlations to the complex mathematical models, representing all phases of the water balance of a river basin (WMO, 1994; Singh, 1995; Beven, 2001). The models were carried out for different purposes as a research or operational tool, for different spatial levels from small to large watershed, and temporal scales from event based models to annual water balances. Although the processes which generate runoff are well understood even within small river basins, it is normally possible to incorporate them into flood forecasting procedures only in a generalised and largely empirical manner because of their spatial and temporal complexity (Smith and Ward, 1998). Simulated runoff as a result of hydrological model can be used further as input into hydraulic models to simulate flood plains what is particularly important for urban areas.

19.2   FLOOD FORECASTING BY HYDROLOGICAL MODELS

Flood forecasting is the use of real-time precipitation and streamflow data in rainfall-runoff and streamflow routing models to forecast flow rates and water levels for periods ranging from a few hours to a few days in advance, depending on the size of the watershed and the period for which the weather is forecasted. The experiences of the WMO trials (WMO, 1975, 1992) and experiences of many authors compared different types of models (Askew, 1989; Bell and Moore, 2000; Kokkonen and Jakeman, 2001; Job et al., 2002) show that the selection of the model is not critical to the success of the forecast. On the other hand, it was shown that the complex models generally do not give better results than the simpler ones. Whilst there are many simulation models in use, the skill is in selecting the right model for the job and balancing data requirements against the cost of model implementation.

It is important to understand the mechanisms which produce flooding in a catchment. Precipitation is a common cause of floods and types of precipitation determines the scale of modelling. Frontal precipitation is associated with large air masses. The precipitation can be prolonged but is usually not very intense. Large volumes of water are produced, but over a period of days. This is associated with large floods in large river basins. Convective precipitation can be very intense and localised. It can cause flash flooding in very short times. Orographic precipitation, unlike the others, is associated with the influence of mountains and is relatively fixed in space.

The relationship between the flood behaviour of the entire river basin is often complex. The downstream characteristics of the flood differ from the upstream characteristics because of lag, routing and scale effects, and changes in geology, physiography and climate from headwaters to the outlet. Flood forecasting is much easier in large basins where the build-up of a flood and the transmission of the flood wave downstream can take days or even weeks (Meuse and Rhine floods of 1995, the Oder floods in 1997; EEA, 2001) and there are more chances for real-time forecasting and mitigation measures such as evacuation and flood protection. In small catchments, with short response times, real-time flood forecasting is much more difficult. The most extreme discharges in such catchments tend to occur as a result of localized convective rainfalls or high-intensity cells within larger synoptic weather systems. Even if there are telemetring raingauges or rainfall radar monitoring over the area, the response times may be too short to issue warnings in real time. The only option is then to issue warnings on the basis of the precipitation forecast, but such warnings tend to be very general. The potential for a flood-producing rainfall might be recognized but it may be very difficult to specify exactly where.

In large catchments both rainfall-runoff modelling and hydraulic modelling of the channels may be involved: the former to determine how much water will contribute to the flood wave; the later to allow predictions of inundation of the flood plain and flooding of property during the event. In small catchments and dispersed settlement like in Slovenia the simulations by using hydrological models are usually suitable for flood prediction.

19.3   EXPERIENCES OF FLOOD FORECASTING IN SLOVENIA

Flood forecasting is in a domain of hydrological services. The organization of a hydrological forecasting service is an internal matter of each country and can differ widely from country to country (WMO, 1994). In Slovenia, the hydrological forecasting service is organized in the frame of the Environmental Agency of the Republic of Slovenia. The service regularly observes the hydrological situations of Slovenian rivers, collects and processes of incoming hydrological information, issues bulletins that relate the current situation about river stages, tendency and any current forecasts or warnings about water levels and discharges, and disseminates the forecasts and warnings to interested users and corresponding authorities. The hydrological reports are issued daily on the basis of available hydrological and meteorological data, meteorological forecast, results of hydrological models and experiences of hydrological forecaster. The forecast is descriptive for 1 to 2 days in advance. In the case of high water events and floods the reports and warnings are delivered frequently during the events. The floods occur quite frequently in Slovenia, but mostly and the heaviest in spring and autumn time. The greatest floods usually occur in autumn when cold front passes central Europe, or by passing of Mediterranean cyclone forms in the bay of Genova (Kobold and Sušely, 2005). The combination of frontal precipitation enhanced by orographic influence plays the most important role in the case of strong events.

Different types of models are used for flood forecasting in Slovenian hydrological forecasting service (Sušnik and Polajnar, 1998; Kobold and Sušnik, 2000). *Regression models* have been developed for predicting the peaks of flood waves, based on statistical analyses of historical high water peak discharges and precipitation events. The regression models for predicting the peaks of flood waves have been developed for some tributaries of the Sava River (Savinja, Ljubljanica, Krka). These models are useful for the quick estimation of peak discharge, but they do not give the information about the time of the peak and the volume of the runoff. In addition to regression models, the conceptual rainfall-runoff models were started to develop for some basins. Among the first, the well known *HEC-1 model* developed by the

US Army Corps of Engineers (Feldman, 1995) has been used for surface runoff simulation. HEC-1 offers many different hydrologic simulation options for each of the main hydrologic processes: precipitation, infiltration/interception, precipitation excess transformation to streamflow, river routing and reservoir routing. This is clearly described by Feldman (1995). The model is not limited to any watershed or river basin size. The subbasin is the smallest areal element in the watershed conceptualization. It can be of any size only limited by the minimum of a one-minute computational interval. Each of the processes is considered to occur uniformly over the subbasin. This is a limitation common to almost all conceptual models. Another limitation of HEC-1 is that the model simulates only single storm events without periods between events because there is no recovery of precipitation loss rates (infiltration, interception, etc.) in the model during periods of no precipitation.

For building up the HEC-1 models the Watershed Modeling System software (WMS, 1997) has been used, which enables to integrate digital terrain analysis and hydrological modelling. The advantage of the software used, is its ability to take digital terrain data for hydrological data development (Kobold and Sušnik, 2000). So far WMS has been implemented on some Slovenian river basins shown in Figure 19.1. Two applications of the HEC-1 model are described. First represents the model set up for the Savinja basin. The Savinja River is the biggest tributary to the Sava River, the main Slovenian watercourse (Figure 19.1), and is

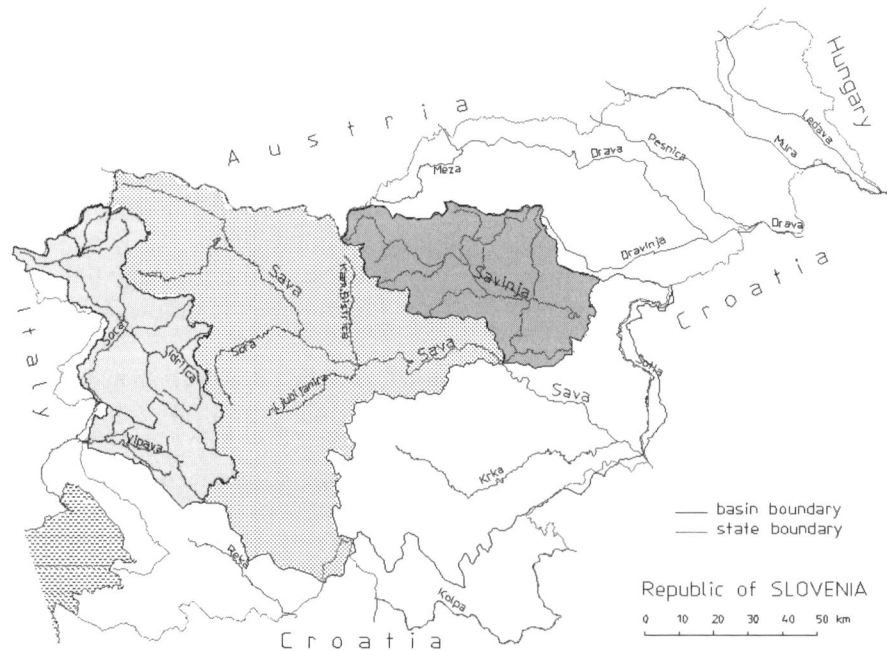

Figure 19.1. River basins in Slovenia covered by the HEC-1 model.

the most flood threatened region in Slovenia. The drainage area of the Savinja basin is $1848\,km^2$. The upper part of the basin is mountainous with altitudes up to 2000 meters (Figure 19.2). The altitudes of the plain area, in the middle reach of the Savinja, are between 200 and 400 meters. Due to its runoff characteristics it has the important influence both on the forming of flood waves and to the forecast of the lower Sava River in Slovenia. Its drainage area can contribute up to 40% of the lower Sava River discharge following extreme meteorological events. The floods, mostly flash floods, are caused by heavy rainfall in headwater mountain areas, especially in autumn. Observed and simulated hydrographs of the last catastrophic flood from November 1998 are presented in Figure 19.2. In that flood event the peak discharge nearly reached the maximum value recorded during the catastrophic flood in 1990 (Kolbezen, 1991). Near the outlet of the catchment at Veliko Širje the peak discharge reached $1458\,m^3/s$ in November 1998, while the value from November 1990 is $1490\,m^3/s$.

Second application refers to the Alpine valley of the Koritnica river which lies in the northwestern part of Slovenia below mountain Mangart (2679 m.a.s.l.) on the border to Italy. The Koritnica catchment covers an area of $87\,km^2$ in headwaters of the Soča River (Figure 19.1). It lies between 400 and over 2000 meters above sea level. The slopes of the catchment are between 60% and 90%. In this part of the Alps the highest number of thunderstorms per year is recorded. More than 400 millimetres of rainfall per day and more than 100 millimetres per hour have been registered there. In autumn 2000 excessive precipitation triggered two landslides, first on November 15 without special damage, and another on November 17 during the night when debris flow killed seven people and destroyed several houses in the village Log pod Mangartom (Mikoš et al., 2002). The period with intense precipitation started in October 2000 and lasted to the end of November 2000. The precipitation from meteorological gauging station Log pod Mangartom (650 m.a.s.l.), located next to the middle of the catchment, and the hydrograph from hydrological gauging station Kal Koritnica, located 1.7 km upstream from the outlet

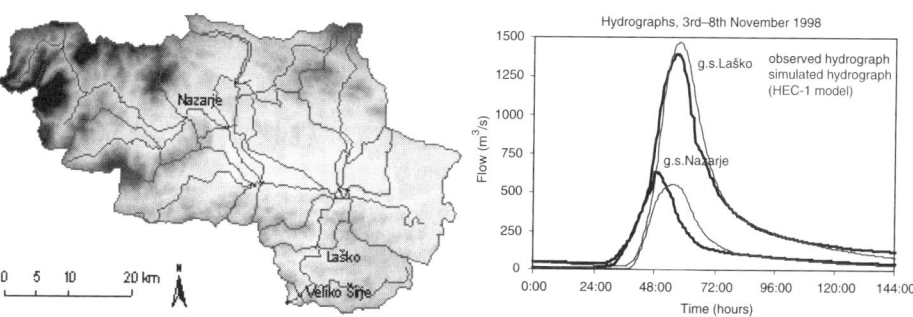

Figure 19.2. The HEC-1 hydrological model of the Savinja basin and flood wave simulation from November 1998.

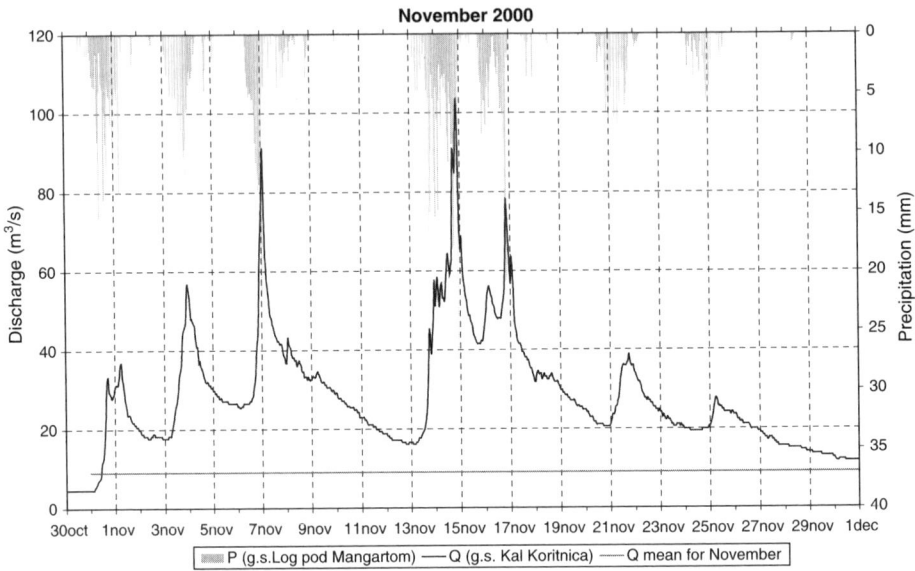

Figure 19.3. River flow hydrograph and precipitation on the Koritnica catchment.

of the catchment, are shown in Figure 19.3. The lag time between precipitation and runoff is less than one hour. About 2000 millimetres of rainfall was registered in the months of October and November 2000 (Mikoš et al., 2002). During the night of the catastrophe about 140 mm of rainfall was registered, common for this area. But monthly amount of precipitation for November 2000 was 1234 mm being more than 100-year event. The maximum peak discharge from November 2000 was 104 m³/s and was below periodical extreme of 311 m³/s. Reconstruction of the event and hydrographs of upstream cross-sections were calculated by HEC-1 model (Figure 19.4). The calculated hydrographs gave the probable discharges regarding the ratios of discharges from their measurements after the event.

The absolute deviation of simulated and observed peak discharges for flood in 1998 (Figure 19.2) is 12.3% for Nazarje and 5.5% for Laško, while the calibration of the HEC-1 model for the Savinja catchment (Kobold and Sušnik, 2000) gave the average absolute deviations of simulated and measured peak discharges between 20% and 30% for smaller subcatchments, but 10% for the whole catchment. The deviations are greater in the mountainous parts of the catchment. The main reason of larger deviations in mountainous parts lies in the insufficient number of recording raingauges (only five recording raingauges) to take into account orographic influences. The average absolute deviation of analysed peaks for the Koritnica catchment is 8%. The analysis of deviations for subcatchments was not possible because there was no any upstream gauging station at the time of the event in November 2000.

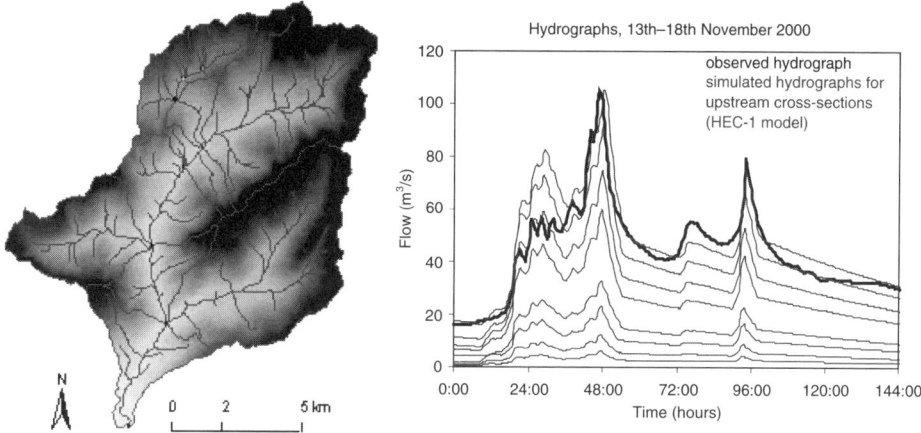

Figure 19.4. The simulation of flood event for the Koritnica catchment by HEC-1 model.

In the year 2003 the Swedish HBV model (Bergström, 1995; Lindström et al., 1997; IHMS, 1999) was tested for runoff simulation on the Savinja basin as a research contribution in EFFS (European Flood Forecasting System) project (EFFS, 2004). The model uses subbasins as primary hydrological units and within these an area-elevation distribution and a classification of land use (forest, open and lakes) can be made. The model consists of subroutines for snow accumulation and melt, soil moisture accounting procedure, routines for runoff generation and a routing procedure. Input data are precipitation, air temperature and potential evapotranspiration. Normally, monthly standard values of potential evapotranspiration are sufficient. The principal output is discharge, however the other output variables relating to water balance components (precipitation, evapotranspiration, soil moisture, water storage) are available from the model (IHMS, 1999). Areal averages of the climatological data are computed separately for each subbasin by a simple weighting procedure. The model has a number of free parameters (Figure 19.5) values of which are found by calibration.

The agreement between observed and computed runoff was evaluated by three main criteria of fit (IHMS, 1999): (a) visual inspection of the computed and observed hydrographs, (b) a continuous plot of the accumulated difference between the computed and the observed hydrographs expressed as

$$Accdiff = \sum (Q_{com} - Q_{obs}) \cdot C_t \qquad (19.1)$$

where Q_{com} is computed discharge, Q_{obs} observed discharge and C_t a constant transforming to mm over the basin in time t, and (c) Nash and Sutcliffe efficiency

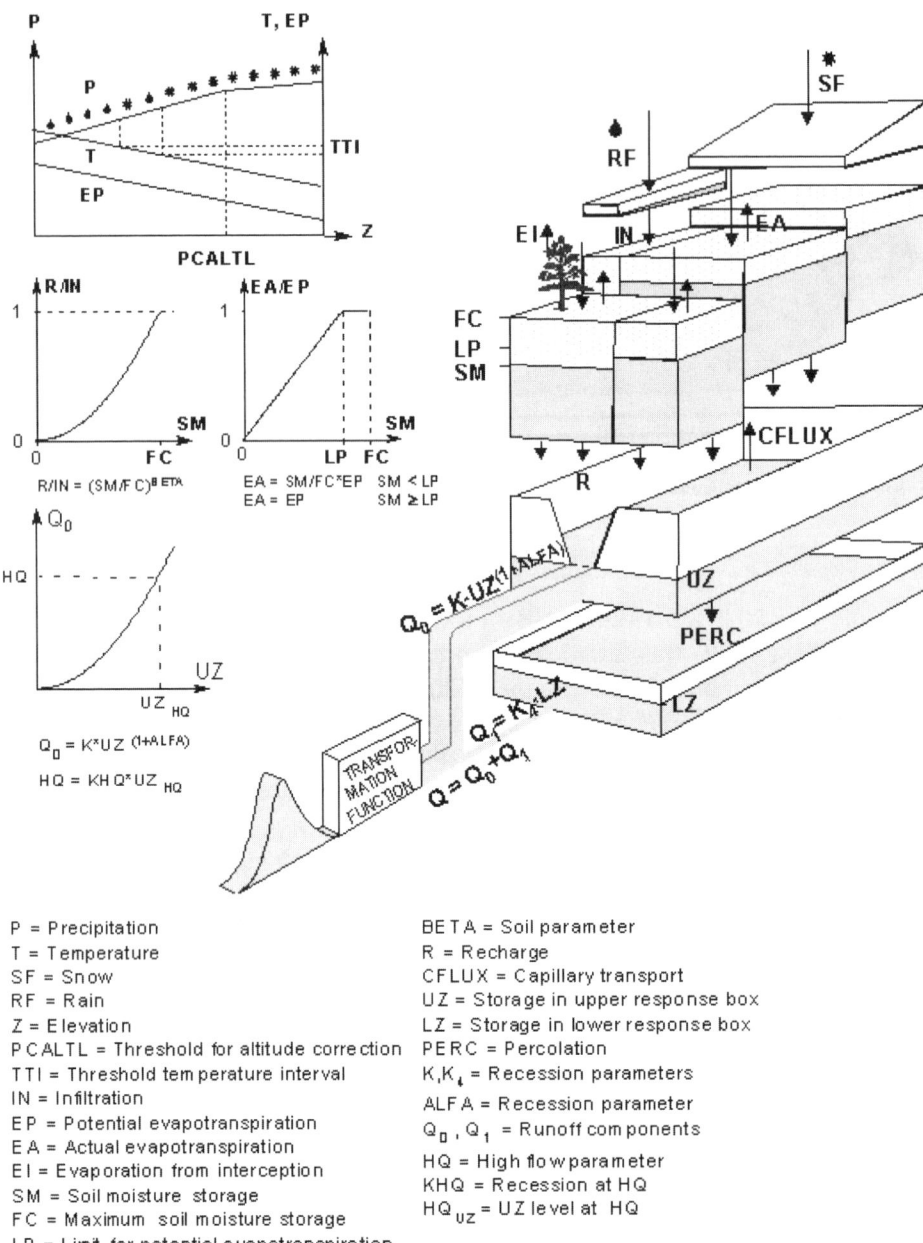

P = Precipitation
T = Temperature
SF = Snow
RF = Rain
Z = Elevation
PCALTL = Threshold for altitude correction
TTI = Threshold temperature interval
IN = Infiltration
EP = Potential evapotranspiration
EA = Actual evapotranspiration
EI = Evaporation from interception
SM = Soil moisture storage
FC = Maximum soil moisture storage
LP = Limit for potential evapotranspiration

BETA = Soil parameter
R = Recharge
CFLUX = Capillary transport
UZ = Storage in upper response box
LZ = Storage in lower response box
PERC = Percolation
K, K_4 = Recession parameters
ALFA = Recession parameter
Q_0, Q_1 = Runoff components
HQ = High flow parameter
KHQ = Recession at HQ
HQ_{UZ} = UZ level at HQ

Figure 19.5. Schematic structure of a subbasin in the HBV-96 model, with routines for snow, soil and response (*Source*: IHMS, 1999).

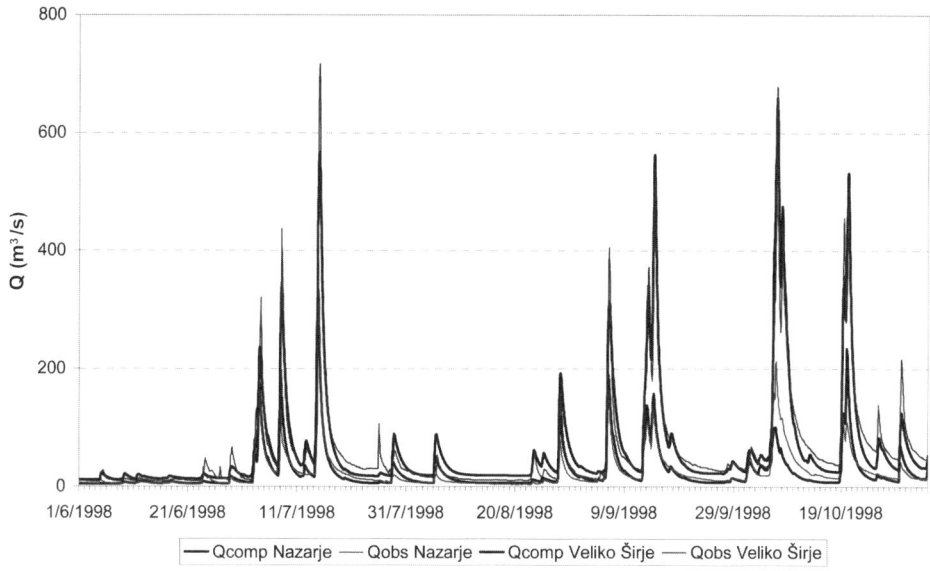

Figure 19.6. A comparison of computed and observed hourly discharges for a part of the year 1998.

criterion (Nash and Sutcliffe, 1970), which is commonly used in hydrological modelling

$$R^2 = 1 - \frac{\sum (Q_{com} - Q_{obs})^2}{\sum (Q_{obs} - \overline{Q_{obs}})^2} \qquad (19.2)$$

A perfect fit would give a value of $R^2 = 1$, but in practice the value above 0.8 means good fit of simulated and measured hydrographs (IHMS, 1999).

The standard HBV model is normally operated on daily time step. To use the model for flash flood forecasting, the version of HBV-96 was applied on the catchment with complex topography like the Savinja basin with time step of one hour (Kobold and Brilly, 2006). For that purpose the hourly values of hydrological and meteorological data for the period 1998–1999 were used in the calibration of the model. Simulated discharges for two water stations Nazarje and Veliko Širje fit with measured ones quite good (Figure 19.6). The R^2 value of 0.78 for g.s. Nazarje and 0.86 for g.s. Veliko Širje was reached by the calibration of hourly data. The criterion of R^2 is smaller for upper sub-catchment, which is mostly mountainous with strong orographic influences. The distribution of precipitation in upper part can vary a lot and areal precipitation over a catchment area is not well defined with available recording raingauges. For the lower sub-catchment the variability of precipitation is not so high and the simulated discharges on the outlet of the basin fit with measured ones quite good. Supposing that the parameters of the HBV-96 model are well defined, the uncertainty of simulated runoff is mainly the result of precipitation uncertainty associated with the average basin precipitation.

19.4   THE SENSITIVITY OF CONCEPTUAL MODELS TO RAINFALL

A peak discharge and time of the peak are the most important information in flood event. The uncertainty of simulated river discharge is mainly the result of precipitation uncertainty associated with the average basin precipitation. Simulation of runoff can give an incorrect result if precipitation is charged with an error what is common not only in estimation of precipitation from raingauges, but also in precipitation forecast (Sattler, 2002; Sattler and Feddersen, 2003; Kobold and Sušelj, 2005). Especially for small catchments the spatial and temporal distribution of precipitation shows a significant influence on peak discharge and total runoff (Faurès et al., 1995; Bell and Moore, 2000).

To find out the measure of runoff uncertainty regarding to precipitation error, the analysis of sensitivity of conceptual models to rainfall error has been performed. The sensitivity of HEC-1 model for the Savinja basin to the rainfall deviation has been carried out by multiflood simulation where the multifloods are computed as ratios from 0.9 to 1.3 of a base precipitation event (Figure 19.7). The deviation in peak discharges for the Savinja River at Laško is shown for more flood

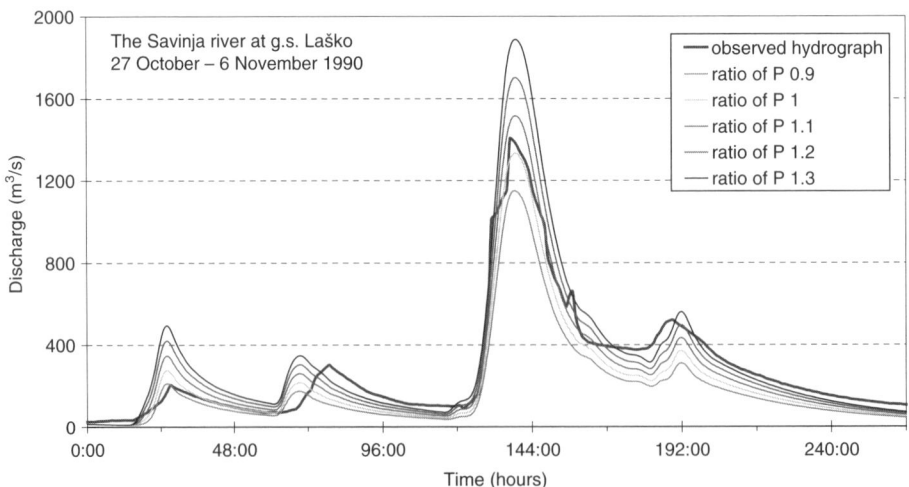

Figure 19.7. Simulated hydrographs by different precipitation scenarios.

events (Figure 19.8). The linear relationship has been obtained and an error in peak discharge is on average 1.6 times greater from an error in rainfall amounts.

Similar, but more detailed analysis was carried out with HBV-96 model for the Savinja basin. The period from beginning of August 1998 to the end of November 1998 (Figure 19.6) was taken in analysis. Precipitation which caused high water events in that period was multiplied by different coefficients from 0.4 to 1.3, representing an error in precipitation. The simulation of each event was performed independent of anothers. That means that only analysed event was charged with an error in precipitation, while the other circumstances remained unchanged. Only peak discharges were analysed in the study ignoring time of the peak. Dimensionless coefficients were calculated for all high water events for water stations Nazarje and Veliko Širje (Figure 19.9), expressed as the ratio of peak discharges calculated by weighted measured precipitation and measured precipitation:

$$k_{Qpeak}^{i,kP} = \frac{Q_{peak}^{i,kP}}{Q_{peak}^{i,P}}$$

(19.3)

where i means high water events, kP weighted precipitation by coefficient k in the range [0.4, 1.3] and P measured precipitation for the event i. The range of peak coefficients is greater in events with low soil moisture, especially in the events occurred after a long dry period when precipitation losses can be very high.

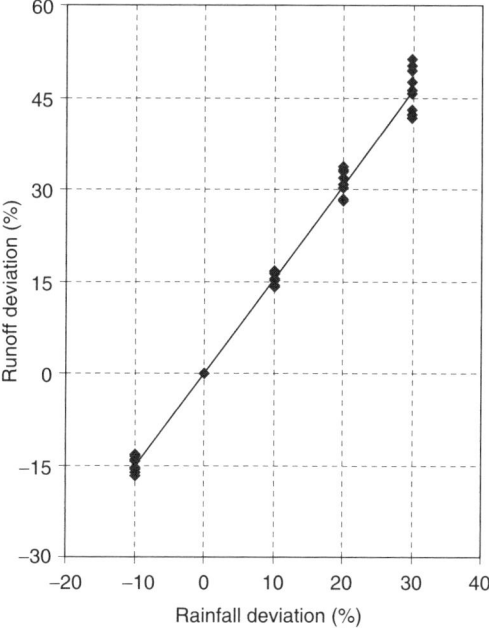

Figure 19.8. Runoff deviation to rainfall deviation for the Savinja at Laško.

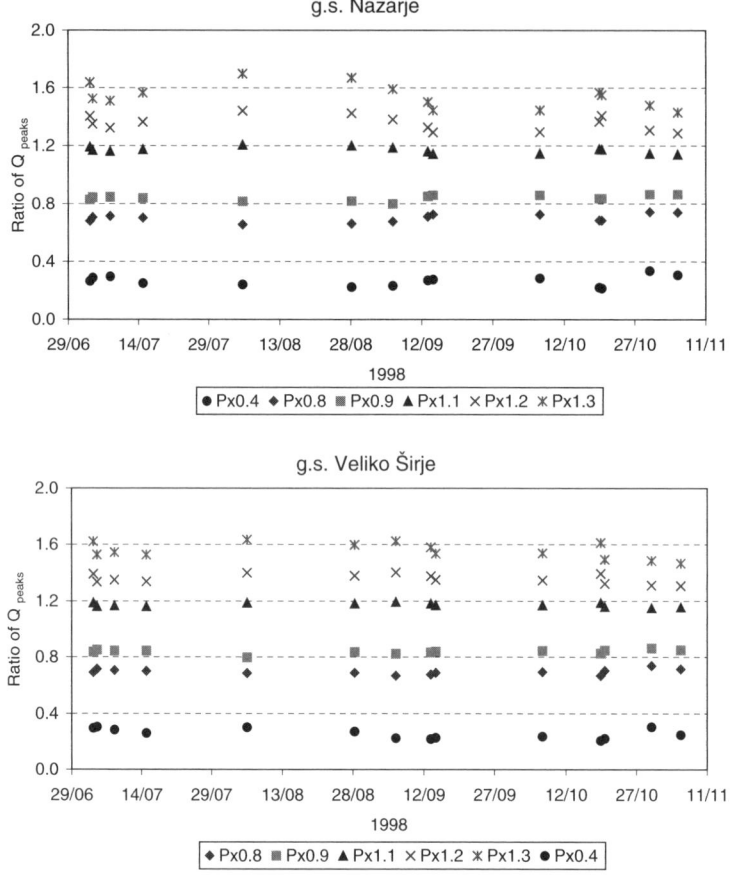

Figure 19.9. Coefficients of peak discharges regarding to an error in precipitation.

The interdependence between coefficients of peak discharges and coefficients of precipitation is shown in Figure 19.10 for both water gauging stations. The coefficients of both stations are mostly covered. The relationship is polynomial and is independent of catchment area. The following polynomial formulae are obtained from Figure 19.10:

$$k_{Qpeak} = 0.65 \cdot k_P^2 + 0.33 \cdot k_P + 0.02 \tag{19.4}$$

for g.s. Nazarje with regression coefficient r^2 of 0.9901, and

$$k_{Qpeak} = 0.67 \cdot k_P^2 + 0.30 \cdot k_P + 0.02 \tag{19.5}$$

for g.s. Veliko Širje with regression coefficient r^2 of 0.9952. k_{Qpeak} is the coefficient of peak discharge and k_P coefficient of precipitation. There are almost no differences in equations.

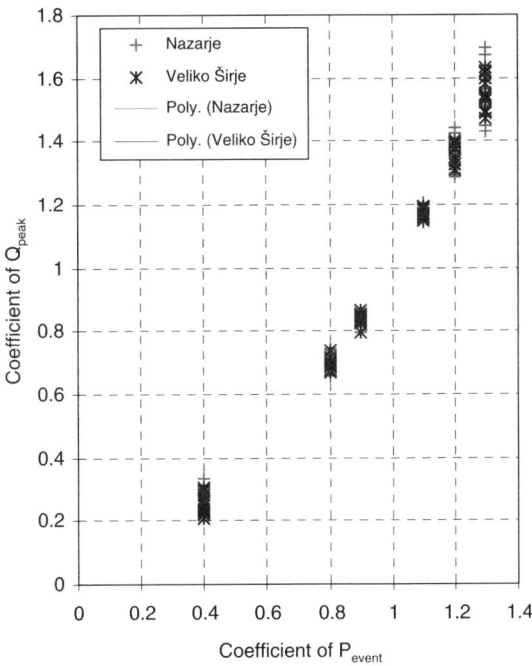

Figure 19.10. Deviation of peak discharges taking into account an error in precipitation, both expressed by coefficients.

Table 19.1. Mean values of coefficients of peak discharges, standard deviations and standard errors resulting from an error in precipitation.

	g.s. Nazarje			g.s. Veliko Širje		
Coefficient of P_{event}	Mean coeff. of Q_{peak}	Standard deviation	Standard error	Mean coeff. of Q_{peak}	Standard deviation	Standard error
0.4	0.27	0.0355	0.0095	0.26	0.0353	0.0094
0.8	0.70	0.0270	0.0070	0.70	0.0196	0.0052
0.9	0.84	0.0200	0.0053	0.84	0.0156	0.0042
1.1	1.17	0.0217	0.0058	1.17	0.0140	0.0037
1.2	1.36	0.0512	0.0137	1.36	0.0325	0.0087
1.3	1.54	0.0836	0.0223	1.56	0.0549	0.0147

Performed analysis has shown that an error in rainfall, which is input into the rainfall-runoff model, can result in great runoff deviation. The mean values of coefficients, standard deviations and standard errors are given for both water stations in Table 19.1. Standard deviation and standard error increase with an error in

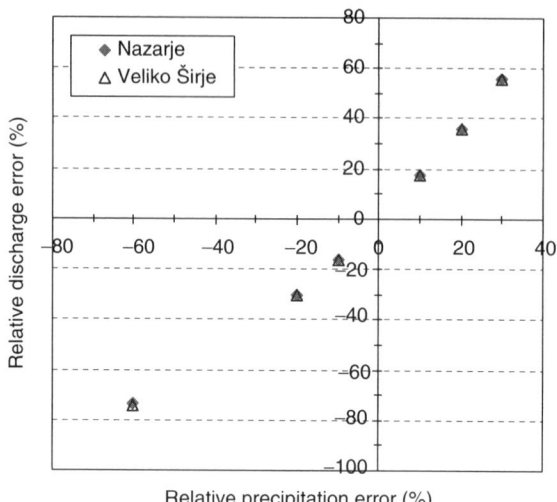

Figure 19.11. The influence of precipitation error on peak discharge error.

precipitation. While the mean coefficient of peak discharge is almost the same for water stations, the standard deviations and standard errors are greater for Nazarje (upper mountainous sub-catchment). The average percentage errors are shown in Figure 19.11.

The results of sensitivity analyses performed with HEC-1 and HBV-96 models are almost identical. Because of selected range of rainfall deviations in HEC-1 model the linear relationship between rainfall and runoff deviations was obtained. It is shown that catchment area does not influence to the interdependence of deviations. Regarding to the results of performed analysis it is very important to assure the accurate precipitation input whether from raingauges or other sources (radar measurements, meteorological forecast). The number of raingauges in the catchment should be densely enough to give proper areal precipitation. However, the accurate representation of precipitation in time and space scale is essential for rainfall-runoff modelling and forecasting.

19.5   FLOOD FORECASTING WITH PREDICTED PRECIPITATION

Numerical modelling is a powerful tool for analysis and forecasting of weather. Currently, quantitative precipitation forecast from four models with very different domain, spatial and time resolution is available at the Agency of the Republic of Slovenia. Results from two global models (ECMWF and DWD/GM) and two limited area models (ALADIN/SI and DWD/LM) are available each day.

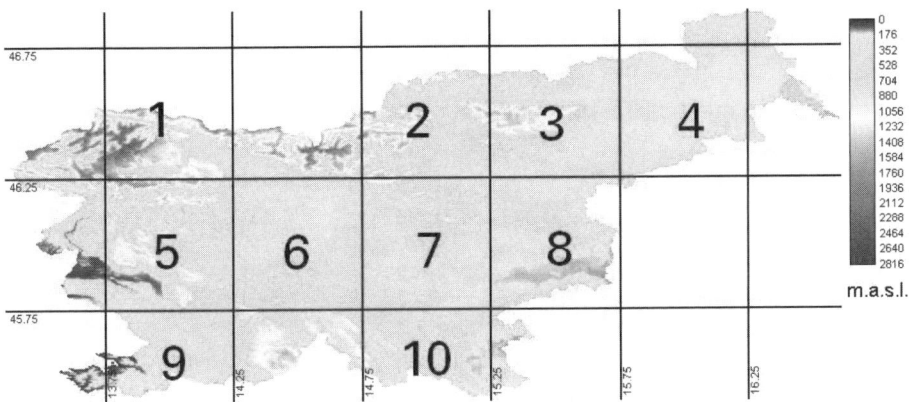

Figure 19.12. ECMWF grid points and relief of Slovenia.

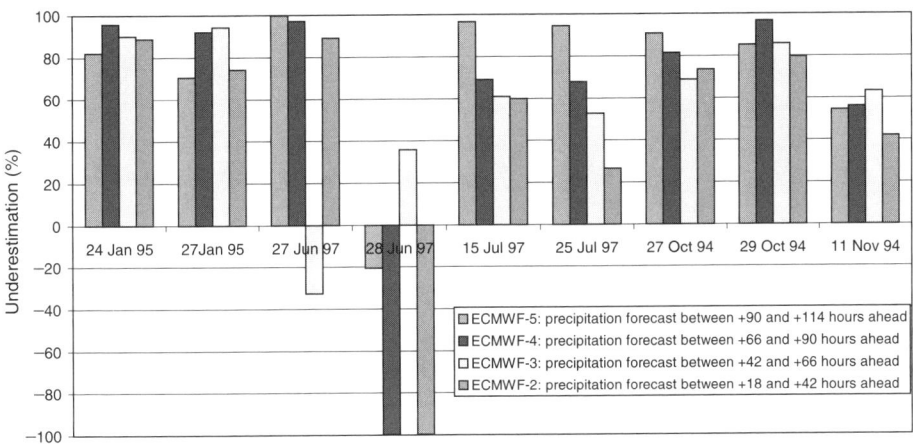

Figure 19.13. Relative difference between ECMWF modelled and measured precipitation for different time intervals for Slovenian territory (Kobold and Sušelj, 2005).

For hydrological forecasting the products of the ECMWF and ALADIN/SI models are generally used.

The European Centre for Medium-Range Weather Forecast (ECMWF) is operationally running global circulation model. It has grid resolution of about 0.5°. Verification has been performed on 10 grid points covering Slovenia. Map is drawn in latitude/longitude projection (Figure 19.12). Results are available for 240 hours in advance. The verification of predicted precipitation was performed for some strong precipitation events (Kobold and Sušelj, 2005). The results show that the ECMWF model underestimates amount of precipitation for Slovenian territory in general except for convective cases (Figure 19.13). The average underestimation

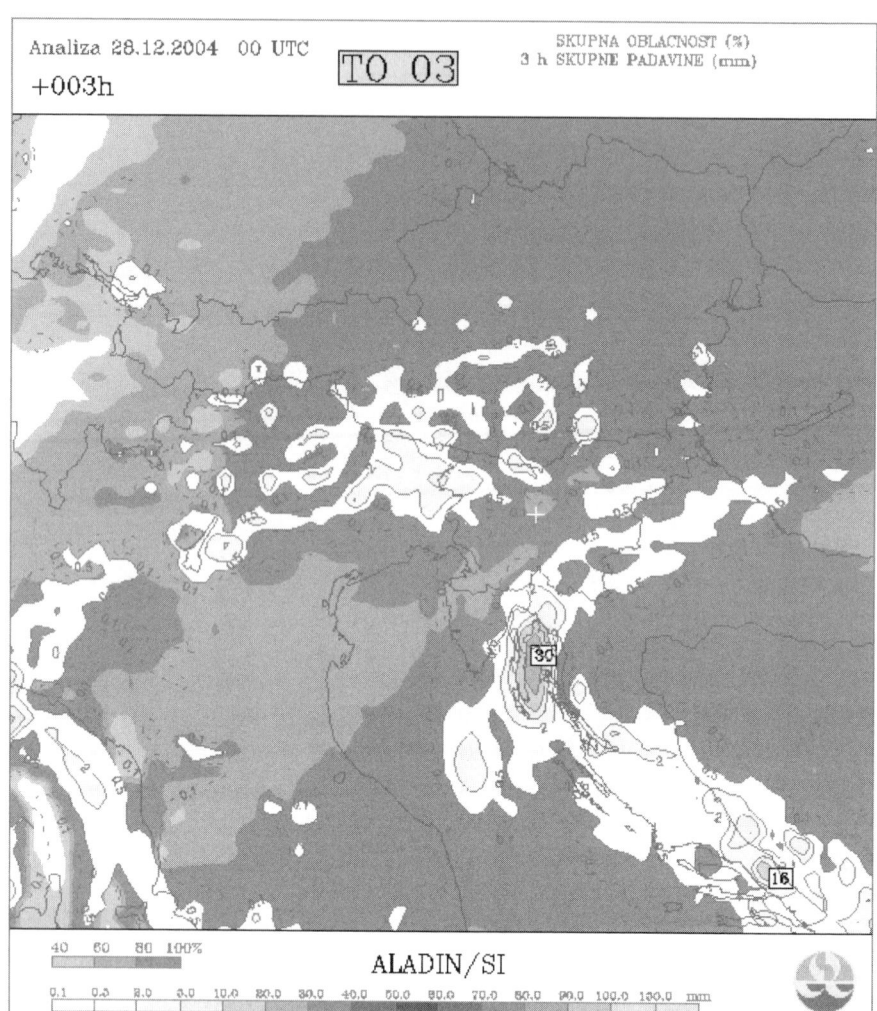

Figure 19.14. Precipitation forecast by the ALADIN/SI model.

is about 60%. The ECMWF model can predict precipitation events correctly, but it is unable to predict the distribution and amount of precipitation correctly.

The forecasts of mesoscale ALADIN/SI model covering Slovenia (Figure 19.14) were tested for operational short-term forecast in the year 2004. In that case the HBV-96 model for the Savinja catchment was run using measured input data of precipitation and temperature for a period before forecast and forecasted precipitation and temperature for a period of forecast. The forecasts of meteorological parameters of the ALADIN/SI model (Vrhovec et al., 1998) are made in spatial resolution of about 9 km and are available up to 48 hours in advance. The model

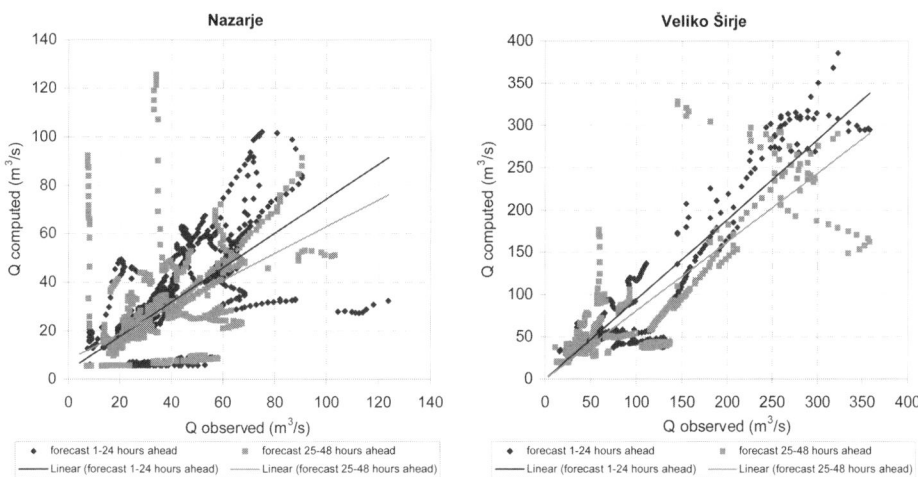

Figure 19.15. The comparison of simulated and observed discharges of operational HBV-96 runs for the Savinja catchment (period from February to August 2004).

is operational and run twice a day. The results of simulations for the period from end of February to August 2004 are presented in Figure 19.15. The comparison of simulated and observed discharges is separated on two parts: forecasts for the first day and forecasts for the second day. The forecasts are better for the first day and they are acceptable for Veliko Širje which is close to the outlet of the catchment. The correlation coefficient r is 0.93 for the forecasts of first day and is 0.82 for the forecasts of second day. Very poor correlation is observed for the upper part of the catchment at Nazarje. The correlation coefficient r is 0.73 for the forecasts of first day and below 0.50 for the forecasts of second day.

The deviations of runoff are mainly the result of uncertainty of ALADIN/SI precipitation. The comparison of areal hourly forecasted precipitation and precipitation calculated from raingauges was made for the Savinja catchment, for the first day (1–24 hours ahead) and for the second day (25–48 hours ahead) of forecast (Figure 19.16). There is no correlation between predicted and measured hourly areal precipitation, the correlation is better for daily areal precipitation. The deviations of predicted daily areal precipitation from measured one are shown in Figure 19.17. The ALADIN/SI model predicts the rainfall event well, but the amount of precipitation can be underestimated or overestimated. The average relative error for precipitation over 3 mm is 46% for the first day of forecast with standard deviation of 34%. For the second day of forecast the deviation is greater; average relative error is 55% with standard deviation of 39% for precipitation over 3 mm. Deviations of precipitations are reflected in deviations of simulated runoff (Figure 19.15).

It is clear that rainfall-runoff models enable the simulations of flash floods and can be used in pre-warning systems. But for the time being the uncertainty of

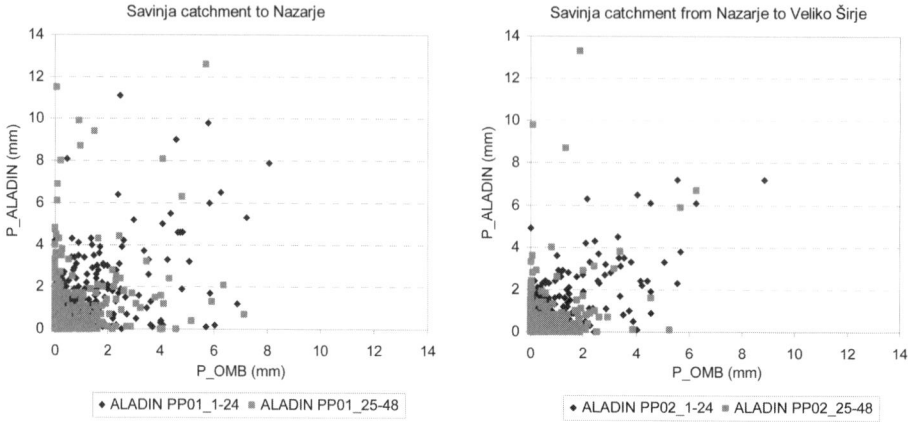

Figure 19.16. The comparison of predicted (P_ALADIN) and measured (P_OMB) hourly precipitation for the Savinja catchment.

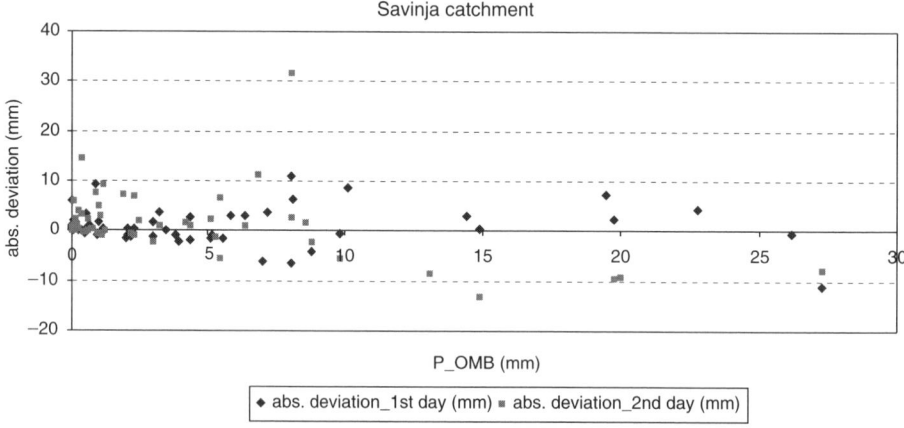

Figure 19.17. Absolute deviation of predicted daily precipitation for the Savinja catchment.

predicted precipitation is still very large and that make flood forecasting and warning much risky. If spatial and time distribution of forecasted precipitation is accurate enough with known uncertainty then hydrological models can aid to give advance warning of potential flooding.

19.6   CONCLUSIONS

Operational hydrological forecasting depends both on measured and forecasted meteorological data. Hydrological forecast is made in real time for the purpose of informing the users (e.g. civil protection, government agencies, the public, private enterprises) and provides a basis for operational decisions in the case of flood. Deficiency of precipitation measurements and uncertainty of precipitation forecast are the main source of uncertainty in rainfall-runoff models and flash flood forecasting.

Rainfall-runoff models are effective tools for flood forecasting. Weather forecasts coupled with information on river basin character and hydrological rainfall-runoff model offer an advance warning of potential flooding. But the uncertainty of predicted precipitation is still very large and on the other hand, the rainfall-runoff models are sensitive to precipitation input. For example, an error of 10% in the amount of overestimated precipitation overestimates the peak of flood wave for about 17%. Uncertainty of precipitation forecast is the main reason for limited use of rainfall-runoff models and it is the primary source of uncertainty in flood forecasting. The accurate precipitation with quantification of uncertainty of predicted precipitation is necessary to increase the reliability of hydrological forecasts and warnings.

Slovenian experiences show that the variability of precipitation in Slovenia is very high. Intense precipitation enhanced by orographic influence causes flash floods, which are characteristics for the most of Slovenian rivers. Those make the flood forecasting more difficult. The resolution of global models like ECMWF is not high enough to predict local and intense precipitation on the territory of Slovenia. The predicted precipitation has very high systematic underestimation. In high precipitation events the model underestimates precipitation amount by average of 60%. Local meteorological ALADIN/SI model has higher resolution, but it can miss the regions of the highest intensities. The model predicts the rainfall event well, but the amount of precipitation is underestimated or overestimated.

REFERENCES

Askew, A.J., 1989. Real time intercomparison of hydrological models. In: New Directions for Surface Water Modelling (Ed. Kavas M.L.), IAHS Pub No. 181, p. 125–132.

Bell, V.A. and Moore, R.J., 2000. The sensitivity of catchment runoff models to rainfall data at different spatial scales. Hydrology and Earth System sciences, 4(4), p. 653–667.

Bergström, S., 1995. The HBV model. In: Computer Models of Watershed Hydrology (ed. by V. P. Sing), 443–476. Water Resources Publication, Colorado, USA.

Beven, K., 2001. Rainfall-runoff modelling. The Primer. John Wiley & Sons, Ltd., England.

EFFS, 2004: A European Flood Forecasting System EFFS. Full Report, edited by: Gouweleeuw, B., Reggiani, P., and de Roo, A., Contract no. EVG1-CT-1999-00011, WL Delft Hydraulics, Netherlands.

Faurès, J.N., Goodrich, D.C., Woolhiser, D.A., Sorooshian, S., 1995. Impact of small-scale spatial rainfall variability on runoff modeling. Journal of Hydrology 173, p. 309–326.

Feldman, A.D., 1995. HEC-1 Flood Hydrograph Package. In: Computer Models of Watershed Hydrology (ed. by V. P. Sing), 119–150. Water Resources Publication, Colorado, USA.

IHMS, 1999. Integrated Hydrological Modelling System. Manual, Version 4.5. Swedish Meteorological and Hydrological Institute, Norrköping, Sweden.

Job, D., Humbel, M., Müller, D. and Sagna, J., 2002. Flood estimation Buenz Valley, Switzerland. Practical use of rainfall-runoff models. In: Proceedings of International Conference on Flood Estimation, CHR Report II-17, Berne, Switzerland, p. 519–527.

Kobold, M. and Sušelj, K., 2005. Precipitation forecasts and their uncertainty as input into hydrological models. EGU, Hydrology and Earth System Sciences, 9(4), p. 322–332.

Kobold, M. and Sušnik, M., 2000. Watershed modelling and surface runoff simulation. In: Tagungspublikation Internationales Symposion INTERPRAEVENT 2000 – Villach, Österreich, Band 2, p. 329–338.

Kokkonen, T.S. and Jakeman, A.J., 2001. A comparison of metric and conceptual approaches in rainfall-runoff modeling and its implications. American Geophsical Union (AGU), Reprinted from Water Resources Research 37, number 9, p. 2345–2352.

Kolbezen, M., 1991. Flooding in Slovenia on November 1, 1990. Ujma 5, Ljubljana, p. 16–18.

Lindström, G., Johansson, B., Persson, M., Gardelin, M. and Bergström, S., 1997. Development and test of the distributed HBV-96 hydrological model. Journal of Hydrology 201, p. 272–288.

Nandakumar, N. and Mein, R.G., 1997. Uncertainty in rainfall-runoff model simulations and the implications for predicting the hydrologic effects of land-use change. Journal of Hydrology 192, p. 211–232.

Nash, J.E. and Sutcliffe, J.V., 1970. River flow forecasting through conceptual models. Part I. A discussion of principles. Journal of Hydrology 10, p. 282–290.

Singh, V. P., 1995. Computer Models of Watershed Hydrology, Water Resources Publication, Colorado, USA.

Smith, K. and Ward, R., 1998. Floods, Physical Processes and Human Impact, John Wiley & Sons Ltd, Chichester, England.

Sušnik, M. and Polajnar, J., 1998. Simple hydrological forecasting models: operational experience. Proceedings of XIXth Conference of the Danube Countries, Osijek, Croatia, p. 31–36.

Vrhovec, T., Žagar, M., Brilly, M. and Šraj, M., 1998. Forecasting of intense precipitation using the ALADIN-SI model and modelling of the runoff. In: Proc. the 17th Goljevscek Memorial Day, Acta hydrotechnica 16/23, Ljubljana, Slovenia, p. 71–84.

WMO, 1975. Intercomparison of conceptual models used in operational hydrological forecasting, Operational Hydrology Report No. 7, Geneva, Switzerland.

WMO, 1992. Simulated real-time intercomparison of hydrological models, Operational Hydrology Report No. 38, Geneva, Switzerland.

WMO, 1994. Guide to hydrological practices, WMO-No. 168, Fifth edition, Geneva, Switzerland.

WMS Manual, 1997. Watershed Modeling System, Reference manual, Version 5.0, Brigham Young University, Engineering Computer Graphics Laboratory, Provo, Utah.

20

An Overview of Flood Protection Barriers

Jean-Luc Salagnac
Centre Scientifique et Technique du Bâtiment, Paris, France

ABSTRACT: An obvious protection principle against floods is to prevent water from reaching buildings or built areas. This is the role of permanent dykes. Temporary equipments have been developed to ensure the same function in case of flood event.

This paper presents an overview of the main types of such equipments. The principle is described and discussion about their performances and limits is introduced. Examples and references are given.

20.1 INTRODUCTION

The most obvious protection principle against floods is to prevent water from reaching buildings or built areas. This of course can be done by choosing construction location out of reach of water.

In flood prone areas, dykes ensure this protection function. These permanent constructions need huge investments and have to be properly maintained to provide the expected protection.

Either for emergency purposes or as temporary protection equipment, various types of removable dykes have been developed. They are an alternative to traditional sand bags.

The following paragraphs describe the principle of such equipments and discuss the performances and the limits of these equipments. References of commercial products are given.

20.2 GENERAL PRINCIPLES

The function of a temporary dyke is similar to the function of a permanent dyke: avoid flood water from running over a given area. There is for instance a specific demand of such equipment for city quarters close to river banks where the absence of permanent dyke allows a good integration of the river in the city landscape.

The principle is to build a structure that will hold water. As such, the requirement for this engineering piece of work is rather simple: resist the thrust of water.

Table 20.1. List of barrier manufacturers (not exhaustive).

Type of barrier	Web sites of manufacturers
Vertical walls	www.bauer.de www.eko-system.de www.msu.fr www.poralu-sa.fr www.interalliance.fr
Folding standby flood defense	www.dutchdam.nl
Dihedral structures	www.bauer.de www.geodesign.se www.sodilor.com
Flexible liners	www.megasecur.com
Water gravity dams	www.jbtt.be www.flooding-agency.com www.beaver-ag.com www.pronal.com www.interalliance.fr
Big bags gravity dams	www.quick-damm.de
Water absorbent bags	www.hydro-bag.com www.sacabso.com

Additional requirements concern the adaptation to local ground conditions, the time needed to erect the barrier, the storage of elements between flood periods and the acceptable water leakage. The cost of the solution takes all these requirements into account.

Different design principles have been proposed to build such temporary dykes that will be analysed in the following paragraphs:

- assembly of elements to a permanent infrastructure in order to form a vertical wall,
- assembly of dihedral elements which form a tilted surface that allows water to press the elements on the ground,
- dihedral shaped flexible liners,
- gravity dams using water reservoirs or big bags,
- water absorption bags.

The effectiveness of such equipments relies on the possibility to obstruct sewers and others underground ducts. If they are not obstructed, flood water may spring from traps situated in the area protected by the barriers.

20.3   VERTICAL WALLS

The general principle is shown on figure 20.1. Horizontal metallic (aluminium) boards are assembled between vertical posts. These posts are first fixed to

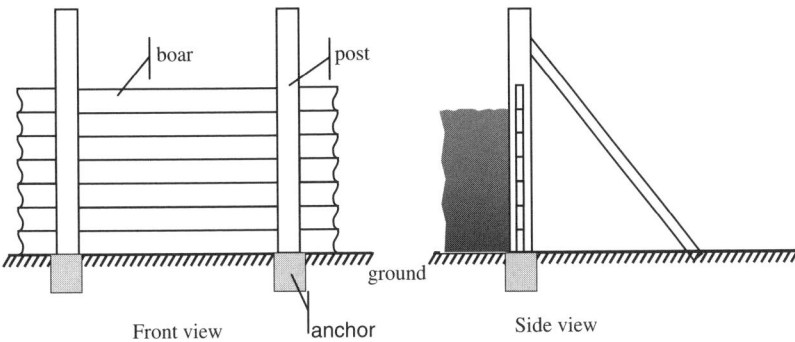

Figure 20.1. Principle of the vertical wall barrier.

permanent anchor plates that form a horizontal base for the system. Steps of this horizontal base are allowed as far as the permanent base line is designed for.

The modular design allows the adaptation of such walls in many circumstances. Straight or bent barriers can be built.

The water tightness between horizontal boards and between boards and posts is ensured by (rubber) joints.

The dimensions of the elements result from engineering as well as from practical considerations (ability of the elements to be (manually) handled, interchangeability of elements ...). Distances between posts are typically from 2 to 5 m. Maximum height of commercially available barriers reaches 5 m. According to design, additional struts may be necessary to ensure the stability of the structure.

20.4   FOLDING STANDBY FLOOD DEFENSE

The general principle is shown on figure 20.2. The barrier is a ready to be erected system which consists of a folding structure. Out of flood periods, this structure is folded flat in a trench that is covered with plates. The erection of the barrier is ensured by opening the plate and unfolding the structure. Stanchions on the back of the dam hold the construction erect. This system allows a fast installation and limits the logistics needed to varry barrier elements on the spot.

20.5   DIHEDRAL STRUCTURES

The general principle is shown on figure 20.3. The dihedral structure is made from the assembly of individual dihedrals. One plane of the (metallic) dihedral structure is in contact with the natural ground. Watertight plates are fixed on the other tilted plane of the structure. The water pressure pushes the structure on the ground.

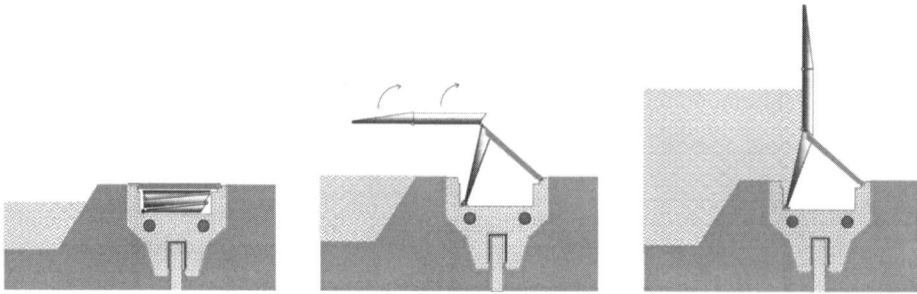

Figure 20.2. Principle of the folding standby flood defense.

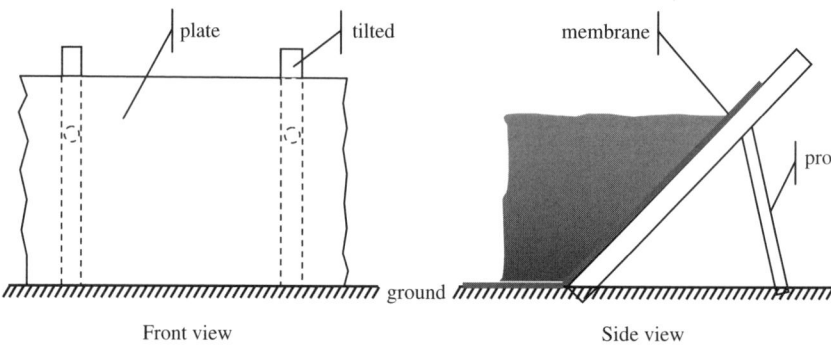

Figure 20.3. Principle of the dihedral barrier.

This action may not be enough to make the barrier watertight: an additional membrane is then often necessary to limit water leakage.

To limit the potential horizontal movement, the extremity of the horizontal part may be pined in the ground.

The installation of such a structure is does not need any permanent anchoring system. The performances are more modest than potential performances of vertical walls. Typical height is about 1 m though some commercially available products claim for more.

Different types of plates are proposed: wood pallets with a membrane, metallic plates, recycled PVC plates.

20.6   FLEXIBLE LINERS

The general principle is shown on figure 20.4. The liner is unrolled on the ground so that the flexible dihedral structure is opened by the water flow and kept open by the water pressure. Would the water level rise, then the structure would open wider.

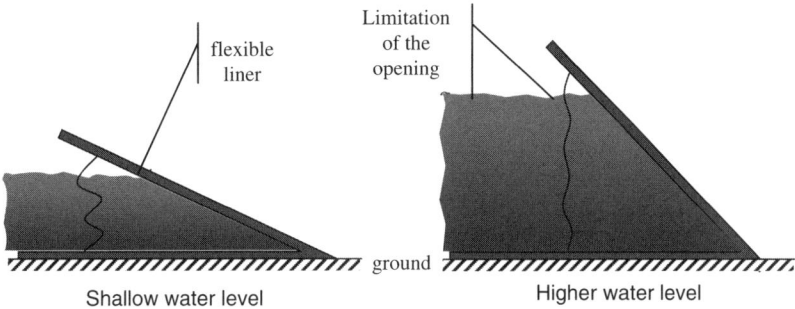

Figure 20.4. Principle of the flexible liner barrier.

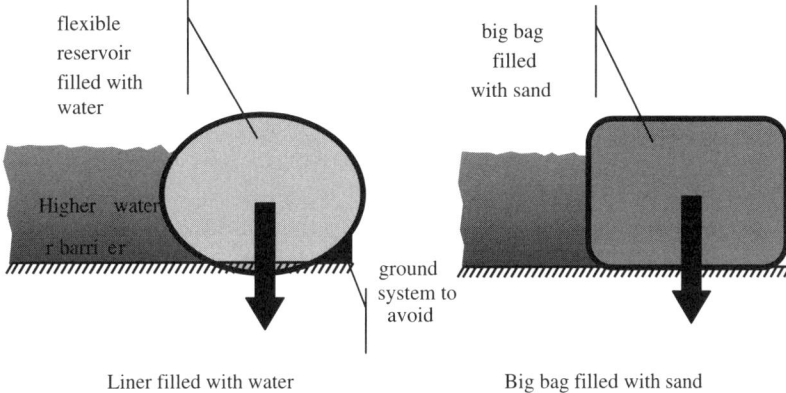

Figures 20.5 and 20.6. Principle of the gravity barrier.

Distributed weights on the membrane in contact with the ground contribute to the watertightness and stability of the system.

Liners can be connected in order to get the required barrier length. The maximum water height is 2 m. As far as the water level is not too high, vehicles can drive over the barrier without affecting the system.

20.7   WATER GRAVITY DAMS

The general principle is shown on figure 20.5. Volumes (tubes, assembly of tubes …) made of liners material are filled with water. The weight of such a structure provides a simple solution to realise a temporary dam. Some products are first installed with compressed air in order to unfold the dam structure.

Volumes can be connected to get the required length. The maximum water height is about 1 m depending on the friction factor of the filled volume on the ground. In the case of tubular structures, two adjacent tubes can be connected in order to avoid the tube-like dam to roll on the ground surface.

20.8   BIG BAGS GRAVITY DAMS

The general principle is shown on figure 20.6. Empty structures made of a metallic frame and of geotextile envelope are unfolded before being filled with about 2 m^3 sand with a loader. The weight of the structure ensures the mechanical resistance to water pressure.

Adjacent bags can form a long temporary dam. One meter high bags can be stacked in order to form higher dams.

20.9   WATER ABSORBENT BAGS

These bags made of textile contain polymers with very high water absorption capacity. They are light when dry and water is absorbed rather quickly. When stacked, they can form a barrier. Different sizes and shapes of bags allow the adaptation of this system in different circumstances. The recommended barrier maximum height by manufacturers is around 1 m.

These bags can only be used once. Then must then be stored in dry places before use. After drying, used bags are processed as garbage.

20.10   SELECTION CRITERIA

An important criterion when selecting a system is the assembly time. The shorter the floods warning time, the shorter the assembly time must be. In practice, most of the described systems that require the assembly of structures are limited to slow plain floods where the water level rises during several days. The systems which are installed by unrolling or unfolding structures show a quicker installation time.

Even if less critical than the previously mentioned criterion, another criterion is the storage volume needed to stack elements. The storage organisation has to be considered together with the logistics organisation. Local authorities have to plan the circulation of trucks under various scenarios. Here again, this performance differs from one system to another.

Cost is also a critical issue. Information is available from systems providers. A balance has to be made from a vulnerability assessment study between the cost of the barrier and the avoided damages.

20.11   PERFORMANCE ASSESSMENT

The indicated performances are based on self declaration of manufacturers.

The performance assessment of the described water barrier systems is not yet very developed. The British Standard Institution published the PAS 1188-2 (Publicly available specification) on flood protection products. This document is not a standard. It *"specifies requirements for the designation testing, factory production control, installation documentation and marking for different configurations of flood protection products intended to be demountable or for temporary installation, for use in the UK or locations with similar exposures, i.e. where there is a temperate climate and advanced warning of flooding is available"*.

The interest of insurance companies for these systems should give an impulse to the development of a third party performance assessment procedure. The result of such a procedure would be a public document containing the description and characteristics of the system, its performances and limits as well as the installation, storage and maintenance conditions.

ACKNOWLEDGEMENT

The reported information is mainly taken from 2004 to 2006 studies carried out in cooperation with the French ministry of construction (Direction Générale de l'Urbanisme, de l'Habitat et de la Construction) and with the French ministry of environment (Direction de la Prévention de la Pollution et des Risques).

21

An Innovative Semi-permanent Flood Protection Structure – Alternative to Sandbags and Supplements to Conventional Earth Embankments

Jarle T. Bjerkholt & Oddvar G. Lindholm
Norwegian University of Life Sciences, Department of Mathematical Sciences and Technology, ÅS, Norway

ABSTRACT: This article reports on innovation and development work of a semi-permanent flood protection structure. This flood protection structure has a preferable use in densely populated urban areas where the access, physical and visual, to a river or other water body should be preserved. Other places it can be utilised are places where it is not possible to build high earth embankments or not possible to build such structures at all. This is also a structure that is easy to use as a closure structure around important objects. The structure developed is described and some of the results from performance tests are reported. Some of the installations that are already done are shown.

21.1 INTRODUCTION

Floods occur naturally all over Norway and world wide and have a naturally and necessary function to keep the balance in aquatic systems. Generally floods in Norway are caused by rain and melting snow, often in combination with saturated or near saturated soil conditions (Eikenæs et al., 2000). Combination of rain and snow melt can cause extreme floods in Norway. In urban areas floods are often caused by intensive rain. The annual cost of damages caused by floods in Norway is on average approximately NOK 200 million, the equivalent of 25 million Euros (NOU, 1996; Sæterbø et al., 1998; Nordskog, 2006). Increased population density, more intensive use of flood prone areas and land use changes are increasing the potential for damages of property. Possible changes in climate can also lead to an increased potential for damages.

Flood protective measurers shall prevent property damages in areas near rivers and lakes or areas close to the sea and shall also protect the water quality and environment of the mentioned water bodies. To achieve these goals it is often necessary to combine different types of protective measures. The most common hydro-technical work for flood protection is earth embankments. Earth embankments normally provide a cost effective and reliable protection against flooding. However, it is not always possible to install this kind of structure in all locations or build them as high as the river engineers recommend. The levees are designed for a maximum height of flooding water given by the return period or the reciprocal annual probability for a flood of a given size to occur. The longer the return period the higher the design flood water level. The higher the earth embankments the more people will have their view reduced or changed.

As already pointed out not all locations are suitable for a conventional earth embankment and the embankments can not always be built to the height given by the wanted return period. Given this it is always necessary to have a certain state of readiness for temporary systems for flood protection. In some cases the temporary or semi-permanent measures may be the only suitable measures. Traditionally, sandbags have been the standard method to build temporary flood barriers. This is a method that is very time consuming to install and also to remove. Due to possible contaminated flood water, sandbags may be regarded as hazardous waste and must be treated as such. Removing sandbags after a flood can therefore also be costly.

In many cases the flood problem is repetitive in the way that some areas are flooded quiet frequent with low water levels, but it is not possible to protect these areas with a permanent structure. Permanent flood protective structures, like earth embankments, are in some cases limited in height by lack of space, the fact that it blocks the view from houses behind the embankment or by other reasons. In many cases the difference between a return period of 100 years and 500 years can be a lot less than 1.0 m. Given this the Norwegian University of Life Sciences started collaboration with the inventor Mr. Thor Olav Rørheim in 1997, with the aim to develop new flood protective structures that could overcome some of the problems associated with sandbags. It was also the intention to develop equipment that could be a supplement to conventional earth embankments.

21.2   OBJECTIVES

The objectives of the work reported here was to develop a method that could be an alternative to sandbags for temporary flood protection and a supplement to earth embankments. Further, the system developed should be easier and faster to install and remove than sandbags. The system should also be reusable and should be at least as cost effective as sandbags. The system should have satisfactory stability against overturning and sliding and the seepage underneath the structure had to be kept at an acceptable level.

21.3   DESCRIPTION OF THE SYSTEM

The semi-permanent flood protection system went through a number of design stages and tests before the system described here was finally patented (Norwegian patent no: NO 314412). The patent is hold by a company called AquaFence and the described system is called AquaFence Semi. The use of the system described need some planning and preinstallations and is by its nature not a direct substitute for sandbags. However, flood safety and preparedness plans to day normally includes sandbags to be utilised if the flood water rises over a certain level, meaning above the height of permanent flood protection structures. In urban areas sandbags has traditionally been the only possible protection of property from rising flood water caused by intensive rain.

The AquaFence Semi system comprises of to main parts; one part that is a permanently installed foundation of concrete and one mobile part that can temporarily be fixed to the foundation. A sketch of the system is shown in figure 21.1. The over ground part of the structure is fixed to the foundation by straps or bolts. This part is made in sections each of 2.0 m and 1.2 m wide and 1.2 m tall, constructed to hold

Figure 21.1. Sketch of the AquaFence semi flood protection system (courtesy of AquaFence AS).

a flood height up to 1.0 m, giving a freeboard of 0.2 m allowing for some differences in the terrain etc. The barrier is made of 15.0 mm plywood. Each section weighs 80 kg. Basically each section comprises of two sheets of plywood design for marine environment, where one is laid flat and the other is supported in upright position whit aluminium members which are fitted into special rapid-lock holders. Each section is joined to the next with a piece of very durable reinforced PVC which is also used to join the two plywood sheets in each section together. The system is intended for use in a well defined place where the flood water is still or flowing slowly.

Seepage underneath the structure or through the ground is not an issue, at least not if installed properly. The foundation is to include a membrane in the ground that is stopping water from seeping through the ground. This is a very important detail due to the small width of the over ground structure. Failing to install the membrane properly may cause a hydraulic exit gradient close to the critical value of approximately 1.0, especially if the soil is held saturated from the flood water for a prolonged period. A hydraulic gradient close to 1.0 may generate ground failure due to washing out of the soil underneath the structure (Janbu, 1989).

To stop the water from passing between the structure and the foundation, a thick flexible polyurethane seal is fitted between the foundation and the over ground structure. The seal is an integrated part of the structure, but can easily be replaced. It is recommended that the seal is replaced after every use to secure a problem-free next time installation and a properly working seal.

The structure is lightweight and easy to install by hand. Normally there should be a minimum of four people to carry out the installation (Rørheim, 2006). The construction allows for small changes in alignment and terrain. Special corner structures are available to allow for sharp corners e.g. to make an enclosure around a building. To make shore that the structure is installed properly the personnel allocated for this job should have some training. However, given that the foundation is in place the over ground structure is fairly fast to install. An instruction for installation will be printed on each element. According to calculations and tests; ten men can put up 100 m of the structure in just one hour.

Initial investments for this structure are high compared to sandbags, mainly due to the need for a secured storage place. This is partly compensated for by the possibility to use the structure several times and lower costs connected to installation, removal and clean-up operations. The AquaFence Semi-system can be delivered in a container, as a self contained unit including all over ground structures. A 40′ container can store up to 400 m of the 1.2 m high flood fence. One should also include in the calculation of cost that the less time needed for installation the less time in advance of an expected flood situation the system has to be installed. This is an advantage to what we can call timeliness costs. The possibility to await the situation longer and see how the flood develops before an installation order is given is a possibility for saving money. This means that it is less possible that an unnecessary installation is carried out.

21.4   ANALYSIS OF STABILITY

To carry out an analysis of stability for a temporary structure like this one must build on many of the same assumptions as for permanent structures. However, some parameters are harder to control for a temporary structure than for permanent installations. The following assumptions are made for these calculations:

- The density of water; $1000 \, \text{kg/m}^3$, the density of concrete; $2400 \, \text{kg/m}^3$.
- The hydrostatic uplift force beneath the structure is assumed zero. The assumption is based on the fact that the presence of a membrane will stop water from seeping through the ground.
- The normal forces are equally distributed over the whole interface between the structure and the ground.
- The shear resistance of the soil is assumed being purely frictional, no cohesion. Practically this is adding some extra safety.
- The weight of the over ground structure is very low and are neglected in the calculations.
- All calculations are made per meter length of structure.

In figure 21.2, a sketch of the structure and the acting forces are shown.

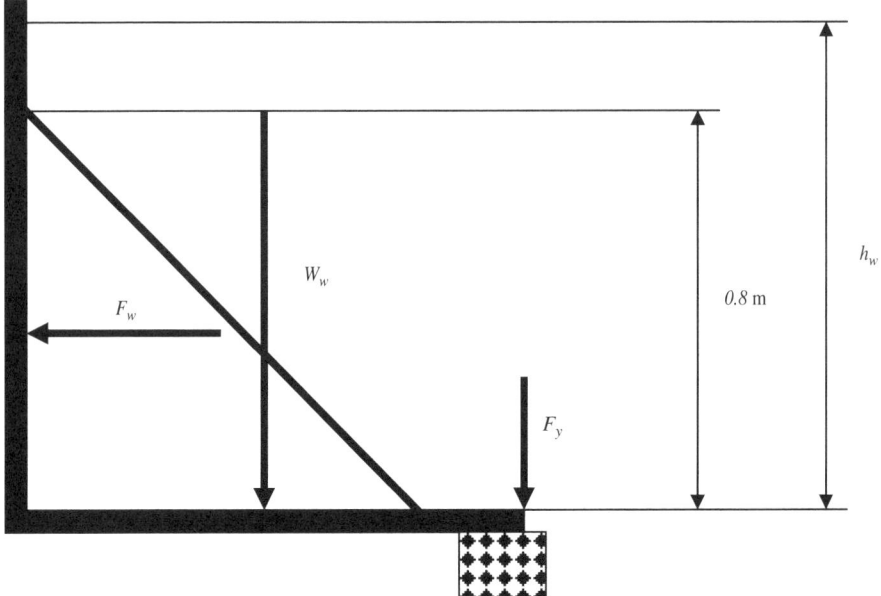

Figure 21.2. Sketch of the flood protection structure and the acting forces, F_w; horizontal force caused by the water pressure acting on the structure, W_w; vertical force caused by the water pressure acting on the structure, h_w; height of flood water, F_y; weight of concrete.

The forces acting on the structure is calculated as follows:

$$F_w = \frac{1}{2}\rho_w g h_w^2$$
$$W_w = \rho_w g h_w b$$

where:

F_w = horizontal force caused by the water pressure acting on the structure
W_w = vertical force caused by the water pressure acting on the structure
h_w = height of flood water
ρ_w = density of water
g = acceleration of gravity

21.5   CALCULATION OF SAFETY AGAINST SLIDING

The factor of safety against sliding is given by the relationship between the forces normal to the ground and the forces parallel to the ground modulated by the factor of proportionality:

$$S_s = \frac{\sum F_n \tan \phi}{\sum F_p}$$

where:

S_s = factor of safety against sliding
$\sum F_n$ = the sum of forces normal to the ground
$\sum F_p$ = the sum of forces parallel to the ground
$\tan \phi$ = coefficient of friction

The flood protection structure can in principal be installed on any type of ground. We have therefore chosen to calculate the minimum coefficient of friction ($\tan \phi$) or angle of friction (ϕ) to decide the minimum requirements for the ground on the place of installation for chosen values of the safety factor. When calculating this, only the weight of the concrete foundation was included, the pull-out forces are put to zero. This is obviously not correct but is an approximation on the safe side. The calculations were done for a flood water height of 1.0 m, which is what the structure is intended for, and for a maximum possible flood water height of 1.2 m corresponding to the height of the structure. The concrete foundation has to be constructed so that it adapts to the actual conditions on site in each case to serve its function. Normally a foundation of 0.4 m by 0.7 m is recommended as minimum size. This gives a minimum weight of the concrete of 6.7 kN/m. Calculations of safety against sliding was done for factors of safety of 1.0 and 1.5. The Norwegian regulation for safety and surveillance of installations in rivers (NVE, 2005) gives a minimum safety factor against sliding of 1.4 for some types of permanent dam structures. For a temporary structure one may be prepared to accept a safety factor that is somewhat lower but always above 1.0 (equilibrium), depending on the

Table 21.1. Minimum friction factors (tan ϕ) and friction angles (ϕ) for safety factors (S_s) of 1.0 and 1.5 against sliding for water depth (h_w) of 1.0 m and 1.2 m.

	Water depth, $h_w = 1.0$ m		Water depth, $h_w = 1.2$ m	
	$S_s = 1.0$	$S_s = 1.5$	$S_s = 1.0$	$S_s = 1.5$
tan ϕ	0.30	0.46	0.41	0.62
ϕ (deg)	18.6	26.9	24.8	35.2

consequences of failure. The results of the calculations of friction factor and friction angles for safety factors of 1.0 and 1.5 are shown in table 21.1.

According to Janbu (1989) we normally find sand and gravel with a porosity of less than 50% having an angle of friction bigger than 35 degrees and silt varying between 25 and 35 degrees. For clayey soils the picture is a bit more complicated since the angle of friction will vary with the water content amongst other factors. However, one should generally avoid clay as building ground for this structure.

21.6   CALCULATION OF SAFETY AGAINST OVERTURNING

With the L-shape of the structure and the same size of the flat laying and the vertical plates, there are no problems whit overturning stability of the structure, assuming that the host structure is stable by itself. The factor of safety against overturning (S_o) is found by calculating the moment equilibrium (relationship between stabilising moment and overturning moment) with respect to the downstream toe of the structure (T_o). This was done for water depths of 1.0 m and 1.2 m.

$$S_o = \frac{M_s}{M_o}$$

where:
   S_o = factor of safety against overturning
   M_s = stabilising moment about toe of the structure
   M_o = overturning moment about toe of the structure

Calculating S_o for water depths of 1.0 m and 1.2 m gives 4.6 and 2.9 respectively. This is in both cases well above a safety factor of 1.4 which is recommended as a minimum by the Norwegian regulation for safety and surveillance of installations in rivers (NVE, 2005).

21.7   TEST METHOD AND RESULTS OF TESTS

There has been carried out several tests of the function of the equipment in a specially constructed test basin at the Norwegian University of Life Sciences. The test

Figure 21.3. AquaFence semi test basin, in the front of the picture we can see the free-standing corner and in the fare end of the picture we can see the termination of the structure against a wall. Test basin is partly filled (Courtesy of AquaFence AS, photograph by Thrana, 2004).

basin was 14 m by 5 m and included free standing corner and a termination of the structure against a concrete wall (figures 21.3 and 21.4, from testing of the proto-type). The test basin has been filled by pumping water from a nearby lake. The speed of filling has been approximately 3 cm per minute. This filling speed is normally way above what one would expect in a normal flood situation and should take care of some of the dynamic forces acting on the construction when the water is rising.

All tests have shown that the equipment is performing well. There has been no leakages observed neither between sections nor between the foundation and the over ground structure. Only minor deformation of the structure has been observed (figure 21.5).

Figure 21.4. Test basin, maximum filling (Courtesy of AquaFence AS, photograph by Thrana, 2004).

21.8   INSTALLATIONS

The AquaFence Semi is a system that is easy to use as a closure structure around important objects. It is also a protection structure that has a preferable use in densely populated urban areas where the access, physical and visual, to a river should be preserved. Other places it can be utilised are places where it is not possible to build high earth embankments or not possible to build such structures at all. Because of the need for a preinstalled concrete foundation, a prerequisite for utilisation of the AquaFence Semi system is that the ways of the flooding water are well known. Some of the installations of the AquaFence system has been installed are shown in figures 21.6–21.8. The cost of the over ground structure of AquaFence 1.2 m is approximately 350 Euros per meter. The total cost of the system will depend on the costs for the foundations and will vary from installation to installation.

21.9   FURTHER WORK

To increase the flexibility of the system, work has been done to make 60 cm high structures that can be installed on hard surfaces, like concrete or asphalt without the need for a pre-installed foundation. A specially designed securing system has been developed for this purpose. Also the possibility of designing the system for

Figure 21.5. Detail of the over ground flood protection structure when partly filled (Courtesy of AquaFence AS, photograph by Thrana, 2004).

more than 1.0 meter of flood water is investigated. The development process is now completely driven by the company AquaFence.

21.10   CONCLUSIONS

The AquaFence Semi is easy to use and has performed well in the tests carried out. Theoretical calculations of safety against sliding and overturning have shown that the structure have a safety factor that is satisfactory on most grounds also when the flood water is at maximum height. The AquaFence Semi is a protection structure that has a preferable use in densely populated urban areas where the access, physical and visual, to a river or other water body should be preserved. Other places it can be utilised are places where it is not possible to build high, or high enough, earth embankments or not possible to build such structures at all.

Figure 21.6. AquaFence Semi, 1.2 m, installed on top of an existing earth embankment at Lillestrøm, a city in southern Norway (Courtesy of AquaFence AS, photograph by Rørheim, 2003).

Figure 21.7. AquaFence Semi, 0.6 m, installed at a TV and radio studio in Oslo, Norway (Courtesy of AquaFence AS, photograph by Rørheim, 2004).

Figure 21.8. AquaFence Semi, 1.2 m installed at a test site in Köln, German (Courtesy of AquaFence AS, photograph by Rørheim, 2004).

REFERENCES

Eikenæs O, Njøs A, Østdahl T and Taugbøl T, Flommen kommer. (Final report from the HYDRA programme, a research programme on floods, in Norwegian with English summary). Norwegian Water Resources and Energy Directorate, Oslo, Norway, 2000.

Janbu N, Grunnlag i Goteknikk. Tapir, Trondheim, 1989.

Nordskog K, Personal communication, Norwegian Natural Perils Pool, 2006.

NOU, Tiltak mot flom (Flood Protection Measures, in Norwegian), Statens forvaltningstjeneste, Oslo, Norway, NOU 1996:16.

NVE, Forskrift om sikkerhet og tilsyn av vassdragsanlegg (Safety regulations, in Norwegian), Norwegian Water Resources and Energy Directorate, Oslo, Norway, 2005.

Rørheim T O, Personal communication, AquaFence, 2006.

Sæterbø E, Syvertsen L and Tesaker E, Vassdragshåndboka. Håndbok i forbygningsteknikk og vassdragsmiljø. (Handbook for construction of flood protection structures, in Norwegian). Tapir forlag. Trondheim, Norway, 1998.

22

The English Planning System and Flood Risk Management

Juliet Richards

Centre for Urban and Regional Ecology (CURE), School of Environment and Development, University of Manchester, Manchester, UK

In England, Local Authorities have certain powers in relation to flood defence provision and drainage. The Land Drainage Act of 1991 essentially 're-enacted powers permitting local authorities to carry out work on [ordinary and critical-ordinary] watercourses. . . . so far as may be necessary in order to prevent or alleviate flooding and to carry out other drainage work' (DoE, 1992, p.27). However, the 'control of floodplain development has been recognised as a non-structural method of flood hazard reduction' (Parker, 1995, p.343).

A hierarchical system of land use and development planning exists in England. At the national level, Planning Policy Guidance (PPG) Notes and the new Planning Policy Statements (PPS) prepared by central government provide the policy framework within which Regional Planning Bodies prepare their Regional Spatial Strategies, which in turn govern the policies and land allocations of Local Planning Authorities in their Local Development Documents. At the time of writing, the current planning policy framework for managing flood risk through the system of land use and development planning in England is contained in Planning Policy Guidance Note 25: Development and Flood Risk (PPG25) (DTLR, 2001). This planning guidance is the latest replacement of a long line of flood-related guidance documents produced by central government, all of which attempted to discourage floodplain development, culminating in the empowerment of local planning departments to refuse permission for floodplain development proposals (Penning-Rowsell, 1976).

The essence of PPG25 is to explain 'how flood risk should be considered at all stages of the planning and development process in order to reduce future damage to property and loss of life' (DTLR, 2001, p.3). The emphasis of its policy concerns is given in the context of sustainable development and the use of the precautionary principle. The key requirements of PPG25 are given in the table below.

Three important features of the guidance are:

(a) the sequential test, which enables LPAs to adopt a risk-based approach to their development plan allocations and development control decisions according to specific flood risk zones. Zones are defined by the annual probability of flooding in that zone, the extent and standard of existing flood defence, the extent of existing development and the nature of development appropriate for that zone. PPG25 requires that any departures from the test must be satisfactorily explained;

(b) it extends the consideration of flood risk in planning decisions to a catchment-wide basis; and

(c) the EA is emphasised as having the lead advisory role on all matters pertaining to flood risk and is advocated as an important consultee on all development proposals where flood risk may be a consideration. The EA is not a statutory consultee[1] in relation to flood risk, however, and LPAs are not obliged to accept the EA's advice.

Key PPG25 requirements

Indicative floodplain maps, plans and other information provided by the EA should be taken into account when preparing development plans

Development plans need to consider flood risk issues at the relevant scale, according to the local significance of flood risk

Development plans should show the areas of flood risk where specific policies are to be applied

Development plans should identify sites where the restoration of natural functions to floodplains could contribute to more sustainable flood management

Development plans should take account of whether an area of flood risk is already defended or not and the standard of that defence

Any necessary flood defence works must be fully funded, including provision for long-term maintenance, as part of the development

The EA has the lead role in advising on flood risk issues at a strategic level and in relation to planning applications

Applications requiring consideration of flood risk issues include those for development:
  i. Within a floodplain or washland shown on the indicative floodplain map prepared by the EA

---

[1] New legislation is currently being drafted which will extend the EA's statutory consultee role to certain developments in flood risk areas (DEFRA, 2005a).

---

ii. Of such a size or nature relative to the receiving watercourse/drainage system that there could be a significant increase in surface runoff

iii. Within or adjacent to any watercourse, particularly where there is potential for flash floods

iv. Includes or adjacent to any flood bank or other flood control structure

v. Situated in an area where the EA has indicated there may be drainage problems

vi. Likely to involve culverting or diverting of any watercourse

Developers should provide the fullest possible information at the earliest possible stage to enable the EA to respond to consultation within 21 days. Developers should carry out a flood risk assessment and submit this with their application

Development plans should include policies which promote the use of sustainable drainage systems

Individual authorities should consider catchment-wide issues in preparing development plans and in determining planning applications

In applying the precautionary principle, planning applications and development plan policies should be tested against the criteria contained in the Sequential Test

---

Source: DTLR (2001).

Central government has monitored the implementation of PPG25 by LPAs in development plans and development control decisions via 'Target 12'[2] reports (see for example, EA and LGA, 2004). There is also an expectation of all LPAs 'to give the matter early and serious attention, regardless of whether their areas have experienced major flooding in recent years' (DTLR, 2001, p.13).

Given that planning policy guidance has been in existence since 1947, suggestions that policy has been ineffective and that the planning profession have failed to address flood risk in their planning decisions abound. Urban development of floodplains was particularly active in the inter-war period, prior to the Town and Country Planning Act of 1947. Since then, development has been piecemeal throughout the

---

[2] The former MAFF produced a series of high level targets for flood and coastal defence provision (MAFF, 1999). Target 12 requires the identification of development plans which do and do not contain flood risk statements or policies and to identify development control or appeal decisions where the EA sustained objections on flood risk grounds, decisions were in line with EA advice and decisions were contrary to EA advice. The findings are reported jointly by the EA and Local Government Association (LGA) each year.

country, particularly where planning control has been less strict. However, it was considered that floodplains could be developed wisely, if the overall benefits of the development to the community as a whole outweigh the costs of flooding, although data to accurately assess the benefits and costs was generally unavailable (Penning-Rowsell, 1976).

Largely in response to the Easter 1998 and autumn 2000 floods, a flurry of literature has emerged, all of which speculate on where the blame should lie for the flood problem. For example, under Hewitt and McGeady's (2001) alluring title 'Who's to blame for the flooding crisis?' (p.14) is the suggestion that the proper policy tools have been in place for well over a decade, but the reason for their failure is due to the local perceptions and practice on the part of the LPAs and the EA. At a Town and Country Planning Association (TCPA) conference in November 2001, evidence was presented concluding that around 40,000 planning applications to develop in the floodplain are made each year, the majority of which were permitted (Young, 2001). In 1996, around 4,000 houses were built in the floodplain compared with around 24,000 being built in 2000 (Crichton, 2001).

According to Thomalla (2001) the EA is frustrated by having to defend areas at flood risk because of unwise planning decisions, that would not otherwise need defending. Flood alleviation objectives have traditionally been focussed on 'economically justifiable defences which are technically sound and environmentally sympathetic' (DoE, 1992, p.28). This emphasis on economic efficiency has perhaps been a driving force behind many planning decisions, particularly given the historical context of benefits and costs described above and the existence of other environmental and physical constraints to development. Indeed, Parker (1995) considers floodplain development may actually be wise, where it achieves wider environmental benefits.

The potential to mitigate the effects of floodplain development has long been recognised by planning professionals. The government approach has traditionally been one of protection, rather than prevention, but floodplain development 'can only be partially protected because structural flood mitigation measures are prone to overtopping and failure' (Parker, 1995, p.342). Indeed, other measures (albeit not always sustainable and sometimes creating new problems downstream, see Sections 2.2.2 and 2.4), such as the raising of ground levels to remove a development site from risk and the provision of compensatory flood storage downstream, have been promoted by the former NRA (Parker, 1995) and by the EA as suitable planning conditions.

Given the EA's feelings on 'unwise' planning decisions and that evidence has shown beyond doubt that some developments have indeed been inappropriate, some arguments suggest this is historical. For example, the majority of properties affected by floods in Lewes and Uckfield, East Sussex, two of the worst hit areas during autumn 2000, were considerably old, many being within the historic town centres. Of all the developments permitted in Lewes during the decade prior to the floods, none were affected. Floodplain development policies were in place and, in reference to one particular planning application, 'considerable thought had been given at the planning stage to siting and levels of the new [development]' (Wilkins, 2001, p.114). The annual High Level Target 12 report for the year 2003/2004 (EA and LGA, 2004) also indicated that nearly all development plans contained some form of flood policy and a survey of LPAs showed that, of those that responded, less than 50% had received a recommendation from the EA to refuse planning permission for development on flood risk grounds and of these, a large majority of LPAs accepted this recommendation. Indeed, the EA appeared to favour conditional consent rather than outright refusal (RTPI, 2001).

Nevertheless, despite removing new developments from risk by implementing certain mitigation measures, the issue remains that downstream effects of development may also arise, putting previously non-risk properties at risk. In this regard, floodplain development has been a focus in that, by reducing the capacity of the floodplain to store water, floodwaters are forced further downstream. However, surface water run-off and inadequate drainage have also created downstream risks, as described in Chapter 2, Section 2.* Notwithstanding the fact that agricultural run-off is effectively beyond the control of LPAs, the contribution of planning decisions to these problems relates primarily to the ability of LPAs to impose planning conditions and obligations and to the difficulties in assessing the cumulative impacts of individual developments.

Planning conditions provide 'considerable scope for the flooding potential of a development to be reduced, or for the adverse flooding effects upon other properties to be mitigated' (Howarth, 2002, p.364). However, 'there may be limitations upon the use of planning conditions insofar as they may not be used to require off-site work to be undertaken. . . . where downstream flood defence works will be needed as a consequence of a proposed development and the developer has no legal power to undertake such works' (p.364). Planning conditions 'must normally relate to matters that are within the boundaries of the proposed development [and] may not involve monetary payments' (p.365) and must 'fairly and reasonably relate to the development that is authorised' (p.364). The validity of requiring on- and off-site flood alleviation measures, including drainage, has been upheld by the courts, however (Howarth, 2002). Despite this, the use of planning conditions to mitigate flood risk appears to be low. For the year 2003/2004, it was reported that, of all the approved planning applications it was consulted on in relation to flood risk, just over half included conditions to mitigate flood risk, with the remaining applications being approved without any flood-related conditions (EA and LGA, 2004). One possible reason for this, is the difficulty in 'enforcing conditions against successive landowners' (Howarth, 2002, p.365) where any measures required as a condition need ongoing or future maintenance.

Planning obligations can be used to overcome this problem, in that they can be transferred to successive landowners. However, the use of planning obligations has been criticised as a means to buy and sell planning permissions and as a means to enable development that would otherwise be considered unacceptable (Howarth, 2002). Furthermore, their use must meet a number of criteria, which Howarth (2002) considered as potentially difficult to balance or indeed, to challenge at appeal given the amount of discretion enjoyed by LPAs and the Secretary of State. Consequently, the use of planning obligations has been far less frequent relative to the use of planning conditions (Cullingworth and Nadin, 2002), potentially because of these uncertainties.

Compounding the uncertainties surrounding the use of planning conditions and obligations, is the difficulty in assessing the cumulative impact of development on flood risk. Whilst PPG25 requires developers to undertake flood risk assessments as a prerequisite to submitting a planning application, a distinct lack of these assessments at the statutory consultation stage of planning process was reported (EA and LGA, 2004). Planning conditions cannot be used to secure flood alleviation measures by one developer, which would be necessary to mitigate the cumulative effect on flood risk resulting from several future developments. Planning obligations, on the other hand, can be used to secure financial contributions towards local infrastructure such as flood defence, but without adequate information on the likely impacts of the proposed development on flood risk, the fairness and reasonableness of any obligation is unlikely to be appropriately balanced.

The EA, as a recommended but not statutory consultee, advises LPAs on flood risk in relation to planning applications. There have been suggestions that EA advice is often misplaced, ignored or not forthcoming and that LPAs have gone against EA advice in granting planning permission to certain applications. Certainly, the EA considered that a review of PPG25 to make it a statutory consultee for flood risk is necessary (Wainscoat, 2005). Whilst LPAs are under no statutory obligation to accord their decisions with EA advice, it is not unreasonable to assume that, given the EA's expertise and advisory role on issues relating to flood risk, LPAs would give serious attention to EA advice. The question arises, therefore, why LPAs do not always follow that advice.

One of the many flood-related activities of the EA is to prepare indicative floodplain maps. These were initially undertaken by the former National Rivers Authority (NRA), in the form of detailed hydrological surveys of certain river sections to show flood risk areas, under Section 105(2) of the Water Resources Act 1991. They were intended as a tool to aid land use and development planning decisions in areas where future development pressure was expected and initially, access to them was generally limited to the NRA and LPAs. Indicative floodplain maps, prepared by the EA, differ from the Section 105 surveys in that they show the estimated extent of a 1 in 100 year flood on all rivers in England and Wales. The maps also include coastal flood limits. Where Section 105 survey data has been available, these have also been incorporated into the EA's indicative floodplain maps (DEFRA and EA, 2003). Confidence in the accuracy of the EA's maps has been low, principally because they 'generally do not delineate floodplain functions, for example routes for passage of floodwater and areas for flood storage' (DEFRA and EA, 2003, p.13) and do not distinguish flood risk areas that benefit from existing flood defences (DTLR, 2001; Crichton, 2003). The fact remains, however, that the maps are intended for guidance only and are not to be used on an individual property basis (Davis, 2004). The maps have since been updated to show different 'flood zones', which are defined in PPG25 (DTLR, 2001), so that the EA can focus the delivery of its advice to LPAs on development proposals in higher risk areas, whilst enabling LPAs to respond more quickly to development proposals in lower risk areas by adopting a standardised response (Davis, 2004). Definition of the zones was based on a combination of computer modelling techniques, aerial surveys, historical flood data and local information from EA staff.

There is a perception that tensions between LPAs and the EA and its predecessors have been ongoing, predominantly due to the historical divorce of water quality planning (the responsibility of the EA) and land use planning (the responsibility of LPAs), both systems having reciprocal impacts (Penning-Rowsell, 1976; Howarth, 2001). Welsh (2001) reported one planning committee councillor as saying 'the Agency blows hot and cold. It moans about the homes and then does nothing about the defences' (p.17). The local knowledge of LPAs about flood prone areas is also considered to be superior to that of the EA (Hewitt and McGeady, 2001; Welsh, 2001) and Hewitt and McGeady (2001) announced that the EA's consultation responses to planning applications were inconsistent, often not forthcoming at all, and generally unfounded. They describe the incredulity expressed by planning officers at the EA's objections to development where flooding is not a problem and EA's non-objection to development where flooding is known to be a problem.

Whilst the LGA confirmed that it was taking action to improve the relationship between LPAs and the EA (Welsh, 2001), it has been suggested that:

"The crucial question is whether the perceived tension between the democratic mandate of local authorities and the technical mandate of the Agency is genuine or not. Whilst it might seem undemocratic to give the, non-elected, Agency power to overturn decisions reached by the, elected, local authority, the reality may be that decisions of local planning authorities are so confined by law and policy guidance that the scope for political considerations to enter into planning determinations is relatively limited. . . . it might be argued that the real tension is between national and local, democratically determined, environmental imperatives."

(Howarth, 2001, p.45)

Certainly, the EA faces the difficulty in 'that its view of the undesirability of a particular development proposal is only one amongst many material considerations that the local planning authority will have to take into account in determining whether to grant the planning permission which is sought' (Howarth, 2001, p.40). This is also evident in the Select Committee on Agriculture's (1998) report:

"On occasions it has to be said that local authorities have acted in accordance with Environment Agency advice but seen their decisions overturned on appeal to the Planning Inspectorate."

(para. 86)

That a number of material considerations exist for local planning officers, it is interesting that Cullingworth and Nadin (2002) comment on national planning policy as follows:

"[National planning policy] advice can be conflicting, perhaps as a result of piece-meal revision at different times. Moreover, as is demonstrated repeatedly at public inquiries, differing interests can 'cherry pick' from the twenty-five PPGs to show how well their arguments meet the official guidance. . . . In applying policies to particular cases, interpretation is required; and often there has to be a balance of conflicting considerations."

(p.49)

Incorporating flood risk management in to the planning system is a compromise between local and national planning concerns coping with the legacy of previous developments and the socio-economic pressures for new development and the precautionary approach to flood risk advocated by the Environment Agency. The advent of the European Water Framework Directive with its emphasis on total catchment management will hopefully lead to integrated decision support tools for United Kingdom planners having to cope with ever increasing risks of flooding.

REFERENCES

Crichton D. (2001) 'An insurance perspective', *speakers notes from the Town and Country Planning Association conference, A Flooded Environment: Planning for Homes*, 8th November.

Crichton D. (2003) *Technical Paper 1 Flood risk and insurance in England and Wales: are there lessons to be learned from Scotland?*, Benfield Greig Hazard Research Centre.

Cullingworth B. and Nadin V. (2002) *Town and Country Planning*, 13th ed., New York: Routledge.

Davis A. (2004) 'New tools for flood risk developments', *Environment Agency document ref: 132/04*, 30th June [online]. Available from: http://www.environment-agency.gov.uk/news/815957?version=1@lang=_e [accessed 11th May, 2005].

Department of the Environment (1992) *Local Government Review: The Functions of Local Authorities in England*, London: HMSO.

Department of Environment, Food and Rural Affairs (2005) 'Making Space for Water: Taking forward a new government strategy for flood and coastal erosion risk management in England', *First response to the autumn 2004 Making Space for Water consultation exercise*, March, 2005, London: DEFRA.

Department of Environment, Food and Rural Affairs and Environment Agency (2003) 'Guide to the Management of Floodplains to Reduce Flood Risks, Stage 1: Development Draft', *Flood and Coastal Defence R&D Programme Report SR599*, February, HR Wallingford.

Department of Transport, Local Government and Regions (2001) *Planning Policy Guidance Note 25: Development and Flood Risk*, Consultation Draft, London: DTLR.

Environment Agency (1997) *Policy and Practice for the Protection of Floodplains*, London: HMSO.

Environment Agency and Local Government Association (2004) 'High Level Target 12: Development and Flood Risk 2003/04', *Joint Report to Department for Environment, Food and Rural Affairs (DEFRA) and the Office of the Deputy Prime Minister (ODPM)*, November.

Hewitt R. and McGeady B. (2001) 'Who's to blame for the flooding crisis?', *Planning*, 2 February: 14–15.

Howarth W. (2001) 'Town and Country Planning and Water Quality Planning', in *Planning and Environmental Protection*, ed. by C. Miller, Oxford: Hart Publishing, pp. 19–45.

Howarth (2002) *Flood Defence Law*, Crayford, Kent: Shaw and Sons.

Parker D.J. (1995) 'Floodplain development policy in England and Wales', *Applied Geography*, 15: 341–363.

Parker D.J (ed.) (2000) *Floods Vols. I and II*. London: Routledge.

Penning-Rowsell E. (1976) 'The effect of flood damage on land use planning', *Geographica Polonica*, 34: 139–153.

RTPI (2001) 'Survey of local planning authorities on flooding', *unpublished questionnaire and results*, February, courtesy of David Barraclough, RTPI.

Select Committee on Agriculture (1998) *Flood and Coastal Defence: Sixth Report*, House of Commons [online]. Available from http://www.publications.parliament.uk/pa/cm199798/cmselect/cmagric/707/70703.htm [accessed 13th August 2004].

Thomalla F. (2001) 'Managing flood risk in the UK', *Town and Country Planning*, April: 111–112.

Wainscoat N. (2005) 'Floods highlight need for policy compliance', *Planning*, 28th January: 9.

Welsh E. (2001) 'Builders ignore warning on floodplain danger', *The Times*, 12 October: 17.

Wilkins B. (2001) 'Towns in flood', *Town and Country Planning*, April: 114.

Young B. (2001) 'Is the planning system able to deliver an appropriate response to current and future flood risk?' *speakers notes from the Town and Country Planning Association conference, A Flooded Environment: Planning for Homes*, 8th November.

23

French Regulations for Urban Flood Management

Nicolas-Gérard Camphuis
*Centre européen de Prévention des Risques d'inondation, Local authorities
and communities in Europe facing flood risk, Orleans, France*

ABSTRACT: The background to the development of flood risk management
arrangements in France has a long history, but has required considerable revision
of the responsibilities and roles of the various stakeholders in response to increasing flood risk. Responsibilities now stretch from individual citizens, through local
communities, with Regional duties and right up to National support and legal
instruments. Current compensatory arrangements for those impacted may be such
that there is too little discouragement to developments in flood plain areas.

23.1 INTRODUCTION

23.1.1 Flood risk exposure of French cities

France has some 280,000 km of waterways with every potential form of flood
behavior. There are 6 major rivers, 3 of which cross national borders. There are
also a large number of coastal waterways. Approximately 14,000 km of waterways
are owned by the State. This either directly ensures their management, monitoring
and maintenance by the State, or these duties can be delegated to a territorial
authority or public body. The other waterways are privately owned and it is the
responsibility of the water bank owner to maintain the portion of the waterway
crossing the property. This may be done in partnership by forming a management
syndicate with other private owners. The territorial authorities have been legally
authorized to replace these arrangements for private landowners since the mid-20th
century, and do so frequently, usually in the form of inter-communal syndicates or
groups constituted from the local drainage basins.

France is partially subject to high sea levels over the entire west and northern
coastlines of the country and also is subject in the southwest to some high water
level flood risk from the Mediterranean Sea. Certain rivers such as the Rhône
and the Loire, are subject to both types of high-water flooding due to their considerable geographical range. The territory has recently suffered from both of

these types of high-water flooding, each causing comparable damage due to its scope.

There is a long history of settlement along the waterways and many major French metropolises are partially built on flood plains of rivers. Some 20–25% of all French communes (8,000) have been identified as partially subject to flooding, although overall, less than 4% of France is subject to flooding. There was major flooding in 1856 and 1910, when floods succeeded one another throughout the year in each French river. This alerted public opinion and mobilized a national solidarity movement at the time. Nonetheless, as a result of economic growth in the 1950 to 1990s, flood-prone urban areas were extensively colonized, contributing to a considerable increase in the population and property exposed to flooding. Today, about 5,000,000 people inhabit an area that is at significant risk of flooding, mainly in the greater Paris area and along the Loire and Rhône rivers.

Several major French cities are particularly exposed to flood risk:

• In the greater Paris area, 900,000 people live in areas prone to flooding from watercourses and this figure rises to 1.3 million if those also exposed to water table flooding are included.

• In Marseille, 80,000 inhabitants are exposed to flood risks; in Orléans, some 50,000 and in Toulouse, some 75,000.

Given the scope of the damage caused by high-water and floods in the last 20 years, France has reacted by combining regulatory measures with detailed analyses of the problems:

• From 1982 to 2002, a large number of regulations were brought in to complete a series of laws originally enacted since 1807. These pertained to four key areas: damage compensation; land use management; preventive information; global water management in drainage basins. In 2003, a Risk law and an amendment to the Water law of 1992 reinforced the work already implemented.

• Since 1995, several major French rivers (Seine, Rhône, Loire and Meuse), as well as some of their estuaries (Saône, Oise-Aisne, Maine) have been studied in detail, combining contingency knowledge with an inventory of human and environmental issues, in order to devise a flood-risk reduction strategy. This has been in collaboration with the Public Basin Establishments; the territorial authorities' groups that have been set up in recent years as representatives to the overall French State.

23.1.2   Administrative context and division of responsibilities

In France, flood risk management is based on:

• Governmental policy that is essentially drafted and implemented by the Ministry in charge of the Environment; it is based on the application of legislation that

is becoming stricter by the year; implemented at local levels by the local State agencies under the authority of the Prefect acting in each of the 100 departments;
- The action of the mayors of each of the 36,000 communes[1] in charge of risk prevention within their territories;
- The responsibility of private owners.

The General Councils elected at the departmental and regional levels of each of the 22 regions, have optional jurisdiction in supporting the communes, which has been provided for by law for over thirty years (Law of 1973 on water defence). They may undertake incentive or mobilization actions, in particular by co-financing operations to reduce the risk of flooding in other responsibilities they conduct under their own authority.

Groups of territorial authorities with the status of Public Basin Establishments, have been formed in a large number of drainage basins to: promote collaboration between municipalities; help to intervene in subsidiarity; raise finances and ensure that indispensable supervisory functions are delivered. These new Establishments could be the main means of delivering significant advances in the way in which flooding is managed in the foreseeable future. Separate Water Agencies manage water quality and water resource management, but do not have any specific mission pertaining to flooding.

In terms of the division of responsibility:
- The government and parliament enact laws, which each Prefect must apply at the level of each department. This relies on mobilization via decentralized State departments and local officials who are designated by Law as responsible for legal application.
- It is the duty of the Ministries to draft a policy, methods and tools to enable the implementation of the laws, as well as to monitor progress on their application and report to government.
- The Law designates the mayor of each of the 36,000 French communes as the primary responsibility for local maintenance of order and safety, land use management and therefore, for exposure to risks due to new construction. In addition to inform the local citizens of the existence of natural and technological risks. In this capacity, the mayor is heavily involved in natural risk prevention, the crisis management involved in high-water alerts and flooding and for providing the population with preventive information to counter such risks.

No single mayor, particularly those of the smallest communes, can assume the responsibilities conferred upon him by Law without the very significant technical support provided by the State departments. State intervention prior to

---

[1] In France, the "commune" is the smallest administrative subdivision. It is either a town or a village and is run by a mayor.

the final application by the mayors also guarantees that a standardised approach is adopted throughout the entire country. It is the State that:
– establishes the risk at the national and departmental levels and brings it to the attention of the Mayor;
– imposes standards on the Mayor in terms of land use management;
– replaces the Mayor in crisis management in the event of a flood when the flooding exceeds the Mayor's jurisdiction due to its geographical scope.
• It is each inhabitant's responsibility to stay informed as to the risks to which they are exposed, to actively seek such information and to protect themselves and their property from flooding.

23.2   THE CURRENT PROGRAMME

The main aspects of French regulation in relation to flood prevention are described in this section.

23.2.1   Regulatory context for flood prevention and remediation

The People living along rivers are solely responsible for their own protection against floods.

The fundamental principle governing French flood protection policy is set out in a law from 1807, still in force, that stipulates that it is each waterway frontage owner's responsibility to protect themselves from flooding, but in terms of overall State control. This includes State financial support. A river frontage owner may not expect or demand that the State protect them from rising water and floods. This law would not exclude all protection but would lead to the construction of works such as dykes or dams primarily falling under the responsibility of groups of private owners subject to State control. The result is that, today, no private owner, nor any mayor, may request national collective support or State protection from flooding.

However, there have been a number of exceptions to this rule along the Loire, the Rhine and the Rhône in particular, in which the State has handled the installation of flood protective works. Since the mid-50s, the law has, among other provisions, authorized territorial authorities to replace owner groups but nonetheless remain under State control. Moreover, they can request that the people who have made the work necessary or who are interested in it contribute to the costs incurred.

23.2.2   Control of urbanization in flood prone areas

French legislation has created a tool to regulate building and development in flood-prone areas. This is called the "Plan de prévention des risques d'inondation" (PPRi) (Flood Prevention Plan). The purpose is to avoid risk to human life, damage to properties and businesses and to preserve floodplains and washland areas as much as possible by controlling construction in areas exposed to flood risk.

The PPRi (flood prevention plan) determines:

- Dangerous areas – areas directly exposed to risk, also known as "red" areas, where, as a general rule, all construction or new installations are prohibited.
- Precautionary areas – areas that are not directly exposed to risks, but in which construction may aggravate existing risks or give rise to new ones, also known as "blue" areas.

The PPRi defines the prevention, protection and mitigation measures that should be applied in each of these areas. It can recommend or control developments and existing construction, facilities or agricultural land. As much as is necessary, it can recommend measures whose purpose is to allow water to drain freely and for the preservation, restoration or expansion of flood plains.

The PPRi can also recommend risk prevention work to private individuals. It can even define regulations relating to public networks and infrastructure, to facilitate evacuation or make the arrival of emergency services easier in a flood. The PPRi is laid down by the Prefect, drafted by the State departments and then ordered by the Prefect, following a consultation period. It is a legal instrument of public interest and is enforceable against third parties. Prevention, protection and mitigation measures laid down by the PPRi can be made obligatory within a 5 year period, which may be reduced in the event of an emergency.

23.2.3 National solidarity and compensation for natural catastrophes

The preamble to the French constitution of 1946, which is still in force, stipulates that "the Nation proclaims the solidarity and equality of all French citizens before the burdens resulting from national calamities". In the early 1980s, a series of high-water incidents and floods, as well as other natural catastrophes, caused serious loss of human life and property, strongly mobilizing national solidarity. The government responded with financial support to ensure compensation for these damages. This resulted in a law on natural catastrophe compensation in 1982.

This law created a blanket national system to compensate for natural catastrophes. This is not a "conventional" insurance system based on market rules, but a system based on community risk-sharing, the working rules of which are determined by law:

- Any citizen who has contracted an insurance policy covering damage and operating losses contributes to the "natural catastrophes" cover, whether or not they are concerned by these natural risks
- The premium rate corresponding to this cover is uniform, equal to 12% of the premiums of contributions related to the basic contracts for all property other than motorized land vehicles, and to 6% of premiums or Fire and Theft contributions for motorized land vehicles.

- The excesses to be paid linked to this cover are fixed and are determined by order.

As soon as flooding has been recognized as a "natural catastrophe" by an Inter-departmental decree, damage is covered by this insurance system. (As a rule, this is the event when the flooding has a return period exceeding 10 years.)

This regime ensures rapid and appropriate compensation for flood victims. However, it has a number of adverse effects as these insurance policies cover only identical reconstruction, without incorporating measures to reduce future vulnerability. Nor does it promote a policy of insurance premium rates based on prevention efforts made by the various parties – insured, commune or State – nor the relocation of existing developments in flood-prone areas. By regulating insurance premiums, it also keeps down the costs to those at risk in flood-prone areas, and as a consequence it does not discourage new construction being scheduled for flood plains.

23.2.4   Preventive information for the community

By law, French "citizens have the right to information on the natural risks to which they are subject in certain areas of the national territory and on their related safety measures. This right applies to foreseeable technological risks and natural risks", particularly flood risks.

For its application, the law stipulates that the Prefects shall establish information documents that they will officially bring to the attention of each communal Mayor governing a territory that is partially or entirely prone to flooding. The Mayor must inform the population of the existence of these documents and make them available to the public. They should be displayed regularly.

The State and its Mayors are responsible for providing citizens with this information, as well as owners and managers of campsites and sellers or landlords of properties.

- The Prefect must draft the "Dossier Départemental des Risques Majeurs" (DDRM) – (Département Report on Major Risks). This report lists all the towns in the département affected by major risks, specifically flooding. Town by town, the DDRM describes the phenomena, their predicted consequences; the historic events that have affected the town, as well as the prevention, protection and mitigation measures planned in the département to reduce these effects.
- The Prefect is also responsible for bringing the risk to the attention of the Mayors of the towns and villages concerned. The DDRM, the existing maps of dangers, and the flood prevention plan (where it exists) must be sent to Mayors. With this aim, ministry decrees impose upon the Prefect the constitution of a "Commission départementale des risques naturels majeurs" – (Department cell for major natural risks).
- In each of the towns and villages cited in the DDRM, the mayor must produce the "document d'information communal sur les risques majeurs" (DICRIM) –

(local information document on major risks). This document is supplementary to the information provided by the Prefect. It indicates the prevention, protection and mitigation measures in response to major risks likely to affect the town. These measures include, as necessary, the security instructions that must be implemented in the event of flooding. The DICRIM should be supported by a security instructions poster campaign.

- In all towns covered by a risk prevention plan, the mayor must provide the citizens with information at least every two years, by a public meeting or any other appropriate means. In flood-prone areas, the Mayor must make an inventory of the historic flood level marks and establish the flood level marks corresponding to the highest ever known.
- In the areas covered by a risk prevention plan, sellers and landlords of both built and planned properties, must, from 1 June 2006, append a "risk status" to the contract as well as a list of the damage the property has sustained.
- In areas subject to a natural risk (including flooding) defined by the Prefect, the owners and managers of campsites and caravan parks must take all possible measures to inform, warn and evacuate to ensure the safety of the occupants. These measures are laid down by the relevant town planning authority.

In addition to these regulatory information tools, since 1994, the government has applied a policy of systematically drafting atlases of flood-prone areas based on an initial pilot study conducted in the early 1990s on the Loire. This entailed mapping the highest known water levels at a scale of 1/25,000 and outlining areas to a 1/25,000 scale according to the significance of the risk to which they were exposed. This also highlighted the drainage paths, as well as areas with severe flooding depths. The risk has been ranked according to the water levels and flow velocities to be expected in the area, based on local topography.

23.2.5   Rescue organization plans and emergency plans

The mayor is responsible for maintaining order and safety in his locality. In the event of an emergency the Mayor will become the "Director of Emergency Operations". The Mayor is responsible for drafting the Local Protection Plan (Plan Communal de Sauvegarde – PCS), which organizes the protection of and aid to the population, in the event of an emergency (particularly flooding). The PCS is mandatory in each town or village covered by an approved flood prevention plan or included within the scope of a specific intervention plan (i.e. concerning a dam).

The aim of the PCS is to organize emergency management (who does what, when and how) and to put this organization down on paper, in order to be in possession of an aide-mémoire when the time comes for an emergency response. The PCS must cover all the known risks, and more specifically flooding. The predecessor to the PCS (Local Protection Plan) was the "Plan Communal de Secours" (Local Emergency Plan) (before 2004). The change in name underlines the fact that the Mayor does not have to organize emergency plans, but only protection plans; in order to inform and protect the population, as well as provide for aid in emergencies.

The "modernization of the civil defence" law of 2004 has created a new instrument for civil mobilization: the local civil defence reserve. The purpose of this organization is to manage the work of volunteers, who generally present themselves, with a view to helping, in the event of an emergency. This reserve organization makes it possible to organize volunteer-work before the emergency, thus complementing the work of the emergency services.

When flooding exceeds the area of a single town, the Prefect takes responsibility for the direction of emergency operations and can trigger all or part of the departmental ORSEC plan (Emergency organization). However, Mayors of the towns concerned continue to be responsible for the safety of the population in their individual town areas.

The ORSEC plan lists the known risks on the scale of the Département, and organizes emergency management and aid. The organization of emergency aid can be divided into main and modular emergency management provisions applicable in any circumstances (core of the Orsec) and specific provisions for certain pre-identified risks, in addition to the main provisions (plans de secours spécialisés (PSS) – (special emergency assistance plans) (i.e. flooding), plans particuliers d'intervention (PPI) – (specific intervention plans), the "red" plan, etc.).

In the event of a large scale emergency, higher levels dealing with the organization of civil defence may be called upon. The Centre Opérationnel de Zone (Area Operational Centre), in each protection area, and the Centre Opérationnel de Gestion Interministériel de Crise (COGIC) – (Interdepartmental Emergency Management Operational Centre) at a national level.

23.2.6   An integrated approach to river basin management

The Law of January 3, 1992 states that "water belongs to the common national heritage" and proclaims that "the protection and enhancement of this resource are in the general interest and [...] includes flood prevention in its water policy". This provides for a management scheme for water harnessing and management (SDAGE) to draft the fundamental outlines defined for each of France's six major hydrographic basins. On the initiative of the State, this scheme is designed for a period of five years by a basin committee (a sort of water parliament) which includes representatives of the State, the territorial authorities and water users. The SDAGE makes a review of the condition of the water resource and aquatic environment. It covers all of the different uses made of existing water resources. It sets the priorities for achieving jointly-defined objectives while considering the protection of the natural aquatic environment, water resource enhancement requirements, foreseeable changes in rural areas, the urban and economic environment and the balance to be maintained between the different water usages (flood prevention can be one of these objectives). It also assesses the economic and financial resources required for its implementation.

SAGE (Scheme for water harnessing and management) can also be created at a sub basin scale. These schemes are drafted by a local water commission (a kind of mini water parliament) that is qualified, in particular, to address flood-related issues.

23.2.6.1   *Public territorial catchment area corporations*

The law dated 30 July 2003 recognized the Etablissements Publics Territoriaux de Bassins (EPTB) (Public Territorial Catchment Area Corporations) as key players in the prevention of flooding, the balanced management of water resources and the preservation and management of wetlands, on a river-basin and sub-river basin scale.

EPTBs are public corporations gathering "collectivités territoriales" (local and territorial authorities). They act in compliance with the principle of subsidiarity, specifically in the area of flood prevention.

23.2.7   Protection: Safety behind dikes

Under the terms of the water regulations, the State must identify dikes threatening public safety and impose specific recommendations (i.e. diagnosis of the dike by the State, which will help define the necessary work to be done) upon their owners. In addition, the mayor has an obligation to warn against "flooding" and the "bursting of dikes", by suitable precautionary means, as well as an obligation to protect citizens in the event of imminent danger.

In 2003, a major flood of the Rhône river caused substantial damage due to dike breaches (in Arles city in particular). Following this disaster, the State decided to make an inventory of all the dikes, and to identify the dikes which were likely to be most significant for public safety. A recurrent problem is that the owners of the various dikes, who should be responsible for their maintenance and safety, do not fulfil their responsibilities. In 10% of cases the dike owner is unknown.

23.2.8   The "Barnier Fund": Fund for the prevention of major natural risks

The French government created a "fund for the prevention of major natural risks", also known as the "Barnier Law", to encourage and give support to the development of measures for the prevention and reduction of vulnerability to natural risks. The Barnier fund can be used to fund the following, with regard to flood prevention:

- Preventative expropriation: In the interest of the public the State can expropriate properties exposed to a foreseeable risk of severe flooding and which seriously threatens human life;
- The costs linked to temporary evacuation and housing of the people exposed to flooding;
- The studies necessary for the development of Plans de Prévention des risques naturels prévisibles (PPRNP) – (Natural Foreseeable Risk Prevention Plans);
- Studies and natural risk prevention work, which "collectivités territoriales" (territorial authorities) must contract. The Finance law 2006, makes provision for the financing of up to 50% of the cost of studies and 25% of the corresponding prevention work using these funds;
- Preventive measures – the purchase of properties exposed to flood risk by private agreement. Also the purchase by private agreement of property which has been more than 50% damaged by flooding in order to make the land unsuitable for further development. Also studies and works made compulsory through a Plan de Prévention des Risques d'Inondation (PPRi) (Flood Prevention Plan), preventative information campaigns on risk).

The Barnier fund comes from:

- A levy on the profits of the additional extra premiums relating to natural catastrophe cover (these extra premiums represent 12% of the comprehensive home or business insurance premium) and the levy is at most 4%.
- Funding from the State.

The fund managed by the Caisse Centrale de Réassurance (CCR) (Central Reinsurer's Fund).

23.2.9   Regulatory forecasting

Forecasting hydrological phenomena is based on the reliable forecast of weather phenomena and the transfer of river water within hydrographic networks. 16,000 km of major French waterways run through 6,300 communes, affecting over 90% of the population living in flood-prone areas. These waterways are monitored by 53 high-water alert departments using automated data gathering networks. This represents over 1,000 automated and re-transmitted measurement stations, 14 ARAMIS radar systems for collecting radar rain data in addition to terrestrial recorded rainfall data. In parallel, METEO France has automated terrestrial pluvio-metric stations equipped with remote-transmission.

High-water forecast models process this data in real-time with alarm systems that are activated during high rainfall and water ponding. The ASPIC decision-making tool developed by Météo France works on this principle and automatically transmits alerts for flood risk observation monitoring to be initiated.

Geographical information systems (GIS) using combined meteorological, hydrological and hydraulic models and mapping test scenarios provide a visual image of likely problematic high rainfall intensities and consequent impacts for better forecasting and prevention. Advanced conduct systems, such as CRISTAL 2 in the Loire basin, are characterized by the interaction and collation of a large amount of data, models and problem-solving processes. A system like CRISTAL 2 for example includes real-time monitoring of the automated hydrographic network and makes simulations of provisional flood risk status, offers decision-making assistance and manages automated information communication that is accessible to the general public (Internet).

In France, flood forecasting is provided by the 22 "Services de Prévision des Crues" (SPC – Flood forecasting departments). A national department, the SCHAPI (Central Hydrometeorology and Flood Prediction Support Department), has been created recently to assist the SPC.

23.2.10   Alert systems

A new system of warning alerts took effect in July 2006. This comprises a national flood warning map and a broadcast national news report. The news report provides advice to assist those at risk to in order for them to adapt to the flood risk.

Flood risk is symbolized with colours:

- green – a normal situation – no risk
- yellow – moderate flood risk, with no significant damage expected
- orange – the foreseeable (or recorded) risk is important, with significant damage expected
- red – major flood risk expected.

The forecast reports are accessible on the Internet (www.vigicrues.ecologie. gouv.fr) and other information about the French regulatory system for flood risk management are available at: The CEPRI website [http://www.cepri.fr/cgloiret/en_ pp_reglementation.htm]; The Ministry in charge of Risk Prevention website [http://www.prim.net/].

23.3   RECENT DEVELOPMENTS IN FLOOD RISK MANAGEMENT IN FRANCE

23.3.1   Feedback from experience

The major floods France has suffered in the last ten years have led the government to systematically organize experience feedback using experts from different ministries. These meet with the various parties involved in the flooding event, including those managing the problems as well as the victims, and, after consultation, the expert group produces a status report which includes recommendations for improvement and a brief proposal for short and mid-term action. In addition, the Inter-ministerial Delegate to Major Risk publishes a yearly report on the damage-causing natural events of the past year, which is made available to the public.

Comparable information, which is updated by the government, is available on the Internet www.prim.net as a contribution to the prevention of major risks for the general public.

23.3.2   Risk education and the organization of rescue measures in scholastic establishments

In 2002, the French Ministry of National Education conducted a campaign to give every school a specific plan for getting pupils to safety when a major risk such as flooding has been identified. The plans must be drafted by the school principal so that, should the establishment become aware of a danger, there is a plan for when and how to give the alert within the school, where and how to get pupils to safety depending upon the nature of the danger, how to manage communication with the authorities, and what instructions to follow at the beginning of the safety drill. The measures should provide awareness and training for all persons involved.

In addition, considerable effort has been deployed by both the competent ministries and civil society to include risk education and behavior in a dangerous situation in school curricula.

23.3.3   Land use and restoring flood plains and washlands

After focusing traditionally on large scale protective infrastructure (dams, impoundments), today's flood risk prevention mainstream policy involves measures to regulate or slow flow into the heads of catchments, while preserving the balance of the associated ecosystems. These complementary measures ensure the mode of land use can maintain its retention capacity, prevent erosion, runoff, and ensure compatibility with their role of flood plains and washlands.

23.4   REINFORCEMENT OF THE PROTECTION OF INHABITED AREAS AND OPERATIONS TO REDUCE THE VULNERABILITY OF PEOPLE AND PROPERTY

Efforts to develop awareness of risks must be coupled with an improvement in the protection of people and property in risk-prone areas. There are incentives in France to urge private and public owners and managers of buildings and facilities to take measures to reduce their vulnerability in order to live with or protect themselves from risk.

Measures to reduce vulnerability may concern: raising floors, moving the most vulnerable equipment above the reach of rising waters, such as electrical and computer installations or engines, and selecting appropriate materials for the flood contingency, etc. Such measures could be imposed as part of a PPR, along with the prohibitive measures the latter contains, and, for the most exposed buildings, support may be allocated to owners in order to relocate them outside of the risk-prone area. The construction work made mandatory by the PPR will qualify for financial aid if carried out by the property owners.

23.4.1   Flood prevention action programs

The State launched a call for projects for flood protection programs in 2002 at a basin or sub-basin scale, with a budget of 130 million euros for actions to be engaged between 2003 and 2008. The purpose of these programmes is to develop local initiatives but also to treat the drainage basins globally and in an integrated manner, within a perspective of sustainable development.

The actions planned for these different programmes are to reinforce and refine:
- The information provision activities amongst officials and the population, facilitating a real awareness of the risk and compelling participatory action;
- High-water forecasting resources;
- The Risk Prevention Plans programme and the control of land use and urban development in high risk areas;
- The restoration of flood plains and washlands and the development of high flow attenuation operations upstream of the drainage basins;
- The protection of inhabited areas and vulnerability-reduction operations;
- The establishment of rescue plans;
- The financial programs for prevention and compensation.

There are some 40 such programmes throughout France.

23.5   EXAMPLES OF FRENCH CITIES LEADING INNOVATIVE ACTIONS OVER AND ABOVE THEIR REGULATORY DUTIES

23.5.1   Paris and "Ile de France" Region: the specialized emergency services flood plan

900,000 people live in flood prone areas in greater Paris. This number rises to 1.3 million if the people exposed to water table flooding are also taken into account. Transport utilities, power, gas and water distribution networks, telecommunications would be affected by a major flood (the last major flood was in 1910).

To avoid a major crisis from flooding by the Seine river, a Specialized Emergency Services Flood Plan (PSSI) is being drafted. The aim of the PSSI is to:
- Analyse the flood risk in the Region of Paris (Ile de France);
- Ensure that the public, social and economic players of Ile de France are informed before, during and after the flood;

– Coordinate emergency flood management measures, ensuring minimum social and economic activities in Ile de France;
– Make provision for a return to normal following flood receding.

To create the PSSI of the Paris Defence Area, the Paris police headquarters devised an action plan with the key players concerned. The action plan is divided into three phases:
  I) Description of flood risk in Ile de France
 II) Action by the business networks and public authorities
III) Coordination of public authorities actions Establishment of the PSSI

Phase I is complete and the description report of flood risks in Ile de France [http://www.prefecture-police-paris.interieur.gouv.fr/prevention/innondation_janvier2006/sommaire.htm#public] can be consulted.

In the framework for phase II, several plans have been made with the operators, which aims to ensure the safety of people and property, and brings together the conditions required for a quick return to normal services (Phase III) in Ile de France :

• RATP (Parisian Transport service) Plan
  This makes provisions for the building of protection walls around air-vents and entries to the RATP underground network (or if necessary covering them with concrete). The plan makes provisions for the human resources, equipment and financing required for this emergency management. It represents €300,000/year and €3M at the time of the alert, avoiding M€3,000 damage.

• An SNCF (Rail company) Plan

• An EDF-GDF (Electricity/Gas) Plan
  This makes provision for prevention measures (protection of high voltage power stations, special connection for prefectures), emergency measures (mobilization of human resources, availability of generators, etc.) and measures for a quick return to normal.

• Telecommunications Plan

• A military action plan has also been drawn up.

There are still a number of plans that need to be devised, such as the management of household refuse, pollution (degradation of drinking water), hygiene (stagnant waters), etc.

These will all be coordinated within the PSSI.

To learn more about the Flood area action plan and the PSSI – (Specialized Emergecy Services Flood Plan): The website of the Paris Police Headquarters [http://www.prefecture-police-paris.interieur.gouv.fr/prevention/innondation_janvier2006/sommaire.htm#public].

23.5.2   Orléans City: a "general interest programme" to help inhabitants reduce the vulnerability of their houses to flooding

Some 50,000 people live in flood prone area In Orleans city. The "communauté d'agglomération d'Orléans" (Agglo-O – the administrative institution which covers Orleans city and the surrounding cities) is assisting inhabitants to reduce the vulnerability of their houses to flooding. A "general interest program" (PIG) began in 2005 for 3 years under which the Agglo-O offers to inhabitants:

• a diagnosis of the vulnerability of their property to flood risk (fully funded by the Agglo-O);
• financial support for the necessary measures required to reduce vulnerability (between 20 and 70% of the cost of works).

The "communauté d'agglomération d'Orléans" has planned to offer 600 to 1,000 diagnosis on the period 2005–2008. 300 households will be assisted. The total budget of this action amounts to around €550,000 (€50,000 for communication, €100,000 for problem analysis, €100,000 for the general organization, €300,000 for the works themselves).

This action has been a major success for the inhabitants, as in less than 6 months, 150 diagnoses have been carried out.

In parallel to the action of the "communauté d'agglomération", the chamber of commerce and industry of the "departement du Loiret" is leading a programme to help industries reducing their vulnerability to flooding.

55% of the industries in the "departement du Loiret" are located in flood-prone areas. A major flood of the Loire river would impact 14,000 employees, and cause €1.5M damage.

The chamber of commerce and industry together in 2006 launched an innovative programme to help company managers to reduce the vulnerability of their industries to flood risk. Realizing that the usual consciousness raising actions were inadequate (public meetings, information documents, etc.), the chamber of commerce and industry decided that it was necessary to provide each company director with direct personal support. A project manager was appointed to meet each company director and provide them with personal advice about the exposure of the company to flood risk and to encourage an overall assessment of the company's vulnerability.

Two diagnoses are introduced. One of these addresses "small industries" and requires half a day of work and is carried out by the directors themselves. The other diagnosis addresses the larger industries. This is made by experts and takes several days to carry out. The aims of these diagnoses are:

• to better understand flood risk;
• to be able to assess direct and indirect damage risks, the company's organisational vulnerability and the potential duration of any disruption that may be caused by flooding;

- to identify measures to be initiated to reduce vulnerability;
- to draft a crisis management plan.

Meetings are planned with 200 company directors.

Feedback on the diagnoses made in Brittany before a major flood occurred showed that some companies had reduced the duration of disruption caused by flooding by a factor of 7, and the amount of damage by a factor of 15.

Author Index